Direct Methods for Stability Analysis of Electric Power Systems

Direct Methods for Stability Analysis of Electric Power Systems

Theoretical Foundation, BCU Methodologies, and Applications

Hsiao-Dong Chiang

WILEY

A John Wiley & Sons, Inc., Publication

Published by John Wiley & Sons, Inc., Hoboken, New Jersey
Published simultaneously in Canada

Library of Congress Cataloging-in-Publication Data is available.

ISBN: 978-0-470-48440-1

oBook ISBN: 978-0-470-87213-0
eBook ISBN: 978-0-470-87212-3

10 9 8 7 6 5 4 3 2 1

Contents

Preface xi

Acknowledgments xiii

1. Introduction and Overview **1**

1.1 Introduction 1
1.2 Trends of Operating Environment 2
1.3 Online TSA 4
1.4 Need for New Tools 5
1.5 Direct Methods: Limitations and Challenges 6
1.6 Purposes of This Book 9

2. System Modeling and Stability Problems **14**

2.1 Introduction 14
2.2 Power System Stability Problem 15
2.3 Model Structures and Parameters 19
2.4 Measurement-Based Modeling 21
2.5 Power System Stability Problems 23
2.6 Approaches for Stability Analysis 25
2.7 Concluding Remarks 27

3. Lyapunov Stability and Stability Regions of Nonlinear Dynamical Systems **29**

3.1 Introduction 29
3.2 Equilibrium Points and Lyapunov Stability 30
3.3 Lyapunov Function Theory 32
3.4 Stable and Unstable Manifolds 34
3.5 Stability Regions 37
3.6 Local Characterizations of Stability Boundary 38
3.7 Global Characterization of Stability Boundary 43
3.8 Algorithm to Determine the Stability Boundary 45
3.9 Conclusion 49

4. Quasi-Stability Regions: Analysis and Characterization **51**

4.1 Introduction 51
4.2 Quasi-Stability Region 51
4.3 Characterization of Quasi-Stability Regions 56
4.4 Conclusions 58

5. Energy Function Theory and Direct Methods **60**

5.1 Introduction 60
5.2 Energy Functions 61
5.3 Energy Function Theory 64
5.4 Estimating Stability Region Using Energy Functions 69
5.5 Optimal Schemes for Estimating Stability Regions 73
5.6 Quasi-Stability Region and Energy Function 75
5.7 Conclusion 78

6. Constructing Analytical Energy Functions for Transient Stability Models **80**

6.1 Introduction 80
6.2 Energy Functions for Lossless Network-Reduction Models 81
6.3 Energy Functions for Lossless Structure-Preserving Models 82
6.4 Nonexistence of Energy Functions for Lossy Models 89
6.5 Existence of Local Energy Functions 92
6.6 Concluding Remarks 93

7. Construction of Numerical Energy Functions for Lossy Transient Stability Models **94**

7.1 Introduction 94
7.2 A Two-Step Procedure 95
7.3 First Integral-Based Procedure 98
7.4 Ill-Conditioned Numerical Problems 105
7.5 Numerical Evaluations of Approximation Schemes 108
7.6 Multistep Trapezoidal Scheme 110
7.7 On the Corrected Numerical Energy Functions 116
7.8 Concluding Remarks 117

8. Direct Methods for Stability Analysis: An Introduction **119**

8.1 Introduction 119
8.2 A Simple System 120
8.3 Closest UEP Method 122
8.4 Controlling UEP Method 123
8.5 PEBS Method 125
8.6 Concluding Remarks 126

9. Foundation of the Closest UEP Method **129**

9.1 Introduction 129
9.2 A Structure-Preserving Model 129
9.3 Closest UEP 132
9.4 Characterization of the Closest UEP 134
9.5 Closest UEP Method 135

9.6 Improved Closest UEP Method 136
9.7 Robustness of the Closest UEP 140
9.8 Numerical Studies 144
9.9 Conclusions 146

10. Foundations of the Potential Energy Boundary Surface Method **148**

10.1 Introduction 148
10.2 Procedure of the PEBS Method 149
10.3 Original Model and Artificial Model 150
10.4 Generalized Gradient Systems 153
10.5 A Class of Second-Order Dynamical Systems 157
10.6 Relation between the Original Model and the Artificial Model 160
10.7 Analysis of the PEBS Method 164
10.8 Concluding Remarks 175

11. Controlling UEP Method: Theory **177**

11.1 Introduction 177
11.2 The Controlling UEP 178
11.3 Existence and Uniqueness 180
11.4 The Controlling UEP Method 181
11.5 Analysis of the Controlling UEP Method 183
11.6 Numerical Examples 188
11.7 Dynamic and Geometric Characterizations 191
11.8 Concluding Remarks 193

12. Controlling UEP Method: Computations **196**

12.1 Introduction 196
12.2 Computational Challenges 197
12.3 Constrained Nonlinear Equations for Equilibrium Points 199
12.4 Numerical Techniques for Computing Equilibrium Points 201
12.5 Convergence Regions of Equilibrium Points 203
12.6 Conceptual Methods for Computing the Controlling UEP 205
12.7 Numerical Studies 207
12.8 Concluding Remarks 212

13. Foundations of Controlling UEP Methods for Network-Preserving Transient Stability Models **215**

13.1 Introduction 215
13.2 System Models 216
13.3 Stability Regions 218
13.4 Singular Perturbation Approach 219
13.5 Energy Functions for Network-Preserving Models 221
13.6 Controlling UEP for DAE Systems 222

13.7 Controlling UEP Method for DAE Systems 224
13.8 Numerical Studies 226
13.9 Concluding Remarks 230

14. Network-Reduction BCU Method and Its Theoretical Foundation 235

14.1 Introduction 235
14.2 Reduced-State System 236
14.3 Analytical Results 237
14.4 Static and Dynamic Relationships 246
14.5 Dynamic Property (D3) 247
14.6 A Conceptual Network-Reduction BCU Method 250
14.7 Concluding Remarks 251

15. Numerical Network-Reduction BCU Method 254

15.1 Introduction 254
15.2 Computing Exit Points 256
15.3 Stability-Boundary-Following Procedure 257
15.4 A Safeguard Scheme 262
15.5 Illustrative Examples 263
15.6 Numerical Illustrations 270
15.7 IEEE Test System 274
15.8 Concluding Remarks 275

16. Network-Preserving BCU Method and Its Theoretical Foundation 279

16.1 Introduction 279
16.2 Reduced-State Model 280
16.3 Static and Dynamic Properties 285
16.4 Analytical Results 288
16.5 Overall Static and Dynamic Relationships 292
16.6 Dynamic Property (D3) 294
16.7 Conceptual Network-Preserving BCU Method 295
16.8 Concluding Remarks 299

17. Numerical Network-Preserving BCU Method 300

17.1 Introduction 300
17.2 Computational Considerations 304
17.3 Numerical Scheme to Detect Exit Points 305
17.4 Computing the MGP 307
17.5 Computation of Equilibrium Points 308
17.6 Numerical Examples 313
17.7 Large Test Systems 319
17.8 Concluding Remarks 325

18. Numerical Studies of BCU Methods from Stability Boundary Perspectives **326**

18.1 Introduction 326
18.2 Stability Boundary of Network-Reduction Models 328
18.3 Network-Preserving Model 334
18.4 One Dynamic Property of the Controlling UEP 339
18.5 Concluding Remarks 342

19. Study of the Transversality Conditions of the BCU Method **345**

19.1 Introduction 345
19.2 A Parametric Study 346
19.3 Analytical Investigation of the Boundary Property 351
19.4 The Two-Machine Infinite Bus (TMIB) System 354
19.5 Numerical Studies 360
19.6 Concluding Remarks 362

20. The BCU–Exit Point Method **365**

20.1 Introduction 365
20.2 Boundary Property 365
20.3 Computation of the BCU–Exit Point 373
20.4 BCU–Exit Point and Critical Energy 376
20.5 BCU–Exit Point Method 378
20.6 Concluding Remarks 379

21. Group Properties of Contingencies in Power Systems **383**

21.1 Introduction 383
21.2 Groups of Coherent Contingencies 385
21.3 Identification of a Group of Coherent Contingencies 386
21.4 Static Group Properties 387
21.5 Dynamic Group Properties 395
21.6 Concluding Remarks 399

22. Group-Based BCU–Exit Method **401**

22.1 Introduction 401
22.2 Group-Based Verification Scheme 402
22.3 Linear and Nonlinear Relationships 403
22.4 Group-Based BCU–Exit Point Method 410
22.5 Numerical Studies 412
22.6 Concluding Remarks 413

23. Group-Based BCU–CUEP Methods **420**

23.1 Introduction 420

23.2 Exact Method for Computing the Controlling UEP 421
23.3 Group-Based BCU–CUEP Method 423
23.4 Numerical Studies 424
23.5 Concluding Remarks 428

24. Group-Based BCU Method **430**

24.1 Introduction 430
24.2 Group-Based BCU Method for Accurate Critical Energy 431
24.3 Group-Based BCU Method for CUEPs 434
24.4 Numerical Studies 438
24.5 Concluding Remarks 445

25. Perspectives and Future Directions **447**

25.1 Current Developments 447
25.2 Online Dynamic Contingency Screening 450
25.3 Further Improvements 452
25.4 Phasor Measurement Unit (PMU)-Assisted Online ATC Determination 453
25.5 Emerging Applications 455
25.6 Concluding Remarks 457

Appendix **458**

A1.1 Mathematical Preliminaries 458
A1.2 Proofs of Theorems in Chapter 9 459
A1.3 Proofs of Theorems in Chapter 10 464

Bibliography 472

Index 483

Preface

Power system instabilities are unacceptable to society. Indeed, recent major blackouts in North America and in Europe have vividly demonstrated that power interruptions, grid congestions, or blackouts significantly impact the economy and society. At present, stability analysis programs routinely used in utilities around the world are based mostly on step-by-step numerical integrations of power system stability models to simulate system dynamic behaviors. This off-line practice is inadequate to deal with current operating environments and calls for online evaluations of changing overall system conditions.

Several significant benefits and potential applications are expected from the movement of transient stability analysis from the off-line mode to the online operating environment. However, this movement is a challenging task and requires several breakthroughs in measurement systems, analytical tools, computation methods, and control schemes. An alternate approach to transient stability analysis employing energy functions is called the direct method, or termed the energy function-based direct method. Direct methods offer several distinctive advantages. For example, they can determine transient stability without the time-consuming numerical integration of a (postfault) power system. In addition to their speed, direct methods can provide useful information regarding the derivation of preventive control and enhancement control actions for power system stability.

Direct methods have a long developmental history spanning six decades. Despite the fact that significant progress has been made, direct methods have been considered impractical by many researchers and users. Several challenges and limitations must be overcome before direct methods can become a practical tool. This book seeks to address these challenges and limitations.

The main purpose of this book is to present a comprehensive theoretical foundation for the direct methods and to develop comprehensive BCU solution methodologies along with their theoretical foundations. In addition, a comprehensive energy function theory, which is an extension of the Lyapunov function theory, is presented along with general procedures for constructing numerical energy functions for general power system transient stability models. It is believed that solving challenging practical problems efficiently can be accomplished through a thorough understanding of the underlying theory, in conjunction with exploring the special features of the practical problem under study to develop effective solution methodologies.

There are 25 chapters contained in this book. These chapters are classified into the following subjects:

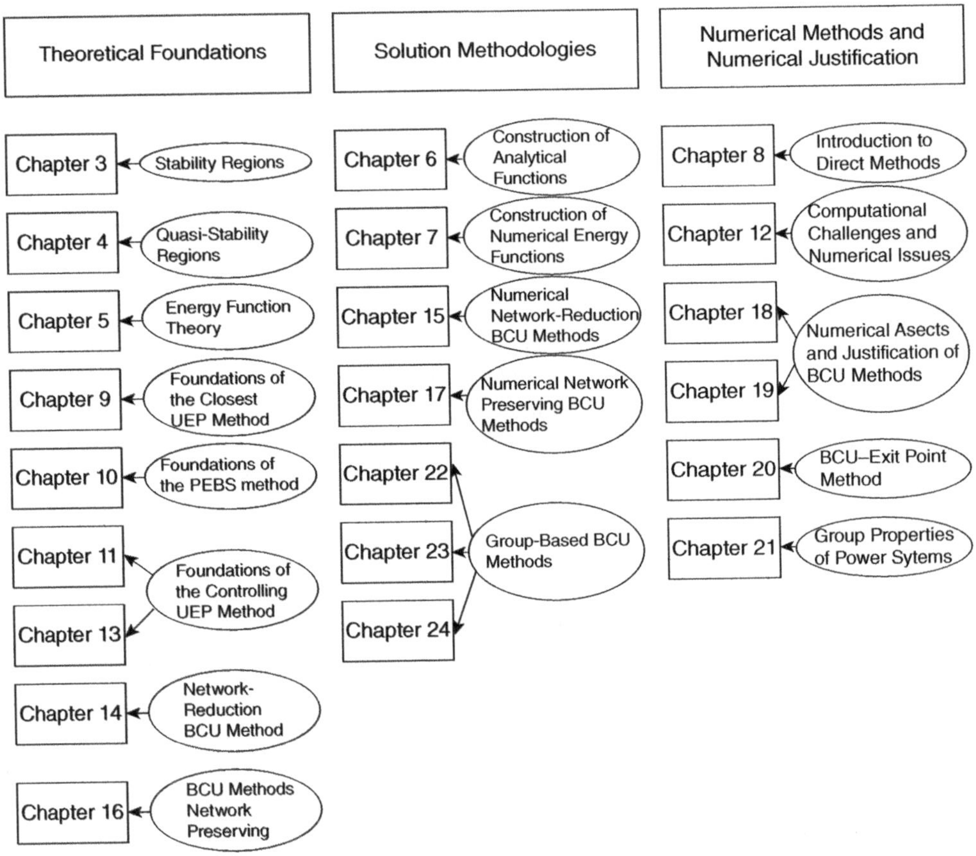

The following stages of research and development can lead to fruitful and practical applications:

Stage 1. Development of theoretical foundations

Stage 2. Development of the solution methodology

Stage 3. Development of reliable methods to numerically implement the solution methodology

Stage 4. Software implementation and evaluation

Stage 5. Industry user interactions

Stage 6. Practical system installation

The first three stages are suitable for university and research institution application, while the last four stages are more suitable for commercial entities. This text focuses on Stages 1 and 2 and touches upon Stage 3. In the following volume, Stage 3 will be more thoroughly explored along with Stages 4 through 6.

Ithaca, New York
May 2010

HSIAO-DONG CHIANG

Acknowledgments

I started my work on direct methods for power system stability analysis while I was a Ph.D. student at the University of California, Berkeley. The advice I received from my advisors, Felix Wu and Pravin Varaiya, I carry with me to this day. Shankar Sastry's instruction on nonlinear systems and Leon Chua's instruction on nonlinear circuits were also very important to my research. In addition, I really appreciate the time Professor Morris Hirsch spent teaching me nonlinear dynamic systems and stability regions. He often spent many hours explaining the world of complex nonlinear phenomena to me, and he was a very inspirational role model.

Several PhD students at Cornell have made significant contributions to the development of the material presented in this book. In particular, I would like to acknowledge Dr. Chia-Chi Chu, Dr. Lazhar Fekih-Ahmed, Dr. Matthew Varghese, Dr. Ian Dobson, Dr. Weimin Ma, Dr. Rene Jean-Jumeau, Dr. Alexander J. Flueck, Dr. Karen Miu, Dr. Chih-Wen Liu, Dr. Jaewook Lee, Mr. Tim Conneen, and Mr. Warut Suampun. Without their hard work, this book would have been incomplete.

Likewise, my former BCU team research associates have made significant contribution to the development of the solution methodologies and the BCU method prototype. I would especially like to acknowledge Dr. Jianzhong Tong, Dr. Chen-Shan Wang, Dr. Yan Zheng, and Dr. Wei Ping. My continual exchange and discussion with Dr. Jianzhong Tong on the general topics of power system dynamic security assessments and control were very enlightening. Furthermore, my joint work with Dr. Hua Li over the past several years has been instrumental to overcoming the challenges of applying the BCU method to practical applications, and he has made significant contribution to the development of group-based BCU methods. My joint work with Dr. Byoung-Kon Choi on the development of new forms of energy functions and the prototype for a new numerical implementation of the BCU method has been very fruitful. Similarly, my discussions with Dr. Bernie Lesieutre, Dr. Zhou Yun, and Dr. Yoshi Suzuki have been very insightful. Dr. Lesieutre and his team's work on the one-parameter transversality condition of the BCU method has been inspirational, and my discussions with Professor Lounan Chen on DAE systems have been invaluable. Lastly, I am greatly indebted to Dr. Luis Fernando Costa Alberto for visiting me every year and for working with me on the areas of stability regions, the BCU method, and direct methods. His insightful and constructive perspective, I believe, will lead to new developments in these areas.

My research associates at the Tokyo Electric Power Company (TEPCO) have been extremely instrumental to the development of TEPCO-BCU and its practical applications in real-world power system models. I would like to express my thanks and appreciation to the following: Dr. Yasuyuki Tada, Dr. Takeshi Yamada, Dr. Ryuya Tanabe, Dr. Hiroshi Okamoto, Dr. Kaoru Koyanagi, Dr. Yicheng Zhou, Mr. Atsushi Kurita, and Mr. Tsuyoshi Takazawa. My working experience with the

TEPCO-BCU team has been truly remarkable. In particular, I am grateful for the continued support, guidance, and vision Dr. Tada has given me all these years. I would also like to thank Mr. Yoshiharu Tachibana and Mr. Kiyoshi Goto, general managers of the R&D center at TEPCO, for their great vision and continued support of my work.

A special thanks goes to my industry friends and associates who have taught me the practical aspects of power system stability problems. Through our joint research and development, I have learned a great deal from them. In particular, I would like to thank Mr. Gerry Cauley, Dr. Neal Balu, Dr. Peter Hirsch, Dr. Tom Schneider, Dr. Ron Chu, Dr. Mani Subramanian, Dr. Dan Sobajic, Dr. Prabha Kundur, Mr. Kip Morison, Dr. Lei Wang, Dr. Ebrahim Vaahedi, Mr. Carson Taylor, Mr. Dave Takash, Mr. Tom Cane, Dr. Martin Nelson, Dr. Soumen Ghosh, Dr. Jun Wu, Mr. Chi Tang, and Mr. William Price. In addition, I would like to thank Mr. Yakout Mansour for his advice on working with 12,000-bus power systems to gain insight into the practical aspects of power systems. His advice has helped shape my research and development these last 15 years.

I am very grateful to Director Chia-Jen Lin and to Director Anthony Yuan-Tian Chen of the Department of System Planning at the Tai-Power Company for their support and for sharing their practical experience with me. My joint research work with China's Electric Power Research Institute (EPRI) in the 1990s was very enjoyable. I would like to thank Mr. Zhou Xiao-Xin, Mr. Zhang Wen-Tao, Mr. Ying Young-Hua, and Mr. Tang Yong. My joint work on the practical application of BCU methods with Si-Fang Automation of Beijing has also been very constructive. In particular, I would like to express my appreciation to Professor Yang Qi-Xun, Professor Wang Xu-Zhao, Mr. Zhang You, Dr. Wu Jing-Tao, Mr. Qi Wen-Bin, and Mr. Sheng Hao.

My academic colleagues have also been a guiding source of support and encouragement. I am very thankful to my colleagues at Cornell University. My working relationship with Professor James S. Thorp and Professor Robert J. Thomas has been very fruitful. In encouraging my work on both the practical and theoretical aspects of power systems, they have inspired my active work on practical applications of nonlinear system theory and nonlinear computation. I thank Professor Peter Sauer for his great advice and guidance over the years and Professor Chen-Ching Liu, who was a great mentor during my early career and who, since then, has become a good friend. Moreover, I would like to thank Professors Anjan Bose, Christ DeMarco, Joe Chow, Robert Fischl, Frank Mercede, David Hill, Ian Hiskens, Vijay Vittal, Aziz Fouad, Maria Pavella, Xia Dao-Zhi, Han Zhen Xiang, Liu Shen, Xue Yu-Shang, Min Yong, Gan Dequing, Li Yinhong, Shi Dong-Yuan, and M. A. Pai for their technical insight into direct methods.

Finally, I would like to thank my family, especially my grandfather Chiang Ah Mu, for their love, sacrifice, and unwavering support.

H-D. C.

Chapter 1

Introduction and Overview

1.1 INTRODUCTION

Power system instabilities are unacceptable to society. Indeed, recent major blackouts in North America and in Europe have vividly demonstrated that power interruptions, grid congestions, or blackouts significantly impact the economy and society. In August 1996, disturbances cascaded through the West Coast transmission system, causing widespread blackouts that cost an estimated $2 billion and left 12 million customers without electricity for up to 8 h. In June 1998, transmission system constraints disrupted the wholesale power market in the Midwest, causing price rises from an average of $30 per megawatt hour to peaks as high as $10,000 per megawatt hour. Similar price spikes also occurred in the summers of 1999 and 2000. In 2003, the Northeast blackout left 50 million customers without electricity and the financial loss was estimated at $6 billion. According to a research firm, the annual cost of power outages and fluctuations worldwide was estimated to be between $119 and $188 billion yearly. Power outages and interruptions clearly have significant economic consequences for society.

The ever-increasing loading of transmission networks coupled with a steady increase in load demands has pushed the operating conditions of many worldwide power systems ever closer to their stability limits. The combination of limited investment in new transmission and generation facilities, new regulatory requirements for transmission open access, and environmental concerns are forcing transmission networks to carry more power than they were designed to withstand. This problem of reduced operating security margins is further compounded by factors such as (1) the increasing number of bulk power interchange transactions and non-utility generators, (2) the trend towards installing higher-output generators with lower inertia constants and higher short circuit ratios, and (3) the increasing amount of renewable energies. Under these conditions, it is now well recognized that any violation of power system dynamic security limits leads to far-reaching consequences for the entire power system.

Direct Methods for Stability Analysis of Electric Power Systems, by Hsiao-Dong Chiang

By nature, a power system continually experiences two types of disturbances: *event disturbances* and *load variations*. Event disturbances (contingencies) include loss of generating units or transmission components (lines, transformers, and substations) due to short circuits caused by lightning, high winds, and failures such as incorrect relay operations, insulation breakdowns, sudden large load changes, or a combination of such events. Event disturbances usually lead to a change in the network configuration of the power system due to actions from protective relays and circuit breakers. They can occur as a single equipment (or component) outage or as multiple simultaneous outages when taking relay actions into account. Load variations are variations in load demands at buses and/or power transfers among buses. The network configuration may remain unchanged after load variations. Power systems are planned and operated to withstand certain disturbances. The North American Electric Reliability Council defines security as the ability to prevent cascading outages when the bulk power supply is subjected to severe disturbances. Individual reliability councils establish the types of disturbances that their systems must withstand without cascading outages.

A major activity in power system planning and operation is the examination of the impact a set of credible disturbances has on a power system's dynamic behavior such as stability. Power system stability analysis is concerned with a power system's ability to reach an acceptable steady state (operating condition) following a disturbance. For operational purposes, power system stability analysis plays an important role in determining the system operating limits and operating guidelines. During the planning stage, power system stability analysis is performed to assess the need for additional facilities and the locations at which additional control devices to enhance the system's static and dynamic security should be placed. Stability analysis is also performed to check relay settings and to set the parameters of control devices. Important conclusions and decisions about power system operations and planning are made based on the results of stability studies.

Transient stability problems, a class of power system stability problems, have been a major operating constraint in regions that rely on long-distance transfers of bulk power (e.g., in most parts of the Western Interconnection in the United States, Hydro-Québec, the interfaces between the Ontario/New York area and the Manitoba/Minnesota area, and in certain parts of China and Brazil). The trend now is that many parts of the various interconnected systems are becoming constrained by transient stability limitations. The wave of recent changes has caused an increase in the adverse effects of both event disturbances and load variations in power system stability. Hence, it is imperative to develop powerful tools to examine power system stability in a timely and accurate manner and to derive necessary control actions for both preventive and enhancement control.

1.2 TRENDS OF OPERATING ENVIRONMENT

The aging power grid is vulnerable to power system disturbances. Many transformers in the grid approach or surpass their design life. The transmission system is often under-invested and overstrained. These result in vulnerable power grids constantly operating near their operating limits. In addition, this operating environment encounters more challenges brought about by dispersed generations whose prime movers can be any renewable energy source such as wind power. As is well recognized, these small-size dispersed generation systems raise even greater concerns of power system stability. Hence, with current power system operating environments, it is increasingly difficult for power system operators to generate all the operating limits for all possible operating conditions under a list of credible contingencies.

At present, most energy management systems periodically perform online power system static security assessment (SSA) and control to ensure that the power system can withstand a set of credible contingencies. The assessment involves selecting a set of credible contingencies and evaluating the system's response to those contingencies. Various software packages for security assessment and control have been implemented in modern energy control centers. These packages provide comprehensive online security analysis and control based almost exclusively on steady-state analysis, making them applicable to SSA and control but not to online transient stability assessment (TSA). Instead, off-line transient stability analysis has been performed for postulated operating conditions. The turn-around time for a typical study can range from hours to days depending on the number of postulated operating conditions and the dynamic study period of each contingency. This off-line practice is inadequate to deal with current operating environments and calls for online evaluations of the constantly changing overall system conditions.

The lack of performing online TSAs in an energy management system can have serious consequences. Indeed, any violation of dynamic security limits has far-reaching impacts on the entire power system and thus on the society. From a financial viewpoint, the costs associated with a power outage can be tremendous. Online dynamic security assessment is an important tool for avoiding dynamic security limit violations. It is fair to say that the more stressed a power system, the stronger the need for online dynamic security assessments.

Several significant benefits and potential applications are expected from the movement of transient stability analysis from the off-line mode to the online operating environment. The first benefit is that a power system can be operated with operating margins reduced by a factor of 10 or more if the dynamic security assessment is based on the actual system configuration and actual operating conditions instead of assumed worst-case conditions, as is done in off-line studies. This ability is especially significant since current environments have pushed power systems to operate with low reserve margins closer to their stability limits. A second benefit to online analysis is that the large number of credible contingencies that needs to be assessed can be reduced to those contingencies relevant to actual operating conditions. Important consequences obtained from this benefit are that more accurate operating margins can be determined and more power transfers among different areas, or different zones of power networks, can be realized. Compared to off-line studies, online studies require much less engineering resources, thereby freeing these resources for other critical activities.

1.3 ONLINE TSA

Online TSA is designed to provide system operators with critical system stability information including (1) TSA of the current operating condition subject to a list of contingencies and (2) available (power) transfer limits at key interfaces subject to transient stability constraints. A complete online TSA assessment cycle is typically in the order of minutes, say, 5 min. This cycle starts when all necessary data are available to the system and ends when the system is ready for the next cycle. Depending on the size of the underlying power systems, it is estimated that, for a large-size power system such as a 15,000-bus power system, the number of contingencies in a contingency list is between 2000 and 3000. The contingency types will include both a three-phase fault with primary clearance and a single line-to-ground fault with backup clearance.

When a cycle of online TSA is initiated, a list of credible contingencies, along with information from the state estimator and topological analysis, is applied to the online TSA program whose basic function is to identify unstable contingencies from the contingency list. An operating condition is said to be transiently stable if the contingency list contains no unstable contingencies; otherwise, it is transiently unstable. The task of online TSA, however, is very challenging.

The strategy of using an effective scheme to screen out a large number of stable contingencies, capture critical contingencies, and apply detailed simulation programs only to potentially unstable contingencies is well recognized. This strategy has been successfully implemented in online SSA. The ability to screen several hundred contingencies to capture tens of the critical contingencies has made the online SSA feasible. This strategy can be applied to online TSA. Given a set of credible contingencies, the strategy would break the task of online TSA into two stages of assessments (Chadalavada et al., 1997; Chiang et al., 1997):

Step 1. Perform the task of dynamic contingency screening to quickly screen out contingencies that are definitely stable from a set of credible contingencies.

Step 2. Perform detailed assessment of dynamic performance for each contingency remaining in Stage 1.

Dynamic contingency screening is a fundamental function of an online TSA system. The overall computational speed of an online TSA system depends greatly on the effectiveness of the dynamic contingency screening, the objective of which is to identify contingencies that are definitely stable and thereby to avoid further stability analysis for these contingencies. It is due to the definite classification of stable contingencies that considerable speedup can be achieved for TSA. Contingencies that are either undecided or identified as critical or unstable are then sent to the time–domain transient stability simulation program for further stability analysis.

Online TSA can provide an accurate determination of online transfer capability constrained by transient stability limits. This accurate calculation of transfer capability allows remote generators with low production cost to be economically dispatched

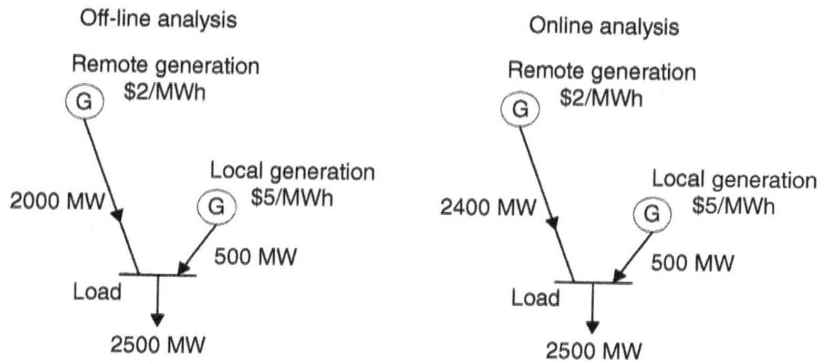

Savings
= (1 h) 400 MW × ($5/MWh − $2/MWh) = $1200
= (24 h) $28,800
= (30 days) $864,000
= (1 year) $10.4 million

Figure 1.1 A hypothetical power system and analysis of financial savings.

to serve load centers. We consider a hypothetical power system containing a remote generator with low production cost, say, a hydro generator of $2 per megawatt hour and a local generator with a high production cost of $5 per megawatt hour that all supply electricity to a load center of 2500 MW (see Figure 1.1). According to the off-line analysis, the transfer capability between the remote generator and the load center was 2105 MW. With a 5% security margin, the output of the remote generator was set to 2000 MW. The local generator then needs to supply 500 MW to the load center to meet the load demand. On the other hand, the actual transfer capability between the remote generator and the load center, according to online TSA, was 2526 MW instead of 2105 MW. With a 5% security margin, the output of the remote generator was set to 2400 MW, while the output of the local generator was set to 100 MW to meet the load demand. By comparing these two different schemes of real power dispatch based on two different transfer capability calculations, the difference in production cost is about $1200 per hour or $28,800 per day. It can be observed that even for such a relatively small load demand of 2500 MW, online TSA allows for significant financial savings amounting to about $10.5 million per year. We recognize that practical power systems may not resemble this hypothetical power system; however, it does illustrate the significant financial benefits of online TSA.

1.4 NEED FOR NEW TOOLS

At present, stability analysis programs routinely used in utilities around the world are based mostly on step-by-step numerical integrations of power system stability models used to simulate system dynamic behaviors. This practice of power system stability analysis based on the time–domain approach has a long history. The stability of the postfault system is assessed based on simulated postfault trajectories. The typical simulation period for the postfault system is 10 s and can go beyond 15 s if multiswing instability is of concern, making this conventional approach rather time-consuming.

The traditional time–domain simulation approach has several disadvantages. First, it requires intensive, time-consuming computation efforts; therefore, it has not been suitable for online application. Second, it does not provide information as to how to derive preventive control when the system is deemed unstable nor how to derive enhancement control when the system is deemed critically stable, and finally, it does not provide information regarding the degree of stability (when the system is stable) and the degree of instability (when the system is unstable) of a power system. This information is valuable for both power system planning and operation.

From a computational viewpoint, online TSA involves solving a large set of mathematical models, which is described by a large set of nonlinear differential equations in addition to the nonlinear algebraic equations involved in the SSA. For a 14,000-bus power system transient stability model, one dynamic contingency analysis can involve solving a set of 15,000 differential equations and 40,000 nonlinear algebraic equations for a time duration of 10–20 s in order to assess the power system stability under the study contingency. Online TSA requires the ability to analyze hundreds or even thousands of contingencies every 5–10 min using online data and system state estimation results. Thus, the traditional time–domain simulation approach cannot meet this requirement.

The computational effort required by online TSA is roughly three magnitudes higher than that of the SSA. This explains why TSA has long remained an off-line activity instead of an online activity in the energy management system. Extending the functions of energy management systems to take into account online TSA and control is a challenging task and requires several breakthroughs in measurement systems, analytical tools, computation methods, and control schemes.

1.5 DIRECT METHODS: LIMITATIONS AND CHALLENGES

An alternate approach to transient stability analysis employing energy functions, called *direct methods*, or termed energy function-based direct methods, was originally proposed by Magnusson (1947) in the late 1940s and was pursued in the 1950s by Aylett (1958). Direct methods have a long developmental history spanning six decades. Significant progress, however, has been made only recently in the practical application of direct methods to transient stability analysis. Direct methods can determine transient stability without the time-consuming numerical integration of a (postfault) power system. In addition to their speed, direct methods also provide a quantitative measure of the degree of system stability. This additional information makes direct methods very attractive when the relative stability of different network configuration plans must be compared or when system operating limits constrained by transient stability must be calculated quickly. Another advantage to direct methods is that they provide useful information regarding the derivation of preventive control actions when the underlying power system is deemed unstable and the derivation of enhancement control actions when the underlying power system is deemed critically stable.

Despite the fact that significant progress has been made in energy function-based direct methods over the last several decades, they have been considered impractical by many researchers and users for power system applications. Indeed, direct methods must overcome several challenges and limitations before they can become a practical tool.

From an analytical viewpoint, direct methods were originally developed for power systems with autonomous postfault systems. As such, there are several challenges and limitations involved in the practical applications of direct methods for power system transient stability analysis, some of which are inherent to these methods while others are related to their applicability to power system models. These challenges and limitations can be classified as follows:

Challenges

- The modeling challenge
- The function challenge
- The reliability challenge

Limitations

- The scenario limitation
- The condition limitation
- The accuracy limitation

The modeling challenge stems from the requirement that there exists an energy function for the (postfault) transient stability model of study. However, the problem is that not every (postfault) transient stability model admits an energy function; consequently, simplified transient stability models have been used in direct methods. A major shortcoming of direct methods in the past has been the simplicity of the models they can handle. Recent work in this area has made significant advances. The current progress in this direction is that a general procedure of constructing numerical energy functions for complex transient stability models is available. This book will devote Chapters 6 and 7 to this topic.

The function limitation stipulates that direct methods are only applicable to first swing stability analysis of power system transient stability models described by pure differential equations. Recent work in the development of the controlling UEP method has extended the first-swing stability analysis into a multiswing stability analysis. In addition, the controlling UEP method is applicable to power system transient stability models described by differential and algebraic equations. This book will devote Chapters 11 through 13 to this topic.

The scenario limitation for direct methods comes from the requirement that the initial condition of a study postfault system must be available and the requirement that the postfault system must be autonomous. It is owing to the requirement of the availability of the initial condition that makes numerical integration of the study fault-on system a must for direct methods. Hence, the initial condition of a study postfault system can only be obtained via the time–domain approach and cannot be available beforehand. On the other hand, the requirement that the postfault system be autonomous imposes the condition that the fault sequence on the system must be well-defined in advance. Currently, the limitation that the postfault system must be an autonomous dynamical system is partially removed. In particular, the postfault system does not need to be a "pure" autonomous system and it can be constituted by a series of autonomous dynamical systems.

The condition limitation is an analytical concern related to the required conditions for postfault power systems: a postfault stable equilibrium point must exist and the prefault stable equilibrium point must lie inside the stability region of the postfault stable equilibrium point. This limitation is inherent to the foundation of direct methods. Generally speaking, these required conditions are satisfied on stable contingencies, while they may not be satisfied on unstable contingencies. From an application viewpoint, this condition limitation is a minor concern and direct methods can be developed to overcome this limitation.

The accuracy limitation stems from the fact that analytical energy functions for general power system transient stability models do not exist. Regarding the accuracy limitation, it has been observed in numerous studies that the controlling UEP method, in conjunction with appropriate numerical energy functions, yields accurate stability assessments. Numerical energy functions are practically useful in direct methods. In this book, methods and procedures to construct accurate numerical energy functions will be presented.

The reliability challenge is related to the reliability of a computational method in computing the controlling UEP for every study contingency. From a theoretical viewpoint, this text will demonstrate the existence and uniqueness of the controlling UEP with respect to a fault-on trajectory. Furthermore, the controlling UEP is independent of the energy function used in the direct stability assessment. Hence, the task of constructing an energy function and the task of computing the controlling UEP are not interrelational. From a computational viewpoint, the task of computing the controlling UEP is very challenging. We will present in Chapter 12 the computational challenges in computing the controlling UEP. A total of seven challenges in computing the controlling UEP will be highlighted. These challenges call into doubt the correctness of any attempt to directly compute the controlling UEP of the original power system stability model. This analysis serves to explain why previous methods proposed in the literature fail to compute the controlling UEP.

The above analysis reveals three important implications for the development of a reliable numerical method for computing controlling UEPs:

1. These computational challenges should be taken into account in the development of numerical methods for computing the controlling UEP.
2. It is impossible to directly compute the controlling UEP of a power system stability model without using the iterative time–domain method.

3. It is possible to directly compute the controlling UEP of an artificial, reduced-state power system stability model without using the iterative time–domain method.

In this book, it will be shown that it is fruitful to develop a tailored solution algorithm for finding the controlling UEPs by exploiting special properties as well as some physical and mathematical insights into the underlying power system stability model. We will discuss in great detail such a systematic method, called the BCU method, for finding controlling UEPs for power system models in Chapters 14 through 17. The BCU method does not attempt to directly compute the controlling UEP of a power system stability model (original model); instead, it computes the controlling UEP of a reduced-state model and relates the computed controlling UEP to the controlling UEP of the original model. This book will devote Chapters 14 through 24 to present the following family of BCU methods:

- The BCU method
- The BCU–exit point method
- The group-based BCU–exit point method
- The group-based BCU–CUEP method
- The group-based BCU method

This book will also explain how to develop tailored solution methodologies by exploring special properties as well as some physical and mathematical insights into the underlying power system stability model. For instance, it will be explained how the group properties of contingencies in power systems are discovered. These group properties will be explored and incorporated into the development of a group-based BCU method. This exploration of group properties leads to a significant reduction in computational efforts for reliably computing controlling UEPs for a group of coherent contingencies and to the development of effective preventive control actions against a set of insecure contingencies and enhancement control actions for a set of critical contingencies.

1.6 PURPOSES OF THIS BOOK

The main purpose of this book is to present a comprehensive theoretical foundation for direct methods and to develop comprehensive BCU solution methodologies along with their theoretical foundations. BCU methodologies have been developed to reliably compute controlling UEPs and to reliably compute accurate critical values, which are essential pieces of information needed in the controlling UEP method. In addition, a comprehensive energy function theory, which is an extension of the Lyapunov function theory, is presented along with a general procedure for constructing numerical energy functions for general power system transient stability models.

This author believes that solving challenging practical problems efficiently can be accomplished through a thorough understanding of the underlying theory, in

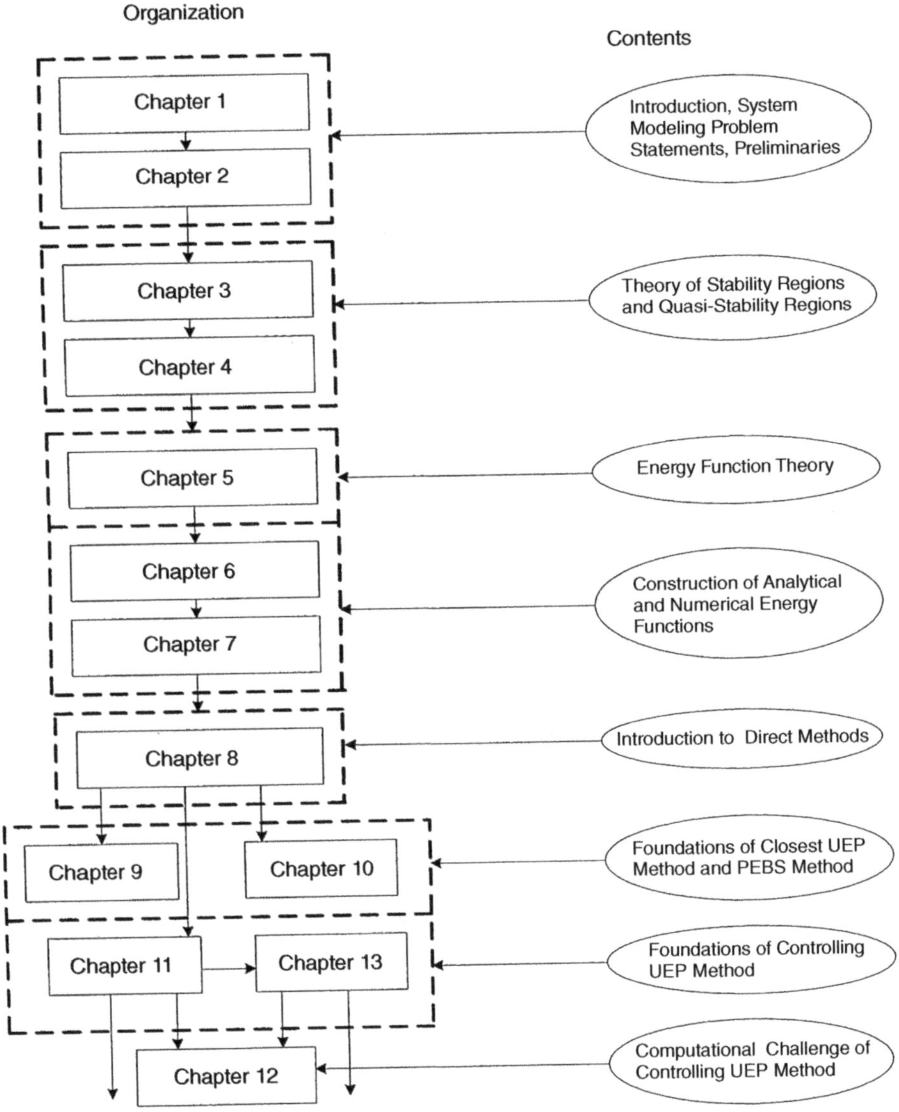

Figure 1.2 An overview of the organization and content of this book.

conjunction with exploring the special features of the practical problem under study, to develop effective solution methodologies. This book covers both a comprehensive theoretical foundation for direct methods and comprehensive BCU solution methodologies.

There are 25 chapters contained in this book. These chapters can be classified into the following (see Figure 1.2):

Chapter 2: System Modeling and Stability Problems

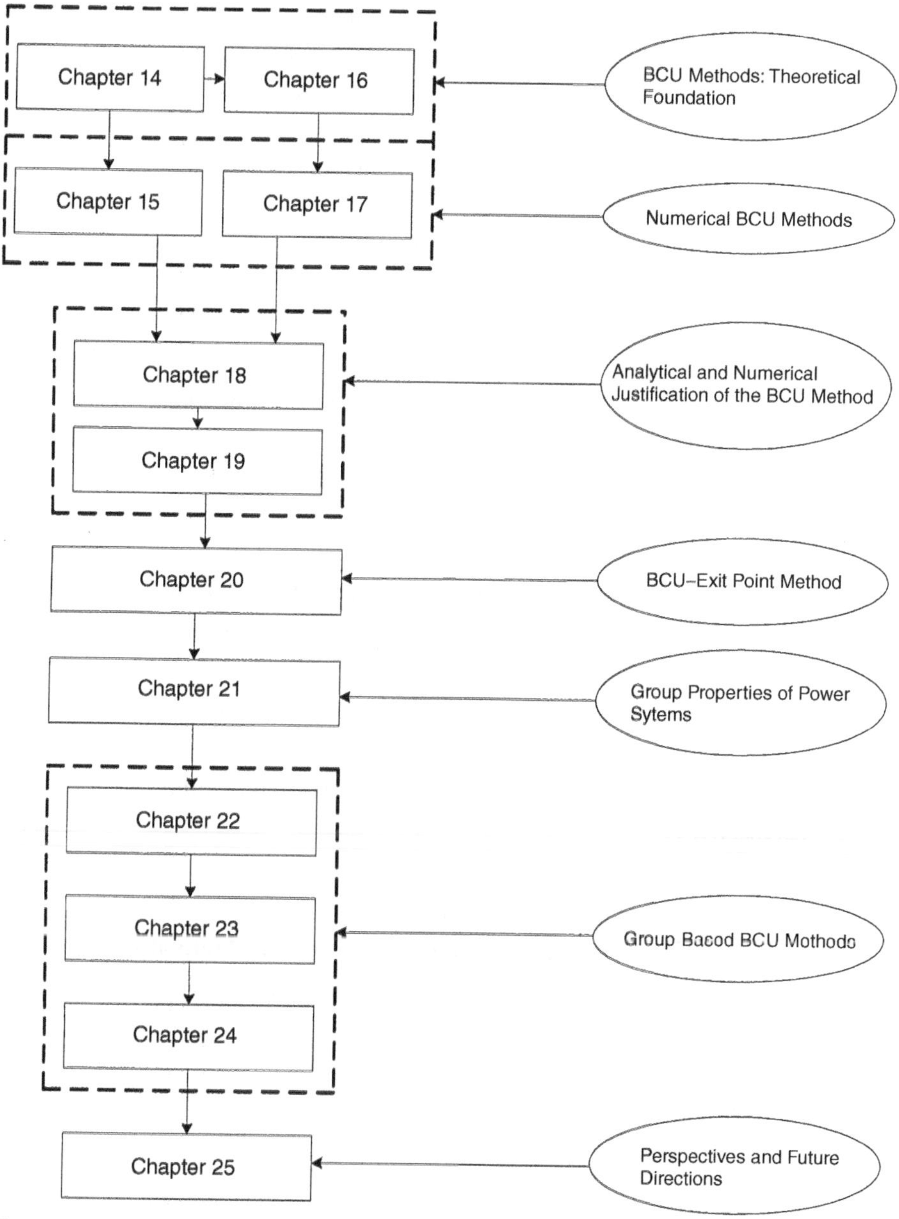

Figure 1.2 Continued

Theory of Stability Regions

Chapter 3: Lyapunov Stability and Stability Regions of Nonlinear Dynamical Systems

Chapter 4: Quasi-Stability Regions: Analysis and Characterization

Energy Functions: Theory and Constructions

Chapter 5: Energy Function Theory and Direct Methods

Chapter 6: Constructing Analytical Energy Functions for Transient Stability Models

Chapter 7: Construction of Numerical Energy Functions for Lossy Transient Stability Models

Direct Methods: Introduction and Foundations

Chapter 8: Direct Methods for Stability Analysis: An Introduction

Chapter 9: Foundation of the Closest UEP Method

Chapter 10: Foundations of the Potential Energy Boundary Surface Method

Controlling UEP Method: Theoretical Foundation and Computation

Chapter 11: Controlling UEP Method: Theory

Chapter 12: Controlling UEP Method: Computations

Chapter 13: Foundations of Controlling UEP Methods for Network-Preserving Transient Stability Models

BCU Methods: Methodologies and Theoretical Foundations

Chapter 14: Network-Reduction BCU Method and Its Theoretical Foundation

Chapter 15: Numerical Network-Reduction BCU Method

Chapter 16: Network-Preserving BCU Method and Its Theoretical Foundation

Chapter 17: Numerical Network-Preserving BCU Method

Chapter 18: Numerical Studies of BCU Methods from Stability Boundary Perspectives

Chapter 19: Study of Transversality Conditions of the BCU Method

Chapter 20: The BCU–Exit Point Method

Group-Based BCU Methods: Group Properties and Methodologies

Chapter 21: Group Properties of Contingencies in Power Systems

Chapter 22: Group-Based BCU–Exit Method

Chapter 23: Group-Based BCU–CUEP Methods

Chapter 24: Group-Based BCU Method

Chapter 25: Perspectives and Future Directions

In summary, this book presents the following theoretical developments as well as solution methodologies with a focus on practical applications for the direct analysis of large-scale power system transient stability; in particular, this book

- provides a general framework for general direct methods, particularly the controlling UEP method;
- develops a comprehensive theoretical foundation for the controlling UEP method, the potential energy boundary surface (PEBS) method, and the closest UEP method;

- presents the BCU methodologies, including the network-reduction BCU method and the network-preserving BCU method;
- presents the theoretical foundation for both the network-reduction BCU method and the network-preserving BCU method;
- develops numerical implementations of both the network-reduction BCU method and the network-preserving BCU method;
- demonstrates the computational procedure of numerical BCU methods using the stability boundary of the original system model and that of the reduced-state model;
- conducts analytical studies of the transversality condition of the BCU method and relates the transversality condition with the boundary condition;
- presents the BCU–exit point method;
- develops group properties of power system contingencies;
- explores the static and dynamic group properties of power system coherent contingencies;
- develops the group-based BCU–exit point method and the group-based BCU–CUEP method; and
- develops group-based BCU methodologies, including the group-based BCU–exit point method, the group-based BCU–CUEP method, and the group-based BCU method.

Chapter 2

System Modeling and Stability Problems

Electric power systems are nonlinear in nature. Their nonlinear behaviors are difficult to predict due to (1) the extraordinary size of the systems, (2) the nonlinearity in the systems, (3) the dynamic interactions within the systems, and (4) the complexity of component modeling. These complicating factors have forced power system engineers to analyze the complicated behaviors of power systems through the process of modeling, simulation, analysis, and validation.

2.1 INTRODUCTION

The complete power system model for calculating system dynamic response relative to a disturbance comprises a set of first-order differential equations:

$$\dot{x} = f(x, y, u), \tag{2.1}$$

describing the internal dynamics of devices such as generators, their associated control systems, certain loads, and other dynamically modeled components. The model is also comprised of a set of algebraic equations,

$$0 = g(x, y, u), \tag{2.2}$$

describing the electrical transmission system (the interconnections between the dynamic devices) and the internal static behaviors of passive devices (such as static loads, shunt capacitors, fixed transformers, and phase shifters). The differential equation (Eq. 2.1) typically describes the dynamics of the speed and angle of generator rotors; the flux behaviors in generators; the response of generator control systems such as excitation systems, voltage regulators, turbines, governors, and boilers; the dynamics of equipment such as synchronous VAR compensators (SVCs), DC lines, and their control systems; and the dynamics of dynamically modeled loads such as induction motors. The stated variables x typically include generator rotor angles, generator velocity deviations (speeds), mechanical powers, field voltages, power

Direct Methods for Stability Analysis of Electric Power Systems, by Hsiao-Dong Chiang

system stabilizer signals, various control system internal variables, and voltages and angles at load buses (if dynamic load models are employed at these buses). The algebraic equations (Eq. 2.2) are composed of the stator equations for each generator, the network equations of transmission networks and loads, and the equations defining the feedback stator quantities. An aggregated representation of each local distribution network is usually used in simulating power system dynamic behaviors. The forcing functions u acting on the differential equations are terminal voltage magnitudes, generator electrical powers, signals from boilers, automatic generation control systems, and so on.

Some control system internal variables have upper bounds on their values due to their physical saturation effects. Let z be the vector of these constrained state variables; then, the saturation effects can be expressed as

$$0 < z(t) \leq \bar{z}. \tag{2.3}$$

For a 900-generator, 14,000-bus power system, the number of differential equations can easily reach as many as 20,000, while the number of nonlinear algebraic equations can easily reach as many as 32,000. The sets of differential equations (Eq. 2.1) are usually loosely coupled (Kundur, 1994; Stott, 1979; Tanaka et al., 1994).

2.2 POWER SYSTEM STABILITY PROBLEM

By nature, a power system continually experiences two types of disturbances: event disturbances and load disturbances (Anderson and Fouad, 2003; Balu et al., 1992). Event disturbances include loss of generating units or transmission components (lines, transformers, and substations) due to short circuits caused by lightning, high winds, failures such as incorrect relay operations or insulation breakdown, or a combination of such events. Event disturbances usually lead to a change in the configuration of power networks. Load disturbances, on the other hand, are the sudden large load changes and the small random fluctuations in load demands. The configuration of power networks usually remains unchanged after load disturbances.

To protect power systems from damage due to disturbances, protective relays are placed strategically throughout a power system to detect faults (disturbances) and to trigger the opening of circuit breakers necessary to isolate faults. These relays are designed to detect defective lines and apparatus or other power system conditions of an abnormal or dangerous nature and to initiate appropriate control actions. Due to the action of these protective relays, a power system subject to an event disturbance can be viewed as going through network configuration changes in three stages: the prefault, the fault-on, and the postfault systems (see Table 2.1).

The prefault system is in a stable steady state; when an event disturbance occurs, the system then moves into the fault-on system before it is cleared by protective system operations. Stated more formally, in the prefault regime, the system is at a known stable equilibrium point (SEP), say (x_S^{pre}, y_S^{pre}). At some time t_0, the system undergoes a fault (an event disturbance), which results in a structural change in the system due to actions from relay and circuit breakers. Suppose the fault duration is

Table 2.1 The Time Evolution, System Evolution, Physical Mechanism, and Mathematical Descriptions of the Power System Stability Problem during the Prefault, Fault-On, and Postfault Stages

Physical mechanism	System is operated around a stable equilibrium point.	A fault occurs on the system that initiates relay actions and circuit breaker actions.	The fault is cleared as the actions of circuit breakers are finished.
System evolution and time evolution	Prefault system $t < t_0$	Fault-on system $t_0 \le t \le t_{cl}$	Postfault system $t > t_{cl}$
Mathematical description	$[x(t), y(t)]$, (x_s, y_s)	$\begin{cases} \dot{x} = f_F^1(x, y) \\ 0 = g_F^1(x, y) \\ t_0 \le t \le t_{F,1} \end{cases}$ . . . $\begin{cases} \dot{x} = f_F^k(x, y) \\ 0 = g_F^k(x, y) \\ t_{F,k-1} \le t \le t_{cl} \end{cases}$	$\dot{x} = f(x, y)$ $0 = g(x, y)$

confined to the time interval $[t_0, t_{cl}]$. During this interval, the fault-on system is described by (for ease of exposition, the saturation effects expressed as $0 < z(t) \le \bar{z}$ are neglected in the following) the following set of differential and algebraic equations (DAEs):

$$\begin{aligned} \dot{x} &= f_F(x, y) \quad t_0 \le t < t_{cl} \\ 0 &= g_F(x, y), \end{aligned} \tag{2.4}$$

where $x(t)$ is the vector of state variables of the system at time t. Sometimes, the fault-on system may involve more than one action from system relays and circuit breakers. In these cases, the fault-on systems are described by several sets of DAEs:

$$\begin{aligned} \dot{x} &= f_F^1(x, y), \quad t_0 \le t \le t_{F,1} \\ 0 &= g_F^1(x, y) \\ \dot{x} &= f_F^2(x, y), \quad t_{F,1} \le t \le t_{F,2} \\ 0 &= g_F^2(x, y) \\ &\cdots \\ \dot{x} &= f_F^k(x, y), \quad t_{F,k} \le t \le t_{cl} \\ 0 &= g_F^k(x, y). \end{aligned} \tag{2.5}$$

The number of sets of DAEs equals the number of separate actions due to system relays and circuit breakers. Each set of DAE depicts the system dynamics due to one action from relays and circuit breakers. Suppose the fault is cleared at time t_{cl} and no additional protective actions occur after t_{cl}. The system, termed the postfault system, is henceforth governed by postfault dynamics described by

$$\begin{aligned} \dot{x} &= f_{PF}(x, y), \quad t_{cl} \le t < \infty \\ 0 &= g_{PF}(x, y). \end{aligned} \tag{2.6}$$

The network configuration may or may not be the same as the prefault configuration in the postfault system. We will use the notation $z(t_{cl}) = (x(t_{cl}), y(t_{cl}))$ to denote the fault-on state at switching time t_{cl}. The postfault trajectory after an event disturbance is the solution of Equation 2.6, with initial condition $z(t_{cl}^{+}) = (x(t_{cl}^{+}), y(t_{cl}^{+}))$ over the postfault time period $t_{cl} \le t < t_{\infty}$.

The fundamental problem of power system stability due to a fault (i.e., a contingency) can be summarized as follows: given a prefault SEP and a fault-on system, will the postfault trajectory settle down to an acceptable steady state? A straightforward approach to assess the power system stability is to numerically simulate the system trajectory and then to examine whether the postfault trajectory settles down to an acceptable steady state. A simulated system trajectory of a large-scale power system transient stability model is shown in Figures 2.1 and 2.2. The simulated system trajectory is composed of the predisturbance trajectory (a SEP) and the fault-on trajectory and the postdisturbance (i.e., the postfault) trajectory. The simulated postfault trajectory settles down to a postfault SEP.

Power system dynamic behaviors after a contingency can be fairly complex. This is because electric power systems are composed of a large number of components (equipment and control devices) interacting with each other, exhibiting nonlinear dynamic behaviors with a wide range of timescales. For instance, the difference between the time constants of excitation systems and those of governors is roughly a couple of orders of magnitude. These physical differences are reflected in the

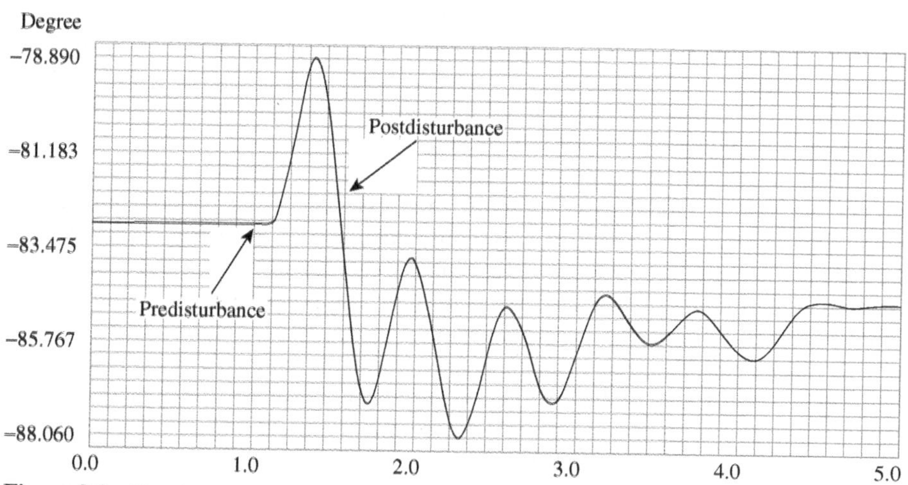

Figure 2.1 The simulated dynamic behavior, prefault, fault-on, and postfault of the generator angle of a large power system model.

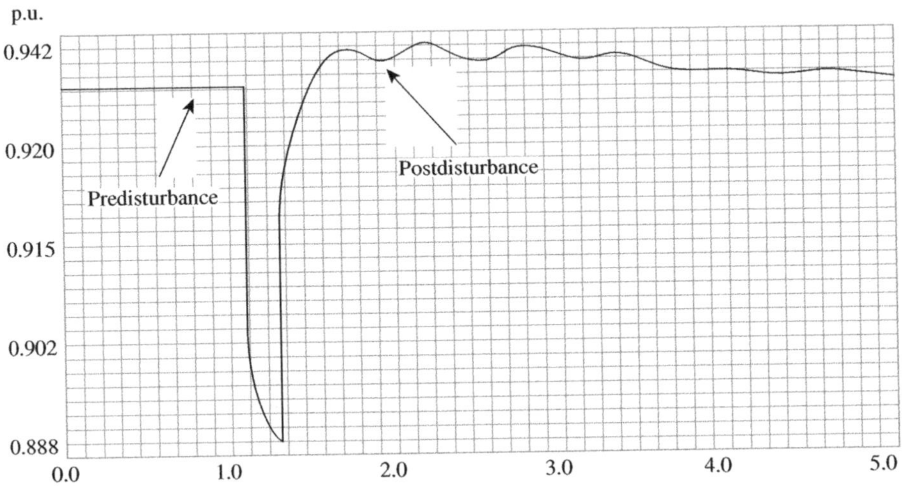

Figure 2.2 The simulated dynamic behavior, prefault, fault-on, and postfault of a voltage magnitude of a large power system model. During the fault, the voltage magnitude drops to about 0.888 p.u.

underlying differential equations, which contain variables of considerably different timescales. The dynamic behavior after a disturbance involves all of the system components, in varying degrees. For instance, a short circuit occurring on a transmission line will trigger the opening of circuit breakers to isolate the fault. This will cause variations in generator rotor speeds, bus voltages, and power flows through transmission lines. Depending on their individual characteristics, voltage variations will activate generator excitation system underload tap changer (ULTC) transformers, SVCs, and undervoltage relays and will cause changes in voltage-dependent loads. Meanwhile, speed variations will activate prime mover governors, underfrequency relays, and frequency-dependent loads. The variations of power flows will activate generation control and ULTC phase shifters. The degree of involvement from each component can be explored to determine the appropriate system model necessary for simulating the dynamic behaviors.

Traditional practice in power system analysis has been to use the simplest acceptable system model, which captures the essence of the phenomenon under study. For instance, the effect of a system component or a control device can be neglected when the timescale of its response is very small or very large compared to the time period of interest. The effects of these components can be roughly taken into account as follows: the dynamic behavior of a system component or a control device can be considered as instantaneously fast if the timescale of its response is very small as compared to the time period of interest. Likewise, the dynamic behavior of a system component or a control device can be considered as a constant if the timescale of its response is very large as compared to the time period of interest. This philosophy has been deemed acceptable because of the severe complexity involved with a full, large-scale power system model (Kundur, 1994; Stott, 1979; Tanaka et al., 1994).

Power system stability models have been divided into three types of stability models with different timescales: (1) a short-term stability model (predominately describing electromechanical transients) on which transient stability analysis is based, (2) a midterm stability model, and (3) a long-term stability model on which long-term stability analysis is based. This division of power system stability models is based on the different timescale involvement of each component and the control device on the overall system's dynamic behaviors (Cate et al., 1984; Kundur, 1994). These three models are described by a set of differential–algebraic equations of the same nature as Equations 2.1 and 2.2, but with different sets of state variables with different time constants. There is, however, a fuzzy boundary distinguishing between the midterm and long-term models. Compared with transient stability analysis, midterm and long-term dynamic behaviors have only come under study relatively recently (Chow, 1982; Kundur, 1994; Stott, 1979; Stubbe et al., 1989; Tanaka et al., 1994).

The time frame of electromechanical oscillations in rotor angle stability typically ranges from a few seconds to tens of seconds. The dynamics of excitation systems, automatic voltage regulators, SVCs, underfrequency load shedding, and undervoltage load shedding are all active in this time frame. These dynamics are called transient (short-term) dynamics, which extend over time intervals on the order of 10 s. The adjective "transient" is added to angle stability to form the term "transient angle stability" when the transient (short-term) power system model is used in a simulation. Similarly, the "adjective transient" is added to voltage stability to form the term "transient voltage stability" when the short-term model is used in voltage stability analysis. When the transient dynamics subside, the system enters the midterm time frame, typically within several minutes, in which the dynamics from such components as ULTC, generator limiters, and load dynamics become active. The time frame following the midterm time frame is the long-term time frame in which turbines, prime mover governors, are active. The adjective "long-term" is added to angle (or voltage) stability to form the term long-term angle (or voltage) stability when the long-term model is used in the simulation.

For transient stability analysis, the assumption of one unique frequency is kept for the transmission network model, but generators have different speeds. Generators are modeled in greater detail, with shorter time constants compared with the models used in long-term stability analysis. Roughly speaking, transient stability models reflect the fast-varying system electrical components and machine angles and frequencies, while the long-term models are concerned with the representation of the slow oscillatory power balance, assuming that the rapid electrical transients have damped out (Kundur, 1994; Tanaka et al., 1994).

2.3 MODEL STRUCTURES AND PARAMETERS

The accuracy of stability analysis has significant impact on power system operational guidelines, operational planning, and design. Accurate stability analysis is necessary to allow for more precise calculations of power transfer capability of transmission grids. The accuracy of stability analysis, however, largely depends on the validity of system models employed in describing power system dynamic behaviors (here, system model refers to the model structure and its associated parameter values). Accurate system models are essential for simulating complex power system behaviors.

In the past, the issue of accurately modeling power system components such as synchronous generators, excitation systems, and loads has received a great deal of attention from the power industry. Standard generator and excitation model structures have been developed (IEEE Standard 421.1, 1986; IEEE Standard 421.5, 1992). The remaining issue is how to derive accurate parameter values for these standard models. This issue is at the heart of parameter estimation in system identification.

Manufacturers develop parameter values for the model structures of generators and their control systems by using an "off-line" approach. In most cases, the parameter values provided by manufacturers are fixed and do not reflect the actual system operating conditions. The effect of nonlinear interaction between the generator (or control system) and the other parts of the system may alter parameter values. For instance, when an excitation system is put into service, its model parameter values may drift due to (1) changes in system operating conditions, (2) the nonlinear interaction between the excitation system and the rest of the power system, and (3) the degree of saturation and equipment aging, and so on. Also, the parameter values of excitation systems provided by manufacturers are typically derived from tests at the plant, before the excitation system is actually put into service, and are often performed by measuring the response of each individual component of the device separately and then by combining those individual components to yield an overall system model. Although adjustments can be made during commissioning, accurate parameter values may not be generally available once the device is installed into the power system. This prompts the use of an online measurement-based approach for developing accurate parameter values.

The measurement-based approach has the advantage of providing reliable data for generators and their integrated control systems by directly measuring the system dynamic behavior as a whole to yield accurate models. For instance, the task of measurement-based generator parameter estimation is to accurately identify the machine's direct and quadrature resistances as well as reactances simultaneously based on measurements without resorting to various off-line tests, such as the open circuit test, in which the machine is isolated from the power system.

Compared to activities in the modeling of generators, loads, and excitation systems, relatively little effort has been devoted to parameter estimation for governors and turbines, whose standard model structures have already been developed (Hannett et al., 1995). This may be explained by the fact that governors and turbines play an important role in power system midterm or long-term stability, but not so much in transient stability, which is much more widely scrutinized. The boiler model, also more relevant in long-term stability studies, is not supported in most current production-grade power system stability programs. In the case of HVDC and some FACTS devices, no standard model is currently available. This is due to one or more of the following reasons: (1) the particular device is new, and standard controls are not well-defined; (2) the device occurs only rarely; and (3) each installation requires a different model.

2.4 MEASUREMENT-BASED MODELING

In the last 20 years, a significant amount of effort has been devoted to measurement-based parameter estimation for synchronous generators, excitation systems, and loads. These efforts are mostly based on the following two classes of methods for estimating these parameters:

- time–domain methods and
- frequency–domain methods.

Historically, frequency–domain methods have appeared to dominate the theory and practice of system identification in control engineering applications. Presently, the literature on system identification is very much dominated by time–domain methods. If the intended use of the model derived from the system identification procedure is to simulate the system or to predict the future outputs of the system, then time–domain methods are most appropriate. Similarly, if the derived model is to be used in conjunction with any state-space/time–domain control system design procedure, then again, time–domain methods are best. However, if the object of system identification is simply to gain general insight into the system, for instance, determining resonances in the response, then frequency–domain methods are probably most appropriate. Most IEEE standard model structures for power components are expressed in the time–domain.

The process of parameter estimation based on measurements can be summarized as follows:

1. Determine a set of input–output data derived from the physical system under study.
2. Estimate its parameters using a suitable method and an estimation criterion.
3. Validate the estimated model using the input–output data.
4. If unsatisfactory, try another model structure and repeat step 2, or try another identification method and another estimation criterion and repeat step 2 until a "satisfactory" model is obtained.

The data from the online measurement are obtained during the occurrence of power system disturbances such as line trippings and faults. The data so acquired reflect the intrinsic characteristics of the system components under normal operating conditions and can be utilized to obtain better parameter values. These improved parameters can in turn be used to improve the modeling and simulation of power system dynamics.

One challenging task in power system modeling is the load modeling. This manifests itself in the unavailability of standard load model structures even though standard generator and excitation model structures have been developed and accepted in the power industry. It is well known that load behaviors have profound impacts on power system dynamic behaviors. Inaccurate load models, for instance, can lead to a power system being operated in modes that result in actual system collapse or separation (CIGRE Task Force 38.02.05, 1990). Simulation studies using simple load models were found to fail in explaining voltage collapse scenarios. Accurate load models are necessary to ensure simulation accuracy in grid operations and planning studies so that more precise stability limits can be determined.

Load models adequate for some types of power system dynamic analysis may be not adequate for others. Hence, representative load models should be developed for certain types, not all types of power system dynamic analysis. For example, voltage stability analysis is more concerned with dynamic behaviors of reactive loads, while transient stability analysis is more concerned with dynamic behaviors of real loads (Liang et al., 1998; Xu and Mansour, 1994). Load models for certain types of power system dynamic analysis were developed in Choi (2006), CIGRE Task Force 38.02.05 (1990), and Ju et al. (1996).

A load model is a mathematical representation of the relationship between a bus voltage (magnitude and frequency) and power (real and reactive) or current flowing into the bus load. At present, the so-called static load model structure (where the load is represented as constant impedance, constant current, constant MVA, or a combination of the three) or voltage frequency-dependent load structure is still commonly used in computer program power system analysis. These static load models are adequate for some types of power system dynamic analysis but not for others. There remains a necessity for the development of accurate dynamic load models. Because of its importance, the subject of load modeling has drawn significant research efforts, for example, those documented in Choi (2006), CIGRE Task Force 38.02.05 (1990), He et al. (2006), IEEE Task Force on Load Representation for Dynamic Performance (1993), and Ju et al. (1996). There are two main time–domain approaches available for developing accurate load models:

- component-based approach and
- measurement-based approach.

The component-based approach builds up the load model from information on dynamic behaviors of all the individual components (Price et al., 1988). Load composition data, load mixture data, and the dynamic behavior of each individual load component of a particular load bus are considered. For a large utility, such surveys of load components can be very difficult and cumbersome tasks.

The measurement-based approach involves placing measurement systems at load buses for which dynamic load models will be developed (CIGRE Task Force 38.02.05, 1990; Craven and Michael, 1983; Hiskens, 2001; Ju et al., 1996). This approach has the advantage of employing direct measurements of the actual load behaviors during system disturbances so that accurate load models can be obtained directly in the form needed for the inputs of existing power system analysis and control programs. These two approaches are complementary to each other. The component-based approach is useful for deriving a suitable model structure for a

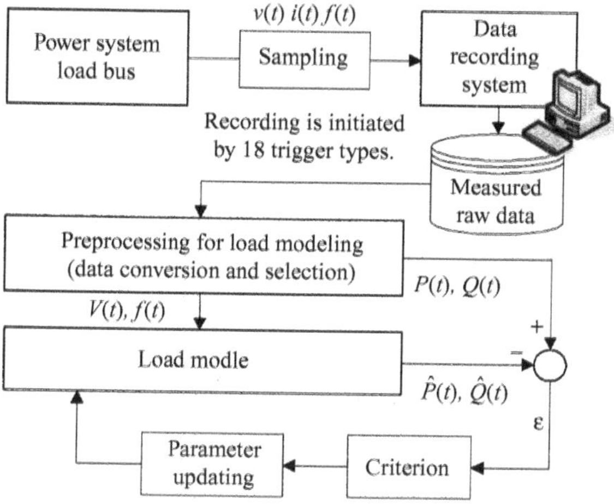

Figure 2.3 A schematic description of the measurement-based load modeling approach.

load bus, whereas the measurement-based approach is appropriate for obtaining values for the associated model parameters.

Figure 2.3 shows a schematic description of the measurement-based load modeling approach. A procedure for identifying a load model using the measurement-based approach is described in the following:

Step 1. Obtain a set of input–output data derived from a set of measurements.

Step 2. Select a load model structure.

Step 3. Estimate its parameters using a suitable method and estimation criterion.

Step 4. Validate the derived model with the parameters obtained in Step 3.

Step 5. If the validation criterion is not met, take remedy actions; for example, try another estimation method or try another model structure and go to Step 3.

2.5 POWER SYSTEM STABILITY PROBLEMS

It is fair to state that any system may always present instabilities when sufficiently large disturbances are introduced. The key point is to find the "proper" disturbance and the appropriate stability condition when a given system or phenomenon is investigated (Hahn, 1967; IEEE TF Report, 1982; Kundur, 1994). A proper disturbance should be relevant to the system and physically meaningful. The "appropriate" stability condition is concerned with the range of deviation in the state space.

There are two types of disturbances in power systems: event disturbances and load variations (Balu et al., 1992). The fundamental problem of power system stability analysis relative to a disturbance (i.e., fault) can be broadly stated as follows:

given a prefault SEP and a fault-on system, will the postfault trajectory settle down to an acceptable SEP (Varaiya et al., 1985)? If the postfault trajectory converges to an equilibrium point, then the system is said to be (rotor) angle stable relative to the fault. Physically, the angle stability represents the ability of synchronous generators of an interconnected power system to remain synchronized after a fault, which initially creates electromechanical oscillations between generators in the system. Furthermore, if the voltage magnitudes both at the equilibrium point and along the postfault trajectory are acceptable, then the system is said to be voltage stable relative to the fault.

Power system instability occurs when the postfault trajectory of a power system subject to a fault does not approach an acceptable steady state (or equilibrium point). Depending on the behaviors of the postfault trajectory, power system instability may be manifested in several different ways. The type of instability may be angle related (an angle instability) or voltage related (a voltage instability), or both. If the voltage magnitudes along the postfault trajectory are unacceptable or the postfault trajectory approaches a steady state with unacceptable voltage magnitudes, then the system is said to suffer from voltage instability relative to the fault. Another type of power system instability is the emergence of oscillatory behaviors such as SSR and low-frequency oscillations. Oscillatory behaviors occur when the postfault trajectory either experiences a prolonged oscillatory transient or approaches a stable periodic solution.

Depending on the size of a disturbance, power system stability can be classified into small-signal (small-disturbance) stability and general (large-disturbance) stability. Small-signal stability refers to the ability of the power system (at an equilibrium point) to maintain synchronism under small disturbances. Given a power system model, a disturbance is called small if the stability of the system can be analyzed based on a linear system model obtained from linearizing the original nonlinear model around the equilibrium point; otherwise, it is called large. Small disturbances, such as small variations in loads and real power generations, occur continually on power systems.

Small-signal stability can be analyzed based on the eigenvalues of the Jacobian matrix at the equilibrium point. Most event disturbances and events involving large variations in loads and real power generations are considered large disturbances. Large-disturbance stability needs to be analyzed based on the postfault trajectory of the original system model. An equilibrium point (operating point) of a power system is small-signal stable if it is an asymptotically stable equilibrium point. From the viewpoint of nonlinear stability analysis, large-disturbance stability, such as angle/voltage stability and transient angle/voltage stability, belongs to the class of asymptotic stability problems.

Two types of basic information are needed in order to perform a power system stability analysis: static data (which is the power flow data) and dynamic data. The power flow data describe the network and its steady-state operating conditions. The dynamic data supply the information needed to compute the response of the modeled devices to a given disturbance. The dynamic data include models of detailed synchronous machines, dynamic load models, induction motor, static VAR compensator, high-voltage DC link, FACTS, and user-defined models. In addition to the above basic information, the following additional information is needed:

- **Description of Disturbances.** This information describes the disturbance to be simulated, for example, fault location and duration, circuit switching, or generation/load rejection. It also specifies the total simulation time.
- **Relay Data.** This data describe the characteristics of the protection devices.
- **Monitoring Information (Optional).** This specifies the variables that are of interest to the user and should be monitored by the program during the simulation.

The accuracy of any simulation depends greatly on the accuracy of the data input into the program. A number of features and facilities are available in some production-grade stability programs to detect questionable data and inconsistencies between the dynamic models and the operating conditions (Electric Power Research Institute, 1992; Stubbe et al., 1989). These features and facilities are provided in the form of error and warning messages and various options to tabulate operating conditions and dynamic data.

2.6 APPROACHES FOR STABILITY ANALYSIS

The complete model for the stability analysis of a typical power system is a set of DAEs numbering on the order of thousands or tens of thousands. At present, stability analysis programs routinely used in utilities around the world are mostly based on step-by-step numerical integrations of the set of equations to simulate system behaviors relative to a given disturbance. A numerical solution of a set of DAEs is a sequence of approximate values of the solution function (curve) at a discrete set of points. A general overview of step-by-step numerical integration methods for power system stability analysis appears in Stott (1979). Several production-grade time-domain simulation packages described in the literature can be found, for example, in Electric Power Research Institute (1992), Kurita et al. (1993), de Mello et al. (1992), Stubbe et al. (1989), and Tanaka et al. (1994).

An alternative approach to transient stability analysis employing energy functions, called direct methods, was originally proposed by Magnusson in 1947 and was pursued in the 1950s and 1960s by several researchers, for example, Aylett (1958), Gless (1966), and El-Abiad and Nagappan (1966). Direct methods received significant research efforts in the 1970s. Direct methods have had a long developmental history spanning six decades, but until recently, many were thought to be impractical for large-scale power systems. Among the direct methods, the classical method of using the concept of closest unstable equilibrium point (UEP) gives a very conservative assessment of stability. The potential energy boundary surface (PEBS) method is fast but may give inaccurate stability assessments. The controlling UEP method is the most viable direct method in terms of its accuracy and slightly conservative nature. The success of the controlling UEP method, however, hinges on the ability to compute the controlling UEP. A great majority of work on the controlling UEP method is based on physical reasonings, heuristics, and simulations without theoretical support. This work has achieved limited success in computing the controlling UEP.

Recent developments of the family of boundary of stability region controlling unstable equilibrium point (BCU) methods have revived the controlling UEP method. The BCU method can reliably compute the controlling UEP. The combination of the controlling UEP method and the BCU method now emerges as a practical means for solving large-scale transient stability analysis problems. Extensive evaluations of the BCU-based controlling UEP method on large-scale power systems such as a 12,000-bus power system were conducted and promising results were reported (Chiang et al., 2006; Tada et al., 2005). Several major installations of BCU methods in modern energy management systems at large power companies have taken place.

The controlling UEP method provides several advantages in stability assessment and control. It can determine transient stability without the time-consuming numerical integration of the (postfault) power system. In addition to its speed, the controlling UEP method also provides an accurate quantitative measure of the degree of system stability. This additional information makes the controlling UEP method very attractive when the relative stability of different network configuration plans must be compared or when system operating limits constrained by transient stability must be calculated quickly. Another advantage of the controlling UEP method is the ability to provide useful information regarding how to derive preventive control actions when the underlying power system is deemed unstable and how to derive enhancement control actions when the underlying power system is deemed critically stable.

We next describe these two different approaches from the state-space viewpoint. The time–domain approach computes the relationship between the prefault stable equilibrium point and the ultimate postfault state explicitly, using step-by-step numerical integrations to simulate the entire system trajectory. On the other hand, direct methods use two steps to solve the stability problem. First, they compute only the relationship between the prefault SEP and the system state at the time of fault clearing using the step-by-step numerical integration of the fault-on system. In the second step, the direct methods directly determine, without numerical integration of the postfault system, whether the initial state of the postfault system lies inside the stability region of a desired SEP. This direct determination of the stability property is based on an energy function (defined for the postfault system) and on a critical energy (relative to the fault-on trajectory). It will be shown that if the energy function value of the initial state of the postfault system is less than the critical energy, then the postfault trajectory will settle down to the desired postfault SEP. This is the analytical basis of direct methods. The great challenge in direct methods is determining the critical energy relative to a fault-on trajectory and deriving energy functions for power system stability models. A comparison between the time–domain simulation approach and the direct methods is summarized in Table 2.2.

Table 2.2 A Comparison of the Time–Domain Approach and the Direct Methods

	Time–domain approach	Direct methods (using energy function)
Prefault system Fault-on system $\dot{x} = f_F(x, y)$ $0 = g_F(x, y)$ $t_0 \leq t \leq t_{cl}$	Prefault SEP 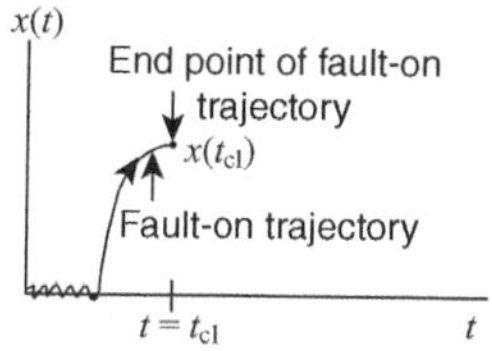	Prefault SEP 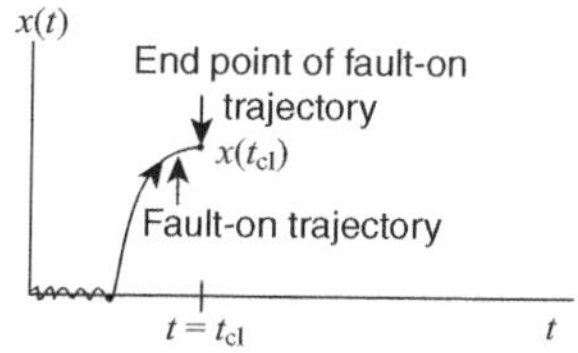
Computation	Numerical integration methods to derive the fault-on trajectory	Numerical integration methods to derive the fault-on trajectory
Postfault system $\dot{x} = f(x)$ $0 = g(x, y)$ $t_{cl} \leq t \leq t_\infty$	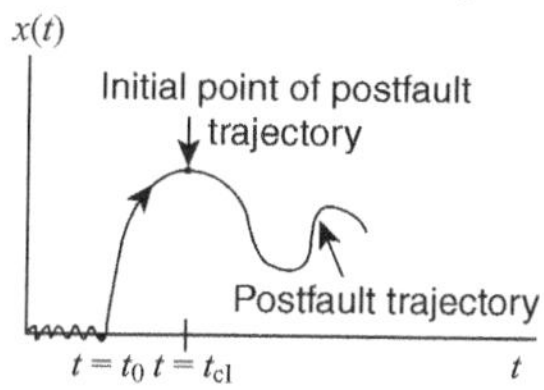	**1.** The postfault trajectory $x(t)$ is not required, while the initial condition $x(t_{cl}^+)$ is required. **2.** The critical energy V_{cr} for the fault-on trajectory is required and the energy function $V(x)$ (for the postfault system) is required.
Computation	Numerical integration methods are used to simulate the postfault trajectory for stability assessment. The simulation time for the postfault system is typically between 10–30 s. If the postfault trajectory settles down, then the postfault system is assessed to be stable; otherwise, it is unstable.	Numerical integration of postfault trajectory is not required. Direct stability assessment of the postfault trajectory is based on the comparison between the energy after the fault and the associated critical energy. In other words, if $V\left(X\left(t_{cl}^+\right)\right) < V_{cr}$, then the postfault trajectory $x(t)$ is stable; otherwise, $x(t)$ may be unstable.

2.7 CONCLUDING REMARKS

After decades of research and developments in the energy function-based direct methods and the time–domain simulation approach, it has become clear that the capabilities of direct methods and that of the time–domain approach complement each other. The current direction of development is to include appropriate direct methods, such as the BCU-based controlling UEP method and a fast time–domain simulation program within the body of overall power system stability simulation programs (Chiang et al., 2007; Jardim et al., 2004; Kim, 1994; Tada and Chiang, 2008). In this development, the BCU-based controlling UEP method provides the advantages of fast computational speed and energy margins, which make it a good complement to the traditional time–domain simulation approach.

The controlling UEP and its functional relations to certain power system parameters, such as power injections, are an effective complement to develop fast calculators for available transfer capability limited by transient stability. In addition, they provide effective information to develop the following control actions:

1. preventive control schemes for credible contingencies that are unstable,
2. enhancement control schemes for credible contingencies that are critically stable, and
3. enhancement control for increasing available transfer capability limited by transient stability.

The analytical basis for direct methods and, in particular, the controlling UEP method, is an understanding of stability regions of nonlinear dynamical systems. We will provide a rigorous theoretical foundation for the direct methods, including the closest UEP method, the PEBS method, and the controlling UEP method in Chapters 8 through 13. The central theme of the theoretical foundation is knowledge of the stability region, which will be developed in the next three chapters.

Chapter 3

Lyapunov Stability and Stability Regions of Nonlinear Dynamical Systems

3.1 INTRODUCTION

Stability is a fundamental subject that unifies engineering and the sciences. This subject has been regarded by many as a fascinating and difficult problem of human culture. As such, there are at least 50 different terms for stability concepts used in the literature. Stability is a very broad subject, and the concept of stability can be formulated in a variety of ways depending on the intended use of stability analysis and design (Alberto and Chiang, 2007; DeCarlo et al., 2000; Hahn, 1967; La Salle and Lefschetz, 1961; May, 1973; Michel et al., 1982). In this chapter, we review some relevant stability concepts from the nonlinear dynamical systems theory. Some of these concepts and their implications will be discussed in detail.

Another important subject related to stability is the stability region of nonlinear dynamical systems. Indeed, the problem of determining stability regions (regions of attraction) of nonlinear dynamical systems is of fundamental importance for many disciplines in engineering and in the sciences (Athay et al., 1984; Chiang et al., 1988; Genesio et al., 1985; Loparo and Blankenship, 1978; Saha et al., 1997; Sastry, 1999; Vu and Liu, 1992). For instance, knowledge of the stability region is essential in the development of direct methods for power system transient stability analysis. Indeed, this problem is at the heart of direct methods. A comprehensive theory of stability regions for the (autonomous) dynamical system will be derived in this chapter. Most of the proofs presented in this chapter are taken from Chiang et al. (1987, 1988) and from Chiang and Thorp (1989b). The topic of estimating stability regions for higher-dimension nonlinear dynamical systems will be discussed in Chapter 5.

Direct Methods for Stability Analysis of Electric Power Systems, by Hsiao-Dong Chiang

3.2 EQUILIBRIUM POINTS AND LYAPUNOV STABILITY

We consider the following (autonomous) nonlinear dynamical system:

$$\dot{x} = f(x), x \in \Re^n. \tag{3.1}$$

It is natural to assume the function (i.e., the vector field) f: $R^n \rightarrow R^n$ satisfies a sufficient condition for the existence and uniqueness of a solution. The solution curve of Equation 3.1 starting from x at $t = 0$ is called the system trajectory starting from x and is denoted by $\phi(x, \cdot)$. The system trajectory starting from x is a function of time; given a specified time, the system trajectory function maps the specified time into a point in the state space. $\bar{x} \in \Re^n$ is said to be an *equilibrium solution* of Equation 3.1 if $f(\bar{x}) = 0$; that is, the equilibrium point is a solution that does not change in time. Hence, equilibrium points are degenerated solution curves that do not move.

The stability property of an equilibrium point will be discussed next. We first review the concepts of (Lyapunov) stability and asymptotic stability.

Definition: Lyapunov Stability
An equilibrium point $\bar{x} \in \Re^n$ of Equation 3.1 is said to be *(Lyapunov) stable* if, for each open neighborhood U of $\bar{x} \in \Re^n$, there exists an open neighborhood V of $\bar{x} \in \Re^n$ such that, for all $x \in V$ and for all $t > 0$, $\phi_t(x) \in U$. Equivalently, $\phi_t(x) \in U$ for all $t > 0$. Otherwise, $\bar{x}$ is unstable.

The concept of (Lyapunov) stability is illustrated in Figure 3.1. Intuitively speaking, an equilibrium point is stable if nearby trajectories stay nearby. In many applications, the requirement of "nearby trajectories stay nearby" is not sufficient; instead, the requirement becomes "nearby trajectories stay nearby and all converge to the equilibrium point." Under this situation, the concept of (Lyapunov) stability can be sharpened into the concept of asymptotic stability as defined in the following.

Definition: Asymptotic Stability
An equilibrium point $\bar{x} \in \Re^n$ of Equation 3.1 is said to be *asymptotically stable* if there exists an open neighborhood U of $\bar{x} \in \Re^n$, such that , for all $x \in U$ and for all

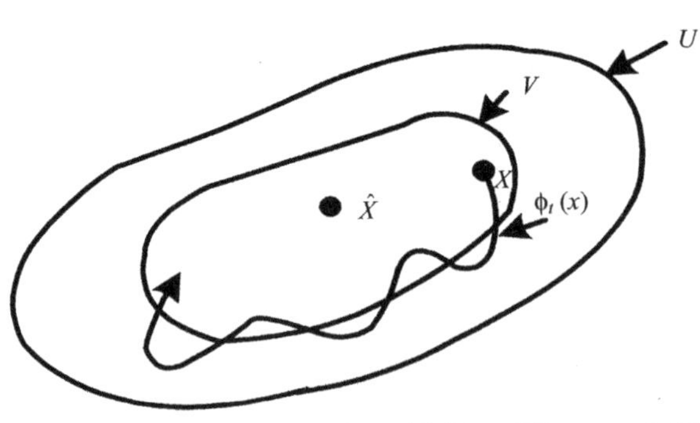

Figure 3.1 An illustration of the definition of (Lyapunov) stability.

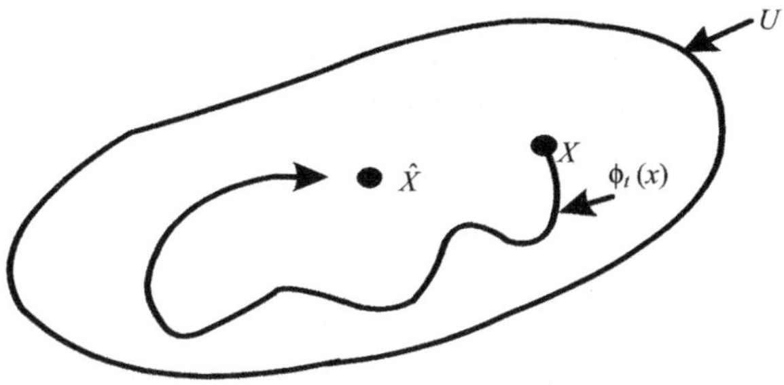

Figure 3.2 An illustration of the definition of asymptotic stability.

$t > 0$, $\phi_t(x) \in U$. In addition, every trajectory $\phi_t(x)$ starting from this neighborhood converges to the equilibrium point $\bar{x}$.

Equivalently,

(i) $\phi_t(x) \in U$, $t > 0$ and

(ii) $\lim_{t\to\infty} \|\phi_t(x) - \bar{x}\| = 0$.

The concept of asymptotic stability is illustrated in Figure 3.2. Intuitively speaking, an equilibrium point is asymptotically stable if it is the sink of nearby trajectories. An asymptotically stable equilibrium point is also termed a sink in the classical literature of nonlinear dynamical systems (Athay et al., 1984; Guckenheimer and Holmes, 1983; Hirsch and Smale, 1974).

In order to determine the stability of $\bar{x}(t)$, we must understand the nature of solutions near $\bar{x}(t)$. Let

$$x(t) = \bar{x}(t) + y(t). \tag{3.2}$$

Substituting Equation 3.2 into Equation 3.1 and Taylor expanding about $\bar{x}(t)$ gives

$$\dot{x}(t) = \dot{\bar{x}}(t) + \dot{y} = f\left(\bar{x}(t) + Df(\bar{x}(t))y + \boldsymbol{O}\left(|y|^2\right)\right) \tag{3.3}$$

where Df is the derivative of f and $|\cdot|$ denotes a norm on $\Re^n$. Using the fact that $\dot{\bar{x}}(t) = f(\bar{x}(t))$, Equation 3.3 becomes

$$\dot{y} = Df\left(\bar{x}(t)y + \boldsymbol{O}\left(|y|^2\right)\right). \tag{3.4}$$

Equation 3.4 describes the evolution of orbits near $\bar{x}(t)$. For stability questions, we are concerned with the behavior of solutions arbitrarily close to $\bar{x}(t)$, so it seems reasonable that this question could be answered by studying the associated *linear system*:

$$\dot{y}(t) = Df(\bar{x}(t))y. \tag{3.5}$$

Therefore, the question of stability of $\bar{x}(t)$ involves the following two steps:

1. Determine if the $y = 0$ solution of Equation 3.5 is stable.
2. Show that stability (or instability) of the $y = 0$ solution of Equation 3.5 implies stability (or instability) of $\bar{x}(t)$ of the system (Eq. 3.1).

It can be shown that if the eigenvalues of the associated linear vector field have nonzero real parts, then the orbit (i.e., trajectory) structure *near an equilibrium solution* of the nonlinear vector field (Eq. 3.1) is essentially the same as that of the linear vector field (Eq. 3.5). This is addressed by the following theorem.

Theorem 3.1: Hartman–Grobman (Hirsch and S. Smale, 1974)

Consider the system (Eq. 3.1) with an equilibrium point $\overline{x}$. If $Df(\overline{x})$ has no zero of purely imaginary eigenvalues, there is a homeomorphism h defined on a neighborhood U of $\overline{x}$ taking orbits of the flow ϕ_t to those of the linear flow $e^{tDf(\overline{x})}$ of Equation 3.5. The homeomorphism preserves the sense of the orbits and is chosen to preserve parameterization by time.

A corollary of Theorem 3.1 and a fundamental linear system theory lead to the following sufficient condition for an equilibrium point to be asymptotically stable.

Theorem 3.2: Asymptotic Stability (Hirsch and Smale, 1974; Sastry, 1999)

Suppose all of the eigenvalues of $Df(\overline{x})$ of the linear system (Eq. 3.5) have negative real parts. Then, the equilibrium solution $x = (\overline{x})$ of the nonlinear system (Eq. 3.1) is asymptotically stable.

This is a well-known stability evaluation method that checks the eigenvalues of the Jacobian matrix at an equilibrium point. The required computational efforts can be immense. An alternative approach is based on the Lyapunov function theory, which is discussed in the next section.

3.3 LYAPUNOV FUNCTION THEORY

One of the most important developments in the stability theory is the Lyapunov function theory. Lyapunov, a Russian mathematician and engineer, laid down the foundation for the Lyapunov theory. Lyapunov stability theorems give sufficient conditions for Lyapunov stability and asymptotic stability. The appealing feature of the Lyapunov function theory is that it derives the stability properties of the equilibrium point without numerically solving the underlying ordinary differential (difference) equations. This is the "spirit of Lyapunov." There are several versions of proof of the Lyapunov function theorem.

During the early period of developing direct methods, the Lyapunov function theorem was applied by many researchers to ensure power system transient stability without time–domian simulation. In this section, we present an overview of a fundamental Lyapunov function theorem.

We denote the following notation as the time derivative of a function $V(x)$, taken along the system trajectory:

$$
\begin{aligned}
\dot{V}(x(t)) &= \frac{\partial V(x(t))^T}{\partial x} \cdot \dot{x}(t) \\
&= \frac{\partial V(x)^T}{\partial_x} \cdot f(x).
\end{aligned}
\tag{3.6}
$$

Since the vector field $f(x)$ and the gradient of the function $V(x)$ are available without the explicit knowledge of the system trajectory, the time derivative of $V(x(t))$ can be performed without knowledge of the system trajectory.

Theorem 3.3: Lyapunov's Stability (Guckenheimer and Holmes, 1983; Hirsch and Smale, 1974)

Let $\hat{x}$ be an equilibrium point of $\dot{x} = f(x)$, where f: $R^n \rightarrow R^n$. Let V: $U \rightarrow R$ be a continuous function defined on a neighborhood of $\hat{x}$, differentiable on $U \rightarrow \hat{x}$, such that

(a) $V(\hat{x}) = 0$ and $V(x) > 0$ if $x \neq \hat{x}$, and $x \in U$.

(b) $\dot{V}(x) \leq 0$ in $U - \hat{x}$.

Then, $\hat{x}$ is stable. Furthermore, if also

(c) $\dot{V}(x) < 0$ in $U - \hat{x}$, then $\hat{x}$ is asymptotically stable.

(d) $\dot{V}(x) < 0$ in $U - \hat{x}$, $U = R^n$, then $\hat{x}$ is asymptotically stable in the large.

Remarks:

1. The Lyapunov function theory asserts not only the stability property of the equilibrium point (a local result) but also that there does not exist any limit cycle (oscillation behavior) or bounded complicated behavior such as an almost periodic trajectory, chaotic motion, and so on, in the subset of the state space where there exists a Lyapunov function.
2. It should be pointed out that the Lyapunov function theory only furnishes sufficient conditions. If for a particular Lyapunov function candidate, V, the required conditions on the derivative of V, that is, $\dot{V}$, are not met, then conclusions regarding the stability or instability of the equilibrium point still cannot be drawn.

There is no systematic way of constructing Lyapunov functions for general nonlinear systems. This is a fundamental drawback of the Lyapunov direct method. Therefore, faced with specific nonlinear dynamical systems, one often has to use experience, intuition, trial and error, and physical insights (e.g., the energy function for electrical and mechanical systems) to search for an appropriate Lyapunov function. In this literature, a number of methods and techniques facilitating the search of Lyapunov functions have been proposed (Khalil, 2002; Michel et al., 1984; Vaahedi et al., 1998; Vidyasagar, 2002).

3.4 STABLE AND UNSTABLE MANIFOLDS

The concepts of an *invariant set*, a *limit set* including an *α-limit set* and an *ω-limit set*, and *stable* and *unstable manifolds* are important to the dynamical system theory. Each of these concepts is defined next. A detailed discussion of these concepts and implications may be found in Guckenheimer and Holmes (1983), in Paganini and Lesieutre (1999), and in Palis and de Melo (1981).

A set $M \in R^n$ is called an *invariant set* of Equation 3.1 if every trajectory of Equation 3.1 starting in M remains in M for all t. A point p is said to be in the ω-limit set of x if, corresponding to each $\varepsilon > 0$ and $T > 0$, there is a $t > T$ with the property that $|\phi(x, t) - p| < \varepsilon$. This is equivalent to saying that there is a sequence $\{t_i\}$ in R, $t_i \to \infty$, with the property that $p = \lim_{i\to\infty}\phi(x, t_i)$. A point p is said to be in the α-limit set of x if, corresponding to each $\varepsilon > 0$ and $T < 0$, there is a $t < T$ with the property that $|\phi(x, t) - p| < \varepsilon$. This is equivalent to saying that there is a sequence $\{t_i\}$ in R, $t_i \to -\infty$, with the property that $p = \lim_{i\to\infty}\phi(x, t_i)$. Hence, the ω-limit set captures the asymptotic behaviors of a bounded trajectory in positive time, while the α-limit set captures the asymptotic behaviors of a bounded trajectory in negative time.

One of the fundamental properties of the limit set is as follows.

Theorem 3.4: Properties of Limit Sets

If a trajectory $\phi(x, t)$ of a system (Eq. 3.1) is bounded for $t \geq 0$ (or $t \leq 0$), then its ω-limit set (or α-limit set) exists; moreover, its limit set is compact, connected, and invariant.

Generally speaking, limit sets can be very complex; they can be equilibrium points, limit cycles (closed orbits), quasi-periodic solutions, and chaos. Stable limit sets are of supreme importance in experimental and numerical settings because they are the only kind of limit sets that can be observed naturally! The concept of stable limit sets is similar to that of equilibrium points.

Definition: Stable Limit Set
A limit set L is said to be *Lyapunov stable* if, for each open neighborhhod U of L, there exists an open neighborhood V of L, such that for all $x \in V$ and for all $t > 0$, $\phi(x) \in U$. Equivalently, $\phi_t(x) \in U$ for all $t > 0$. Otherwise, L is unstable.

Definition: Asymptotically Stable Limit Set
A limit set L is *asymptotically stable* if there exists an open neighborhood V of L such that the ω-limit set of every point in V is L. Equivalently,

(i) $\phi_t(x) \in V$, $t > 0$ and

(ii) $\lim_{t\to\infty} \operatorname*{infimum}_{y\leftarrow L} \|\phi_t(x) - y\| = L.$

We next review the concept of stable and unstable manifolds of limit sets. We start from the simplest limit set—the equilibrium point. Let $\hat{x}$ be an equilibrium point and $U \subset R^n$ be a neighborhood of $\hat{x}$. We define the local stable manifold of $\hat{x}$ as follows:

$$W^S_{loc}(\hat{x}) := \{x \in U : \phi_t(x) \to \hat{x} \text{ as } t \to \infty\}.$$

The local unstable manifold of $\hat{x}$ is defined as

$$W^u_{loc}(\hat{x}) := \{x \in U : \phi_t(x) \to \hat{x} \text{ as } t \to -\infty\}.$$

Note that $W^S_{loc}(\hat{x})$ is a positive invariant set, while $W^u_{loc}(\hat{x})$ is a negative invariant set. They may not be manifolds when $\hat{x}$ is nonhyperbolic. We next present a fundamental theorem of stable and unstable manifolds for a hyperbolic equilibrium point. Recall that an equilibrium point is *hyperbolic* if the corresponding Jacobian matrix has no eigenvalues with zero real parts; otherwise, it is a nonhyperbolic *equilibrium point.*

Theorem 3.5: Unstable–Stable Manifold Theorem (Guckenheimer and Holmes, 1983; Sastry, 1999)

Suppose that the nonlinear system (Eq. 3.1) has a hyperbolic equilibrium point $\bar{x}$. Then the sets $W^S_{loc}(\bar{x})$ and $W^u_{loc}(\bar{x})$, often referred to as the local stable and unstable manifolds, are manifolds of the same dimensions n_s, n_u as those of the stable and unstable eigenspaces E^s, E^u of the linearized system (Eq. 3.5). These manifolds are also tangent to E^s, E^u at $\bar{x}$. $W^S_{loc}(\bar{x})$ and $W^u_{loc}(\bar{x})$ are as smooth as the vector field $f(x)$ of Equation 3.1 (see Figure 3.3).

The stable manifold $W^S(\bar{x})$ and the unstable manifold $W^u(\bar{x})$ are obtained by letting points in $W^S_{loc}(\bar{x})$ flow backward in time and points in $W^u_{loc}(\bar{x})$ flow forward in time:

$$W^S(\bar{x}) := \bigcup_{t \leq 0} \phi_t(W^S_{loc}(\bar{x})) \text{ and} \tag{3.7}$$

$$W^u(\bar{x}) := \bigcup_{t \geq 0} \phi_t(W^u_{loc}(\bar{x})). \tag{3.8}$$

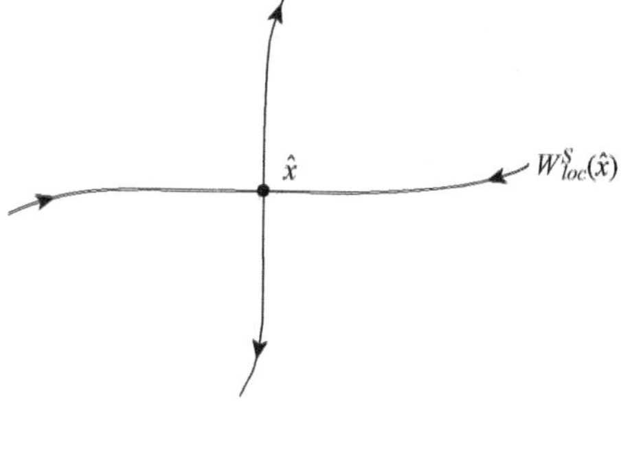

Figure 3.3 The local stable and unstable manifolds of an equilibrium point.

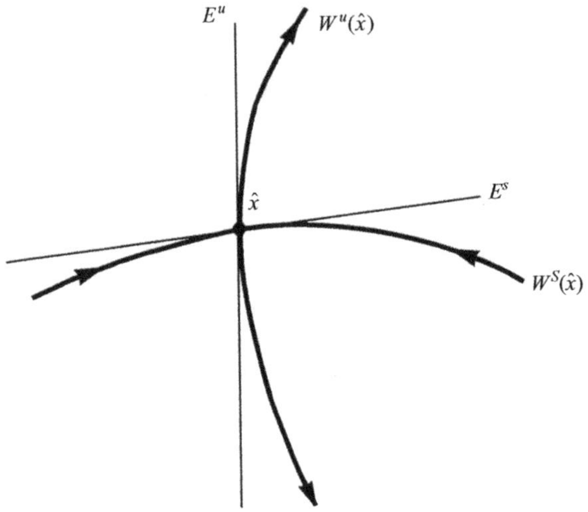

Figure 3.4 The relationship between a stable eigenspace, an unstable eigenspace, and stable and unstable manifolds at a hyperbolic equilibrium point.

Remarks:

1. Clearly, $\hat{x}$ is the ω-limit set of every point in $W^S(\hat{x})$, as well as the α-limit set of every point in $W^u(\hat{x})$. For a hyperbolic equilibrium point, the dimension of the stable manifold $W^S(\hat{x})$ equals the number of eigenvalues of the Jacobian $J_f(\hat{x})$ with negative real parts. The sum of the dimension of $W^S(\hat{x})$ and that of $W^u(\hat{x})$ equals the dimension of the state space (see Figure 3.4).
2. Stable and unstable manifolds are invariant sets. Every trajectory in the stable manifold $W^S(\hat{x})$ converges to $\hat{x}$ as time goes to positive infinity, while every trajectory in the stable manifold $W^u(\hat{x})$ converges to $\hat{x}$ as time goes to negative infinity.
3. The above concepts and definitions of stable and unstable manifolds are applicable to other classes of hyperbolic limit sets such as periodic solutions, quasi-periodic solutions and chaos.

Motivated by Theorem 3.5, we define the type of equilibrium point based on the number of eigenvalues of the corresponding Jacobian matrix with a positive real part.

Definition: Type of Equilibrium Point
The *type* of hyperbolic equilibrium point p of a general nonlinear system (Eq. 3.1) is defined as the number of eigenvalues of $(\partial f/\partial x)$ (p) with a positive real part. If $(\partial f/\partial x)$ (p) has exactly one eigenvalue with a positive real part, we call p a *type-one equilibrium point*. Similarly, p is called a type-k equilibrium point if $(\partial f/\partial x)$ (p) has exactly a k eigenvalue with a positive real part.

Type-one equilibrium points play a dominating role in the characterization of a stability boundary and a quasi-stability boundary. Hence, type-one equilibrium points and stable equilibrium points (SEPs) are important equilibrium points used in this text.

The idea of transversality is basic in the study of dynamical systems (Palis, 1969; Palis and de Melo, 1981; Smale, 1967). If A and B are injectively immersed manifolds in M, we say they satisfy the *transversality condition* if either (1) at every point of intersection $x \in A \cap B$, the tangent spaces of A and B span the tangent space of M at x; that is,

$$T_x(A)+T_x(B)=T_x(M) \quad \text{for } x \in A \cap,$$

or (2) they do not intersect at all. One of the most important features of a hyperbolic equilibrium point $\hat{x}$ is that its stable and unstable manifolds intersect transversely at $\hat{x}$. This transverse intersection is important because it persists under perturbation of the vector field.

3.5 STABILITY REGIONS

For a SEP, say, x_s, there exists a number $\delta > 0$ such that every point in the set, $\|x_0 - x_s\| < \delta$ implies that the trajectory starting from the initial point x_0 converges to the SEP x_s; that is, $\phi_t(x_0) \to \hat{x}$ as $(t \to \infty)$. If δ is arbitrarily large, then $\hat{x}$ is called a *global SEP*. There are many physical systems containing SEPs but not globally stable equilibrium points. A useful concept for these kinds of systems is that of the *stability region* (also called the *region of attraction*). The stability region of a SEP x_s is the set of all points x such that

$$\lim_{t\to\infty} \phi_t(x) \to x_s. \tag{3.9}$$

We will denote the stability region of x_s by $A(x_s)$ and its closure by $\bar{A}(x_s)$, respectively; hence,

$$A(x_s) := \{x \in R^n : lim_{t\to\infty}\phi_t(x) = x_s\}. \tag{3.10}$$

When it is clear from the context, we write A for $A(x_s)$, and so on. Alternatively, the stability region can be expressed as

$$A(x_s) = \{x \in R^n : \omega(x) = x_s\}, \tag{3.11}$$

where $\omega(x)$ denotes the ω-limit set of x. From a topological point of view, the stability region $A(x_s)$ is an open, invariant, and connected set (Hirsch, 1976). The boundary of the stability region $A(x_s)$ is called the *stability boundary* (also called *separatrix*) of x_s and will be denoted by $\partial A(x_s)$ (see Figure 3.5).

We next discuss some topological properties of the stability region and stability boundary. Since the stability region of a SEP is, in fact, the stable manifold, the following topological properties of stability regions are based on the properties of the stable manifold of x_s (Hirsch and Pugh, 1970).

Theorem 3.6: Topological Property

The stability region $A(x_s)$ is an open, invariant set that is diffeomorphic to R^n.

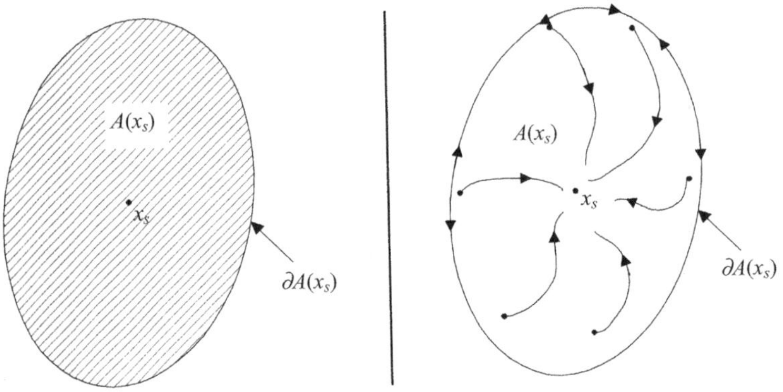

Figure 3.5 As time increases, every trajectory in the stability region $A(x_s)$ converges to the SEP. x_s and every trajectory on the stability boundary evolve on the boundary.

The above result asserts that every trajectory in a stability region lies entirely in the stability region and that the dimension of the stability region is n. Since the boundary of an invariant set is also invariant and the boundary of an open set is a close set, the set of the stability boundary has the following property.

Theorem 3.7: Topological Property

The stability boundary $\partial A(x_s)$ is a closed invariant set of dimension $<n$. If $A(x_s)$ is not dense in R^n, then $\partial A(x_s)$ is of dimension $n - 1$.

If there are at least two SEPs, then the dimension of each stability boundary is $n - 1$; in particular, stability boundaries are nonempty in this case. Characterization of stability regions can be achieved via characterization of the stability boundary. We will develop a comprehensive theory for stability regions of nonlinear dynamical systems (Eq. 3.1) in the next section.

3.6 LOCAL CHARACTERIZATIONS OF STABILITY BOUNDARY

Our aim is to present a comprehensive theory of stability regions for the (autonomous) dynamical system (Eq. 3.1). In this section, several dynamic and topological properties of a stability boundary will be also derived. A complete characterization for two fundamental limit sets of general nonlinear dynamical systems (i.e., equilibrium points and limit cycles) lying on the stability boundary will be derived.

Our approach starts from a local characterization of the stability boundary and moves towards a global characterization of the stability boundary. We first derive a complete characterization for an equilibrium point to lie on the stability boundary, which is a key step in the characterization of the stability region $A(x_s)$. We do this in two steps. First, we impose only one assumption on the dynamical system, namely, that equilibrium points are hyperbolic, and derive conditions for an equilibrium point to be on the stability boundary in terms of both its stable and unstable manifolds. Additional conditions are then imposed on the dynamical system and the results are further sharpened. We also derive the characterizations of closed orbits on the stability boundary. We use the notation $A - B$ to denote those elements that belong to A but not to B. We use the term critical element to denote equilibrium points and limit cycles.

Let x be a hyperbolic critical element. Let U be a neighborhood of x in $W^s(x)$ whose boundary ∂U is transversal to the vector field f. We call ∂u a *fundamental domain* of $W^s(x)$. A cross section $V \subset R^n$ of a vector field f is a manifold V of dimension $n - 1$, which need not be a hyperplane but must be in a manner such that the flow of f is everywhere transversal to it. Any cross section of f containing ∂U and transversal to $W^s(x)$ is the so-called *fundamental neighborhood* $G(x)$ associated with $W^s(x)$. It follows that $W^s(x) = \cup_{t\in R}\Phi_t(\partial U) \cup \{x\}$ and $\cup_{t\geq 0}\Phi_t(G(x)) \cup W^u(x)$ contains a neighborhood of x.

Theorem 3.8: Characterization of Equilibrium Point on the Stability Boundary (Chiang et al., 1988)

Let A be the stability region of a SEP x_s. Let $x \neq x_s$ be a hyperbolic equilibrium point. Then,

(i) if $\{W^u(\hat{x}) - \hat{x}\} \cap \bar{A} \neq \phi$, then $\hat{x} \in \partial A$. Conversely, if $\hat{x} \in \partial A$, then $\{W^u(\hat{x})\} \cap \bar{A} \neq \phi$;

(ii) Suppose $\hat{x}$ is not a source (i.e., $\{W^S(\hat{x}) - \hat{x}\} \neq \phi$). Then $\hat{x} \in \partial A$ if and only if $\{W^S(\hat{x}) - \hat{x}\} \cap \partial\bar{A} \neq \phi$.

Proof: (i) If $y \in W^u(x) \cap \bar{A}$, then

$$\lim_{t\to\infty} \Phi_{-t}(y) = \hat{x}$$

But since $\bar{A}$ is invariant, we have

$$\Phi_{-t}(y) \in \bar{A}.$$

It follows that

$$\hat{x} \in \bar{A}.$$

Since $\hat{x}$ cannot be in the stability region, $\hat{x}$ is on the stability boundary.

Suppose, conversely, that $\hat{x} \in \partial A$. Let $G \subset \{W^u(\hat{x}) - \hat{x}\}$ be a fundamental domain of $W^u(\hat{x})$; this means that G is a compact set, such that

$$\cup_{t\in R} \Phi_t(G) = \{W^u(\hat{x}) - \hat{x}\}.$$

Let $G\varepsilon$ be the ε-neighborhood of G in R^n. Then $\cup_{t<0}\Phi_t(G_\varepsilon)$ contains a set of the form $\{U - W^S(\hat{x})\}$, where U is a neighborhood of $\hat{x}$. Since $\hat{x} \in \partial A$, it follows that $U \cap A \neq \phi$. But, by assumption, $\hat{x} \in \partial A$, so $W^S(\hat{x}) \cap A = \phi$. Therefore, we have

$$\{U - W^S(\hat{x})\} \cap A \neq \phi$$

or

$$\bigcup_{t<0} \Phi_t(G_\varepsilon) \cap A \neq \phi.$$

This implies that $G_\varepsilon \cap \Phi_t(A) \neq 0$ for some t. Since A is invariant under the flow, it follows that

$$G_\varepsilon \cap A \neq \phi.$$

Since $\varepsilon > 0$ is arbitrary and G is a compact set, we conclude that G contains at least a point of $\bar{A}$. The proof of (ii) is similar to the proof of (i), thus completing the proof.

The above characterization of an equilibrium point lying on the stability boundary can be extended to another critical element, that is, closed orbit (limit cycle). By a *closed orbit* of a dynamical system we mean the image of a nonconstant periodic solution of Equation 3.1; that is, a trajectory γ is a closed orbit if γ is not an equilibrium point and $\Phi_t(x) = x$ for some $x \in \gamma$, $t \neq 0$. A closed orbit γ is said to be *hyperbolic* if, for any $p \in \gamma$, $n - 1$ of the eigenvalues of the Jacobian of $\Phi_t(\gamma)$ at p have a modulus not equal to 1 (one eigenvalue must always be 1). A *critical element* of the vector field f is either a closed orbit or an equilibrium point.

The stable and unstable manifolds of a hyperbolic closed orbit γ are defined as the following:

$$W^S(\gamma) = \{x \in M : \Phi_t(x) \to \gamma \text{ as } t \to \infty\} \text{ and}$$
$$W^u(\gamma) = \{x \in M : \Phi_t(x) \to \gamma \text{ as } t \to \infty\}.$$

A characterization of the closed orbit (limit cycle) on the stability boundary is as follows.

Theorem 3.9: Characterization of a Closed Orbit on the Stability Boundary (Chiang et al., 1988)

Let A be the stability region of a SEP. Let γ be a hyperbolic closed orbit. Then,

(i) $\gamma \subseteq \partial A$ if and only if $\{W^u(\gamma) - \gamma\} \cap \bar{A} \neq \phi$;
(ii) Suppose $\{W^S(\gamma) - \gamma\} \neq \phi$. Then, $\gamma \subseteq \partial A$ if and only if $\{W^S(\gamma) - \gamma\} \cap \partial A \neq \phi$.

As a corollary to Theorem 3.8, if $\{W^u(\hat{x}) - \hat{x}\} \cap A \neq \phi$, then $\hat{x}$ must be on the stability boundary. Since any trajectory in $A(x_s)$ approaches x_s, we see that a sufficient condition for $\hat{x}$ to be on the stability boundary is the existence of a trajectory in $W^u(\hat{x})$, which approaches x_s. The nice thing about this condition is that it can be checked numerically. From a practical point of view, we would like to see when this condition is necessary. We will show that this condition becomes necessary under two additional assumptions.

So far, we have assumed only that the critical elements are hyperbolic. This is a generic property for dynamical systems. Roughly speaking, we say a property is generic for a class of systems if that property is true for *almost all* systems in the class. A formal definition is given in Hirsch (1976). It has been shown in Palis (1969) that among $C^r(r \geq 1)$ vector fields, the following properties are generic: (1) all equilibrium points and closed orbits are hyperbolic, and (2) the intersections of the stable and unstable manifolds of critical elements satisfy the transversality condition. Theorem 3.8 can be sharpened under two conditions, one of which is generic for a nonlinear dynamical system (Eq. 3.1). This is the transversality condition. The other condition requires that every trajectory on the stability boundary approaches one of the critical elements.

The following proposition (Hirsch, 1976) is used in the proofs of the next two theorems.

Lemma 3.10

Let x_i and x_j be hyperbolic critical elements of the nonlinear dynamical system (Eq. 3.1). Suppose that the intersection of the stable and unstable manifolds of x_i, x_j satisfy the transversality condition and $\{W^u(x_i) - x_i\} \cap \{W^S(x_j) - x_j\} \neq \phi$. Then, dimension $W^u(x_i) \geq \dim W^u(x_j)$, where the equality sign is true only when x_i is an equilibrium point and x_j is a closed orbit.

The following Lemma, which is a weak version of the λ-lemma (Chiang et al., 1988), is useful in the proof of the next theorem. Recall that the type of an equilibrium point is the dimension of its unstable manifold. An m-disk is a disk of dimension m.

Lemma 3.11

Let $\hat{\upsilon}$ be a hyperbolic critical element of the nonlinear dynamical system (Eq. 3.1) with dimension $W^u(\hat{\upsilon}) = m$. If $\hat{\upsilon}$ is an equilibrium point, let D be an m-disk in $W^u(\hat{\upsilon})$. If $\hat{\upsilon}$ is a closed orbit, let D be an $(m-1)$-disk in $W^u(\hat{\upsilon}) \cap S$, where S is a cross section at $p \in \hat{\upsilon}$. Let N be an m-disk (if $\hat{\upsilon}$ is an equilibrium point) or $(m-1)$-disk (if $\hat{\upsilon}$ is a closed orbit) having a point of transversal intersection with $W^S(\hat{\upsilon})$. Then D is contained in the closure of the set $\cap_{t \geq 0} \Phi_t(N)$.

Now, we present the key theorem of this section, which characterizes an equilibrium point being on the stability boundary in terms of both its stable and unstable manifolds. From the practical point of view, this result is more useful in the numerical verification of equilibrium points on the stability boundary than Theorem 3.8.

Theorem 3.12: Characterization of an Equilibrium Point on the Stability Boundary (Chiang et al., 1988)

Let A be the stability region of a SEP of the nonlinear dynamical system (Eq. 3.1). Let $\hat{x}$ be an equilibrium point. Assume the following.

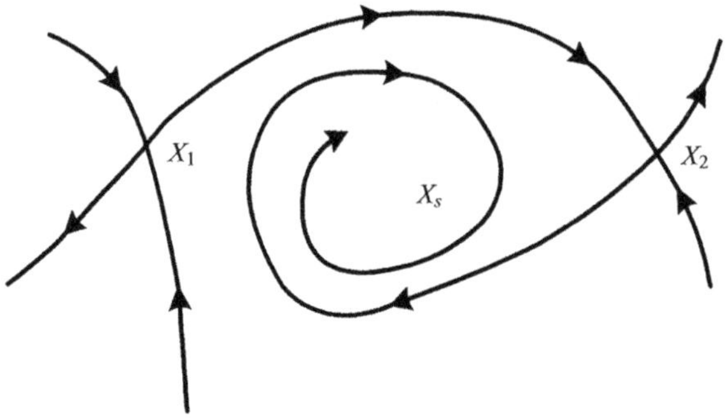

Figure 3.6 The intersection between the unstable manifold of x_1 and the stable manifold of x_2 does not satisfy the transversality condition.

(A1) All the equilibrium points on ∂A are hyperbolic.

(A2) The stable and unstable manifolds of equilibrium points on ∂A satisfy the transversality condition.

(A3) Every trajectory on ∂A approaches one of the equilibrium points as $t \to \infty$.

Then,

(1) $\hat{x} \in \partial A$ if and only if $W^u(\hat{x}) \cap A \neq \phi$ and

(2) $\hat{x} \in \partial A$ if and only if $W^S(\hat{x}) \subseteq \partial A$.

To show that the transversality condition is needed in Theorem 3.12, let us consider the example taken from Tsolas et al. (1985). In Figure 3.6, the transversality condition is not satisfied because the intersection of the unstable manifold of x_1 and the stable manifold of x_2 is a portion of the manifold whose tangent space has a dimension of 1. Note that the unstable manifold of x_1 intersects with the stability boundary (see Theorem 3.8) but not the stability region (see Theorem 3.12). A part of the stable manifold of x_1 (upper part in Figure 3.6) is not in the stability boundary (see Theorem 3.12).

Theorem 3.13 below extends the result of Theorem 3.12 to accommodate closed orbits on the stability boundary.

Theorem 3.13: Characterization of a Critical Element on the Stability Boundary (Chiang et al., 1988)

Let A be the stability region of a SEP of the nonlinear dynamical system (Eq. 3.1). Let r be a critical element. Assume the following.

(B1) All the critical elements on ∂A are hyperbolic.

(B2) The stable and unstable manifolds of critical elements on ∂A satisfy the transversality condition.

(B3) Every trajectory on ∂A approaches one of the critical elements as $t \to \infty$.

Then,

(1) $\hat{r}$ is on the stability boundary ∂A if and only if $W^u(\hat{r}) \cap A \neq \phi$ and

(2) $\hat{r}$ is on the stability boundary ∂A if and only if $W^S(\hat{r}) \subseteq \partial A$.

The next result concerns the number of equilibrium points on the stability boundary. We say that $S \subset R^n$ is a *smooth* manifold of dimension s if, for each point $p \in S$, there exist a neighborhood $U \subset S$ of p and a homeomorphism $h{:}U \rightarrow V$, where V is an open subset of R^s, such that the inverse homeomorphism $h^{-1}{:}V \rightarrow U \subset R^n$ is an immersion of class C^1.

Theorem 3.14: Number of Equilibrium Points on the Stability Boundary

If the stability boundary ∂A of a SEP is a smooth compact manifold and all the equilibrium points on ∂A are hyperbolic, then the number of equilibrium points on ∂A is even.

Proof: The proof is based on the following fact (Hirsch, 1976, p. 139, Exercise 7): the Euler characteristic of the boundary of a compact manifold is even. From the Poincaré–Hopf index theorem (Guillemin and Pollack, 1974, p. 134), it follows that the sum of the indices of equilibrium points of f on the smooth, compact stability boundary ∂A are even. But the index of f at a hyperbolic equilibrium point is either +1 or −1 (Milnor, 1965, p. 37). Consequently, Theorem 3.14 follows.

3.7 GLOBAL CHARACTERIZATION OF STABILITY BOUNDARY

In this section, we characterize the stability boundary for a fairly large class of the nonlinear dynamical system (Eq. 3.1) whose stability boundary is nonempty. We make the following assumptions concerning the vector field:

(A1) All the equilibrium points on the stability boundary are hyperbolic.

(A2) The stable and unstable manifolds of equilibrium points on the stability boundary satisfy the transversality condition.

(A3) Every trajectory on the stability boundary approaches one of the equilibrium points as $t \rightarrow \infty$.

Remark: Assumption (A1) is a generic property of C^1 dynamical systems and can be checked for a particular system by direct computation of the eigenvalues of the corresponding Jacobian matrix of the vector field. Assumption (A2) is also a generic property; however, it is not easy to check. Assumption (A3) is not a generic property, but in many systems, it can be verified by several means.

Theorem 3.15 asserts that if Assumptions (A1)–(A3) are satisfied, then the stability boundary is the union of the stable manifolds of the equilibrium points on the stability boundary (Zaborszky et al., 1988a).

Theorem 3.15: Characterization of the Stability Boundary

For a nonlinear autonomous dynamical system (Eq. 3.1) which satisfies Assumptions (A1)–(A3), let x_i, $i = 1, 2, \ldots$ be the equilibrium points on the stability boundary ∂A of the SEP. Then,

[1] $x_i \in \partial A$ if and only if $W^u(x_i) \cap A \neq \phi$ and

[2] $\partial A = \bigcup_i W^S(x_i)$.

Proof: Part [1] is shown in Theorem 3.12. We prove part [2]. Let x_i, $i = 1, 2, \ldots$ be the equilibrium points on the stability boundary ∂A. Theorem 3.12 implies the following:

$$\partial A \supseteq \bigcup_i W^S(x_i). \tag{3.12}$$

Assumption (A3) implies

$$\partial A \subseteq \bigcup_i W^S(x_i). \tag{3.13}$$

Combining Equations 3.12 and 3.13, we complete the proof for part [2].

Theorem 3.15 can be generalized to allow closed orbits to exist on the stability boundary. The following theorem gives an interesting result on the structure of the equilibrium points on the stability boundary. Moreover, it presents a necessary condition for the existence of certain types of equilibrium points on a *bounded* stability boundary.

Theorem 3.16: Structure of Equilibrium Points on the Stability Boundary

For the nonlinear autonomous dynamical system (3.1) containing two or more SEPs, if it satisfies Assumptions (A1)–(A3), then the stability boundary must contain at least one type-one equilibrium point. If, furthermore, the stability region is bounded, then ∂A must contain at least one type-one equilibrium point and one source.

Proof: Since there are at least two SEPs including, say, x_s, it follows that the dimension of $\partial A(x_s)$ is $(n - 1)$. Since $\partial A(x_s) = \cup W^S(x_j)$, where $x_j \in \partial A(x_s)$, at least one of the x_j must be a type-one equilibrium point, say, x_1, so that the dimensions of $\cup W^S(x_j)$ is $(n - 1)$. Repeating the same argument, if $\partial W^S(x_1)$ is nonempty, then the dimension of $\partial W^S(x_1)$ is $\leq(n - 2)$, say, $(n - k)$. The application of Theorem 3.15 yields $\partial W^S(x_1) = \cup W^S(x_j)$, $x_j \in \partial W^S(x_1)$. In order for $\cup W^S(x_j)$ to have dimension $(n - k)$, at least one of the x_j must be a type-k equilibrium point. If the stability region is bounded, the same argument can be repeated until we reach a type-n equilibrium point (a source). This completes the proof.

The contrapositive of Theorem 3.16 leads to the following corollary, which is useful in predicting the unboundedness of the stability region.

Theorem 3.17: Sufficient Condition for an Unbounded Stability Region

Consider the nonlinear autonomous dynamical systems (Eq. 3.1) with a SEP x_s whose stability boundary is nonempty. If Assumptions (A1)–(A3) are satisfied and if $\partial A(x_s)$ contains no source, then the stability region $A(x_s)$ is unbounded.

3.8 ALGORITHM TO DETERMINE THE STABILITY BOUNDARY

Theorem 3.15 leads to the following conceptual algorithm for determining the stability boundary of a SEP of the nonlinear autonomous dynamical systems (Eq. 3.1) that satisfy Assumptions (A1)–(A3).

Algorithm: To determine the stability boundary $\partial A(x_s)$,

Step 1. Find all the equilibrium points.

Step 2. Identify those equilibrium points whose unstable manifolds contain trajectories approaching the SEP x_s.

Step 3. The stability boundary of x_s is the union of the stable manifolds of the equilibrium points identified in Step 2.

Step 1 in the algorithm involves finding all the solutions of $f(x) = 0$. Step 2 can be accomplished numerically. The following procedures are suggested:

1. Find the Jacobian at the equilibrium point (say, $\hat{x}$).
2. Find many of the generalized unstable eigenvectors of the Jacobian having unit length.
3. Find the intersection of each of these normalized, generalized, and unstable eigenvectors (say, y_i) with the boundary of an ε-ball of the equilibrium point (the intersection points are $\hat{x} + \varepsilon y_i$ and $\hat{x} - \varepsilon y_i$).
4. Integrate the vector field backward (reverse time) from each of these intersection points up to some specified time. If the trajectory remains inside this this ε-ball, then go to next step. Otherwise, we replace the value ε by $\alpha\varepsilon$ and also the intersection points $\hat{x} \pm \varepsilon y_i$ by $\hat{x} \pm \alpha\varepsilon y_i$ where $0 < \alpha < 1$. Repeat this step.
5. Numerically integrate the vector field starting from these intersection points.
6. Repeat steps 3–5. If any of these trajectories approach x_s, then the equilibrium point is on the stability boundary.

For a planar system, the equilibrium point on the stability boundary is either a type-one equilibrium point or a type-two equilibrium point, which is a source. The stable manifold of a type-one equilibrium point in this case has dimension one, which can easily be determined numerically as follows:

1. Find a normalized stable eigenvector y of the Jacobian at the equilibrium point $\hat{x}$.
2. Find the intersection of this stable eigenvector with the boundary of an ε-ball of the equilibrium point $\hat{x}$ (where the intersection points are $\hat{x} + \varepsilon y_i$ and $\hat{x} - \varepsilon y_i$).
3. Integrate the vector field from each of these intersection points after some specified time. If the trajectory remains inside this ε-ball, then go to next step. Otherwise, we replace the value ε by $\alpha\varepsilon$ and also the intersection points $\hat{x} \pm \varepsilon y_i$ by $\hat{x} + \alpha\varepsilon y_i$, where $0 < \alpha < 1$. Repeat this step.
4. Numerically integrate the vector field backward (reverse time) starting from these intersection points.
5. The resulting trajectories are the stable manifold of the equilibrium point.

For higher-dimensional systems, the numerical procedure similar to the one above can only provide a set of trajectories on the stable manifold. Finding the stable and unstable manifolds of an equilibrium point is a nontrivial problem, and advanced numerical methods for computing stable and unstable manifolds are needed (Ushiki, 1980).

The method for a complete determination of a stability region will be illustrated on two simple examples. In each example, two figures will be derived: one compares the estimated stability region by the previous methods and the present method, while the other gives the phase portrait of the system to verify the results of this method. Throughout these examples, we *assume* the transversality condition Assumption (A2) is satisfied.

Example 1: This is an example studied in Genesio and Vicino (1984b) and in Michel et al. (1982):

$$\begin{aligned} \dot{x}_1 &= -2x_1 + x_1x_2 \\ \dot{x}_2 &= -x_1 + x_1x_2. \end{aligned} \tag{3.14}$$

There are two equilibrium points: (0.0, 0.0) is a SEP and (1, 2) is a type-one equilibrium point. Assumption (A1) is satisfied. The trajectory on the unstable manifold of (1, 2) converges to the SEP (0.0, 0.0); hence, (1, 2) is on the stability boundary (see Theorem 3.8). Next, we check Assumption (A3). Consider the following function:

$$V(x_1, x_2) = x_1^2 - 2x_1x_2 + x_2^2.$$

The derivative of $V(x_1, x_2)$ along the trajectory of Equation 3.14 is

$$\begin{aligned} \dot{V}(x_1, x_2) &= \frac{\partial V}{\partial x_1}\dot{x}_1 + \frac{\partial V}{\partial x_2}\dot{x}_2 \\ &= -2(2x_1 - x_2)(x_1 - x_2). \end{aligned}$$

Hence,

$$\dot{V}(x_1, x_2) < 0 \quad \text{for } (x_1, x_2) \in B^C := R^2 - B,$$

where $B := \{(x_1, x_2): 2x_1 - x_2 \geq 0 \text{ and } x_1 - x_2 \leq 0\}$.

Define the following sets:

$$\tilde{B} := B_1 \cup B_2 \cup B_3,$$

where

$$B_1 = \{(x_1, x_2): x_1 < 1, x_2 < 2\},$$
$$B_2 = \{(x_1, x_2): x_1 \geq 1, x_2 \leq 2\} \cap B, \text{ and}$$
$$B_3 = \{(x_1, x_2): x_1 > 1, x_2 > 2\}.$$

Since in the set B_1 both $|x_1(t_i)|$ and $|x_2(t_i)|$ are strictly decreasing sequences, we conclude that B_1 is inside the stability region (0, 0). In other words, the stability boundary $\partial A(0, 0)$ cannot lie in B_1. On the other hand, every trajectory of Equation 3.14 in the set B_3 is unbounded as $t \to \infty$; therefore, the stability boundary $\partial A(0, 0)$ cannot lie in B_3 either. However, by checking the vector field of Equation 3.14 in B_2, we find that every trajectory in B_2 will either enter into B_1 or B_3 or converge to the point (1, 2). Hence, we have shown that the stability boundary $\partial A(0, 0)$ cannot be in B_1 nor B_3; the part of the stability boundary $\partial A(0, 0)$ in B_2 must converge to (1, 2). Next, we will show that the part of the stability boundary $\partial A(0, 0)$ in $R^2 - \tilde{B}$ also converges to (1, 2). Then, we may claim that Assumption (A3) is satisfied. Note that $\dot{V}(x_1, x_2) < 0$ for $(x_1, x_2) \in R^2 - \tilde{B} \subseteq B^C$ and the function $V(x_1, x_2)$ is a proper map in $R^2 - \tilde{B}$. Thus, every trajectory of $\partial A(0, 0)$ in $R^2 - \tilde{B}$ is bounded and, if it converges in $R^2 - \tilde{B}$, it must converge to an equilibrium point in $R^2 - \tilde{B}$. However, there is no equilibrium point in $R^2 - \tilde{B}$. So, the part of the stability boundary $\partial A(0, 0)$ in $R^2 - \tilde{B}$ must enter the set $\tilde{B}$. However, we have shown that the stability boundary $\partial A(0, 0)$ in $\tilde{B}$ converges to (1, 2). Therefore, the trajectories on the stability boundary $\partial A(0, 0)$ converge to (1, 2), and Assumption (A3) is shown to be satisfied.

It follows that the stability boundary is the stable manifold of (1, 2) (see Theorem 3.15), which is curve C in Figure 3.7. Because there is no source, the stability region is unbounded (Theorem 3.17). Curves A and B in Figure 3.7 are

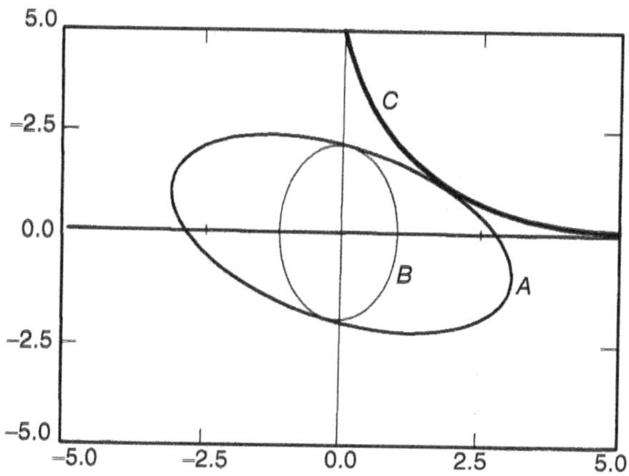

Figure 3.7 Predictions of the stability region of Example 1 by different methods. Curves A and B are obtained by the methods in Michel et al. (1982) and in Genesio and Vicino (1984b). Curve C is obtained by the present method.

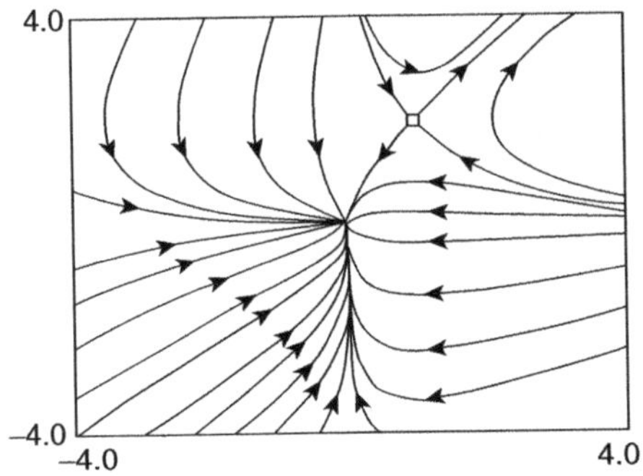

Figure 3.8 The phase portrait of this system. Note that all the points inside curve C converge to the stable equilibrium point, which verified that curve C is the exact stability boundary.

obtained by the methods in Michel et al. (1982) and in Genesio and Vicino (1984b), respectively. The approximately true stability boundary mentioned in Genesio and Vicino (1984b) seems to agree with curve C. Figure 3.8 plots the phase portrait of this system, which confirms that curve C represents the exact stability boundary.

Example 2: The following system is close to a power system transient stability model and is considered in Athay et al. (1979):

$$\dot{x}_2 = 0.301 - \sin(x_1 + 0.4136) + 0.138 \sin 2(x_1 + 0.4136) - 0.279 x_2. \quad (3.15)$$

It is straightforward to verify that Assumption (A3) is satisfied with this simple example. The equilibrium points of Equation 3.15 are periodic on the subspace $\{x_1, x_2 | x_2 = 0\}$ and the Jacobian matrix of Equation 3.15 at (x_1, x_2) is

$$J(x) = \begin{bmatrix} 0 & 1 \\ a & -0.279 \end{bmatrix}, \quad (3.16)$$

where $a = -\cos(x_1 + 0.4136) + 0.276 \cos 2(x_1 + 0.4136)$. Let λ_1, λ_2 be the eigenvalue of $J(x)$:

$$\lambda_1 + \lambda_2 = -0.279$$

$$\lambda_1 \times \lambda_2 = -a.$$

The following observations are immediate:

1. Assumption (A1) is satisfied.
2. At least one of the eigenvalues must be negative, which implies there is no source in the system (Eq. 3.15). By Theorem 3.17 we conclude that the stability region (with respect to any SEP) is unbounded.
3. The SEPs and the type-one equilibrium points are located alternately on the x_1-axis.

It can be shown that (6.284098, 0.0) is a SEP of Equation 3.15. Let us consider its stability region. The application of Theorem 3.8 shows that the type-one equilibrium points (2.488345, 0.0) and (8.772443, 0.0) are on the stability boundary. The

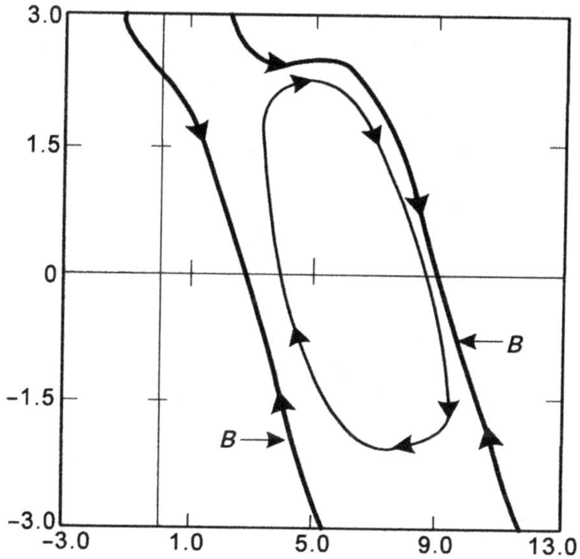

Figure 3.9 Estimations of the stability region of Example 2 by different methods. Curve A is the estimated stability boundary from the methods in Michel et al. (1982) (after a shift in coordinates). Curve B is the estimated stability boundary by the present method.

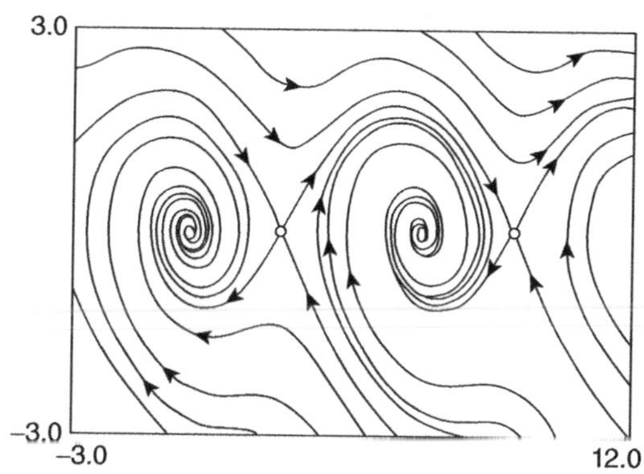

Figure 3.10 The phase portrait of this system, which confirms that curve B in Figure 3.9 is the exact stability boundary.

stability region is again unbounded due to the absence of a source. The stability boundary obtained by the present method is curve B shown in Figure 3.9, which is the union of stable manifolds of the equilibrium points (2.488345, 0.0) and (8.772443, 0.0). Curve A is the estimated stability boundary obtained by the method described in Michel et al. (1982) (after a shift in coordinates). It is clear from the phase portrait in Figure 3.10 that the trajectories of the points inside curve B converge to the SEP, which verifies that curve B is the exact stability boundary.

3.9 CONCLUSION

Depending on the intended use of stability analysis and design, stability can be formulated in a variety of different ways. In this chapter, we have reviewed some relevant stability concepts from nonlinear dynamical systems theory. One of the most important developments in stability theory is the Lyapunov function theory. An overivew of the Lyapunov function theory has been presented. In terms of application, characterization and estimation of the stability region of a SEP are very important. In this chapter, a comprehensive theory of stability regions of SEPs has been presented.

A complete dynamic characterization of the stability boundary of a fairly large class of nonlinear autonomous dynamical systems has been derived. Practical numerical methods are needed to accurately estimate the stability regions of nonlinear dynamical systems and are important to many disciplines in the sciences and engineering (Genesio and Vicino, 1984b; Loccufier and Noldus, 2000; Luyckx et al., 2004; Miyagi and Yamashita, 1986; Vittal and Michel, 1986). Presently, few methods are computationally practical and also do not yield accurate results for general large-scale nonlinear dynamical systems are available.

The analytical characterizations of equilibrium points on the stability boundary and the complete characterization of the stability boundary are very useful in the development of theoretical foundations for direct methods of power system transient stability analysis. Furthermore, these analytical results in combination with explorations of the structures of power system transient stability models will lead to the development of practical direct methods for large-scale power systems. These developments will be presented in later chapters.

Chapter 4

Quasi-Stability Regions: Analysis and Characterization

4.1 INTRODUCTION

The stability boundary structure of general nonlinear dynamical systems can be very complex. A simple three-dimensional example given in Zaborszky et al. (1988b) shows that the closure of stability regions may contain subsets of the stability boundary. The stability boundary of a simple swing equation may have a truncated fractal structure (Varghese and Thorp, 1988). There are several factors that contribute to the complexity of the stability boundary. One of them is the presence of critical elements (i.e., the equilibrium point and limit cycle) in the interior of the closure of the stability region. This motivates a study of the notions of the quasi-stability boundary and region. Indeed, from an engineering viewpoint, the quasi-stability region is a "practical" stability boundary while the quasi-stability boundary is less complex than the stability boundary.

We present in this chapter the quasi-stability boundary and develop some characterizations of the quasi-stability region that are useful for the analysis of direct methods. The closest unstable equilibrium point (UEP) and controlling UEP will then be characterized by using the characterization of equilibrium points on the quasi-stability region developed in this chapter. The topic of optimal estimation of the quasi-stability region is also presented. Most of the proofs presented in this chapter are taken from Chiang and Fekih-Ahmed (1996a,b) and from Chiang and Chu (1996).

4.2 QUASI-STABILITY REGION

To illustrate the concept of the quasi-stability boundary, consider the following 2-D system, which is a reduced system of the system presented in Zaborszky et al. (1988b):

Direct Methods for Stability Analysis of Electric Power Systems, by Hsiao-Dong Chiang

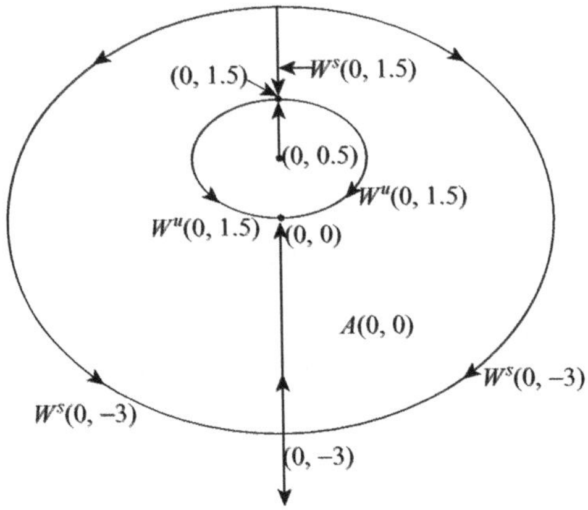

Figure 4.1 The stability boundary of the stable equilibrium point (0, 0) of the system (Eq. 4.1) is composed of the stable manifold of (0, −3) and the stable manifold of (0, 1.5). On the other hand, the quasi-stability boundary of the stable equilibrium point (0, 0) of the system (Eq. 4.1) is composed only of the closure of the stable manifold of (0, −3). In other words, the quasi-stability boundary of the stable equilibrium point (0, 0) of the system (Eq. 4.1) is the union of the stable manifold of (0, −3) and the unstable equilibrium point (0, 3).

$$\begin{aligned}\dot{x} &= \left\{\left(\sqrt{x^2+y^2}-3\right)\left[x^2+y^2+(y-2)\sqrt{x^2+y^2}-2y+1.5\right]+y\right\}x \\ \dot{y} &= \left\{\left(\sqrt{x^2+y^2}-3\right)\left[x^2+y^2+(y-2)\sqrt{x^2+y^2}-2y+1.5\right]y\right\}-x^2.\end{aligned} \tag{4.1}$$

It can be easily verified that (0, 0) is an asymptotically stable equilibrium point of the system. Its stability region A is shown in Figure 4.1. There are four equilibrium points on the stability boundary, namely, (0, 0.5), (0, 3), (0. 1.5), and (0, −3). The first two equilibrium points are sources, while the remaining two are type-one equilibria. The phase portrait of the system shows that the stability boundary is composed of two parts. One consists of the stable manifold of (0, −3) and the other consists of the stable manifold of (0, 1.5). The two parts are connected to each other by the source (0, 3). Observe that the stable manifold of (0, 1.5) lies in the interior of $\bar{A}$. A slight perturbation of any trajectory lying on the stable manifold of (0, 1.5) may cause this trajectory to eventually converge to the origin. Thus, this part of the stability boundary effectively behaves as if it were part of the stability region. This leads us to believe that this part of the boundary should be practically considered part of the stability region. The second part of the boundary, which consists of the stable manifold of (0, −3), divides the state space into two regions. One region consists of the points whose trajectories will most likely converge to the origin. We will call this region the quasi-stability region. The second region consists of points whose trajectories will move away from the closure of the stability region. The curve that separates these two regions will be termed the quasi-stability boundary.

The above 2-D system suggests a preliminary intuitive definition of a quasi-stability boundary by looking at the phase portrait of the system. A suitable definition of a quasi-stability boundary for high-dimension nonlinear systems must be given. For the 2-D system example, we note that this system satisfies Assumptions (A1)–(A3). Thus, by Theorem 3.15, for an equilibrium point to be on the stability boundary, its unstable manifold must intersect the stability region. This property is obvious for the two sources on the boundary because their unstable manifolds are 2-D. Furthermore, we can see in Figure 4.1 that the two saddle equilibrium points on the boundary have this property also. There is, however, a substantial difference between these two saddle points. The unstable manifold of (0, 1.5) is included in the closure of the stability region A; that is, $W^u\ (0, 1.5) \cap (\bar{A})^c = \varnothing$. In contrast, the unstable manifold of (0, −3) can be viewed as being composed of two parts. One part of the unstable manifold is contained in the closure of A, while the other part is contained in $(\bar{A})^c$.

The above example leads to a desirable intuitive definition of a critical element on ∂A_p shown as follows.

Definition
A critical element σ is on the quasi-stability boundary ∂A_p if and only if the following two conditions are met:

(i) $\sigma \in \partial A$.

(ii) $W^u\ (\sigma) \cap (\bar{A})^c \neq \varnothing$.

In order to show that the above definition is consistent, one needs to show that the (topological) boundary of the quasi-stability region equals the quasi-stability boundary ∂A_p. This will be shown in the following.

We consider the following (autonomous) nonlinear dynamical system:

$$\dot{x} = f(x), x \in \Re^n. \tag{4.2}$$

It is natural to assume the function (i.e., the vector field) f: $R^n \rightarrow R^n$ satisfies a sufficient condition for the existence and uniqueness of the solution. We make the following assumptions concerning the vector field:

(A1) All the equilibrium points on the stability boundary are hyperbolic.

(A2) The stable and unstable manifolds of the equilibrium points on the stability boundary satisfy the transversality condition.

(A3) Every trajectory on the stability boundary approaches one of the equilibrium points as $t \rightarrow \infty$.

The next analytical result gives a necessary condition for a critical element to be on the quasi-stability boundary.

Theorem 4.1: A Necessary Condition

Let A be the stability region of a stable equilibrium point of the nonlinear dynamical system (Eq. 4.2) satisfying Assumptions (A1)–(A3). Let σ be a hyperbolic critical element, then the fact that the element lies on the stability boundary implies that the stable manifold of the element is contained in the complement of the stability region; that is, $\sigma \in \partial A \Rightarrow W^s(\sigma) \subset (A)^c$.

Proof: Suppose $\sigma \in \partial A_p$. By definition, let $D \subset W^u(\sigma) \cap (A)^c$ be an m-disk, $m = \dim W^u(\sigma)$. Let $y \in W^s(\sigma)$ be arbitrary. For any $\varepsilon > 0$, let N be an m-disk transversal to $W^s(\sigma)$ at y, contained in the ε-neighborhood of y. In Lemma 3.11, there exists a $t > 0$ such that $\phi_t(N)$ is so close to D that $\phi_t(N)$ contains a point $p \in (A)^c$.

Thus, $\phi_{-t}(p) \in N$. Since $(A)^c$ is an invariant set, $N \cap (\bar{A})^c = \varnothing$. As $\varepsilon \to 0$, we get $y \in (A)^c$. Thus $W^s(\sigma) \subset (A)^c$. This completes the proof.

We next establish the relationship between the critical elements belonging to the sets ∂A_p and $\partial \bar{A}$.

Theorem 4.2: A Relation between ∂A_p and $\partial \bar{A}$

Let A be the stability region of a stable equilibrium point of the nonlinear dynamical system (Eq. 4.2). If we let σ be a hyperbolic critical element, then $\sigma \in \partial A_p \Rightarrow \sigma \in \partial \bar{A}$.

Proof: From definition, we have $\sigma \in \partial A_p$ if and only if (i) $\sigma \in \partial A$ and (ii) $W^u(\sigma) \cap (\bar{A})^c \neq \varnothing$. If we let $y \in \{W^u(\sigma) - \sigma\} \cap (\bar{A})^c$, then $\lim_{t \to -\infty} \phi_t(y) = \sigma$. Since $\overline{(\bar{A})^c}$ is invariant, we have $\sigma \in \overline{(\bar{A})^c}$. Also, since $\sigma \in \partial A$, $\sigma \in \bar{A}$. Thus, $\sigma \in \overline{(\bar{A})^c} \cap \bar{A}$, which is the definition of $\partial \bar{A}$. This completes the proof.

In the following, we present two useful results that characterize a critical element on $\partial \bar{A}$. These results provide insight into the definition of the quasi-stability boundary. It will also be shown that $\partial A_p = \partial \bar{A}$.

Theorem 4.3: Characterization of Critical Elements on $\partial \bar{A}$

Let A be the stability region of a stable equilibrium point of the nonlinear dynamical system (Eq. 4.2). If σ is a hyperbolic critical element, then $\sigma \in \partial \bar{A}$ implies $\{W^u(\sigma) - \sigma\} \cap \overline{(\bar{A})^c} \neq \varnothing$. Conversely, if $\{W^u(\sigma) - \sigma\} \cap \overline{(\bar{A})^c} \neq \varnothing$ and $\{W^u(\sigma) - \sigma\} \cap \bar{A} \neq \varnothing$, then $\sigma \in \partial \bar{A}$.

Proof: Let $G \subset \{W^u(\sigma) - \sigma\}$ be a fundamental domain of $W^u(\sigma)$; that is, G is compact and $\bigcup_{t<0} \phi_t(G) = \{W^u(\sigma) - \sigma\}$. Let G_ε be an ε-neighborhood of G in $\Re^n$. Then $\bigcup_{t<0} \phi_t(G_\varepsilon)$ contains a set of the form $\{U - W^s(\sigma)\}$, where U is a neighborhood of σ. Since $\sigma \in \partial \bar{A}$ is equivalent to $\sigma \in \partial(\bar{A})^c$, $U \cap (\bar{A})^c = \varnothing$. But by the topological inclusion $\partial \bar{A} \subset \partial A$, we get $W^s(\sigma) \subset \partial A$; hence, $W^s(\sigma) \cap (\bar{A})^c = \varnothing$. Therefore, we have $\{U - W^s(\sigma)\} \cap (\bar{A})^c = \varnothing$. Equivalently, $U_{t<0} \phi_t(G_\varepsilon) \cap (\bar{A})^c = \varnothing$. This implies that $G_\varepsilon \cap \phi_t[(\bar{A})^c] = \varnothing$ for some t.

Since $(\bar{A})^c$ is an invariant open set under the flow, we conclude that $G_\varepsilon \cap (\bar{A})^c = \varnothing$. Since $\varepsilon > 0$ is arbitrarily chosen and G is compact, G contains a point in $\overline{(\bar{A})^c}$; thus, $\{W^u(\sigma) - \sigma\} \cap \overline{(\bar{A})^c} \neq \varnothing$. For the second part, let $y \in W^u(\sigma) \cap \overline{(\bar{A})^c}$. Then $\lim_{t \to -\infty} \phi_t(y) = \sigma$, and hence, $\sigma \in \overline{(\bar{A})^c}$. Also, from the assumptions, $\sigma \in \bar{A}$; therefore, $\sigma \in \partial \bar{A}$. This completes the proof.

Theorem 4.4: Characterization of Critical Elements on $\partial \bar{A}$

Let A be the stability region of a stable equilibrium point of the nonlinear dynamical system (Eq. 4.2) satisfying Assumptions (A1)–(A3). If we let σ be a hyperbolic critical element, then $\sigma \in \partial \bar{A}$ implies that $\{W^u(\sigma) - \sigma\} \cap (\bar{A})^c \neq \varnothing$.

Proof: We shall prove this proposition for an equilibrium point $\ddot{x}$. The proof is similar for the case of a closed periodic orbit. From Theorem 4.3 we have $\{W^u(\hat{x})-\hat{x}\}\cap\overline{(\bar{A})^c}\neq\varnothing$. Let $n_u(\ddot{x})$ denote the type of equilibrium point $\ddot{x}$. From Assumption (A1), one has $n_u(\ddot{x})\geq 1$ for any $\ddot{x}\in\partial A$. Call y a point in $\{W^u(\hat{x})-\hat{x}\}\cap\overline{(\bar{A})^c}$. If $y\in(\bar{A})^c$, then the proof is complete. If $y\in\overline{(\bar{A})^c}=\partial\bar{A}$, then there exists an equilibrium point $\hat{z}\in\partial A$ and $y\in\{W^s(\hat{z})-\hat{z}\}$. Let $h=n_u(\ddot{x})$, and $m=n_u(\ddot{z})$, where h and m are positive integers. By Assumption (A2), $W^u(\hat{x})$ intersects $W^s(\hat{z})$ transversally at y. Thus, $h>m$. Two cases may arise:

(a) If $h=1$, then $m=0$, which is a contradiction. Hence, $\{W^u(\hat{x})-\hat{x}\}\cap(\bar{A})^c\neq\varnothing$.

(b) If $h>1$, then $m\leq h-1$. Assume inductively that $\{W^u(\hat{x})-\hat{x}\}\cap(\bar{A})^c\neq\varnothing$. $W^u(\hat{x})$ contains an m-disk N at y and is transverse to $W^s(\hat{z})$. By Lemma 3.11, we have $\phi_t(N)\cap(\bar{A})^c\neq\varnothing$ for some $t>0$. Since $(\bar{A})^c$ is an invariant set, we have $U\cap(\bar{A})^c\neq\varnothing$. This completes the proof.

We are now in a position to derive the following main result, which provides greater insight into the concept of quasi-stability regions.

Theorem 4.5: Relation between ∂A_p and $\partial\bar{A}$

Consider a nonlinear dynamical system that satisfies Assumptions (A1)–(A3). If we let σ be a critical element, then $\sigma\in\partial A_p$, if and only if $\sigma\in\partial\bar{A}$.

Proof: From Theorem 4.2, we have $\sigma\in\partial A_p$, which implies that $\sigma\in\partial\bar{A}$. Now, if $\sigma\in\partial\bar{A}$, then $\sigma\in\partial A$ because $\partial\bar{A}\subset\partial A$. Moreover, Theorem 4.4 implies that $\{W^u(\sigma)-\sigma\}\cap(\bar{A})^c\neq\varnothing$. In other words, $\sigma\in\partial A_p$. This completes the proof.

We have so far developed the concept of a quasi-stability boundary via critical elements. To define the quasi-stability boundary completely, we need to come up with a definition that includes points other than equilibrium points and periodic solutions. Theorem 4.5, in conjunction with the following topological fact:

$$\partial\overline{int\bar{A}}\subset\partial int\bar{A}\subset\partial\bar{A}\subset\partial A, \tag{4.3}$$

suggests that either $\partial\overline{int\bar{A}}$ or $\partial int\bar{A}$ or $\partial\bar{A}$ can be chosen as ∂A_p. Theorem 4.5 asserts that the last relationship is probably the most appropriate choice and thus the following formal definition of a quasi-stability region is derived.

Definition: Quasi-Stability Boundary

Let A be the stability region of a stable equilibrium point of the nonlinear dynamical system (Eq. 4.2). The *quasi-stability boundary* ∂A_p is $\partial\bar{A}$ and the *quasi-stability* region A_p is the open set $int\bar{A}$.

In order to have a consistent definition, one needs to show that the boundary of A_p equals ∂A_p. In the next analytical result, which is based on a known topology property that for any open set A, it is true that $\partial int\bar{A}=\partial\bar{A}$, which shows that the quasi-stability region and the quasi-stability boundary are well-defined. A graphical illustration of the difference between stability region and quasi-stability region is shown in Figure 4.2.

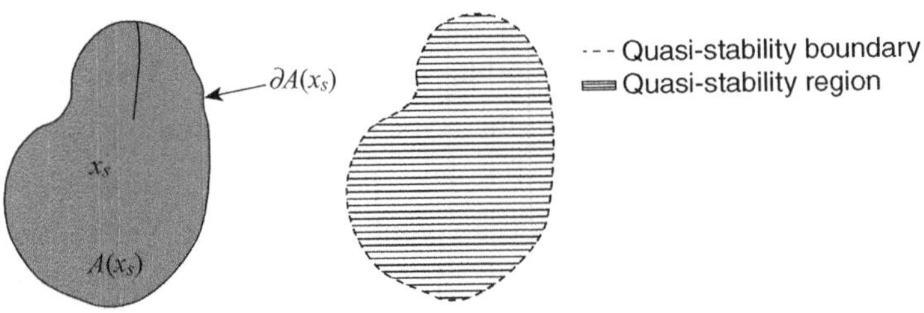

Figure 4.2 A graphical illustration of the difference between the stability region and the quasi-stability region.

Theorem 4.6: Topological Property

Let A be the stability region of a stable equilibrium point of the nonlinear dynamical system (Eq. 4.2), which satisfies Assumptions (A1)–(A3). Let the quasi-stability boundary be denoted as ∂A_p and the quasi-stability region be denoted as A_p, then $\partial int\bar{A} = \partial A_p$.

Based on the definition of quasi-stability regions, we have the following.

Proposition 4.7: Topological Property

The quasi-stability region A_p is an open invariant set, which is diffeomorphic to $\Re^n$, and the quasi-stability boundary ∂A_p is a closed invariant set. Moreover, if $\bar{A} \neq \Re^n$, then ∂A_p has dimension $n - 1$.

Proof: The first part follows from the facts that $A \subset A_p$, A is an invariant set diffeomorphic to $\Re^n$, and $\partial\bar{A}$ is an invariant set. To prove the second part, observe that if $A = \Re^n$, then $\partial A_p = \partial\overline{int\bar{A}} = \partial\Re^n = \varnothing$. And if $A \neq \Re^n$, then $\partial A_p = \partial int\bar{A}$. Since $int\bar{A}$ is an open set and $\overline{int\bar{A}} \neq \Re^n$, ∂A_p has dimension $n - 1$. This completes the proof.

4.3 CHARACTERIZATION OF QUASI-STABILITY REGIONS

In this section, we will derive a complete characterization of the quasi-stability boundary for a fairly large class of nonlinear dynamical systems. To this end, we will first derive conditions that an equilibrium point (or closed orbit) must possess to be on the quasi-stability boundary. We will then show that the quasi-stability boundary is the union of stable manifolds of all the critical elements on the quasi-stability boundary. Moreover, the quasi-stability boundary is the union of the closure of the stable manifolds of all the type-one critical elements on the quasi-stability boundary.

We first present conditions that an equilibrium point (or limit cycle) must satisfy to be on the quasi-stability boundary $\partial A_p(x_s)$. This is a key step in the complete characterization of $\partial A_p(x_s)$.

Theorem 4.8: Critical Elements on the Quasi-Stability Boundary

Consider the general nonlinear dynamical system (Eq. 4.2). Let A_p be the quasi-stability region of x_s and A be the stability region of x_s. Let $\sigma \neq x_s$ be a hyperbolic critical element. If Assumptions (A1)–(A3) are satisfied, then the following analytical results hold:

(1) $\sigma \in \partial A_p$ if and only if $W^u(\sigma) \cap A \neq \varnothing$ and $W^u(\sigma) \cap (\bar{A})^c \neq \varnothing$.

(2) $\sigma \in \partial A_p$ if and only if $W^s(\sigma) \subset \partial A_p$.

Proof: The proof of Part (1) follows from definition and Theorem 4.6. We only prove Part (2). It suffices to show that $\sigma \in \partial\bar{A} \Leftrightarrow W^s(\sigma) \subset \partial\bar{A}$.

Sufficiency: If $W^s(\sigma) \subset \partial\bar{A}$, then $\sigma \in \partial A_p$ since $\sigma \in W^s(\sigma)$.

Necessity: Suppose $\sigma \in \partial A_p$, it follows that $W^s(\sigma) \subset \overline{(\bar{A})^c}$. But Theorem 4.1 implies that $W^s(\sigma) \subset \partial A$; therefore, $W^s(\sigma) \not\subset (\bar{A})^c$. Now, $\overline{(\bar{A})^c} = (\bar{A})^c \cup \partial(\bar{A})^c = (\bar{A})^c \cup \partial\bar{A}$; thus, $W^s(\sigma) \subset \partial\bar{A}$. This completes this proof.

We next develop a complete characterization of the quasi-stability boundary for a fairly large class of nonlinear dynamical systems whose stability boundary is nonempty.

Theorem 4.9: Characterization of the Quasi-Stability Boundary

Consider a nonlinear dynamical system described by Equation 4.2, which satisfies Assumptions (A1)–(A3). Let σ_i, $i = 1, 2, \ldots$ be the critical elements on the quasi-stability boundary $\partial A_p(x_s)$ of the stable equilibrium point x_s. Then,

$$\partial A_p(x_s) = \bigcup_{\sigma_i \in \partial A_p(x_s)} W^s(\sigma_i).$$

Proof: We deduce from Theorem 4.8 that

$$\bigcup_{\sigma_i \in \partial A_p(x_s)} W^s(\sigma) \subset \partial A_p(x_s).$$

Since $\partial A_p(x_s)$ is an invariant set, Assumption (A3) implies that every point of $\partial A_p(x_s)$ must be on some $W^s(\sigma)$ for some critical element $\sigma \in \partial A_p(x_s)$. Thus, $\partial A_p(x_s) \subset \bigcup_{\sigma_i \in \partial A_p(x_s)} W^s(\sigma)$. Hence, the proof is completed.

We next study a topological relationship between the critical elements on the quasi-stability boundary. In particular, type-one critical elements and their stable manifolds play a dominant role in the characterization of the quasi-stability boundary. The proof of the following theorem can be found in Chiang and Fekih-ahmed (1996).

Theorem 4.10: A Topological Relation

Consider a nonlinear dynamical system described by Equation 4.2, which satisfies Assumptions (A1)–(A3). Let A_p be the quasi-stability region of x_s and let $\sigma \in \partial A_p$ be a hyperbolic critical element whose type is greater than one. Then, there exists a type-one critical element $\sigma_1 \in \partial A_p$ such that $\sigma \in W^s(\sigma_1)$.

In fact, it can be further shown by using arguments similar to those of Theorem 4.1 that $W^u(\sigma) \cap W^s(\sigma) \neq \varnothing$. Let x be a point at which $W^u(\sigma)$ and $W^u(\sigma_1)$ meet transversally and consider a closed disk $D \subset W^s(\sigma_1)$ containing x. Since the intersection is transversal, by Lemma 4.6, we have $W^s(\sigma) \subset \overline{\cup_{t \leq 0} \phi_t(D)} \subset \overline{W^s(\sigma_1)}$. By Theorem 4.9, it follows that

$$\partial A_p \subset \bigcup_{\sigma_i \in \Sigma_1 \cap \partial A_p} \overline{W_s(\sigma_i)},$$

where Σ_1 is the set of type-one critical elements of the system. Since ∂A_p is a closed and invariant set, Assumption (A3) implies the following relationship:

$$\bigcup_{\sigma_i \in \Sigma_1 \cap \partial A_p} \overline{W_s(\sigma_i)} \subset \partial A_p.$$

Thus, we have the following result that is a refined version of Theorem 4.9.

Theorem 4.11: Characterization of Quasi-Stability Boundary

Consider a nonlinear dynamical system described by Equation 4.2, which satisfies Assumptions (A1)–(A3). Let σ_i^1, $i = 1, 2, \ldots$ be the type-one critical elements on the quasi-stability boundary $\partial Ap(x_s)$ of stable equilibrium point xs. Then,

$$\partial A_p(x_s) \subset \bigcup_{\sigma_i^1 \in \partial A_p(x_s)} \overline{W^s(\sigma_i^1)}.$$

Theorem 4.11 asserts that the quasi-stability boundary equals the union of closure of the stable manifolds of all type-one critical elements on the boundary. Applying this theorem of direct methods for power system transient stability, it follows that the closest UEP and the controlling UEP in direct methods are all type-one equilibrium points. In addition, the unstable manifolds of the closest UEP and the controlling UEP all converge to the stable equilibrium point.

4.4 CONCLUSIONS

The concept of quasi-stability regions for general nonlinear dynamical systems has been introduced and characterizations of the quasi-stability boundary have been presented. The quasi-stability boundary is shown to be the union of stable manifolds of equilibrium points and limit cycles on the boundary. In addition, the quasi-stability boundary is shown to be the union of the closure of stable manifolds of type-one equilibrium points and limit cycles on the boundary. Dynamic and topological properties of quasi-stability regions have also been derived. In general, a stability region is always a subset of the quasi-stability region. Finally, the class of nonlinear dynamical systems whose stability regions equal their quasi-stability regions has been characterized.

The analytical results presented in this chapter are applicable to power system transient stability models. For example, it will be shown that the closest UEP and controlling UEP in direct methods are all type-one equilibrium points. In addition, the unstable manifolds of the closest UEP and the controlling UEP all converge to the stable equilibrium point.

Chapter 5

Energy Function Theory and Direct Methods

5.1 INTRODUCTION

Power system stability analysis is concerned with whether the postfault trajectory will settle to an acceptable operating condition. Direct methods for stability analysis use algorithmic procedures to assess, without integrating the postfault system, the stability property of the postfault trajectory by comparing the system energy at the initial state of the postfault trajectory to a critical energy value. Direct methods not only avoid the time-consuming procedure of numerical integration of the postfault system, but they also provide a quantitative measure of the degree of system stability. Given a power system transient stability model with specified fault-on systems and a specified postfault system, direct methods for transient stability analysis are composed of the following steps:

Step 1. Numerically simulate the fault-on trajectory.

Step 2. Compute the initial point of the postfault system.

Step 3. Construct an energy function for the postfault power system.

Step 4. Compute the energy function value at the initial point of the postfault system.

Step 5. Compute the critical energy for the fault-on trajectory.

Step 6. Perform direct stability analysis by comparing the system energy at the initial state of the postfault system (computed at Step 4) with the critical energy (computed at Step 5). If the former is smaller than the latter, then the postfault trajectory will be stable; otherwise, it may be unstable.

Steps 1 and 2 require efficient numerical methods for simulating the fault-on trajectory. The construction of an energy function for a postfault power system (needed in Step 3) will be discussed in the next two chapters. Step 4 is a direct

Direct Methods for Stability Analysis of Electric Power Systems, by Hsiao-Dong Chiang

calculation of the energy function value at the initial condition obtained in Step 2. In Step 6, direct methods directly determine whether a postfault power system will remain stable by comparing the system energy computed in Step 4 to the critical energy computed in Step 5. Hence, it is crucial that an energy function capturing the system energy value be used to compute the system energy and an accurate critical energy value (for the fault-on trajectory) be calculated in direct methods. The topic of how to construct an energy function for a postfault power system will be discussed in this chapter and in the next chapter. The various methods for computing critical energy values along with the theoretical foundation of these methods will be presented in later chapters.

The theoretical basis of direct methods for the stability assessment of a postfault power system is the knowledge of the stability region of the postfault power system; if the initial condition of the postfault system lies inside the stability region of a desired postfault stable equilibrium point (SEP), then one can ensure, without performing numerical integrations, that the ensuing postfault trajectory will converge to the desired point. Therefore, knowledge of a stability region plays an important role in the theoretical foundation of direct methods.

In this chapter, a comprehensive energy function theory for general nonlinear autonomous dynamical systems will be presented. In addition, the theoretical basis for using energy functions to estimate the stability regions of large-scale nonlinear systems will be presented. The energy function theory has been applied to power system transient stability models to develop a theoretical foundation for direct methods. We will present this development in later chapters.

5.2 ENERGY FUNCTIONS

In this section, we study energy functions that can be viewed as extension of the Lyapunov functions. Energy functions are useful for global analysis of system trajectories and for estimating stability regions and quasi-stability regions, among others. The energy function theory for general nonlinear autonomous dynamical systems will be presented in the next section.

We consider a general nonlinear autonomous dynamical system described by the following equation:

$$\dot{x}(t) = f(x(t)). \tag{5.1}$$

We say a function, $V: \mathrm{R}^n \to R$, is an energy function for the system (Eq. 5.1) if the following three conditions are satisfied:

(i) The derivative of the energy function $V(x)$ along any system trajectory, $x(t)$, is nonpositive; that is,

$$\dot{V}(x(t)) \leq 0.$$

(ii) If $x(t)$ is a nontrivial trajectory (i.e., $x(t)$ is not an equilibrium point), then along the nontrivial trajectory $x(t)$, the set

$$\{t \in R : \dot{V}(x(t)) = 0\}$$

has a measure zero in R.

(iii) That a trajectory $x(t)$ has a bounded value of $V(x(t))$ for $t \in R^+$ implies that the trajectory $x(t)$ is also bounded. Stating this in brief, that $V(x(t))$ is bounded, implies that $x(t)$ itself is also bounded.

Property 1 indicates that the energy function is nonincreasing along its trajectory but does not imply that the energy function is strictly decreasing along its trajectory. There may exist a time interval $[t_1, t_2]$ such that $\dot{V}(x(t)) = 0$ for $t \in [t_1, t_2]$. Properties 1 and 2 imply that the energy function is strictly decreasing along any system trajectory. Property 3 states that the energy function is a proper map along any system trajectory but need not be a proper map for the entire state space. Recall that a proper map is a function $f: X \to Y$ such that for each compact set $D \in Y$, the set $f^{-1}(D)$ is compact in X. It will become clear that property 3 can be viewed as a "dynamic" proper map. From the above definition of the energy function, it is obvious that an energy function may not be a Lyapunov function.

As an illustration of the energy function, we consider the following classical transient stability model and derive an energy function for the model. Consider a power system consisting of n generators. Let the loads be modeled as constant impedances. Under the assumption that the transfer conductance of the reduced network after eliminating all load buses is zero, the dynamics of the ith generator can be represented by the equations

$$\begin{aligned} \dot{\delta}_i &= \omega_i \\ M_i\dot{\omega}_i &= P_i - D_i\omega_i - \sum_{\substack{j=1 \\ j\neq i}}^{n+1} V_iV_jB_{ij}\sin(\delta_i - \delta_j), \end{aligned} \tag{5.2}$$

where the voltage at node $i + 1$ is served as the reference; that is, $\delta_{i+1} := 0$. The set of equations (Eq. 5.2) for $i = 1, 2, \ldots, n$ describes the power system and can be written in matrix form:

$$\begin{aligned} \dot{\delta} &= \omega \\ M\dot{\omega} &= P - D\omega - f(\delta, V), \end{aligned} \tag{5.3}$$

where $\delta = (\delta_1, \ldots, \delta_n)^T$, $\omega = (\omega_1, \ldots, \omega_n)^T$, $P = (P_1, \ldots, P_n)^T$, $V = (V_1, \ldots, V_n)^T$, and $f = (f_1, \ldots, f_n)^T$ are n-vectors; $M = \text{diag}(M_1, \ldots, M_n)$, and $D = \text{diag}(D_1, \ldots, D_n)$ are $n \times n$ positive diagonal matrices; and

$$f_i(\delta, V) = \sum_{\substack{j\neq i \\ j=1}}^{n+1} V_iV_jB_{ij}\sin(\delta_i - \delta_j).$$

The equations for the prefault, fault-on, and postfault systems have the same form as Equation 5.2 except that the B_{ij} is different due to the change in network topology. This is a version of the so-called *classical model* of the power system. We next show that there exists an energy function, $V(\delta, \omega)$, for the classical model (Eq. 5.3).

Consider the following function:

$$V(\delta,\omega)=\frac{1}{2}\sum_{i=1}^{n}M_i\omega_i^2-\sum_{i=1}^{n}P_i(\delta_i-\delta_i^s)-\sum_{i=1}^{n}\sum_{j=i+1}^{n+1}V_iV_jB_{ij}\left\{\cos(\delta_i-\delta_j)-\cos(\delta_i^s-\delta_j^s)\right\}, \tag{5.4}$$

where $x^s = (\delta^s, 0)$ is the SEP under consideration.

Differentiating V along the trajectory $(\delta(t), \omega(t))$ of Equation 5.2 gives

$$\dot{V}(\delta(t),\omega(t))=\sum_{i=1}^{n}\left(\frac{\partial V}{\partial\delta_i}\dot{\delta}_i+\frac{\partial V}{\partial\omega_i}\dot{\omega}_i\right)=-\sum_{i=S_n}D_i\omega_i^2\le 0. \tag{5.5}$$

For the trajectory starting at an equilibrium point, $(\delta^s, 0)$, we have $\omega(t) = 0$; hence,

$$\dot{V}(\delta(t),\omega(t))=0. \tag{5.6}$$

Suppose that there is an interval, $t \in [t_1, t_2]$, such that

$$\dot{V}(\delta(t),\omega(t))=0 \quad t\in[t_1,t_2]. \tag{5.7}$$

Hence,

$$\omega(t)=0 \quad t\in[t_1,t_2]. \tag{5.8}$$

But this implies $\omega(t) = 0$ and $\delta(t) =$ constant for $t \in [t_1, t_2]$. It then follows from Equation 5.2 that

$$P_i-\sum_{\substack{j\neq i\\ j=1}}^{n+1}V_iV_jB_{ij}\sin(\delta_i-\delta_j)=0, \tag{5.9}$$

which are precisely the equations for the equilibrium point of Equation 5.2. Therefore, $(\delta(t), \omega(t))$ $t \in [t_1, t_2]$ must be on an equilibrium point.

Let us first integrate Equation 5.2 for ω_i:

$$\omega_i(t)=e^{-D_i/M_i t}\omega_i(0)+\int_0^t e^{-D_i/M_i(t-s)}\left\{P_i-\sum_{\substack{j\neq i\\ j=1}}^{n+1}V_iV_jB_{ij}\sin(\delta_i(s)-\delta_j(s))\right\}ds. \tag{5.10}$$

The term in the bracket is uniformly bounded, say, by a_i; that is,

$$\left|P_i-\sum_{\substack{j\neq i\\ j=1}}^{n+1}V_iV_jB_{ij}\sin(\delta_i(s)-\delta_j(s))\right|\le a_i. \tag{5.11}$$

Since D_i and M_i are positive numbers, we have from Equation 5.10

$$|\omega_i(t)| \le |\omega_i(0)| + a_i \frac{M_i}{D_i}; \tag{5.12}$$

that is, $\omega_i(t)$ is bounded by $b_i := |\omega_i(0)| + a_i M_i / D_i$.

We next show that condition (iii) of the energy function is satisfied, suppose $V(\delta(t), \omega(t))$ is bounded below and above, say, by c_1 and c_2, respectively. Then, we have

$$c_1 - \frac{1}{2}\sum_{i=1}^{n} M_i b_i^2 < -\sum_{i=1}^{n} P_i(\delta_i - \delta_i^s) - \sum_{i=1}^{n}\sum_{j=i-1}^{n+1} V_i V_j B_{ij}\{\cos(\delta_i - \delta_j) - \cos(\delta_i^s - \delta_j^s)\} < c_2. \tag{5.13}$$

But the second term on the right-hand side is uniformly bounded, say, by c; that is,

$$\left|\sum_{i=1}^{n}\sum_{j=i-1}^{n+1} V_i V_j B_{ij}\{\cos(\delta_i - \delta_j) - \cos(\delta_i^s - \delta_j^s)\}\right| < c. \tag{5.14}$$

Substituting Equation 5.14 into Equation 5.13, we get

$$c_1 - \frac{1}{2}\sum_{i=1}^{n} M_i b_i^2 - c < -\sum_{i=1}^{n} P_i(\delta_i - \delta_i^s) < c_2 + c. \tag{5.15}$$

Hence, the term $P^T\delta$ is bounded. It is shown in Arapostathis et al. (1982) that $P^T\delta$ being bounded implies $\delta(t)$ is bounded. This result, together with Equation 5.12, asserts that the trajectory $((\delta(t), \omega(t)))$ on the stability boundary is bounded. Hence, condition (iii) of the energy function is satisfied.

5.3 ENERGY FUNCTION THEORY

The theoretical basis of direct methods for power system transient stability assessment is the knowledge of the stability region. We present analytical results showing that the stability regions of a class of nonlinear systems can be completely characterized. For this class of nonlinear systems, which have energy functions, a complete characterization of the stability boundary will be derived and optimal schemes to estimate its stability regions via an energy function will be presented.

We first examine the behaviors of system trajectories of nonlinear systems having energy functions. In general, the dynamic behaviors of trajectories of general nonlinear systems can be very complicated. The asymptotic behaviors (i.e., the ω-limit set) of trajectories can be quasi-periodic trajectories or chaotic trajectories. However, as shown below, if the underlying dynamical system has some special properties, then the system may admit only simple trajectories. For instance, every trajectory of the system (Eq. 5.1) having an energy function has only two modes of behaviors: its trajectory either converges to an equilibrium point or goes to infinity (becomes unbounded) as time increases or decreases. This result is explained in the following theorem.

Theorem 5.1: Global Behavior of Trajectories

If there exists a function satisfying conditions (i) and (ii) of the energy function for the system (Eq. 5.1), then every bounded trajectory of the system (Eq. 5.1) converges to one of the equilibrium points.

Theorem 5.1 asserts that there does not exist any limit cycle (oscillation behavior) or bounded complicated behavior such as an almost periodic trajectory and a chaotic motion in the system. Applying this result to power system models, it indicates that for a power system model having an energy function, there is no complicated behavior such as closed orbits (limit cycle), quasi-periodic trajectories, and chaotic motions in the state space of the model.

We next show a sharper result, asserting that every trajectory on the stability boundary must converge to one of the unstable equilibrium points (UEPs) on the stability boundary.

Theorem 5.2: Trajectories on the Stability Boundary

If there exists an energy function for the system (Eq. 5.1), then every trajectory on the stability boundary $\partial A(x_s)$ converges to one of the equilibrium points on the stability boundary $\partial A(x_s)$.

The significance of this theorem is that it offers an effective way to characterize the stability boundary. In fact, Theorem 5.2 asserts that the stability boundary $\partial A(x_s)$ is contained in the union of stable manifolds of the UEPs on the stability boundary. One corollary of Theorem 5.2 shown next will be extended to develop a theoretical foundation for direct methods, in particular, the closest UEP method and the controlling UEP method.

Corollary 5.3: Energy Function and Stability Boundary

If there exists an energy function for the system (Eq. 5.1), which has an asymptotically stable equilibrium point, x_s (but not globally asymptotically stable), then the stability boundary $\partial A(x_s)$ is contained in the set, which is the union of the stable manifolds of the UEPs on the stability boundary $\partial A(x_s)$; that is,

$$\partial A(x_s) \subseteq \bigcup_{x_i \in \{E \cap \partial A(x_s)\}} W^s(x_i).$$

The following two theorems give interesting results on the structure of the equilibrium points on the stability boundary. Moreover, it presents a necessary condition for the existence of certain types of equilibrium points on a *bounded* stability boundary. The proof is similar to that of Theorem 3.16 and is therefore omitted.

Theorem 5.4: Structure of Equilibrium Points on the Stability Boundary

If there exists an energy function for the system (Eq. 5.1), which has an asymptotically stable equilibrium point, x_s (but not globally asymptotically stable), then the stability boundary $\partial A(x_s)$ must contain at least one type-one equilibrium point. If, furthermore, the stability region is bounded, then the stability boundary $\partial A(x_s)$ must contain at least one type-one equilibrium point and one source.

The contrapositive of Theorem 5.4 leads to the following corollary, which is useful in predicting the unboundedness of a stability region.

Theorem 5.5: Sufficient Condition for Unbounded Stability Region

If there exists an energy function for the system (Eq. 5.1), which has an asymptotically stable equilibrium point, x_s (but not globally asymptotically stable), and if $\partial A(x_s)$ contains no source, then the stability region $A(x_s)$ is unbounded.

We next address the issue of whether it is possible that two equilibrium points of the system (Eq. 5.1) have the same energy function value $V(.)$ (i.e., $V(x_1) = V(x_2)$ for $x_1, x_2 \in E$) In the context of direct methods, this issue is related to the uniqueness of the closest UEP and the uniqueness of the controlling UEP. Note that all the potential energy $V_p(x)$ in the (total) energy functions $V(.)$ for existing power system transient stability models have the property that $\nabla V_p(x) = 0$ at any equilibrium point. To answer this question, we first observe the conditions for two equilibrium points x_1 and x_2 to have the relationships that $V_p(x_1) = V_p(x_2)$ and $\nabla V_p(x_1) = \nabla V_p(x_2)$. This is a set of $(2n + 1)$ nonlinear algebraic equations with $2n$ unknowns (x_1, x_2); it is unlikely that there exists a solution. A rigorous justification of this observation is presented in Theorem 5.6. This theorem states that generically, all the equilibrium points of the system (Eq. 5.1) have distinct energy function values. We recall the concept of a generic property. Let X be a complete metric space and let $P(x)$ be a statement about points x in X. We say that $P(x)$ is a *generic property* if the set of points where it holds true contains a countable intersection of open dense sets.

The following result can be proved via Thom's famous transversality theorem (Hirsch, 1976; Munkres, 1975).

Theorem 5.6: Energy Functions and Equilibrium Points

Let $V(.)$ be an energy function for the system (Eq. 5.1) and $\nabla V(x) = 0$ at the equilibrium points of the system (Eq. 5.1). Then, the property that all the equilibrium points of the system (Eq. 5.1) have distinct values of energy function $V(.)$ is generic.

We next apply the analytical results derived in this section to the classical transient stability model (Eq. 5.2), which admits an energy function for the model. First, the hyperbolicity of equilibrium points of the classical power system stability model (Eq. 5.3) is shown.

Proposition 5.7: Hyperbolic Equilibrium Points

Consider the classical power system stability model (Eq. 5.3). If the Jacobian matrix $\left(\frac{\partial f}{\partial \delta}\right)$ is nonsingular at an equilibrium point, then the equilibrium point is hyperbolic.

Proof: Let $\hat{x} = (\hat{\delta}, \hat{w})$ be an equilibrium point of Equation 5.3. The Jacobian at $\hat{x}$ is given by

$$J(\hat{x}) = \begin{bmatrix} 0 & I \\ -M^{-1}F(\hat{\delta}) & -M^{-1}D \end{bmatrix}, \tag{5.16}$$

where the ijth element of the matrix $F(\hat{\delta})$ is $\partial f_i/\partial \delta_j(\hat{\delta})$. Let λ be an eigenvalue of $J(\hat{x})$ and $x = (x_1, x_2)$ be a corresponding eigenvector:

$$\begin{bmatrix} 0 & I \\ -M^{-1}F(\hat{\delta}) & -M^{-1}D \end{bmatrix} \begin{bmatrix} x_1 \\ x_2 \end{bmatrix} = \lambda \begin{bmatrix} x_1 \\ x_2 \end{bmatrix} \tag{5.17}$$

or

$$x_2 = \lambda x_1 \text{ and} \tag{5.18}$$

$$-M^{-1}F(\hat{\delta})x_1 - M^{-1}Dx_2 = \lambda x_2. \tag{5.19}$$

Substituting Equations 5.18 into Equation 5.19 and premultiplying by $x_1^T M$, we get

$$\lambda^2 x_1^T M x_1 + \lambda x_1^T D x_1 + x_1^T F(\hat{\delta}) x_1 = 0. \tag{5.20}$$

We are going to show that $\lambda \neq 0$ and $\lambda \neq j\alpha$; hence, the equilibrium point is hyperbolic. The two cases are considered separately.

1. Suppose $\lambda = 0$; then, Equation 5.20 becomes

   $$x_1^T F(\hat{\delta}) x_1 = 0,$$

   which contradicts the assumption that $F(\hat{\delta})$ is nonsingular.

2. Suppose $\lambda = j\alpha$; then, Equation 5.20 becomes

   $$\alpha^2 x_1^T M x_1 = x_1^T F(\hat{\delta}) x_1 \text{ and} \tag{5.21}$$

$$j\alpha x_1^T D x_1 = 0. \tag{5.22}$$

Again this is a contradiction because D is assumed to be nonsingular. This completes the proof.

The hyperbolicity of the equilibrium points of the classical power system stability model (Eq. 5.3) implies that every equilibrium point is isolated. We next apply Theorem 5.2 to conclude that every trajectory on the stability boundary of the classical power system stability model (Eq. 5.3) converges to one of the equilibrium points, and the following characterization of the stability boundary for the classical power system stability model (Eq. 5.3) is derived.

Proposition 5.8: Characterization of Stability Boundary

Let x_s be an asymptotically stable equilibrium point of the classical power system stability model (Eq. 5.3), then the stability boundary $\partial A(x_s)$ is contained in the set, which is the union of the stable manifolds of the UEPs on the stability boundary $\partial A(x_s)$; that is,

$$\partial A(x_s) \subseteq \bigcup_{x_i \in \{E \cap \partial A(x_s)\}} W^s(x_i).$$

The stability region of the classical power system stability model (Eq. 5.3) is shown to be unbounded as stated in the following.

Proposition 5.9: Unbounded Stability Region

The stability boundary $\partial A(x_s)$ of an asymptotically stable equilibrium point of the classical power system stability model (Eq. 5.3) is unbounded.

Proof: We will show that the equilibrium points of Equation 5.3 do not contain a source. Specifically, we will show that the Jacobian $J(\hat{x})$ always has eigenvalues with negative real parts. This is done by the application of the inertia theorem (Wimmer, 1974), which states that if H is a nonsingular Hermitian matrix and A has no eigenvalues on the imaginary axis (i.e., $n_c(A) = n_c(H) = 0$), then $AH + HA^* \geq 0$ implies that

$$In(A) = In(H). \tag{5.23}$$

Now we choose H to be the symmetric matrix:

$$H = \begin{bmatrix} -F(\hat{x})^{-1} & 0 \\ 0 & -M^{-1} \end{bmatrix}. \tag{5.24}$$

We have

$$J(\hat{x})H + HJ(\hat{x})^T = 2\begin{bmatrix} 0 & 0 \\ 0 & M^{-1}DM^{-1} \end{bmatrix}. \tag{5.25}$$

Hence,

$$In(J(\hat{x})) = In\begin{bmatrix} -F(\hat{x})^{-1} & 0 \\ 0 & -M^{-1} \end{bmatrix}. \tag{5.26}$$

Since M is a diagonal matrix with positive elements, Equation 5.26 implies that, at any equilibrium point $\hat{x}$, the Jacobian $J(\hat{x})$ has at least n eigenvalues with negative real parts. This completes the proof.

Remark: It should be emphasized that Proposition 5.9 asserts that the stability region is unbounded for the power system model under consideration. In practice, of course, constraints from devices not modeled here would prevent the actual system from operating with a large deviation of frequency and angle differences.

5.4 ESTIMATING STABILITY REGION USING ENERGY FUNCTIONS

In this section, we focus on how to characterize, via an energy function, the stability region of a high-dimension nonlinear system such as a power system. This characterization will prove important in the direct methods. We consider the following set:

$$S_v(k) = \{x \in R^n : V(x) < k\}, \tag{5.27}$$

where $V(.):R^n \to R$ is an energy function. Sometimes, we drop the subscript v of $S_v(k)$, simply writing $S(k)$, when it is clear from the context. We shall call the boundary of the set (Eq. 5.27) $\partial S(k) := \{x \in R^n : V(x) = k\}$ the *level set* (or *constant energy surface*) and k the *level value*. If k is a regular value (i.e., $\nabla V(x) \neq 0$, for all $x \in V^{-1}(k)$), then by the inverse function theorem, $\partial S(k)$ is a $C^r(n-1)$-dimensional submanifold of R^n (Hurewicz and Wallman, 1948, p. 46). Moreover, if $r > n - 1$, then by the Morse–Sard theorem, the set of regular values of V is residual; in other words, "almost all" level values are regular. In particular, for almost all values of k, the level set $\partial S(k)$ is a $C^r(n-1)$-dimensional submanifold.

Generally speaking, this set $S(k)$ can be very complicated with several connected components even for the two-dimensional case. Let

$$S(k) = S^1(k) \cup S^2(k) \cup \ldots \cup S^m(k), \tag{5.28}$$

where $S^i(k) \cap S^j(k) = \phi$ when $i \neq j$. That is, each of these components is connected and disjoint from each other. Since $V(.)$ is continuous, $S(k)$ is an open set. Because $S(k)$ is an open set, the level set $\partial S(k)$ is of $(n - 1)$ dimensions. Furthermore, each component of $S(k)$ is an invariant set.

In spite of the possibility that a constant energy surface may contain several disjoint connected components, there is an interesting relationship between the constant energy surface and the stability boundary. This relationship is that at most, one connected component of the constant energy surface $\partial S(r)$ has a nonempty intersection with the stability region $A(x_s)$. This relationship is established in Theorem 5.10.

Theorem 5.10: Constant Energy Surface and Stability Region

Let x_s be a SEP of the system (Eq. 5.1) and $A(x_s)$ be its stability region. Then, the set $S(r)$ contains only one connected component, which has a nonempty intersection with the stability region $A(x_s)$ if and only if $r > V(x_s)$.

Motivated by Theorem 5.10, we shall use the notation $S_{x_s}(r)$ to denote the connected set of $S(r)$ (whose level value is r) containing the SEP x_s. We drop the subscript x_s of $S_{x_s}(r)$ when it is clear from the context. In Figure 5.1, the relation between the constant energy surfaces at different level values and the stability region $A(x_s)$ is shown. It is observed from this figure that the connected set $S_{x_s}(r)$ with a level value r smaller than the critical value is very conservative in the approximation of the stability boundary $\partial A(x_s)$. As the set $S_{x_s}(r)$ is expanded by increasing the level value r, the approximation gets improved until this constant energy surface hits the stability boundary $\partial A(x_s)$ at some point. This point will be shown to be a UEP. We call this point the *closest UEP* of the SEP x_s with respect to the energy function $V(.)$.

Furthermore, as we increase the level value r, the connected set $S_{x_s}(r)$ would contain points that lie outside the stability region $A(x_s)$. It is therefore inappropriate to approximate the stability region $A(x_s)$ by the connected set $S_{x_s}(r)$ with a level value higher than that of the closest UEP. From this figure, it becomes obvious that among the several connected sets of the constant energy surface, there is only one connected component of the constant energy surface that has a nonempty intersection with the stability region $A(x_s)$.

Topological and dynamic characterizations for the point on the stability boundary with the minimum value of an energy function will be derived. These characterizations are useful in identifying the point with the minimum value of an energy function over the stability boundary.

Theorem 5.11: Topological Characterization

Let x_s be a SEP and $A(x_s)$ be the stability region of the system (Eq. 5.1), which has an energy function. If the stability region $A(x_s)$ is not dense in R^n, then the point with the minimum value of the energy function over the stability boundary $\partial A(x_s)$ exists, and it must be a UEP.

Proof: By the conditions of energy functions, x_s is a SEP if and only if x_s is a local minimum of any energy function $V(x)$ associated with the system (Eq. 5.1). Hence, given any energy function $V(x)$, there exists a positive number n such that the set $\partial S(n)$, which is defined to be the connected component of the level set $\{x{:}V(x_s) \le V(x) \le n\}$ containing the SEP x_s, is closed and bounded.

Next, we investigate how the set $\partial S(n)$ changes as the level value n is varied. Suppose that n is a regular value of $V(.)$; then, as a consequence of the implicit function theorem, the set $SC_n(x_s)$ is a manifold. Furthermore, as we increase the level value n to the new value, say, m with $m > n$, such that there are no critical

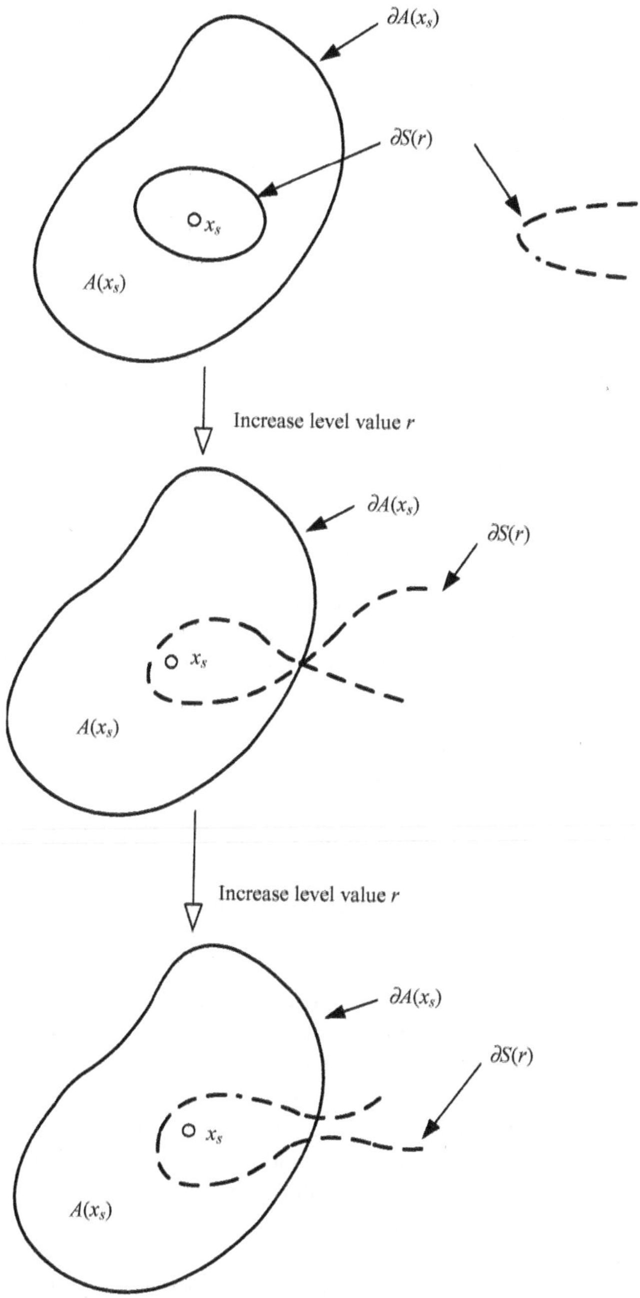

Figure 5.1 The relation between the constant energy surfaces $S(r)$ at different level values and the stability region $A(x_s)$.

points in the set $\{S(m) - S(n)\}$, then it can be shown (Franks, 1980) that the set $\partial S(m)$ is diffeomorphic to $\partial S(n)$. In particular, the set $\partial S(m)$ is closed and bounded.

Therefore, as we increase the level value n, the level set $\partial S(n)$ remains to be bounded until it hits a critical point, which is an equilibrium point of the system (Eq. 5.1). Now we argue that this equilibrium point, say, $\hat{x}$, must be lying on the stability boundary $\partial(x_s)$. By contradiction, suppose that $V(\hat{x}) = \overline{m}$ and this equilibrium point $\hat{x} \in \overline{A(x_s)}^c$, here, $\overline{A(x_s)}^c$ denotes the complement of $\overline{A(x_s)}$. It is easy to see that the set $S\overline{m}$ has a nonempty intersection with the stability boundary $\partial A(x_s)$. Let $S := S(\overline{m}) \cap \partial A(x_s)$. We will show that S must contain at least an equilibrium point and thus contradicts the condition that $\hat{x}$ is the only equilibrium point lying in $S(\overline{m})$. We note the following:

1. $\partial A(x_s)$ is a closed and invariant set of the system (Eq. 5.1) and
2. $SC_{\hat{m}}(x_s)$ is a compact and positive invariant set of the system (Eq. 5.1).

Hence, the set S is a compact, positive invariant set. Since any compact, positive invariant set must contain at least an ω-limit set and the ω-limit set of the system (Eq 5.1) consists of equilibrium points, the set S must contain at least one equilibrium point, which is a contradiction, and we complete the proof.

Theorem 5.11 asserts that the point with the minimum value over the stability boundary must be a UEP. Theorem 5.12 gives a dynamic characterization of this UEP in terms of its unstable manifold. Note that this theorem holds without the transversality condition.

Theorem 5.12: Dynamic Characterization

Let x_s be a SEP and $A(x_s)$ be the stability region of the system (Eq. 5.1), which has an energy function. If the stability region $A(x_s)$ is not dense in R^n and $\hat{x}$ is the point with the minimum value of the energy function over the stability boundary $\partial A(x_s)$, then $W^u(\hat{x}) \cap A(\hat{x}) \neq \phi$.

Proof: We prove this theorem by contradiction. Suppose that the unstable manifold of $\hat{x}$ does not converge to the SEP x_s. Theorem 3.8 states that if x_s is a SEP and $\hat{x}$ is a hyperbolic equilibrium point, then $\hat{x}$ is on the stability boundary $\partial A(x_s)$ if and only if its unstable manifold has a nonempty intersection with the closure of the stability region; that is, $W^u(\hat{x}) \cap \bar{A}(x_s) \neq \phi$. Therefore, there exists a trajectory $x(t)$ in the unstable manifold $W^u(\hat{x})$ such that $x(t) \in \partial A(x_s)$ and $\lim_{t \to -\infty} x(t) = \hat{x}$. In addition, the energy function value is strictly decreasing along any nontrivial trajectory of the system (Eq. 5.1). These two facts imply that there exist other points on the stability boundary with a lower energy function value than $V(\hat{x})$. This contradicts the fact that $\hat{x}$ possesses the lowest energy function value within the stability boundary $\partial A(x_s)$. This completes the proof.

We have analyzed some behaviors of the constant energy surface and the relationship between the constant energy surface and the stability region. The next subject is approximating a stability region via a constant energy surface in an optimal way. Indeed, given an energy function, the central point in the energy function approach for estimating the stability region of a SEP is the determination of the critical level value for the energy function.

5.5 OPTIMAL SCHEMES FOR ESTIMATING STABILITY REGIONS

We discuss in this section how to optimally determine the critical level value of an energy function for estimating the stability boundary $\partial A(x_s)$. We use the notation $\bar{A}^c(x_s)$ to denote the complement of the set $\bar{A}_p(x_s)$. We use the notation $S_k(x_s)$ to denote the only component of the several disjoint connected components of S_k that contains the SEP x_s.

Theorem 5.13: Optimal Estimation

Consider the nonlinear system (Eq. 5.1) that has an energy function $V(x)$. Let x_s be an asymptotically stable equilibrium point whose stability region $A(x_s)$ is not dense in R^n. Let E_1 be the set of equilibrium points and $\hat{c} = \min_{x_i \in \partial A(x_s) \cap E_1} V(x_i)$, then

[1] $S_{\hat{c}}(x_s) \subset A(x_s)$ and

[2] the set $\{(S_b(x_s) \cap \bar{A}^c(x_s)\}$ is nonempty for any number $b > \hat{c}$.

The above theorem asserts that the connected component $S_{\hat{c}}(x_s)$ is the best candidate to approximate the stability region $A(x_s)$. Part 2 of Theorem 5.13 asserts that the scheme of choosing $\hat{c} = \min_{x_i \in \partial A(x_s) \cap E_1} V(x_i)$ is optimal because the estimated stability region characterized by the corresponding constant energy surface is the largest one within the entire stability region.

Based on Theorems 5.11–5.13, we propose the following scheme to estimate the stability region $A(x_s)$ of a nonlinear dynamical system (Eq. 5.1) via an energy function $V(.)$.

Scheme: Optimal Estimation of the Stability Region $A(x_s)$ via an Energy Function $V(.)$

A. Determining the critical level value of the energy function

Step A1. Find all the equilibrium points.

Step A2. Order these equilibrium points whose corresponding values, $V(.)$, are greater than $V(x_s)$.

Step A3. Of these, identify the one with the lowest energy function value and whose unstable manifold converges to the SEP x_s. (Let this one be $\hat{x}$.)

Step A4. The value of the energy function at $\hat{x}$ gives the critical level value of this energy function (i.e., $V(\hat{x})$).

B. Estimating the stability region $A(x_s)$

Step B1. The connected component of $\{x|V(x) < V(\hat{x})\}$ containing the SEP x_s gives the estimated stability region.

Remark: The computation associated with Step A1 could be very involved. Efficient numerical methods in conjunction with utilizing special properties of the system under study are needed to implement this step. The above scheme is general and applicable to a nonlinear dynamical system (Eq. 5.1) that has an energy function. The computation efficiency of this scheme, however, can be improved by focusing on type-one equilibrium points if the quasi-stability region is to be estimated. In this situation, instead of finding all of the equilibrium points in Steps A1 and A2, one merely needs to find and check all the type-one equilibrium points.

For the purpose of illustration, we consider the following simple example:

$$\begin{aligned}\dot{x}_1 &= -\sin x_1 - 0.5\sin(x_1 - x_2) + 0.01\\ \dot{x}_2 &= -0.5\sin x_2 - 0.5\sin(x_2 - x_1) + 0.05.\end{aligned} \tag{5.29}$$

It is easy to show that the following function is an energy function for the system (Eq. 5.29):

$$V(x_1, x_2) = -2\cos x_1 - \cos x_2 - \cos(x_1 - x_2) - 0.02x_1 - 0.1x_2. \tag{5.30}$$

The point $x^s := (x_1^s, x_2^s) = (0.02801, 0.06403)$ is the SEP whose stability region is to be estimated. Applying the optimal scheme to the system (Eq. 5.29), we have the following:

Steps A1 and A2. There are two type-one equilibrium points within the region $\{(x_1, x_2): x_1^s - \pi < x_1 < x_1^s + \pi, x_2^s - \pi < x_2 < x_2^x + \pi\}$.

Step A3. The type-one equilibrium point (0.04667, 3.11489) is the one with the lowest energy function value among all the UEPs on the stability boundary $\partial A(x^s)$, and its unstable manifold converges to the SEP (0.02801, 0.06403).

Step A4. The critical level value for the system (Eq. 5.29) is –0.31329 (see Table 5.1).

Curve *A* in Figure 5.2 is the exact stability boundary $\partial A(x^s)$ of the system (Eq. 5.29) according to Corollary 5.3. Curve *B* is the stability boundary estimated by the

Table 5.1 The Coordinates of Two Type-One Equilibrium Points and Their Energy Function Values

Type-one equilibrium point	x_1	x_2	$V(.)$
1	0.04667	3.11489	–0.31329
2	–3.03743	0.33413	2.04547

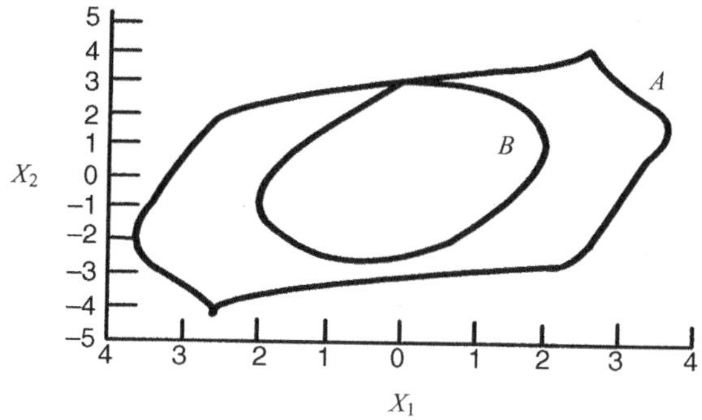

Figure 5.2 Curve A is the exact stability boundary $\partial A(x^s)$ of the system (Eq. 5.29) according to Corollary 5.3. Curve B is the stability boundary estimated by the constant energy surface (with a level value of −0.31329) of the energy function. The optimality of this estimation is also shown in this figure.

connected component (containing the SEP x^s) of the constant energy surface of the energy function (Eq. 5.30) passing through −0.31329. It can be seen from Figure 5.2 that the critical level value −0.31329 determined by the proposed scheme is indeed the optimal value of the energy function (Eq. 5.30) for estimating the stability region $A(x^s)$. The estimate by curve B, however, is still conservative compared to the exact stability boundary.

5.6 QUASI-STABILITY REGION AND ENERGY FUNCTION

In this section, we will establish a relationship between the structures of a quasi-stability boundary and constant energy surfaces at different level values. Recall that only one component of the constant energy surface can intersect the quasi-stability region. The constant energy surface with a small level value is very conservative in approximating the quasi-stability boundary. As the constant energy surface is expanded with the increasing level value, the approximation becomes more accurate until this constant energy surface intersects the stability boundary at the closest UEP (see Figure 5.3). Two scenarios can occur: the first scenario is that the closest UEP lies in the quasi-stability region and not on the quasi-stability boundary (see Figure 5.3), and the second scenario is that the closest UEP lies on the stability boundary but not on the quasi-stability region.

If the closest UEP does not lie on the quasi-stability boundary, then it must lie on the quasi-stability region. Even as we increase the level value, the new constant energy surface would still lie inside the quasi-stability region $A_p(x_s)$ and it is a connected but not simply connected set, as shown in the following.

Proposition 5.14: Bounded Property

Consider the nonlinear dynamical system described by Equation 5.1, which admits an energy function $V(x)$. Let A, ∂A, and A_p be the stability region, the stability boundary, and the quasi-stability region, respectively, of x_s. Let $c = \min_{x \in \partial A \cap E} V(x)$, where

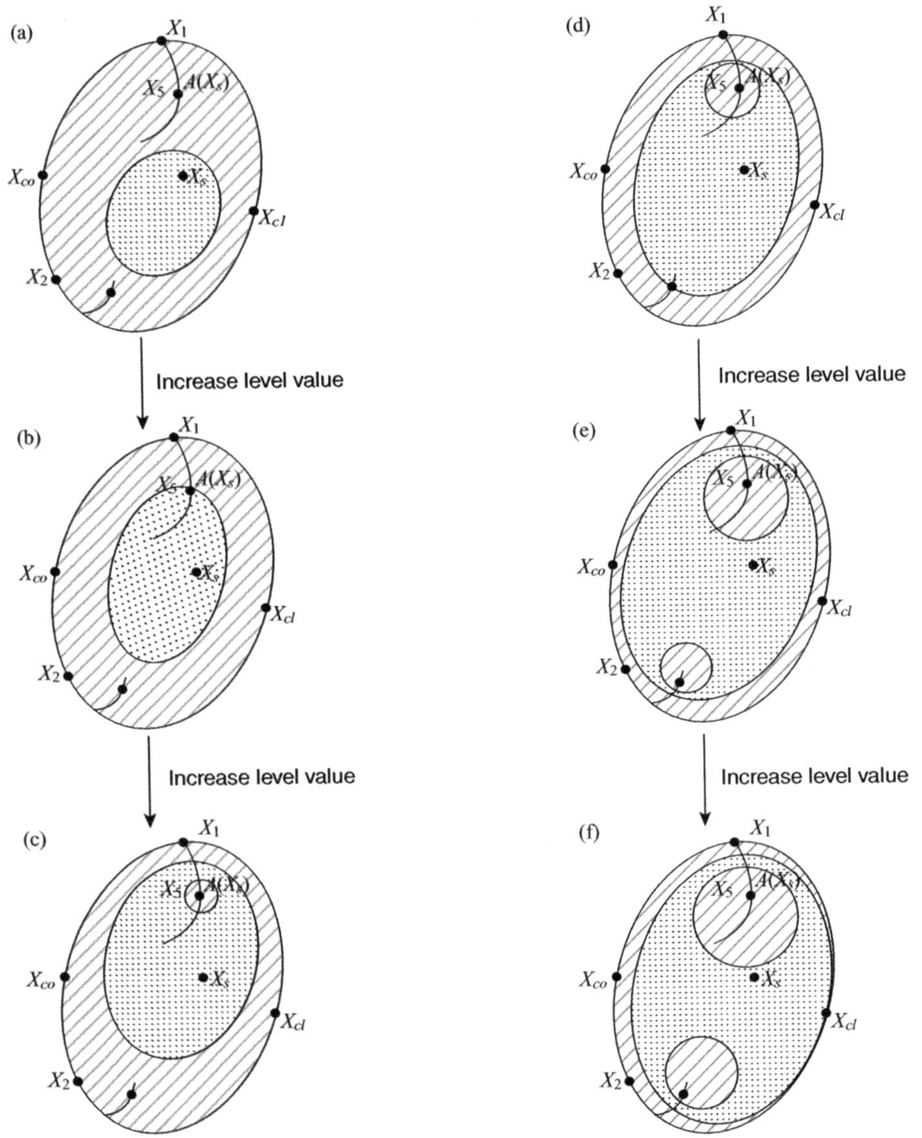

Figure 5.3 The structure of a constant energy surface changes with the increase of its level value. The constant energy surface contains several connected components; some of them are simply connected, while others are not simply connected.* In (a), only one connected constant energy surface intersects with the quasi-stability region and the constant energy surface is conservative in approximating the quasi-stability boundary. In (b), the constant energy surface intersects with the closest UEP, which lies inside the quasi-stability region. In (c), as the level value increases, there is a "hole" in the connected constant energy surface, which is then not simply connected. In (d), the constant energy surface intersects with another UEP, which lies inside the quasi-stability region. In (f), with the increase in the level value, the constant energy surface eventually intersects with a type-one UEP lying on the quasi-stability boundary.

*A set is said to be connected if the line connecting two points in the set lies in the set. A set is said to be simply connected if the set is connected and the (topological) boundary of the set is also connected.

E is the set of equilibrium points of Equation 5.1. If the value c is attained at an equilibrium point $\hat{x} \in A_P$, then for any $\varepsilon > 0$ sufficiently small, the set $\overline{S_{c+\varepsilon}(x_s)}$ is contained in the quasi-stability region; that is, $\overline{S_{c+\varepsilon}(x_s)} \subset A_p$.

Proof: Let k be the type of $\hat{x}$. Since $\hat{x}$ is a hyperbolic equilibrium point, by the Morse lemma (Milnor, 1973), there exists local coordinates $(x_1, \ldots, x_k, y_1, \ldots, y_{n-k})$ defined in the neighborhood U centered at $\hat{x}$, such that $V(x, y) = c - |x|^2 + |y|^2$ in these coordinates. In these local coordinates, let $H(\varepsilon)$ be a rectangular neighborhood of $\hat{x}$ defined as

$$H(\varepsilon) = \left\{(x, y) \in U : |y|^2 \le \varepsilon \text{ and } |x|^2 \le 2\varepsilon\right\}.$$

The set H is called a handle of equilibrium point $\hat{x}$. It can be shown that $\overline{S_{c+\varepsilon}(x_s)}$ is homeomorphic to $\overline{S_{c-\varepsilon}(x_s)} \cup H(\varepsilon)$ (Franks, 1980). Note that $\overline{S_{c-\varepsilon}(x_s)}$ and $H(\varepsilon)$ are compact sets. Thus, $\overline{S_{c+\varepsilon}(x_s)}$ is also compact because the image of a compact set under a homeomorphism is compact. Hence, $S_{c+\varepsilon}(x_s)$ is bounded.

Now, since $\hat{x} \in A_P = int\bar{A}$ for ε small enough, it follows that $H(\varepsilon) = int\bar{A}$. Moreover, $\overline{S_{c-\varepsilon}(x_s)} \subset int\bar{A}$. This implies that $H(\varepsilon) \cup \overline{S_{c-\varepsilon}(x_s)} = \overline{S_{c+\varepsilon}(x_s)} \subset int\bar{A}$. This completes the proof.

Proposition 5.14 asserts that if one keeps on increasing the level value from $c + \varepsilon$ to a higher value, say, c', without hitting $\partial\bar{A}$, then $S_{c'}(x_s)$ is still inside A_p and is bounded. In addition, it can be shown that if $a < b$ are regular values and there are no equilibrium points in $S_b(x_s) - S_a(x_s)$, then $S_a(x_s)$ and $S_b(x_s)$ are diffeomorphic. Hence, the stability region estimated by a suitable constant energy surface always has a nice geometrical structure because it is always homeomorphic to an open n-dimensional sphere. The quasi-stability region estimated by a constant energy surface, however, can exhibit a very complicated shape, as we can see in the following proposition.

Proposition 5.15: Connected but Not Simply Connected Property

Consider the nonlinear dynamical system described by Equation 5.1, which has an energy function $V(x)$. Let A, ∂A, and A_p be the stability region, the stability boundary, and the quasi-stability region, respectively, of x_s. Let $c = \min_{x \in \partial A \cap E} V(x)$, where E is the set of equilibrium points of Equation 5.1. If the value c is attained at an equilibrium point $\hat{x} \in A_p$, then $S_{c+\varepsilon}(x_s)$ is not simply connected.

Proof: We will first review the concept of deformation retraction (Massey, 1967). Let B be a subset of X. A deformation reaction of X onto B is a continuous map $F: X \times I \to X$, where I is the unit interval, such that $F(x, 0) = x$ for $x \in X$, $F(x, 1) \in B$ for $x \in X$, and $F(b, t) = b$ for $b \in B$. If such an F exists, then B is called a deformation retract of X. One can visualize a deformation retraction as a gradual collapsing of the space X onto the subspace B, such that each point of B remains fixed during the shrinking process. This concept is used in the following lemma.

Lemma 5.16:

Let n_u be the type of $\hat{x}$ and let $W_\varepsilon^u(\hat{x})$ be an n_u-dimensional neighborhood of $\hat{x}$ in $W^u(\hat{x}) \cap H(\varepsilon)$, where $H(\varepsilon)$ is the rectangular neighborhood. Then, $S_{c+\varepsilon}(x_s)$ has $S_{c-\varepsilon}(x_s) \cup W_\varepsilon^u(\hat{x})$ as a strong deformation retract.

Lemma 5.16 says that in changing the level value from c to $c + \varepsilon$, $S_{c+\varepsilon}(x_s)$ can be continuously transformed into $S_{c-\varepsilon}(x_s) \cup W_\varepsilon^u(\hat{x})$. Therefore, any topological invariant of the first set is preserved after the collapsing.

Now we will prove the proposition. Since $\hat{x} \in A_P = int\bar{A}$, it follows that $\{W^u(\hat{x}) - \hat{x}\} \subset A$, which implies $W^u(\hat{x}) \subset int\bar{A}$ Hence, $W_\varepsilon^u(\hat{x}) \subset int\bar{A}$. From Lemma 5.16, it follows that since $S_{c+\varepsilon}(x_s)$ has $S_{c-\varepsilon}(x_s) \cup W_\varepsilon^u(\hat{x})$ as a strong deformation retract, $S_{c+\varepsilon}(x_s)$ must self-intersect. If not, then $W_\varepsilon^u(\hat{x})$ would connect $S_{c-\varepsilon}(x_s)$, which lies in $int\bar{A}$ to another component $C \subset (\bar{A})^c$ of the level set corresponding to the level value $c - \varepsilon$. So before reaching the level value $c - \varepsilon$, the component C must intersect $\partial\bar{A}$ and $S_{c-\varepsilon}(x_s) \cap C = \emptyset$. This contradicts the fact that, for a given level value, exactly one component of the level set intersects $\bar{A}$, and this component must be $S_{c+\varepsilon}(x_s)$. Since $S_{c-\varepsilon}(x_s)$ is homeomorphic to an n-dimensional disk, D^n, it can be retracted to a line segment. $W_\varepsilon^u(\hat{x})$ would connect $S_{c-\varepsilon}(x_s)$, which must self-intersect. So, we conclude that $S_{c-\varepsilon}(x_s) \cup W_\varepsilon^u(\hat{x})$ can be retracted to a circle. Since the circle is not simply connected and $S_{c+\varepsilon}(x_s)$ has the same homotopy type as the circle (Massey, 1967), we conclude that $S_{c+\varepsilon}(x_s)$ is not simply connected. This completes the proof.

5.7 CONCLUSION

In this chapter, we have presented a comprehensive energy function theory for general nonlinear autonomous dynamical systems. This comprehensive theory includes the following analytical results:

- global behavior of trajectories,
- global behaviors of trajectories on the stability boundary,
- characterizations of the stability boundary,
- structures of equilibrium points on the stability boundary,
- sufficient conditions for an unbounded stability boundary, and
- distinct energy function values of equilibrium points.

In addition, the theoretical basis for using energy functions to estimate stability regions of large-scale nonlinear systems has been presented. The relationship between the structures of the quasi-stability boundary and the constant energy surfaces at different level values has been established. Optimal schemes for estimating the stability region and quasi-stability regions have been presented, and these schemes form the basis for the closest UEP method. The energy function theory can be applied to power system transient stability models to develop a theoretical foundation for direct methods. We will present this development in later chapters.

One important extension of energy functions is to allow derivatives along system trajectories to be positive in some bounded sets instead of being nonpositive in the state space. We term this extension of energy functions as a generalized energy function. To this end, the extended invariance principle seems promising (Rodrigues et al., 2000, 2001). If the extension is successful, then it is expected that generalized energy functions are applicable to more complicated power system stability models than current transient stability models.

Chapter 6

Constructing Analytical Energy Functions for Transient Stability Models

6.1 INTRODUCTION

The task of constructing an energy function for a (postfault) transient stability model is essential to direct methods. The role of the energy function is to make feasible a direct determination of whether a given point (such as the initial point of a postfault power system) lies inside the stability region of a postfault stable equilibrium point (SEP) without performing numerical integration. An energy function is an extension of the Lyapunov function in that it must satisfy the three conditions described in the previous chapter. A Lyapunov function may not be an energy function.

Two different classes of power system models for direct transient stability analysis have been proposed: network-reduction models and network-preserving models. Traditionally, direct methods have been developed for the network-reduction models where all the loads are modeled as constant impedances and the entire network representation is reduced to the generator internal buses. Network-preserving models were proposed in the 1980s to overcome some of the shortcomings of network-reduction models. In each class of model, a model is termed lossless if the nonzero transfer conductance terms are neglected; otherwise, it is termed lossy, such as lossy network-reduction models and lossy network-preserving models.

Practical power system stability models are lossy, where the losses are either from the transmission system and loads or from the transfer conductances in the reduced system admittance matrix after the elimination of load buses. Considerable efforts have concentrated on the construction of analytical energy functions for lossy power system stability models (Guadra, 1975; Henner, 1976; Machias, 1986; Pai and Murthy, 1973). These efforts, however, have been in vain.

Direct Methods for Stability Analysis of Electric Power Systems, by Hsiao-Dong Chiang

On the other hand, most of the existing energy functions for power systems with losses are based on unjustified approximations (Athay et al., 1979; Hartman, 1973; Kakimoto et al., 1978; Pai and Varwandkar, 1977; Uemura et al., 1972). These functions are either not well-defined or the derivative of these functions along system trajectories is not negative. Thus, the practical merits of these energy functions for the stability analysis of power systems with losses are unclear. Indeed, it has been shown that the energy function in Uemura et al. (1972) may lead to an estimate of critical clearing time higher than its actual value obtained through step-by-step methods.

Consequently, these efforts raise the issue of the existence of energy functions for power systems with losses. Regarding this issue, the existence of local Lyapunov functions for power systems with transfer conductances has been shown in Caprio (1986), Kwatny et al. (1985), Narasimhamurthi and Musavi (1984), and Saeki et al. (1983). Unfortunately, these local Lyapunov functions only serve to determine the stability property of an equilibrium point and cannot be directly applied to the determination of the stability region. As a result, these local Lyapunov functions cannot be employed in direct methods for transient stability analysis.

The issue therefore has been only partly resolved: there is a local energy function for power systems with losses. However, does a global energy function (or Lyapunov function) for lossy power system stability models exist? This question was addressed in Chiang (1989) by demonstrating the nonexistence of a general energy function for power systems with losses. While this may be a disappointing message, it has also been shown that under certain conditions, there is an energy function (defined in a compact set of the state space) that can be used in direct methods for the stability analysis of power systems with certain losses. One implication of these results is that any general procedure attempting to construct an analytical energy function for a power system with losses must include a step that checks its existence. These theoretical results confirm the current practice of using numerical energy functions for direct transient stability analysis.

This chapter is focused on the methodology of constructing (analytical) energy functions for lossless power system stability models. In addition, a theoretical study of the nonexistence of energy functions for lossy power system stability models will be presented. The construction of (numerical) energy functions for lossy power system stability models will be presented in the next chapter.

6.2 ENERGY FUNCTIONS FOR LOSSLESS NETWORK-REDUCTION MODELS

Many existing lossless network-reduction transient stability models can be written in the following compact form (Chu and Chiang, 1999):

$$\begin{aligned} T\dot{x} &= -\frac{\partial U}{\partial x}(x, y) \\ \dot{y} &= z \\ M\dot{z} &= -Dz - \frac{\partial U}{\partial y}(x, y), \end{aligned} \tag{6.1}$$

where $x \in R^n$, y, $z \in R^m$, T, M, and D are positive diagonal matrices. $U(x, y)$ is a smooth function. A sufficient condition for the existence of an energy function for this compact representation is the following (Chu and Chiang, 1999):

$$W(x, y, z) = K(z) + U(x, y) = \frac{1}{2} z^T M z + U(x, y), \tag{6.2}$$

where the function $W: R^{n+2m} \to R$. If, along every nontrivial trajectory $(x(t), y(t), z(t))$ with a bounded function value $W(x, y, z)$, $x(t)$ is also bounded, then $W(x, y, z)$ is an energy function for the system.

These classical lossless, network-reduction transient stability models are seldom used in practical applications because of the underlying unwarranted simplifications in the formulation of generators and loads. This type of model is known to have the following shortcomings: (1) it precludes the models for nonlinear load behaviors; (2) reduction of the network leads to loss of network topology and hence precludes the study of energy shifts among different components of the network; (3) neglect of transfer conductances leads to a bias in stability assessments of unknown magnitude and direction (overestimates or underestimates). Structure-preserving models have been proposed to overcome some of the shortcomings of network-reduction models. We discuss in the next section a general procedure to derive analytical energy functions for network-preserving transient stability models.

6.3 ENERGY FUNCTIONS FOR LOSSLESS STRUCTURE-PRESERVING MODELS

We construct analytical energy functions for network-preserving lossless transient stability models. Several existing network-preserving transient stability models can be rewritten as a set of general differential and algebraic equations (DAEs) of the following compact form (Bergen and Hill, 1981; Bergen et al., 1986; Chu and Chiang, 2005; Padiyar and Ghosh, 1989; Padiyar and Sastry, 1987):

$$\begin{aligned}
0 &= -\frac{\partial}{\partial u} U(u, w, x, y) \\
0 &= -\frac{\partial}{\partial w} U(u, w, x, y) \\
T\dot{x} &= -\frac{\partial}{\partial x} U(u, w, x, y) \\
\dot{y} &= z \\
M\dot{z} &= -Dz - \frac{\partial}{\partial y} U(u, w, x, y),
\end{aligned} \tag{6.3}$$

where $u \in R^k$ and $w \in R^l$ are instantaneous variables and $x \in R^n$, $y \in R^n$, and $z \in R^n$ are state variables. T is a positive definite matrix, and M and D are diagonal positive definite matrices. Here, differential equations describe generator and/or load dynamics, while algebraic equations express the power flow equations at each bus. Also, the function $U(u, w, x, y)$ satisfies the following conditions:

(D1) $\nabla U(u, w, x, y) = \nabla U(u_1 + 2m_1 \pi, \ldots, u_k + 2m_k \pi, w, x, y_1 + 2l\pi, \ldots, y_n + 2l_n \pi)$ for all $m_i, l_j \in Z$.

(D2) For $i = 1, \ldots, n$, $k = 1, \ldots, n$, and $j = 1, \ldots, m$, there exist four polynomials with positive coefficients $p_{1j}(u_1, \ldots, u_k, x_1, \ldots, x_m)$, $p_{2ik}(u_1, \ldots, u_k, x_1, \ldots, x_m)$, $p_{3ij}(u_1, \ldots, u_k, x_1, \ldots, x_m)$, and $p_{4ij}(u_1, \ldots, u_k, x_1, \ldots, x_m)$, such that

$$\left|\frac{\partial}{\partial y_i} U(u, w, x, y)\right| \leq p_{1i}(|u_1|, \ldots, |u_k|, |x_1|, \ldots, |x_m|),$$

$$\left|\frac{\partial}{\partial^2 y_i y_k} U(u, w, x, y)\right| \leq p_{2ik}(|u_1|, \ldots, |u_k|, |x_1|, \ldots, |x_m|),$$

$$\left|\frac{\partial}{\partial w_j \partial y_i} U(u, w, x, y)\right| \leq p_{3ij}(|u_1|, \ldots, |u_k|, |x_1|, \ldots, |x_m|), \text{ and}$$

$$\left|\frac{\partial}{\partial x_j \partial y_i} U(u, w, x, y)\right| \leq p_{4ij}(|u_1|, \ldots, |u_k|, |x_1|, \ldots, |x_m|),$$

respectively.

Due to their intricate and complicated dynamics near singular surfaces, DAE systems are usually difficult to analyze. To avoid an awkward analysis on the DAE representation, various methods have been presented in the literature. Here, we will treat the algebraic equations as singularly perturbed differential equations (Chiang et al., 1994). The application of the singularly perturbed systems in network-preserving power system models was initiated by Sastry, Desour, and Varaiya (Sastry and Desoer, 1981; Sastry and Varaiya, 1980) and was considered later by Arapostathis et al. (1982), Bergen et al. (1986), De Marco and Bergen (1984, 1987), Chiang et al. (1992), and Zou et al. (2003).

With the aid of singularly perturbed systems, the compact representation of the network-preserving, lossless, transient stability model becomes

$$\begin{aligned}
\varepsilon_1 \dot{u} &= -\frac{\partial}{\partial u} U(u, w, x, y) \\
\varepsilon_1 \dot{w} &= -\frac{\partial}{\partial w} U(u, w, x, y) \\
T\dot{x} &= -\frac{\partial}{\partial x} U(u, w, x, y) \\
\dot{y} &= z \\
M\dot{z} &= -Dz - \frac{\partial}{\partial y} U(u, w, x, y),
\end{aligned} \tag{6.4}$$

where ε_1 and ε_2 are sufficiently small positive numbers.

We next construct an analytical energy function for the lossless network-preserving transient stability model.

Theorem 6.1: Existence of an Analytical Energy Function

For the compact representation of the singularly perturbed network-preserving power system model (Eq. 6.4), consider the function $W: R^{5n} \to R$:

$$W(u, w, x, y, z) = K(z) + U(u, w, x, y) = \frac{1}{2} z^T M z + U(u, w, x, y). \tag{6.5}$$

Suppose that along every nontrivial trajectory $(u(t), w(t), x(t), y(t), z(t))$ of the system (Eq. 6.4) with bounded $W(. , . , . , . , .)$, $(u(t), w(t), x(t))$ is also bounded. Then, $W(u, w, x, y, z)$ is an energy function for the system (Eq. 6.4).

Proof: We will check conditions (i)–(iii) of the energy function:

(i) Differentiating $W(u(t), w(t), x(t), y(t), z(t))$ along the trajectory yields the following:

$$\begin{aligned}\dot{W}(u(t), w(t), x(t), y(t), z(t)) &= \frac{\partial W^T}{\partial u}\dot{u} + \frac{\partial W^T}{\partial w}\dot{w} + \frac{\partial W^T}{\partial x}\dot{x} + \frac{\partial W^T}{\partial y}\dot{y} + \frac{\partial W^T}{\partial z}\dot{z} \\ &= -\frac{\partial U^T}{\partial u}\varepsilon_1^{-1}\frac{\partial U}{\partial u} - \frac{\partial U^T}{\partial w}\varepsilon_2^{-1}\frac{\partial U}{\partial w} - \frac{\partial U^T}{\partial x}T^{-1}\frac{\partial U}{\partial x} \\ &= -z^T D z \le 0.\end{aligned}$$

This inequality shows that condition (i) of the energy function is satisfied.

(ii) Suppose that there is an interval $t \in [t_1, t_2]$ such that $\dot{W}(u(t), w(t), x(t), y(t), z(t)) = 0$. Hence, it follows that $z(t) = 0$ and $\dot{u}(t) = \dot{w}(t) = \dot{x}(t) = 0$ for $t \in [t_1, t_2]$. But this also implies that $y(t)$ is a constant, which implies that the system is at an equilibrium point. Hence, condition (ii) of the energy function also holds.

(iii) Because $u(t)$, $w(t)$, and $x(t)$ are bounded along the nontrivial trajectory with bounded $W(. , . , . , . , .)$, only components $y(t)$ and $z(t)$ will be examined for condition (iii). We then apply Barbalat's lemma and the covering map to conclude that $u(t)$, $w(t)$, $x(t)$, $y(t)$, and $z(t)$ are bounded. From conditions (i)–(iii) we conclude that $W(u(t), w(t), x(t), y(t), z(t))$ is indeed an energy function. This completes the proof.

For an illustrational purpose, we next derive an analytical energy function for the two-axis generator with first-order exciter dynamics and a dynamic load model with the following characteristics:

1. Each generator is modeled as a two-axis generator plus a first-order exciter system.

2. Each load demand is a hybrid of constant susceptance, constant complex power, and a frequency-dependent dynamic load model.
3. For an internal generator bus, the two-axis generator model is written as follows: For $i = 1, \ldots, n$,

$$\begin{aligned}
\dot{\delta}_i &= \omega_i \\
M_i\dot{\omega}_i &= -D_i\omega_i + P_{mi} - P_{ei} \\
T_{doi}\dot{E}'_{qi} &= -\frac{x_{di}}{x'_{di}}E'_{qi} + \frac{x_{di} - x'_{di}}{x'_{di}}V_i\cos(\delta_i - \theta_i) + E_{fi} \\
T_{qoi}\dot{E}'_{di} &= -\frac{x_{qi}}{x'_{qi}}E'_{di} - \frac{x_{qi} - x'_{qi}}{x'_{qi}}V_i\sin(\delta_i - \theta_i),
\end{aligned} \tag{6.6}$$

where

$$P_{ei} = \frac{1}{x'_{di}}E'_{qi}V_i\sin(\delta_i - \theta_i) + \frac{1}{x'_{qi}}E'_{di}V_i\cos(\delta_i - \theta_i) + \frac{x'_{di} - x'_{qi}}{2x'_{di}x'_{qi}}V_i^2\sin[2(\delta_i - \theta_i)].$$

4. Exciter dynamics: various exciter dynamic models have been proposed in the past. For simplicity and to avoid the nonsmooth dynamics of hard limits in the detailed exciter models, the following first-order approximation exciter dynamic is considered:

$$T_{vi}\dot{E}_{fi} = -E_{fi} - \mu_i k_i V_i\cos(\delta_i - \theta_i) + l_i. \tag{6.7}$$

5. For an external generator bus, the power flow equations at each generator can be described in the following equations. For $i = 1, \ldots, n$,

$$\begin{aligned}
0 = &\sum_{j\neq i}^{n+1} V_iV_j(B_{ij}\sin(\theta_i - \theta_j)) + G_{ij}\cos(\theta_i - \theta_j) \\
&+ \sum_{l=n+2}^{n+m+1} V_iV_l(B_{il}\sin(\theta_i - \theta_l) + G_{il}\cos(\theta_i - \theta_l)) \\
&- \frac{1}{x'_{di}}E'_{qi}V_i\sin(\delta_i - \theta_i) - \frac{1}{x'_{di}}E'_{di}V_i\cos(\delta_i - \theta_i) \\
&- \frac{x'_{di} - x'_{qi}}{2x'_{di}x'_{qi}}V_i^2\sin[2(\delta_i - \theta_i)] \text{ and}
\end{aligned} \tag{6.8}$$

$$\begin{aligned}
0 = &\frac{V_i^2}{x'_{di}} - \frac{1}{x'_{di}}E'_{qi}V_i\cos(\delta_i - \theta_i) - \frac{1}{x'_{qi}}E'_{di}V_i\sin(\theta_i - \delta_i) \\
&- \frac{x'_{di} - x'_{qi}}{2x'_{di}x'_{qi}}V_i^2[\cos 2(\delta_i - \theta_i) - 1] \\
&- \sum_{j=1}^{n+1} V_iV_j(B_{ij}\cos(\theta_i - \theta_j)) + G_{ij}\sin(\theta_i - \theta_j) \\
&- \sum_{l=n+2}^{n+m+1} V_iV_l(B_{il}\cos(\theta_i - \theta_l) + G_{il}\sin(\theta_i - \theta_l)).
\end{aligned} \tag{6.9}$$

6. Load bus: because the portion of the constant impedance load is incorporated into the network admittance matrix Y, the load demand at each load bus is a hybrid combination of the following three terms: (1) frequency-dependent loads—$D_k\dot{\theta}_k$; (2) constant power $P_k^d + JQ_k^d(V_k)$; and (3) constant current injection $I_{Lk}\angle\phi_k$, where both I_{Lk} and ϕ_k are constants. The bus voltage angle θ will be changed during the transient stage. So, the power factor of the constant current injection will not remain a constant during the transient response. Hence, the complex power demand at each node is of the form $P_k^d + V_k I_{Lk}\cos(\theta_k - \phi_k) + j[Q_k^d(V_k) + V_k I_{Lk}\sin(\theta_k - \phi_k)]$. The power flow equations at the load bus can be represented in the following equations: for $k = n + 2, \ldots, n + m + 1$,

$$P_k^d + V_k I_{Lk}\cos(\theta_k - \phi_k) - D_k\dot{\theta}_k = \sum_{j=1}^{n+m+1} V_k V_j (B_{kj}\sin(\theta_k - \theta_j) + G_{kj}\cos(\theta_k - \theta_j))$$

$$Q_k^d(V_k) + V_k I_{Lk}\sin(\theta_k - \phi_k) = -\sum_{j=1}^{n+m+1} V_k V_j (B_{kj}\cos(\theta_k - \theta_j) + G_{kj}\sin(\theta_k - \theta_j)). \tag{6.10}$$

We consider the function (Chu and Chiang, 2005)

$$U(\delta, \theta, V, E_q', I_L) = \sum_{i=1}^{14} U_i, \tag{6.11}$$

where

$$U_1 = -\sum_{i=1}^{n} P_{mi}\delta_i,\ U_2 = -\sum_{i<j}^{n+1} V_i V_j B_{ij}\cos(\theta_i - \theta_j),\ U_3 = -\sum_{i=1}^{n+1}\sum_{k=n+2}^{n+m+1} V_i V_k B_{ik}\cos(\theta_i - \theta_k)$$

$$U_4 = -\sum_{k=n+2}^{n+m+1} P_k^d \delta_k,\ U_5 = -\sum_{k<1}^{n+m+1} V_k V_l B_{kl}\cos(\theta_k - \theta_l),\ U_6 = -\sum_{k=n+2}^{n+m+1} V_{kk}^2 B_{kk}$$

$$U_7 = -\sum_{k=n+2}^{n+m+1}\int_{V_k^0}^{V_k}\frac{Q_k^d(V_k)}{V_k}dV_k,\ U_8 = -\frac{1}{2}\sum_{i=1}^{n+1} B_{ii} + \frac{x_{di}' - x_{qi}}{2x_{qi}x_{di}'}[\cos[2(\delta_i - \theta_i)] - 1]V_i^2$$

$$U_9 = -\sum_{i=1}^{n+1}\frac{1}{x_{qi}'}E_{qi}'V_i\cos(\delta_i - \theta_i),\ U_{10} = \frac{1}{2}\sum_{i=1}^{n+1} E_{qi}'^2\frac{x_{di}}{x_{di}'(x_{di} - x_{di}')}$$

$$U_{11} = \sum_{i=1}^{n}\frac{1}{2x_{qi}'}(E_{di}' + 2E_{di}'V_i\sin(\delta_i - \theta_i)),\ U_{12} = -\sum_{i=1}^{n}\frac{V_i^2}{2x_{qi}'},\ U_{13} = \sum_{i=1}^{n}\frac{E_{di}'^2}{2(x_{qi} - x_{qi}')}$$

$$U_{14} = -\sum_{k=n+2}^{n+m+1} V_k I_{Lk}\sin(\theta_i - \phi_i).$$

We next express the overall dynamic equations in terms of partial derivatives of $U(\delta, \theta, V, E_q', E_d', I_L)$. This expression can facilitate the construction of energy functions:

$$0 = -\frac{\partial U}{\partial \phi_i}(\delta, \theta, V, E'_q, E'_d, I_L)$$

$$0 = -V_i \frac{\partial U}{\partial V_i}(\delta, \theta, V, E'_q, E'_d, I_L)$$

$$0 = -\frac{\partial U}{\partial \phi_k}(\delta, \theta, V, E'_q, E'_d, I_L)$$

$$0 = -V_i \frac{\partial U}{\partial V_k}(\delta, \theta, V, E'_q, E'_d, I_L)$$

$$\dot{\delta}_i = \omega_i$$

$$M_i \dot{\omega}_i = -D_i \omega_i - \frac{\partial U}{\partial \delta_i}(\delta, \theta, V, E'_q, E'_d, I_L)$$

$$T_{doi} \dot{E}'_{qi} = E_{fdi} - (x_{di} - x'_{di}) \frac{\partial U}{\partial E'_{qi}}(\delta, \theta, V, E'_q, E'_d, I_L)$$

$$T_{qoi} \dot{E}_{di} = -(x_{qi} - x'_{qi}) \frac{\partial U}{\partial E'_{di}}(\delta, \theta, V, E'_q, E'_d, I_L)$$

$$T_{vi} \dot{E}_{fdi} = -E_{fdi} - \mu_i k_i V_i \cos(\theta_i - \phi_i) + l_i.$$

In the preceding set of equations, only the last equation does not contain the analytical expression of the partial derivative of $U(\delta, \theta, V, E'_q, E'_d, I_L)$. We next add terms into $U(\delta, \theta, V, E'_q, E'_d, I_L)$ in order to make every equation contain the partial derivative of U. Note that

$$\begin{aligned} \frac{\partial U}{\partial E'_{qi}}(\delta, \theta, V, E'_q, E'_d, I_L) &= \frac{x_{di} E'_{qi}}{x'_{di}(x_{di} - x'_{di})} - \frac{V_i \cos(\theta_i - \phi_i)}{x'_{di}} \\ &= \frac{1}{x_{di} - x'_{di}}(T_{doi} \dot{E}'_{qi} - E_{fdi}). \end{aligned} \tag{6.12}$$

We next define two extra terms:

$$\begin{aligned} U_{15} &= -\frac{1}{2} E^T B^{-1} A E \\ U_{16} &= -C^T E, \end{aligned} \tag{6.13}$$

where

$$\begin{aligned} E &= [E_1, \ldots, E_n], \\ T &= \text{blockdiag}[T_1, \ldots, T_n], \\ A &= \text{blockdiag}[A_1, \ldots, A_n], \\ B &= \text{blockdiag}[B_1, \ldots, B_n], \\ C &= [C_1, \ldots, C_n]^T, \end{aligned}$$

$$E_i = [E_{qi}, E_{fdi}]^T,\ T_i = \begin{bmatrix} T'_{doi} & 0 \\ 0 & T'_{vi} \end{bmatrix},\ A_i = \begin{bmatrix} 0 & 1 \\ -\dfrac{\mu_i k_i x'_{di}}{x_{di} - x'_{di}} & -1 \end{bmatrix},\ B_i = \begin{bmatrix} \dfrac{x_{di} - x'_{di}}{x'_{di}} & \alpha_i \\ -\mu_i k_i & \beta_i \end{bmatrix},$$

and

$C_i = [0, l_i]^T$.

It follows that

$$\frac{\partial U}{\partial E} = \frac{\partial U}{\partial E} - B^{-1}AE - B^{-1}C, \tag{6.14}$$

and the voltage dynamics E can be written as the following equation:

$$T\dot{E} = -B\frac{\partial U}{\partial E}.$$

We next add these two terms into Equation 6.11:

$$U(\delta, \theta, V, E'_q, E'_d, E_{fd}, I_L) = U(\delta, \theta, V, E'_q, E'_d, I_L) + U_{15} + U_{16} \tag{6.15}$$

The overall system can thus be rewritten as the following compact set:

$$\begin{aligned}
0 &= -\frac{\partial U}{\partial \phi_i}(\delta, \theta, V, E'_q, E'_d, E_{fd}, I_L) \\
0 &= -\frac{\partial U}{\partial \phi_k}(\delta, \theta, V, E'_q, E'_d, E_{fd}, I_L) \\
0 &= -V_i\frac{\partial U}{\partial V_i}(\delta, \theta, V, E'_q, E'_d, E_{fd}, I_L) \\
0 &= -V_k\frac{\partial U}{\partial V_k}(\delta, \theta, V, E'_q, E'_d, E_{fd}, I_L) \\
\dot{\delta}_i &= \omega_i \\
M_i\dot{\omega}_i &= -D_i\omega_i - \frac{\partial U}{\partial \delta_i}(\delta, \theta, V, E'_q, E'_d, E_{fd}, I_L) \\
T'_{qoi}\dot{E}'_{di} &= -(x_{qi} - x'_{qi})\frac{\partial U}{\partial E'_{di}}(\delta, \theta, V, E'_q, E'_d, E_{fd}, I_L) \\
T\dot{E} &= -B\frac{\partial U}{\partial E}(\delta, \theta, V, E'_q, E'_d, E_{fd}, I_L).
\end{aligned} \tag{6.16}$$

Clearly, the new model can be written as the compact form of Equation 6.3, where

$u = [\theta, \phi]^T$, $w = \log V$, $x = [E'_d, E'_q, E_{fd}]$, $y = \delta$, $z = \omega$, and

$$T = \text{blockdiag}\left[I_n, \frac{T'_{q0}}{x_q - x'_q}, B^{-1}T\right].$$

We next summarize a key result for constructing the above network-preserving transient stability model.

Theorem 6.2: Existence of the Analytical Energy Function (Chu and Chiang, 2005)

Consider the new class of network-preserving power system models described in Equations 6.8–6.12. Define

$$W(\omega, \delta, \theta, V, E'_q, E'_d, I_L, E_{fd}) = k(\omega) + U(\delta, \theta, V, E'_q, E'_d, I_L, E_{fd}) = \frac{1}{2}\omega^T M\omega + \sum_{i=1}^{16} U_i, \tag{6.17}$$

where U_i are defined in Equations 6.11 and 6.13, respectively. Then, $W(\omega, \delta, \theta, V, E'_q, E'_d, I_L, E_{fd})$ is an (analytical) energy function for the new class of network-preserving power system models.

A physical interpretation of each term of the energy function (Eq. 6.17) along with a comparison with other energy functions proposed for different transient stability models, such as the models proposed in BH (Bergen and Hill, 1981), the NM model (Narasimhamurthi and Musavi, 1984), and the TAV model (Tsolas et al., 1985), are summarized in Table 6.1.

6.4 NONEXISTENCE OF ENERGY FUNCTIONS FOR LOSSY MODELS

In this section, the nonexistence of a general energy function for power systems with losses will be shown. While this may be a disappointing message, it is also shown that, under certain conditions, there is an energy function (defined in a compact set of the state space) that can be used in direct methods for the stability analysis of power systems with a certain degree of losses. One implication of these results is that any general procedure attempting to construct an energy function for a power system with losses must include a step that serves to check for the existence of an energy function.

Consider a lossy classical power system stability model. Let the loads be modeled as constant impedances. The dynamics of the ith generator can be represented by the following equations:

$$\begin{aligned} \dot{\delta}_{in} &= \omega_i - \omega_n \\ M_i\dot{\omega}_i &= P_i - D_i\omega_i - \sum_{j\neq i, j=1}^{n} E_iE_jB_{ij}\sin(\delta_i - \delta_j) - \sum_{j\neq i, j=1}^{n} E_iE_jG_{ij}\cos(\delta_i - \delta_j), \end{aligned} \tag{6.18}$$

where node n serves as the reference node. E_i is the constant voltage behind direct axis transient reactance and M_i is the generator's moment of inertia. The damping

Table 6.1 Various Network-Preserving Power System Models and Their Corresponding Potential Energy Functions

Potential energy	Formula	BH model	NM model	TAV model	CC model
Mechanical power injections	U_1	Y	Y	Y	Y
Transfer between generator terminal buses	U_2	Y	Y	Y	Y
Transfer between generator terminal buses and load buses	U_3	Y	Y	Y	Y
Active power demand at load buses	U_4	Y	Y	Y	Y
Transfer between load buses	U_5	Y	Y	Y	Y
Loss at each load bus	U_6		Y	Y	Y
Reactive demand at load buses	U_7		Y	Y	Y
Transfer between generator internal buses and terminal buses	U_8			Y	Y
Quadrature axis voltage variations at generator internal buses	U_9			Y	Y
Quadrature axis voltage variations at generator internal buses actions of exciters at generator internal buses	U_{10}			Y	Y
D-axis voltage variations	U_{11}				Y
Damper winding on the Q-axis	U_{12}				Y
Damper winding on the Q-axis	U_{13}			Y	Y
Constant current loads	U_{14}				Y
Interactions between Q-axis voltage variations and exciters	U_{15}				Y
Exciters at generator internal buses	U_{16}				Y

CC, Chu and Chiang.

constant D_i is assumed to be positive. The G_{ij} term represents the transfer conductance of the *i*–*j* element in the reduced admittance matrix of the system (Eq. 6.18). $P_i = P_{mi} - E_i^2 G_{ii}$, where P_{mi} is the mechanic power. Furthermore, when uniform mechanical damping is assumed, that is, $D_i/M_i = C$, $i = 1, 2, \ldots, n$, the dynamics of the *i*th generator can be represented by the following equations:

$$\begin{aligned}\dot{\delta}_{in} &= \omega_{in}, i = 1, 2, \ldots, n-1 \\ \dot{\omega}_{in} &= (P_{mi} - P_{ei}) M_i^{-1} + (P_{mn} - P_{en}) M_n^{-1} - C\omega_{in} \\ &= -C\omega_{in} + f_{in}(\delta).\end{aligned} \tag{6.19}$$

We first consider the following system, which describes one machine connected to an infinite bus through a lossy transmission line:

$$\begin{aligned}\dot{\delta} &= \omega \\ M\dot{\omega} &= P - D\omega - EB\sin(\delta) - EG\cos(\delta),\end{aligned} \tag{6.20}$$

where δ is the phase angle of the machine with respect to the infinite bus. Consider the following function for (Eq. 6.20):

$$V(\delta,\omega)=\frac{1}{2}M\omega^2-P\delta-EB\cos(\delta)+EG\sin(\delta). \tag{6.21}$$

The derivative of Equation 6.21 along the trajectory of the system (Eq. 6.20) is given by

$$\dot{V}(\delta,\omega)=-D\omega^2\leq 0.$$

Hence, the function (Eq. 6.21) is a general energy function for the one-machine infinite bus system (Eq. 6.20).

The extension of the energy function (Eq. 6.21) to multimachine power systems represents a great challenge. In their pioneer paper (Abiad and Nagappan, 1966), El-Abiad and Nagappan proposed the following function for the multimachine power systems with losses:

$$\begin{aligned}V_1(\delta,\omega)=&\frac{1}{2}\sum_{i=1}^{n}M_i\omega_i^2-\sum_{i=1}^{n}P_i(\delta_i-\delta_j^s)\\&-\sum_{i=1}^{n}\sum_{j=i+1}^{n+1}E_iE_jB_{ij}\left[\cos(\delta_i-\delta_j)-\cos(\delta_i^s-\delta_j^s)\right]\\&-\sum_{i=1}^{n}\sum_{j=i+1}^{n+1}E_iE_jG_{ij}\left[\sin(\delta_i-\delta_j)-\sin(\delta_i^s-\delta_j^s)\right].\end{aligned} \tag{6.22}$$

Differentiating the function V_1 $(\delta,\ \omega)$ along the trajectories of the system (Eq. 6.18) gives

$$\begin{aligned}\dot{V}_1(\delta,\omega)&=\frac{\partial V}{\partial\delta}\dot{\delta}+\frac{\partial V}{\partial\omega}\dot{\omega}\\&=-\sum_{i=1}^{n}D_i\omega_i^2-2\sum_{i=1}^{n}\sum_{j=i+1}^{n+1}E_iE_jG_{ij}\omega_i\cos(\delta_i-\delta_j).\end{aligned} \tag{6.23}$$

This function V_1 $(\delta,\ \omega)$ is therefore not an energy function for the system (Eq. 6.18). Note that $\dot{V}_1$ will be negatively definite for large values of ω_i.

Many attempts have since been made to derive general energy functions for the system (Eq. 6.18) but without success. Another approach pursued by Uemura et al. (1972), Athay et al. (1979), Kakimoto et al. (1978), and Guindi and Mansour (1982) is based on numerical approximations accommodating the transfer conductance effects. It should be stressed that these approximations lead to functions that do not possess the properties of analytical energy functions.

At present, there is no general energy function for multimachine lossy stability models. The following key result has been shown in (Chiang, 1989).

Theorem 6.3: Nonexistence of an Energy Function

There is no general energy function for the power system (Eq. 6.18) with losses, where $n\geq 2$ is the number of generators.

Remarks:

1. This main result shows that there is no general analytical energy function for multimachine power systems with losses (Eq. 6.18). The usefulness of existing local, analytical energy functions in direct methods is not clear. It should be noted that in order to be useful, these local energy functions must be at least well-defined in the stability region of the postfault system.
2. For the simple system of one-machine infinite bus through a lossy transmission line, it can be shown (by direct analysis of the associated differential equation) that there is no limit cycle in the state space. The existence of a global analytical energy function (e.g., Eq. 6.21) for these one-machine infinite bus systems also confirms this point.
3. One key implication is that any general procedure attempting to construct an analytical energy function must include a step serving to check for the existence of an analytical energy function. This step essentially plays the same role that the Lyapunov equation plays in determining the stability of an equilibrium point.

6.5 EXISTENCE OF LOCAL ENERGY FUNCTIONS

We will show the existence of an energy function defined on any compact set of the state space of the system (Eq. 6.19) with small losses. It is well known that, under the assumption that the transfer conductances of the system (Eq. 6.19) are zero, there exists an energy function for this system. Indeed, with $G_{ij} = 0$, we define a function, $V(\delta, \omega)$, as follows:

$$\begin{aligned} V(\delta,\omega) &= \sum_{i=1}^{n-1}\sum_{j=i+1}^{n}\left\{\frac{1}{2}M_iM_j\omega_{ij}^2 - (P_iM_j - P_jM_i)(\delta_{ij}-\delta_{ij}^s)\right. \\ &\quad \left. + \left(\sum_{i}^{n} M_i\right)E_iE_jB_{ij}(\cos\delta_{ij} - \cos\delta_{ij}^s)\right\} \\ &= V_k(\omega) + V_p(\delta). \end{aligned} \tag{6.24}$$

At this point, some concepts from the mathematical dynamical system theory are recalled. Let $\Phi^f(.)$ (resp. $\Phi^g(.)$) be the flow of a vector field f (resp. g) and Ω (f) (resp. Ω (g)) be its nonwandering set. We say f is Ω-stable if, for every small C^r-close ($r > 0$) vector field of f (i.e., every vector field close to f in the C^r topology), say, g, there exists a homeomorphism h: $\Omega(f) \to \Omega(g)$, which is orbit preserving (Nitecki and Shub, 1975; Pugh and Shub, 1970). The next analytical result shows that the system (Eq. 6.19) with transfer conductance $G = 0$ is Ω-stable.

Theorem 6.4: Ω-Stable

If the system (Eq. 6.19) with $G = 0$ satisfies the assumption of hyperbolic equilibrium points, then this system is Ω-stable.

Theorem 6.4 implies that, under the assumption of hyperbolicity for the power system (Eq. 6.19) without losses, there exists a positive number, α, such that if the transfer conductance of the system (Eq. 6.19) satisfies $|G| < \alpha$, then the nonwandering set of the system (Eq. 6.19) with a certain transfer conductance also consists entirely of equilibrium points. This property, in combination with an analytical result regarding the fine filtration (Nitecki and Shub, 1975), leads to the following result: for any compact set S of the state space of the system (Eq. 6.19), there exists a positive number, α, such that if the transfer conductance of the system (Eq. 6.19) satisfies $|G| < \alpha$, then there exists an energy function defined on this compact set S.

The above analytical results confirm the current practice of using numerical energy functions in the area of direct methods. The issues, however, are the reliability and the performance of these numerical energy functions in stability analysis. Further investigation is needed, although these numerical energy functions seem to have performed very well in direct methods.

6.6 CONCLUDING REMARKS

The existence of analytical energy functions for lossless power system transient stability models, including the network-reduction and network-preserving models, has been shown. A methodology for deriving the analytical energy functions has been presented. Physical explanations of each term in the derived analytical energy functions have been provided.

The nonexistence of general analytical energy functions for lossy power system transient stability models has been shown. One key implication is that any general procedure attempting to construct an energy function for a lossy power system transient stability model must include a step that checks for the existence of an energy function. This step essentially plays the same role as the Lyapunov equation in determining the stability of an equilibrium point.

The existence of a local analytical energy function defined on any compact set of the state space of a power system transient stability model with small losses has been shown. This analytical result is local, and it is not clear whether a local analytical energy function is useful or not. The analytical results derived in this chapter confirm the current practice of using numerical energy functions in the area of direct methods. The topic of how to construct numerical energy functions will be discussed in the next chapter.

Chapter 7

Construction of Numerical Energy Functions for Lossy Transient Stability Models

7.1 INTRODUCTION

The existing numerical energy functions for lossy power system stability models are not analytical in the sense that they are not well-defined functions. They all contain two major terms: the analytical terms (i.e., path-independent terms) and the path-dependent terms. The path-dependent terms are not well-defined functions and require numerical approximations to become well-defined. The so-called ray approximation scheme and the trapezoidal approximation scheme are popular schemes for numerically approximating the path-dependent terms (Fouad and Vittal, 1991; Pai, 1989; Pavella and Murthy, 1994; Qiang and Zhong, 2005; Sasaki, 1979).

In this chapter, methodologies for constructing numerical energy functions for lossy power system stability models will be presented. At present, only two methods are available. One is based on the so-called first integral principle, while the other is based on a two-step procedure. Numerical energy functions derived by these two methods all contain path-dependent terms which need to be evaluated by numerical approximation schemes.

It will be shown that the so-called ray approximation scheme leads to a numerical ill-conditioned problem during the numerical approximation of path-dependent terms. A scheme to eliminate this numerical ill-conditioning will be presented. However, even after this elimination, the ray approximation scheme can give unacceptable results in the critical clearing time (CCT) estimate. It is recommended that the trapezoidal rule be used to evaluate the path-dependent terms of the numerical energy function. Numerical studies of these two approximation schemes will be conducted and the numerical results will be compared.

Finally, improved numerical approximation schemes for evaluating the path-dependent terms will be presented. These improved numerical approximation schemes are based on multipoint integrations. It will be numerically shown that the improved schemes are superior to the two popular approximation schemes: ray approximation and trapezoidal approximation schemes.

7.2 A TWO-STEP PROCEDURE

The existing (numerical) energy functions for lossy power system stability models are not analytical in the sense that they are not well-defined functions. Numerical energy functions were introduced to resolve the difficulty of constructing (analytical) energy functions via, for example, the inclusion of the first integral of the vector field introduced by the transfer conductances.

A general two-step procedure to derive a numerical energy function for a lossy transient stability model is described as follows:

Step 1. Construct an analytical energy function for the underlying model without the lossy terms, such as transfer conductance.

Step 2. Construct a numerical energy function by adding the first integral of the vector field introduced by the transfer conductances to the analytical energy function constructed in Step 1.

The above two-step procedure leads to a numerical energy function, which has two terms: an analytic energy function and a path-dependent energy function related to lossy terms.

We next give an illustration of the above two-step procedure. Many existing lossy network-reduction transient stability models can be written in the following compact form (Chu and Chiang, 1999):

$$\begin{aligned} T\dot{x} &= -\frac{\partial U}{\partial x}(x, y) + g_1(x, y) \\ \dot{y} &= z \\ M\dot{z} &= -Dz - \frac{\partial U}{\partial y}(x, y) + g_2(x, y), \end{aligned} \tag{7.1}$$

where $x \in R^n$, y, and $z \in R^m$, T, M, and D are positive diagonal matrices; $g_1(x, y)$ and $g_2(x, y)$ represent the transfer conductance of the reduced network. $U(x, y)$ is a smooth function. We apply Step 1 of the procedure to derive an analytical energy function; that is, $W(x, y, z) = K(z) + U(x, y) = \frac{1}{2}z^T Mz + U(x, y)$ for the following lossless model:

$$T\dot{x} = -\frac{\partial U}{\partial x}(x, y)$$

$$\dot{y} = z$$

$$M\dot{z} = -Dz - \frac{\partial U}{\partial y}(x, y).$$

By adding the first integral of the vector field introduced by the transfer conductances to the analytical energy function constructed in Step 1, it leads to a numerical energy function that has two terms: an analytical energy function, $W_{ana}(x, y, z) = \frac{1}{2}z^T Mz + U(x, y)$, and a path-dependent potential energy, $U_{path}(x, y, z)$:

$$\begin{aligned} W_{num}(x,y,z) &= W_{ana}(x, y, z) + U_{path}(x, y) \\ &= K(z) + U(x, y) + U_{path}(x, y). \end{aligned} \tag{7.2}$$

The above numerical energy function contains path-dependent terms, which can be approximately evaluated using two trajectory approximation schemes: (1) ray approximation and (2) trapezoidal approximation.

We next explain how to evaluate path-dependent terms using ray approximation. For example, we evaluate the following path-dependent term:

$$U_1 = \sum_{i=1}^{n} \int G_{ii} V_i^2 d\theta_i, \tag{7.3}$$

in which the state variables involved in the integration can be approximated by the following ray:

$$\begin{aligned} V_i &= V_{io} + \lambda \Delta V_i \\ \theta_i &= \theta_{io} + \lambda \Delta \theta_i. \end{aligned} \tag{7.4}$$

By substituting Equation 7.3 into Equation 7.4, it follows that

$$\begin{aligned} U_1 &\cong \sum_{i=1}^{n} \int_0^1 G_{ii} (V_{io} + \lambda \Delta V_i)^2 \Delta\theta_i d\lambda \\ &= \frac{G_{ii}\Delta\theta_i}{3}(V_i^2 + V_i V_{io} + V_{io}^2). \end{aligned} \tag{7.5}$$

For another example, we evaluate the following path-dependent term:

$$U_2 = \sum_{i=1}^{n} \sum_{\substack{j=1 \\ j \neq i}}^{n} G_{ij} \int V_i V_j \cos\theta_{ij} d\theta_i. \tag{7.6}$$

By using the ray approximation, U_2 can be rewritten as follows:

$$U_2 = \sum_{i=1}^{n} \sum_{j=i+1}^{n} G_{ij} \int V_i V_j \cos\theta_{ij} d(\theta_i + \theta_j)$$

$$\cong \sum_{i=1}^{n} \sum_{j=i+1}^{n} G_{ij}(\Delta\theta_i + \Delta\theta_j) \begin{bmatrix} \frac{V_{io}V_{jo}}{\Delta\theta_{ij}}(\sin\theta_{ij} - \sin\theta_{ijo}) \\ + \frac{V_{io}\Delta V_j + V_{jo}\Delta V_i}{\Delta\theta_{ij}^2}(\cos\theta_{ij} - \cos\theta_{ijo} + \Delta\theta_{ij}\sin\theta_{ij}) \\ + \frac{\Delta V_i \Delta V_j}{\Delta\theta_{ij}^3}(2\Delta\theta_{ij}\cos\theta_{ij} + (\Delta\theta_{ij}^2 - 2)\sin\theta_{ij} + 2\sin\theta_{ijo}) \end{bmatrix} \text{ and} \tag{7.7}$$

$$U_3 = \sum_{i=1}^{n} \sum_{\substack{j=1 \\ j\neq i}}^{n} G_{ij} \int V_j \sin\theta_{ij} dV_i$$

$$\cong \sum_{i=1}^{n} \sum_{\substack{j=1 \\ j\neq i}}^{n} G_{ij}\Delta V_i \begin{bmatrix} \frac{V_{jo}}{\Delta\theta_{ij}}(\cos\theta_{ijo} - \cos\theta_{ij}) \\ + \frac{\Delta V_j}{\Delta\theta_{ij}^2}(\sin\theta_{ij} - \sin\theta_{ijo} - \Delta\theta_{ij}\cos\theta_{ij}) \end{bmatrix}. \tag{7.8}$$

Another scheme to approximate the path-dependent terms is based on the trapezoidal rule, which is a special case of the closed Newton–Cotes formulae for numerical integration. This rule approximates the integration via the following equation:

$$\int_a^b f(x)dx \cong \frac{b-a}{2}(f(a) + f(b)):$$

Applying the trapezoidal rule to the following term, it follows that the term at the kth time step becomes

$$U_4(k) = \sum_{i=1}^{n} \int G_{ii} V_i^2 d\theta_i$$

$$\cong \sum_{i=1}^{n} \frac{G_{ii}}{2}\left(V_i(k)^2 + V_i(k-1)^2\right)(\theta_i(k) - \theta_i(k-1)) + U_4(k-1), \tag{7.9}$$

where $U_4(0) = 0$.

Applying the trapezoidal rule to the following term, it follows that

$$U_5(k) = \sum_{i=1}^{n} \sum_{\substack{j=1 \\ j\neq i}}^{n} G_{ij} \int V_i V_j \cos\theta_{ij} d\theta_i$$

$$\cong \sum_{i=1}^{n} \sum_{\substack{j=1 \\ j\neq i}}^{n} \frac{G_{ij}}{2}(V_i(k)V_j(k)\cos\theta_{ij}(k) + V_i(k-1)V_j(k-1)\cos\theta_{ij}(k-1))$$

$$\times(\theta_i(k) - \theta_i(k-1)) + U_5(k-1), \tag{7.10}$$

where $U_5(0) = 0$.

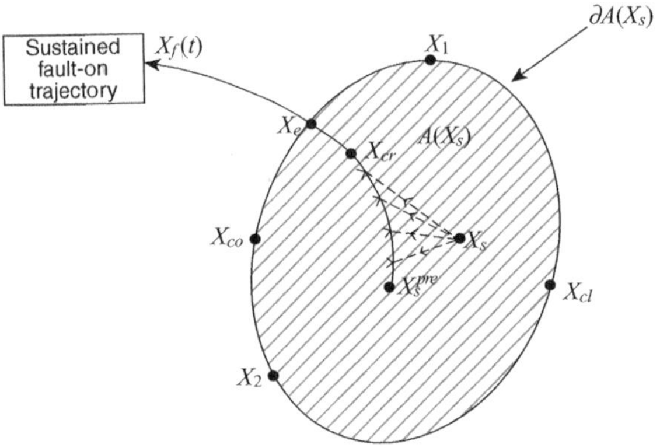

Figure 7.1 Illustration of calculating numerical energy function along the fault-on trajectory using ray approximation.

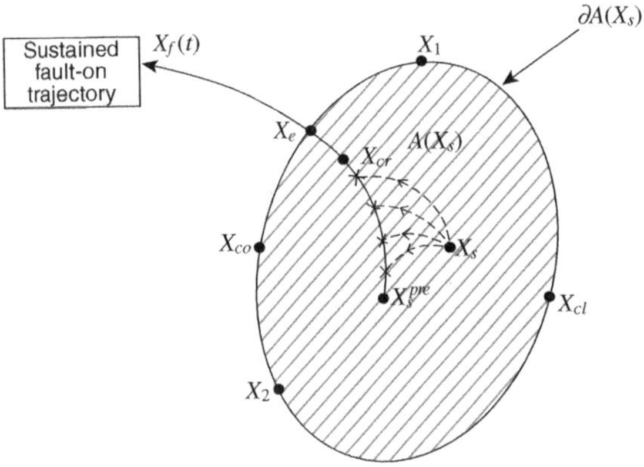

Figure 7.2 Illustration of calculating numerical energy function along the fault-on trajectory using trapezoidal approximation.

Hence, the numerical energy function contains an analytical energy function, which is a path-independent term, and a nonanalytical function, which depends on the integration path. The path-dependent term can be approximated by applying the trapezoidal rule or by the ray approximation scheme. The differences between these two approximations can be small or noticeable depending on the nature of the fault-on trajectory. An illustration of these two approximation schemes in evaluating path-dependent terms is shown in Figures 7.1 and 7.2, respectively.

7.3 FIRST INTEGRAL-BASED PROCEDURE

The purpose of this section is to present a procedure based on the first integral principle to derive an energy function for a general lossy transient stability model. In order to build the complete transient stability model in a systematic manner, the bus number sequence and relevant interfacing variables are defined as follows:

- **Network Bus.** Generator buses are numbered $1, 2, \ldots, n$. Load buses are numbered $n + 1, n + 2, \ldots, n + m$.

- **Bus Voltages.** The voltage magnitudes and phase angles are represented as $V_1, V_2, \ldots, V_{n+m}$ and $\theta_1, \theta_2, \ldots, \theta_{n+m}$, respectively.
- **Power Generations.** The generations from the generator terminal buses into the network are represented by $P_{Gi} + jQ_{Gi}$ for $i = 1, \ldots, n$. If generator resistance is neglected, then P_{Gi} is equal to the electrical power P_{ei}. And $P_{Gi} = Q_{Gi} = 0$ for $i = n + 1, \ldots, n + m$.
- **Static VAR Compensation.** The reactive power supplied by the static VAR compensator installed at bus i is represented as $Q_{SVCi} = B_{SVCi}V_i^2$, where B_{SVCi} is the equivalent shunt susceptance, and $B_{SVCi} = 0$ if there is no SVC device at bus i.
- **Loads.** Bus loads are represented by $P_{Li} + jQ_{Li}$ for $i = 1, \ldots, n + m$. Both P_{Li} and Q_{Li} are voltage-dependent ZIP models; that is,

$$P_{Li} = P_{Li}^{(0)}\left[\alpha_{Pi} + \beta_{Pi}\left(\frac{V_i}{V_i^{(0)}}\right) + \gamma_{Pi}\left(\frac{V_i}{V_i^{(0)}}\right)^2\right] \quad \alpha_{Pi} + \beta_{Pi} + \gamma_{Pi} = 1$$

$$Q_{Li} = Q_{Li}^{(0)}\left[\alpha_{Qi} + \beta_{Qi}\left(\frac{V_i}{V_i^{(0)}}\right) + \gamma_{Qi}\left(\frac{V_i}{V_i^{(0)}}\right)^2\right] \quad \alpha_{Qi} + \beta_{Qi} + \gamma_{Qi} = 1$$

  for $i = 1, \ldots, n + m$.
- **Admittance Matrix.** The original network admittance matrix is denoted as $Y = G + jB$, where $G, B \in R^{(n+m)\times(n+m)}$, or $Y_{ij} = G_{ij} + jB_{ij}$, $i = 1, \ldots, n + m$, $j = 1, \ldots, n + m$.

  It should be noted that the generator internal reactances are not included in the above network admittance matrix.
- **Network Equation.** The network equation is represented by the power flow equation:

$$\sum_{j=1}^{n+m} V_iV_j(G_{ij}\cos\theta_{ij} + B_{ij}\sin\theta_{ij}) + P_{Li} - P_{Gi} = 0$$

$$\sum_{j=1}^{n+m} V_iV_j(G_{ij}\sin\theta_{ij} - B_{ij}\cos\theta_{ij}) + Q_{Li} - Q_{Gi} - Q_{SVCi} = 0.$$

  $i = 1, \ldots, n + m$. It should be noted that for $i = n + 1, \ldots, n + m$, $P_{Gi} = Q_{Gi} = 0$. And $Q_{SVCi} = 0$ if bus i is not installed with an SVC device.

The complete transient stability model is composed of mathematical equations describing the system components such as generators, control systems, and transmission networks and loads. Therefore, the energy function W to be derived can be viewed as a combination of several terms related to each component in the model:

- W_{KE}: the kinetic energy associated with generator rotating inertias and speeds,
- U_{GEN}: the potential energy associated with generator circuit equations,

- U_{NET}: the potential energy associated with network equations and loads,
- U_{AVR}: the potential energy associated with automatic voltage regulation systems,
- U_{PSS}: the potential energy associated with power system stabilizers,
- U_{GOV}: the potential energy associated with generator speed governors, and
- U_{SVC}: the potential energy associated with the static VAR compensators.

In summary, the energy function can be written as

$$W = W_{KE} + U_{GEN} + U_{NET} + U_{AVR} + U_{PSS} + U_{GOV} + U_{SVC}.$$

Because the interaction between generators and loads is closely related to the network equations, we derive W_{KE}, U_{GEN}, and U_{NET} simultaneously and denote their sum as $W_{G\text{–}N}$; that is,

$$W_{G\text{–}N} = W_{KE} + U_{GEN} + U_{NET}.$$

Hence, the total energy function can be rearranged as

$$W = W_{G\text{–}N} + U_{AVR} + U_{PSS} + U_{GOV} + U_{SVC}.$$

This arrangement is beneficial because it would produce more analytical terms. Based on the first integral principle, the various parts of this energy function will be derived in the following. We next illustrate a procedure to derive an energy function associated with generators and network equations. This procedure is based on the first integral principle.

7.3.1 Energy Function Associated with Swing Equations

The transient energy function associated with the swing equation is calculated as follows:

$$\begin{aligned} W_s &= \int \sum_{i=1}^{n} \left(M_i \frac{d\omega_i}{dt} \left(-P_{mi} + \frac{M_i}{M_T} P_{col} \right) + P_{Di} \right) \frac{d\delta}{dt} dt \\ &= \sum_{i=1}^{n} \frac{1}{2} M_i \omega_i^2 - \sum_{i=1}^{n} \int P_{mi} d\delta_i + \sum_{i=1}^{n} \int P_{ei} d\delta_i + \sum_{i=1}^{n} \int D_i \omega_i d\delta_i . \end{aligned}$$

The first term represents the kinetic energy, and the other three terms represent the potential energy due to mechanical, electrical and damping powers.

7.3.2 Energy Function Associated with Power Flow Equation

Multiplying the real power flow equation by $\frac{d\theta_i}{dt}$ and summing over all the $n + m$ equations, we get

$$\sum_{i=1}^{n+m}\sum_{j=1}^{n+m} V_iV_j(G_{ij}\cos\theta_{ij}+B_{ij}\sin\theta_{ij})\frac{d\theta_i}{dt}+\sum_{i=1}^{n+m}P_{Li}(V_i)\frac{d\theta_i}{dt}-\sum_{i=1}^{n}P_{Gi}\frac{d\theta_i}{dt}=0.$$

Multiplying both sides of the reactive power flow equation by $\frac{1}{V_i}\frac{dV_i}{dt}$ and summing over all the $n+m$ equations, we obtain

$$\sum_{i=1}^{n+m}\sum_{j=1}^{n+m} V_j(G_{ij}\sin\theta_{ij}-B_{ij}\cos\theta_{ij})\frac{dV_i}{dt}+\sum_{i=1}^{n+m}\frac{Q_{Li}}{V_i}\frac{dV_i}{dt}-\sum_{i=1}^{n}\frac{Q_{Gi}}{V_i}\frac{dV_i}{dt}=0.$$

Integrating the above two equations and summing them over them, we get

$$\begin{aligned}W_{PF}&=\int\sum_{i=1}^{n+m}\sum_{j=1}^{n+m}[V_iV_j(G_{ij}\cos\theta_{ij}+B_{ij}\sin\theta_{ij})d\theta_i+V_j(G_{ij}\sin\theta_{ij}-B_{ij}\cos\theta_{ij})dV_i]\\&+\int\sum_{i=1}^{n+m}\left(P_{Li}d\theta_i+\frac{Q_{Li}}{V_i}dV_i\right)-\int\sum_{i=1}^{n}\left(P_{Gi}d\theta_i+\frac{Q_{Gi}}{V_i}dV_i\right).\end{aligned}$$

By carrying out this integration, we have the following equation:

$$\begin{aligned}W_{pF}&=-\frac{1}{2}\sum_{i=1}^{n+m}B_{ii}\left(V_i^2-V_i^{0^2}\right)-\sum_{i=1}^{n+m}\sum_{j=i+1}^{n+m}\left(B_{ij}\left(V_iV_j\cos\theta_{ij}-V_i^{(0)}V_j^{(0)}\cos\theta_{ij}^{(0)}\right)\right)\\&+\sum_{i=1}^{n+m}\int G_{ij}V_i^2d\theta_i+\sum_{i=1}^{n+m}\sum_{j=i+1}^{n+m}G_{ij}\int\left(V_iV_j\cos\theta_{ij}d(\theta_i+\theta_j)+\sin\theta_{ij}(V_jdV_i-V_idV_j)\right)\\&+\int\sum_{i=1}^{n+m}(P_{Li}d\theta_i)+\left(\frac{Q_{Li}}{V_i}\right)dV_i-\sum_{i=1}^{n}\int\left(P_{Gi}d\theta_i+\frac{Q_{Gi}}{V_i}dV_i\right).\end{aligned}$$

It follows that

$$\begin{aligned}W_s+W_{PF}&=\frac{1}{2}\sum_{i=1}^{n}M_i\omega_i^2-\frac{1}{2}\sum_{i=1}^{n+m}B_{ii}\left(V_i^2-V_i^{(0)^2}\right)\\&-\sum_{i=1}^{n+m-1}\sum_{j=i+1}^{n+m}B_{ij}\left(V_iV_j\cos\theta_{ij}-V_i^{(0)}V_j^{(0)}\cos\theta_{ij}^{(0)}\right)-\sum_{i=1}^{n}\int P_{mi}d\delta_i\\&+\sum_{i=1}^{n}\int P_{ei}d\delta_i+\sum_{i=1}^{n}\int D_i\omega_id\delta_i+W_{Gii}+W_{Gij}+W_{Load}\\&-\sum_{i=1}^{n}\int\left(P_{Gi}d\theta_i+\frac{Q_{Gi}}{V_i}dV_i\right).\end{aligned}$$

We define $W_{Gen}=\sum_{i=1}^{n}\int P_{ei}d\delta_i-\sum_{i=1}^{n}\int\left(P_{Gi}d\theta_i+\frac{Q_{Gi}}{V_i}dV_i\right)$, then by carrying out the integration terms, we have

$$
\begin{aligned}
W_{gen} &= \frac{1}{4}\sum_{i=1}^{n}\frac{x''_{di}+x''_{qi}}{x''_{di}x''_{qi}}\left(V_i^2 - V_i^{0^2}\right) \\
&- \frac{1}{4}\sum_{i=1}^{n}\left(\frac{1}{x''_{qi}}-\frac{1}{x''_{di}}\right)\left(V_i^2\cos 2(\delta_i-\theta_i) - V_i^{0^2}\cos 2(\delta_i^0-\theta_i^0)\right) \\
&- \sum_{i=1}^{n}\int\left(\frac{x''_{di}-x_{ei}}{x''_{di}(x'_{di}-x_{ei})}e'_{gi} + \frac{(x_{di}-x'_{di})(x'_{di}-x''_{di})+(x'_{di}-x_{ei})^2}{x''_{di}(x_{di}-x_{ei})(x'_{di}-x_{ei})}e''_{qi}\right)d(V_i\cos(\delta_i-\theta_i)) \\
&+ \sum_{i=1}^{n}\int\left(\frac{x''_{qi}-x_{ei}}{x''_{qi}(x'_{qi}-x_{ei})}e'_{di} + \frac{(x_{qi}-x'_{qi})(x'_{qi}-x''_{qi})+(x'_{qi}-x_{ei})^2}{x''_{qi}(x_{qi}-x_{ei})(x'_{qi}-x_{ei})}e''_{di}\right)d(V_i\sin(\delta_i-\theta_i)).
\end{aligned}
$$

7.3.3 Energy Function Associated with Rotor Circuit Equations

An energy function associated with the rotor circuit equations can be obtained by the following procedure:

Step 1. Multiply both sides of $\frac{de'_{qi}}{dt}$ equation by $\frac{1}{x_{di}-x'_{di}}\left(\frac{de'_{gi}}{dt}\right)$.

Step 2. Multiply both sides of $\frac{de'_{gi}}{dt}$ equation by

$$\frac{(x_{di}-x'_{di})(x'_{di}-x''_{di})+(x'_{di}-x_{ei})^2}{(x'_{di}-x''_{di})(x_{di}-x_{ei})^2}\left(\frac{de''_{qi}}{dt}\right).$$

Step 3. Multiply both sides of $\frac{de'_{gi}}{dt}$ equation by

$$\frac{(x_{di}-x'_{di})(x'_{di}-x''_{di})+(x'_{di}-x_{ei})^2}{(x'_{di}-x''_{di})(x_{di}-x_{ei})^2}\left(\frac{de''_{qi}}{dt}\right).$$

Step 4. Multiply both sides of $\frac{de'_{di}}{dt}$ equation by $\frac{1}{x_{qi}-x'_{qi}}\left(\frac{de'_{di}}{dt}\right)$.

Step 5. Multiply both sides of $\frac{de'_{di}}{dt}$ equation by

$$\frac{(x_{qi}-x'_{qi})(x'_{qi}-x''_{qi})+(x'_{qi}-x_{ei})^2}{(x'_{qi}-x''_{qi})(xq-x_{ei})^2}\left(\frac{de''_{di}}{dt}\right).$$

By summing over all the equations and integrating them, we get (after a tedious procedure) the energy function associated with generator and network equations, which can be summarized as follows:

$$W = W_s + W_{PF} + W_{Rotor}.$$

The above energy function can be rearranged as follows:

$$W = W_k + \sum_{i=1}^{23} U_i$$

$$W_k = \text{kinetic energy} = \frac{1}{2}\sum_{i=1}^{n} M_i\omega_i^2$$

$$U_1 = \text{potential energy due to network reactance} = -\frac{1}{2}\sum_{i=1}^{n+m} B_{ii}\left(V_i^2 - V_i^{0^2}\right) - \sum_{i=1}^{n+m-1}\sum_{j=i+1}^{n+m} B_{ij}\left(V_iV_j\cos\theta_{ij} - V_i^0V_j^0\cos\theta_{ij}^0\right)$$

$$U_2 = \text{due to generator subtransient reactance} = \frac{1}{4}\sum_{i=1}^{n}\frac{x''_{di}+x''_{qi}}{x''_{di}x''_{qi}}\left(V_i^2 - V_i^{(0)^2}\right)$$

$$U_3 = \text{due to } x''_{di} \neq x''_{qi} = -\frac{1}{4}\sum_{i=1}^{n}\left(\frac{x''_{di}-x''_{qi}}{x''_{di}x''_{qi}}\right)\left(V_i^2\cos 2(\delta_i-\theta_i) - V_i^{(0)^2}\cos 2\left(\delta_i^{(0)}-\theta_i^{(0)}\right)\right)$$

$$U_4 = \text{due to } e'_{qi}V_i = -\sum_{j=1}^{n}\int\frac{x''_{di}-x_{ei}}{x''_{di}(x'_{di}-x_{ei})}\left(e'_{qi}V_i\cos(\delta_i-\theta_i) - e'^{(0)}_{qi}V_i^{(0)}\cos\left(\delta_i^{(0)}-\theta_i^{(0)}\right)\right)$$

$$U_5 = \text{due to } e''_{qi}V_i = -\sum_{i=1}^{n}\int\frac{x''_{di}-x_{ei}}{x''_{di}(x'_{di}-x_{ei})}\left(1+\frac{(x_{di}-x'_{di})(x'_{di}-x''_{di})}{(x'_{di}-x_{ei})^2}\right)\left(e''_{qi}V_i\cos(\delta_i-\theta_i) - e''^{(0)}_{qi}V_i^{(0)}\cos\left(\delta_i^{(0)}-\theta_i^{(0)}\right)\right)$$

$$U_6 = \text{due to } e'^2_{qi} = \sum_{i=1}^{n}\frac{1}{2}\left(\frac{1}{x_{di}-x'_{di}} + \frac{x'_{di}-x''_{di}}{(x'_{di}-x_{ei})^2} + \frac{(x''_{di}-x_{ei})^2}{x''_{di}(x'_{di}-x_{ei})^2}\right)\left(e'^2_{qi} - \left(e'^{(0)}_{qi}\right)^2\right)$$

$$U_7 = \text{due to } e''^2_{qi} = \frac{1}{2}\sum_{i=1}^{n}\frac{x'_{di}}{x''_{di}}\frac{(x'_{di}-x_{ei})^2}{(x'_{di}-x''_{di})(x_{di}-x_{ei})^2}\left(1+\frac{(x_{di}-x'_{di})(x'_{di}-x''_{di})}{(x'_{di}-x_{ei})^2}\right)^2\left(e''^2_{qi} - \left(e''^{(0)}_{qi}\right)^2\right)$$

$$U_8 = \text{due to } e'_{qi}e''_{qi} = -\sum_{i=1}^{n}\frac{x_{ei}}{e''_{di}(x_{di}-x_{ei})}\left(1+\frac{(x_{di}-x'_{di})(x'_{di}-x''_{di})}{(x'_{di}-x_{ei})^2}\right)\left(e'_{qi}e''_{qi} - e'^{(0)}_{qi}e''^{(0)}_{qi}\right)$$

$$U_9 = \text{due to } e'_{di}V_i = \sum_{i=1}^{n}\frac{x''_{qi}-x_{ei}}{x''_{qi}(x'_{qi}-x_{ei})}\left(e'_{di}V_i\sin(\delta_i-\theta_i) - e'^{(0)}_{di}V_i^{(0)}\sin\left(\delta_i^{(0)}-\theta_i^{(0)}\right)\right)$$

$$U_{10} = \text{due to } e''_{di}V_i$$
$$= \sum_{i=1}^{n} \frac{(x'_{qi} - x_{ei})}{x''_{qi}(x_{qi} - x_{ei})} \left(1 + \frac{(x_{qi} - x'_{qi})(x'_{qi} - x''_{qi})}{(x'_{qi} - x_{ei})^2}\right)$$
$$\left(e''_{di}V_i \sin(\delta_i - \theta_i) - e''^{(0)}_{di}V_i^{(0)} \sin\left(\delta_i^{(0)} - \theta_i^{(0)}\right)\right)$$

$$U_{11} = \text{due to } e'^2_{di}$$
$$= \sum_{i=1}^{n} \frac{1}{2}\left(\frac{1}{x_{qi} - x'_{qi}} + \frac{x'_{qi} - x''_{qi}}{(x'_{qi} - x_{ei})^2} + \frac{(x''_{qi} - x_{ei})^2}{x''_{qi}(x'_{qi} - x_{ei})^2}\right)\left(e'^2_{di} - \left(e'^{(0)}_{di}\right)^2\right)$$

$$U_{12} = \text{due to } e''^2_{di}$$
$$= \frac{1}{2}\sum_{i=1}^{n} \frac{x'_{qi}}{x''_{qi}} \frac{(x'_{qi} - x_{ei})^2}{(x'_{qi} - x''_{qi})(x_{qi} - x_{ei})^2}\left(1 + \frac{(x_{qi} - x'_{qi})(x'_{qi} - x''_{qi})}{(x'_{qi} - x_{ei})^2}\right)^2\left(e''^2_{di} - \left(e''^{(0)}_{di}\right)^2\right)$$

$$U_{13} = \text{due to } e'_{di}e''_{di}$$
$$= -\sum_{i=1}^{n} \frac{x_{ei}}{x''_{qi}(x_{qi} - x_{ei})}\left(1 + \frac{(x_{qi} - x'_{qi})(x'_{qi} - x''_{qi})}{(x'_{qi} - x_{ei})^2}\right)\left(e'_{di}e''_{di} - e'^{(0)}_{di}e''^{(0)}_{di}\right)$$

$$U_{14} = \text{due to conductance}$$
$$= \sum_{i=1}^{n+m} G_{ii} \int V_i^2 d\theta_i + \sum_{i=1}^{n+m-1}\sum_{j=i+1}^{n+m} G_{ij} \int \left(V_iV_j \cos\theta_{ij} d(\theta_i + \theta_j) + \sin\theta_{ij}(V_j dV_i - V_i dV_j)\right)$$

$$U_{15} = \text{due to active load} = \sum_{i=1}^{n+m} \int (P_{Li} d\theta_i)$$

$$U_{16} = \text{due to active load} = \sum_{i=1}^{n+m} \int \frac{Q_{Li}}{V_i} dV_i$$

$$U_{17} = \text{due to mechanical power} = -\sum_{i=1}^{n} \int P_{mi} d\delta_i$$

$$U_{18} = \text{due to damping} = \sum_{i=1}^{n} D_i \int \omega_i d\delta_i$$

$$U_{19} = \text{due to excitation} = -\sum_{i=1}^{n} \frac{1}{x_{di} - x'_{di}} \int E_{fdi} dE'_{gi}$$

$$U_{20} = \text{due to } \frac{de'_{qi}}{dt} \text{ variation} = \sum_{i=1}^{n} \frac{T'_{doi}}{x_{di} - x'_{di}} \int \left(\frac{de'_{gi}}{dt}\right)^2 dt$$

$$U_{21} = \text{due to } \frac{de''_{qi}}{dt}$$
$$= \sum_{i=1}^{n} \frac{T''_{dki}}{x'_{di} - x''_{di}} \frac{(x_{di} - x'_{di})(x'_{di} - x''_{di}) + (x'_{di} - x_{ei})^2}{(x_{di} - x_{ei})^2} \left(\frac{de''_{qi}}{dt}\right)^2 dt$$

$$U_{22} = \text{due to } \frac{de'_{di}}{dt} = \sum_{i=1}^{n} \frac{T'_{qoi}}{x_{qi} - x'_{qi}} \int \left(\frac{de'_{di}}{dt} \right)^2 dt$$

$$U_{23} = \text{due to } \frac{de''_{di}}{dt} = \sum_{i=1}^{n} \frac{T''_{qki}}{x'_{qi} - x''_{qi}} \frac{(x_{qi} - x'_{qi})(x'_{qi} - x''_{qi}) + (x'_{qi} - x_{ei})^2}{(x_{qi} - x_{ei})^2} \left(\frac{de''_{di}}{dt} \right)^2 dt.$$

Note that $U_1 - U_{13}$ are path-independent terms that have analytical forms; $U_{14} - U_{23}$ are path-dependent terms for which numerical integration is required by proper path assumptions.

Because the derivation of the above energy function is based on the first integral, therefore, we have $\frac{dW}{dt} = 0$ along the postfault trajectory. If the terms $U_{18} \cdot U_{20} \cdot U_{21} \cdot U_{22} \cdot U_{23}$ are not included in the energy function, then along the postfault trajectory, we have

$$\begin{aligned}
\frac{dW}{dt} = &-\sum_{i=1}^{n} D_i \omega_i^2 - \sum_{i=1}^{n} \frac{T'_{doi}}{x_{di} - x'_{di}} \int \left(\frac{de'_{gi}}{dt} \right)^2 - \sum_{i=1}^{n} \frac{T'_{qoi}}{x_{qi} - x'_{qi}} \int \left(\frac{de'_{gi}}{dt} \right)^2 \\
&- \sum_{i=1}^{n} \frac{T''_{dki}}{x'_{di} - x''_{di}} \frac{(x_{di} - x'_{di})(x'_{di} - x''_{di}) + (x'_{di} - x_{ei})^2}{(x_{di} - x_{ei})^2} \left(\frac{de''_{qi}}{dt} \right)^2 \\
&- \sum_{i=1}^{n} \frac{T''_{qki}}{x'_{qi} - x''_{qi}} \frac{(x_{qi} - x'_{qi})(x'_{qi} - x''_{qi}) + (x'_{qi} - x_{ei})^2}{(x_{qi} - x_{ei})^2} \left(\frac{de''_{di}}{dt} \right)^2 \leq 0.
\end{aligned}$$

7.4 ILL-CONDITIONED NUMERICAL PROBLEMS

A numerical ill-conditioning problem can occur in the numerical energy function by the ray approximation scheme. By observing Equations 7.7 and 7.8, it is clear that an ill-conditioning problem will occur if $\Delta\theta_{ij}$ is close to zero. We illustrate this ill-conditioning problem on the following nine-bus system, shown in Figure 7.3. We consider the case of nonuniform damping: Di/Mi = [0.1, 0.2, 0.3] with the constant impedance load (see Table 7.1).

We summarize the postfault stable equilibrium point (SEP) and controlling unstable equilibrium point (UEP) for Contingency 1 in Table 7.2. The angle differences between the controlling UEP and the postfault SEP are needed for evaluating the energy margin in the controlling UEP method. The angle differences between the exit point and the postfault SEP needed for evaluating the energy margin in the potential energy boundary surface (PEBS) method are listed in Table 7.3. It is noticed that the angle difference, $\Delta\theta_{45} = \Delta\theta_4 - \Delta\theta_5$, is almost zero, which will cause a numerical ill-conditioning problem in the computation of potential energy related to transfer conductance based on the ray approximation scheme.

We summarize the postfault SEP and controlling UEP for Contingency 2 in Table 7.4. The angle differences between the controlling UEP and the postfault SEP are needed for evaluating the energy margin in the controlling UEP method. The angle differences between the exit point and the postfault SEP needed for evaluating

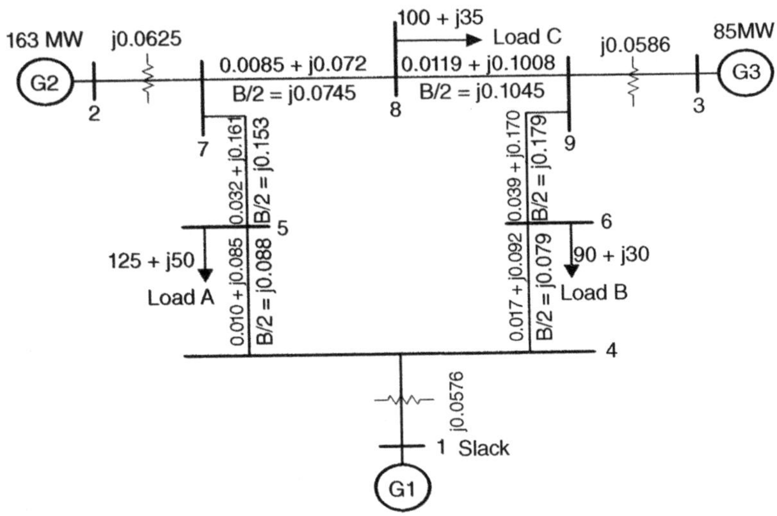

Figure 7.3 A three-generator, nine-bus test system.

Table 7.1 Contingency List and Postfault SEPs

Contingency number	Faulted bus	Fault clearing type	Description from bus to bus		Postfault SEPs (δ_1, δ_2, δ_3)
1	7	Line trip	7	5	(−0.1736, 0.5199, 0.2576)
2	7	Line trip	8	7	(−0.0987, 0.3980, −0.0708)
4	5	Line trip	5	4	(−0.0590, 0.1786, 0.0839)
5	4	Line trip	4	6	(−0.0592, 0.1957, 0.0492)
7	9	Line trip	6	9	(−0.1322, 0.3365, 0.3230)
8	9	Line trip	9	8	(−0.0764, 0.2031, 0.1682)

Table 7.2 Postfault SEP and Controlling UEP for Contingency 1

	(δ_1, δ_2, δ_3)	(V_1, V_2, ..., V_9)	(θ_1, θ_2, ..., θ_9)
x_{sep}	(−0.1736, 0.5199, 0.2576)	(1.0141, 1.0067, 0.9955, 0.9748, 0.9257, 0.9544, 0.9973, 0.9796, 0.9930)	(−0.2065, 0.3426, 0.1142, −0.2404, −0.3383, −0.1834, 0.2456, 0.1319, 0.0669)
x_{cuep}	(−0.7657, 2.0560, 1.6419)	(0.8276, 0.8436, 0.6418, 0.6106, 0.5799, 0.3227, 0.7490, 0.6381, 0.5272)	(−0.7641, 1.8889, 1.4835, −0.7616, −0.8596, −0.3516, 1.7669, 1.6004, 1.3850)

the energy margin in the PEBS method are listed in Table 7.5. We notice that the angle difference, $\Delta\theta_{89} = \Delta\theta_8 - \Delta\theta_9$, is almost zero, which will cause a numerical ill-conditioning problem in the computation of potential energy related to transfer conductance based on the ray approximation scheme.

Table 7.3 Angle Differences for Contingency 1

	$\Delta\delta_1$	$\Delta\delta_2$	$\Delta\delta_3$	$\Delta\theta_1$	$\Delta\theta_2$	$\Delta\theta_3$	$\Delta\theta_4$	$\Delta\theta_5$	$\Delta\theta_6$	$\Delta\theta_7$	$\Delta\theta_8$	$\Delta\theta_9$
$\theta_{cuep} - \theta_{sep}$	−0.5921	1.5360	1.3843	−0.557	1.5462	1.3693	−0.5212	−0.5212	−0.1682	1.5213	1.4685	1.3181
$\theta_{exit} - \theta_{sep}$	−0.5404	1.8362	0.3400	−0.508	1.7496	0.4448	−0.4763	−0.4563	−0.3134	1.5708	1.1733	0.4645

The angle difference, $\Delta\theta_{45} = \Delta\theta_4 - \Delta\theta_5$, is almost zero, which will cause a numerical ill-conditioning problem.

Table 7.4 Postfault System SEP and Controlling UEP for Contingency 2

	$[\delta_1, \delta_2, \delta_3]$	$[V_1, V_2, \ldots, V_9]$	$[\theta_1, \theta_2, \ldots, \theta_9]$
x_{sep}	[−0.0987, 0.3980, −0.0708]	[1.0289, 1.0145, 0.9893, 1.0038, 0.9647, 0.9835, 1.0092, 0.9476, 0.9851]	[−0.1331, 0.2200, −0.2174, −0.1674, −0.1342, −0.2530, 0.1237, −0.3576, −0.2662]
x_{cuep}	[−0.4879, 2.0301, −0.4842]	[0.8697, 0.6270, 0.8816, 0.6926, 0.3936, 0.7334, 0.4405, 0.8093, 0.8413]	[−0.4934, 1.7938, −0.6074, −0.5013, −0.2621, −0.6115, 1.4984, −0.7469, −0.6555]

We summarize the postfault SEP and the controlling UEP for Contingency 4 in Table 7.6. The angle differences between the controlling UEP and the postfault SEP are needed for evaluating the energy margin in the controlling UEP method. The angle differences between the exit point and the postfault SEP needed for evaluating the energy margin in the PEBS method are listed in Table 7.7. We note that the angle difference, $\Delta\theta_{57} = \Delta\theta_5 - \Delta\theta_7$, is almost zero, which will cause a numerical ill-conditioning problem in the computation of potential energy related to transfer conductance based on the ray approximation scheme.

The numerical ill-conditioning problem using the ray approximation scheme can be eliminated by using the following rule:

$$\text{If } \Delta\theta_{ij} = \theta_{ij} - \theta_{ij0} < 1.0e^{-6}, \text{ then assume } \Delta\theta_{ij} = 0.0.$$

Then, the path-dependent terms U_2 and U_3 can be rewritten by the modified ray approximation in the case that $\Delta\theta_{ij} \approx 0.0$:

$$U_2 \approx \sum_{i=1}^{n-1}\sum_{j=i+1}^{n} G_{ij}\left(\Delta\theta_i + \Delta\theta_j\right)\cos\theta_{ij0}\left[V_{i0}V_{j0} + \frac{1}{2}\left(V_{i0}\Delta V_j + V_{j0}\Delta V_i\right) + \frac{1}{3}\Delta V_i \Delta V_j\right] \tag{7.14}$$

$$U_3 \approx \sum_{i=1}^{n-1}\sum_{j=i+1}^{n} G_{ij}\Delta V_i \sin\theta_{ij0}\left(\frac{V_j + V_{j0}}{2}\right). \tag{7.15}$$

7.5 NUMERICAL EVALUATIONS OF APPROXIMATION SCHEMES

To evaluate the numerical energy function using the ray approximation scheme and that using the trapezoidal approximation scheme, we apply the controlling UEP method to these two numerical energy functions to direct the estimation of CCTs. The estimated CCTs are then compared to the exact CCTs based on the time–domain simulation. A summary of the comparison between the exact CCTs and the estimated CCTs is presented in Tables 7.8 and 7.9.

Table 7.8 lists for each contingency the potential energy at the exit point used as the critical energy by the PEBS method, the potential energy at the controlling UEP used as the critical energy by the controlling UEP method, the CCT based on the controlling UEP method with the numerical energy function evaluated via the

Table 7.5 Angle Differences for Contingency 2

	$\Delta\delta_1$	$\Delta\delta_2$	$\Delta\delta_3$	$\Delta\theta_1$	$\Delta\theta_2$	$\Delta\theta_3$	$\Delta\theta_4$	$\Delta\theta_5$	$\Delta\theta_6$	$\Delta\theta_7$	$\Delta\theta_8$	$\Delta\theta_9$
$\theta_{cuep} - \theta_{sep}$	−0.3892	1.6321	−0.4134	−0.3603	1.5738	−0.3900	−0.3339	−0.1279	−0.3585	1.3747	−0.3892	−0.3892
$\theta_{exit} - \theta_{sep}$	−0.5790	1.8280	0.6604	−0.4863	1.8447	0.4377	−0.3529	−0.1629	−0.0919	1.7284	0.3203	0.3203

$\Delta\theta_{89} = \Delta\theta_8 - \Delta\theta_9$ is almost zero, which will cause a numerical ill-conditioning problem in the computation of potential energy related to transfer conductance based on the ray approximation scheme.

Table 7.6 Postfault System SEP, Point, and Controlling UEP (CUEP) for Contingency 4

	$[\delta_1, \delta_2, \delta_3]$	$[V_1, V_2, \ldots, V_9]$	$[\theta_1, \theta_2, \ldots, \theta_9]$
SEP	[−0.0590, 0.1786, 0.0839]	[1.0499, 0.9950, 1.0146, 1.0444, 0.8798, 1.0208, 0.9797, 0.9836, 1.0180]	[−0.0895, 0.0008, −0.0551, −0.1187, −0.2714, −0.1638, −0.0983, −0.1384, −0.0999]
CUEP	[−0.8563, 2.2192, 2.0069]	[0.8265, 0.8137, 0.5975, 0.6084, 0.6333, 0.2638, 0.7052, 0.5974, 0.4703]	[−0.8564, 2.0440, 1.8339, −0.8566, 1.7351, −0.6079, 1.9082, 1.8060, 1.7133]

modified ray approximation, the exact CCT based on the time–domain simulation, and the associated error. As observed from the error information provided in this table, the large errors suggest that the ray approximation scheme is not acceptable for constructing numerical energy functions.

Table 7.9 lists for each contingency the potential energy at the exit point used as the critical energy by the PEBS method, the potential energy at the controlling UEP used as the critical energy by the controlling UEP method, the CCT based on the controlling UEP method with the numerical energy function evaluated via the trapezoidal approximation scheme, the exact CCT based on the time–domain simulation, and the associated error. As observed from the error information provided in this table, the small errors suggest that the trapezoidal approximation scheme is acceptable for constructing numerical energy functions from which the trapezoidal approximation scheme seems acceptable in this numerical study.

Even though the modified ray approximation scheme can eliminate the numerical ill-conditioning problem, the estimated CCTs obtained using the modified ray approximation are significantly different from the exact CCTs. In addition, it is observed that in most contingencies of the test system, the numerical energy function based on the modified ray approximation scheme leads to overestimates in CCTs. On the other hand, the trapezoidal approximation scheme does not suffer from the numerical ill-conditioned problem, and this scheme leads to conservative estimates of CCTs for all contingency cases in this numerical study, which shows that the trapezoidal approximation scheme outperforms the ray approximation scheme in approximating path-dependent terms. To further improve the trapezoidal approximation scheme, an improved trapezoidal approximation scheme will be presented in the next section.

7.6 MULTISTEP TRAPEZOIDAL SCHEME

An improved approximation scheme for evaluating numerical energy functions is presented in this section. The basic idea of this improved scheme is the following: instead of evaluating the energy function value at the controlling UEP by applying the trapezoidal approximation scheme to the two state vectors, the controlling UEP and the postfault SEP, the evaluation is performed by applying the trapezoidal rule

Table 7.7 Angle Differences for Contingency 4

	$\Delta\delta_1$	$\Delta\delta_2$	$\Delta\delta_3$	$\Delta\theta_1$	$\Delta\theta_2$	$\Delta\theta_3$	$\Delta\theta_4$	$\Delta\theta_5$	$\Delta\theta_6$	$\Delta\theta_7$	$\Delta\theta_8$	$\Delta\theta_9$
$\theta_{cuep} - \theta_{sep}$	–0.7973	2.0406	1.9230	–0.7670	2.0433	1.8890	–0.7380	2.0065	–0.4441	2.0065	1.9444	1.8132
$\theta_{exit} - \theta_{sep}$	–0.7919	2.1944	1.5537	–0.7435	2.1346	1.5821	–0.6859	2.0377	–0.3282	2.0377	1.8762	1.5487

$\Delta\theta_{57} = \Delta\theta_5 - \Delta\theta_7$ is almost zero, which will cause a numerical ill-conditioning problem in the computation of potential energy related to transfer conductance based on the ray approximation scheme.

Table 7.8 CCT Estimates Using the Numerical Energy Function Evaluated Via the Modified Ray Approximation

Contingency number	Potential energy at exit point	Potential energy at CUEP	CCT estimate	Exact CCT	Error (%)
1	1.8234	3.5802	0.262	0.170	–54.12
2	1.8975	–0.2992	0.0000	0.187	100.0
4	9.2757	9.5050	0.520	0.424	–22.64
5	7.8330	7.6405	1.000	0.320	–212.5
7	0.9867	2.1678	0.259	0.224	–15.63
8	2.6840	1.0988	0.174	0.245	28.98

The large errors suggest that the ray approximation scheme is not acceptable for constructing numerical energy functions.

Table 7.9 CCT Estimates Using the Numerical Energy Function Evaluated Via the Trapezoidal Approximation Scheme

Contingency number	Potential energy at exit point	Potential energy at CUEP	CCT estimate	Exact CCT	Error (%)
1	1.0625	0.7663	0.158	0.170	7.06
2	1.7909	0.9465	0.155	0.187	17.11
4	2.3545	2.2614	0.411	0.424	3.07
5	2.4229	2.1744	0.304	0.320	5.00
7	1.9412	1.0476	0.194	0.224	13.40
8	2.6094	1.7112	0.207	0.245	15.51

The small errors suggest that the trapezoidal approximation scheme is acceptable for constructing numerical energy functions.

to multiple points between the controlling UEP and the postfault SEP (see Figure 7.4).

For instance, the three-step trapezoidal approximation scheme applies the trapezoidal rule to the controlling UEP, x_1, x_2, and the postfault SEP as the "end points" to evaluate the path-dependent integration via the following three numerical integrations:

$$\int_{x_s^{post}}^{x_{co}} f(x)dx \cong \frac{b-a_1}{2}(f(a_1)+f(b))+\frac{a_1-a_2}{2}(f(a_1)+f(a_2)) + \frac{a_2-a}{2}(f(a_2)+f(a)) \tag{7.16}$$

where $b = f(x_{co})$, $a_1 = f(x_1)$, $a_2 = f(x_2)$, $a = f\left(x_s^{post}\right)$.

This multistep trapezoidal approximation scheme provides a more accurate numerical energy function evaluation. When applied to the nine bus system with six contingencies, it also leads to more accurate CCT estimates for each contingency.

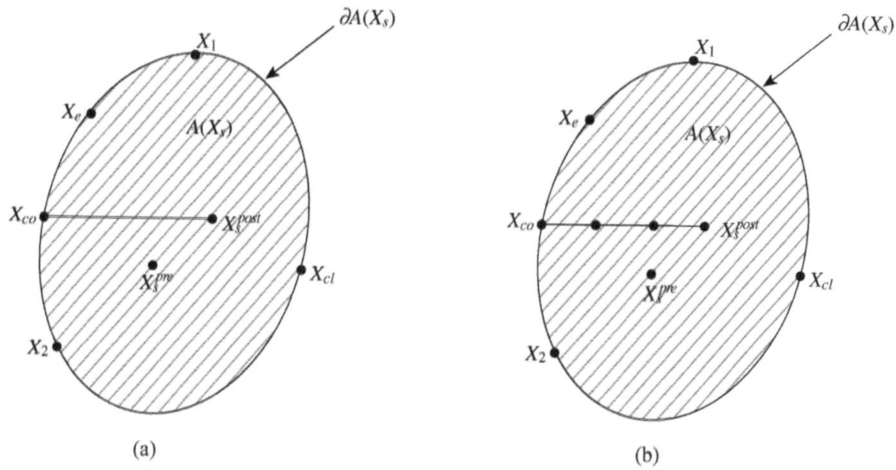

Figure 7.4 (a) The traditional trapezoidal approximation scheme applies the trapezoidal rule to the controlling UEP and the postfault SEP as the "end points" to evaluate the path-dependent integration. (b) The three-step trapezoidal approximation scheme applies the trapezoidal rule to the controlling UEP, x_1, x_2, and the postfault SEP as the "end points" to evaluate the path-dependent integration via three numerical integrations.

A summary of the comparison among the exact CCTs using the time–domain simulation, the estimated CCTs using the trapezoidal approximation scheme, and the (new) estimated CCTs using the multistep trapezoidal approximation scheme are presented in Table 7.9. The errors of estimated CCTs using the multistep trapezoidal approximation scheme, as compared with the exact CCTs shown in the table, are considerably smaller than the errors using the traditional trapezoidal approximation scheme. For example, the error for Contingency 1 is reduced from 7.06% to 1.18%, and from 3.07% to 0.24% for Contingency 4. The error for Contingency 5 is reduced from 5.00% to 1.88% and from 15.51% to 9.8% for Contingency 8. Hence, the multistep trapezoidal approximation scheme leads to accurate estimated CCTs as compared with the traditional trapezoidal approximation scheme for each contingency.

The number of steps used in the multistep trapezoidal approximation scheme impacts on the energy function value at the point of evaluation such as at the controlling UEP. As the number of steps increases, the energy function value at the point of evaluation also improves at the expense of minor computational efforts. As the number of steps increases to a threshold value, the energy function value at the point of evaluation saturates. As listed in Table 7.10, the CCT estimate is based on the traditional one, while the new CCT estimate is based on the multistep one. The error is associated with the multistep trapezoidal approximation scheme. For instance, the two-step trapezoidal approximation scheme yields the numerical energy function value of 0.8385 at the controlling UEP, while the corresponding CCT estimate is 0.165 s.

Table 7.10 CCT Estimates Using the Numerical Energy Function Evaluation Via the Multistep Trapezoidal Approximation Scheme

Contingency number	Potential energy at exit point	Improved potential energy at CUEP	CCT estimate	New CCT estimate	Exact CCT	Error (%)
1	1.0625	0.8733	0.158	0.168	0.170	1.18
2	1.7909	1.1251	0.155	0.167	0.187	10.7
4	2.3545	2.3934	0.411	0.425	0.424	0.24
5	2.4229	2.3528	0.304	0.314	0.320	1.88
7	1.9412	1.1908	0.194	0.204	0.224	8.93
8	2.6094	2.0343	0.207	0.221	0.245	9.80

Table 7.11 The Number of Point Evaluation, the Variation of Potential Energy Related to Path-Dependent Terms, the Variation of Potential Energy at CUEP, and CCT Estimates for Contingency 1

Number of integration steps	Path-dependent potential energy	Total potential energy	CCT estimate (s)
1	0.7238	0.7663	0.158
2	0.796	0.8385	0.165
3	0.8149	0.8574	0.166
5	0.8251	0.8676	0.167
7	0.8279	0.8704	0.168
10	0.8294	0.8719	0.168
15	0.8302	0.8727	0.168
20	0.8305	0.873	0.168
30	0.8307	0.8732	0.168
50	0.8308	0.8733	0.168
80	0.8308	0.8733	0.168
100	0.8308	0.8733	0.168

The improvement in estimated CCTs and in critical energy evaluation increases with the number of steps in the evaluation.

We next perform numerical studies for evaluating the impacts of numbers of steps in the multistep trapezoidal approximation scheme on CCT estimations. The number of steps starts from 1, and then 2, and then 3, 5, 7, 10, 15, 20, 30, 50, 80, and finally 100. The corresponding estimated CCTs are listed in Tables 7.11–7.14 for four different contingencies, respectively. As shown in Table 7.11, the traditional trapezoidal approximation scheme (i.e., the one-step scheme) yields the numerical energy function value of 0.7663 at the controlling UEP, and the corresponding CCT estimate is 0.158 s. The two-step trapezoidal approximation scheme yields the

Table 7.12 The Number of Point Valuation, the Variation of Potential Energy Related to Path-Dependent Terms, the Variation of Potential Energy at CUEP, and CCT Estimates for Contingency 2

Number of integration steps	Path-dependent potential energy	Total potential energy	CCT estimate (s)
1	−0.3974	0.9465	0.155
2	−0.2752	1.0686	0.163
3	−0.2446	1.0993	0.165
5	−0.2281	1.1157	0.166
7	−0.2235	1.1203	0.167
10	−0.221	1.1228	0.167
15	−0.2197	1.1241	0.167
20	−0.2193	1.1246	0.167
30	−0.2189	1.1249	0.167
50	−0.2188	1.1251	0.167
80	−0.2187	1.1251	0.167
100	−0.2187	1.1251	0.167

The improvement in estimated CCTs and in critical energy evaluation increases with the number of steps in the evaluation, but this improvement saturates when the number of steps equals seven and beyond.

Table 7.13 The Number of Point Evaluation, the Variation of Potential Energy Related to Path-Dependent Terms, the Variation of Potential Energy at CUEP, and CCT Estimates for Contingency 4

Number of integration steps	Path-dependent potential energy	Total potential energy	CCT estimate (s)
1	2.7657	2.2614	0.411
2	2.8428	2.3385	0.419
3	2.8731	2.3687	0.422
5	2.8893	2.385	0.424
7	2.8936	2.3892	0.424
10	2.8958	2.3914	0.425
15	2.8969	2.3926	0.425
20	2.8973	2.3929	0.425
30	2.8976	2.3932	0.425
50	2.8977	2.3934	0.425
80	2.8978	2.3934	0.425
100	2.8978	2.3934	0.425

The improvement in estimated CCTs and in critical energy evaluation increases with the number of steps in the evaluation, but this improvement saturates when the number of steps equals five and beyond.

Table 7.14 The Number of Point Evaluation, the Variation of Potential Energy Related to Path-Dependent Terms, the Variation of Potential Energy at CUEP, and CCT Estimates for Contingency 5

Number of integration steps	Path-dependent potential energy	Total potential energy	CCT estimate (s)
1	2.3019	2.1744	0.304
2	2.4075	2.2799	0.31
3	2.4479	2.3204	0.312
5	2.4693	2.3418	0.313
7	2.4749	2.3473	0.314
10	2.4777	2.3502	0.314
15	2.4792	2.3517	0.314
20	2.4797	2.3522	0.314
30	2.4801	2.3526	0.314
50	2.4803	2.3528	0.314
80	2.4803	2.3528	0.314
100	2.4803	2.3528	0.314

The improvement in estimated CCTs and in critical energy evaluation increases with the number of steps in the evaluation, but this improvement saturates when the number of steps equals seven and beyond.

numerical energy function value of 0.8385 at the controlling UEP and the corresponding CCT estimate is 0.165 s. The three-step trapezoidal approximation scheme yields the numerical energy function value of 0.8574 at the controlling UEP and the corresponding CCT estimate is 0.166 s. The improvement in the estimated CCTs and in the critical energy values increases with the number of steps in the evaluation, but this improvement saturates when the number of steps equals seven and beyond.

Similar improvements also occur with other contingencies as summarized in Tables 7.12–7.14, respectively. The claim that the improvement in the estimated CCTs and in the critical energy evaluation increases with the number of steps in the evaluation but that this improvement saturates when the number of steps reaches a threshold still holds. These threshold values vary among different contingencies. For example, the threshold value is 7 for Contingencies 1 and 2, 5 for Contingency 4, and 7 for Contingency 5. The determination of this threshold value seems contingency dependent.

7.7 ON THE CORRECTED NUMERICAL ENERGY FUNCTIONS

Since there may not exist an analytical energy function for lossy transient stability models, the issue of verifying numerical energy functions and of correcting these numerical energy functions when they fail to provide accurate energy margins arises.

This issue has not received serious treatment. In the literature, one attempt to correct energy margin by using physical arguments to adjust the kinetic energy of those generators that contribute to the system separation was proposed. The accuracy of this heuristic method is still unclear. Because there are presently no better alternatives, the numerical energy functions constructed via explicit trajectory approximations are still in widespread use.

There are some other numerical difficulties brought about by these path-dependent terms in the potential energy function. From a theoretical viewpoint, energy function values should be evaluated with respect to a reference point—the postfault SEP. However, the simulated trajectory between the postfault SEP and the point in the state space whose energy value is to be evaluated requires a full time–domain simulation, which is time-consuming. One alternative is to compute and store energy values along the fault-on trajectory. Once the critical energy is determined, which is the energy value at the controlling UEP using the linear path from prefault SEP to the controlling UEP, the estimated CCT can be obtained by comparing the computed energy margin to the stored energy values along the fault-on trajectory. However, the accuracy of this CCT computing procedure can be severely damaged by the path-dependent terms in the energy function. This is because the critical energy margin is computed by using the linear path trajectory (i.e., the ray approximation) from the prefault SEP to the controlling UEP, while the energy values along the fault-on trajectory are computed using the actual fault-on trajectory. The problem is that the fault-on trajectory may be way off the linear path from the prefault SEP and the fault clearing point, making the error of the resulting numerical energy function unacceptable. Apparently, some correction to the computed energy margin must be made to the goal that the corrected energy margin is accurate and conservative.

It is believed that a corrective scheme to be added into the current procedure of computing energy margins can be developed. The purpose of the corrective scheme is to compensate the effect of the assumed linear path or the quadratic path in computing the energy margin. One approach to the development of the corrective scheme is the usage of the actual fault-on trajectory instead of the linear path. This approach deserves further investigation.

7.8 CONCLUDING REMARKS

Two general methods for deriving numerical energy functions have been presented in this chapter. One method is based on the so-called first integral principle, while the other is a two-step procedure. The existing numerical energy functions all contain two parts: the analytical terms (i.e., path-independent terms) and the path-dependent terms. The path-dependent terms are computed based on either the so-called ray approximation scheme or the trapezoidal approximation scheme. These two approximation schemes have been examined.

In this chapter, it has been shown that a numerical ill-conditioned problem may arise in the evaluation of path-dependent potential energy using the ray approximation scheme. A modified ray approximation scheme has been proposed to overcome the problem. However, the estimated CCTs obtained using this modified approximation are still significantly different from the exact CCTs. It is observed from numerical studies that the modified ray approximation scheme can lead to overestimates in CCTs. On the other hand, the trapezoidal approximation scheme does not suffer from the ill-conditioned problem, and this scheme leads to conservative estimates of CCTs for all contingency cases in this numerical study. Hence, the trapezoidal approximation scheme seems to outperform the ray approximation scheme in the evaluation of path-dependent potential energy.

To further improve the trapezoidal approximation scheme, a multistep trapezoidal approximation scheme has been presented. One significant feature of the improved scheme is that the error of the estimated CCTs using the multistep trapezoidal approximation scheme, when compared to the exact CCTs, is considerably smaller than the error from the traditional trapezoidal approximation scheme. The improvement in the estimated CCTs and in the critical energy evaluation increases with the number of steps in the multistep trapezoidal approximation scheme, but this improvement plateaus when the number of steps reaches a threshold value. The determination of this threshold value is contingency dependent.

Since there are presently no better alternatives, the numerical energy functions constructed via explicit trajectory approximations are still in widespread use. To correct the possible error brought about by numerical energy functions, corrections to the computed energy functions must be made such that the corrected energy function leads to accurate and conservative stability assessment results. The idea of adding a corrective scheme to the current procedure for computing energy margins seems promising. The purpose of this corrective scheme is to compensate the effect of an assumed linear or quadratic path when computing energy functions. One effective approach to the development of this corrective scheme is the usage of the actual fault-on trajectory instead of the linear path since the actual fault-on trajectory is available in direct methods. Nevertheless, this approach deserves further investigation.

Chapter 8

Direct Methods for Stability Analysis: An Introduction

8.1 INTRODUCTION

The conventional time–domain approach numerically integrates both the fault-on system and the postfault system. The stability of the postfault system is assessed based on the simulated postfault trajectory. The typical simulation period for the postfault system is 10 s and can go beyond 15 s if multiswing instability is of concern, making this conventional approach rather time-consuming. This time-consuming nature makes the conventional time–domain approach infeasible for practical application in power system online stability assessments (Chiang, 1996, 1999; El-kady et al., 1986; Groom et al., 1996). By contrast, direct methods only integrate the fault-on system and determine, without integrating the postfault system, whether or not the postfault system will remain stable once the fault is cleared by comparing the system energy (when the fault is cleared) to a critical energy value. Direct methods not only avoid the time-consuming numerical integration of the postfault system but they also provide a quantitative measure of the degree of system stability/instability (Chiang, 1991; Chiang et al., 1995; Fouad and Vittal, 1988; Gibescu et al., 2005; Hiskens and Hill, 1989).

The fundamental problem of transient stability is the following: starting from the postfault initial state $X(t_{cl})$, will the postfault system settle down to the steady state condition X_s? In other words, the purpose of power system stability analysis is to determine whether the initial point of the postfault trajectory is located inside the stability region (domain of attraction) of an acceptable stable equilibrium point (SEP) (acceptable steady state).

The problem of direct stability analysis can be translated into the following: given a set of nonlinear equations with an initial condition, determine whether or not the ensuing trajectories will settle down to a desired steady state without resorting to explicit numerical integrations. It is assumed in the area of direct methods that the following condition is satisfied:

Direct Methods for Stability Analysis of Electric Power Systems, by Hsiao-Dong Chiang

Assumption (A1): The prefault SEP, X_s^{pre}, lies inside the stability region of a desired postfault SEP, X_s.

At present, the most popular method for analyzing transient stability is to numerically simulate the postfault dynamic behaviors via numerical integrations of the postfault system model. On the other hand, direct methods first assume that the postfault SEP, X_s, which satisfies (power system) operational constraints (an acceptable steady state), exists. If this point does not exist, then the direct method cannot solve the problem and must resort to the time–domain simulation for verification. Direct methods next determine whether the initial point of the postfault trajectory lies inside the stability region of the acceptable SEP. If it does, direct methods then declare that the resulting postfault trajectory will converge to X_s without knowing the dynamics of the postfault trajectory.

The basis of direct methods for the stability assessment of a postfault system is knowledge of the stability region: if the initial condition of the postfault system lies inside the stability region of a desired postfault SEP, then one can ensure, without performing any numerical integration, that the ensuring postfault trajectory will converge to the desired SEP. Therefore, knowledge of the stability region plays an important role in direct methods.

In the next three sections, a heuristic introduction to direct methods, including the closest UEP method, the potential energy boundary surface (PEBS) method, and the controlling UEP method will be presented.

8.2 A SIMPLE SYSTEM

Heuristic arguments for the applicability of direct methods can be derived from the classical equal-area criterion. We consider one-machine infinite bus systems described by the following equations:

$$\begin{aligned} \dot{\delta} &= \omega \\ M\dot{\omega} &= -D\omega - P_0 \sin\delta + P_m. \end{aligned} \tag{8.1}$$

There are three equilibrium points lying within the range of $\{(\delta, \omega) = -\pi < \delta < \pi, \omega = 0\}$, and they are $(\delta_s, 0) = (\sin^{-1}(P_m/P_0), 0)$, which is a SEP, and $(\delta_1, 0) = (\pi - \sin^{-1}(P_m/P_0), 0)$ and $(\delta_2, 0) = (-\pi - \sin^{-1}(P_m/P_0), 0)$, which are unstable equilibrium points (UEPs). We consider the following function, an energy function for this simple system:

$$E(\delta, \omega) = \frac{1}{2}M\omega^2 - P_m\delta - P_0\cos\delta. \tag{8.2}$$

The energy function can be divided into kinetic energy $K(\omega)$ and potential energy functions $U(\omega)$:

$$E(\delta, \omega) = K(\omega) + U(\delta), \tag{8.3}$$

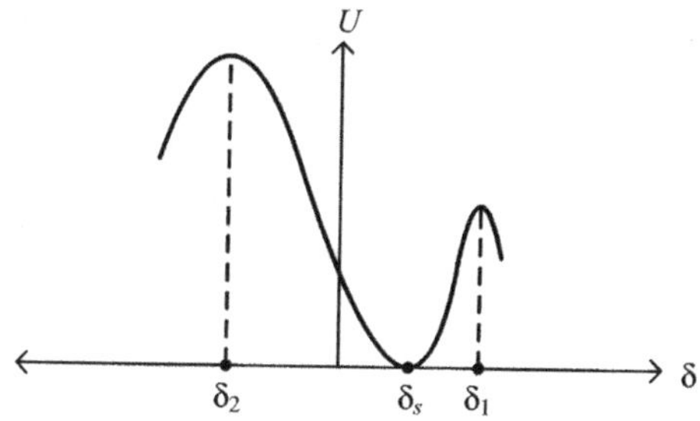

Figure 8.1 The potential energy function $U(\delta)$ is a function of δ only and reaches its local maximum at UEPs δ_1 and δ_2.

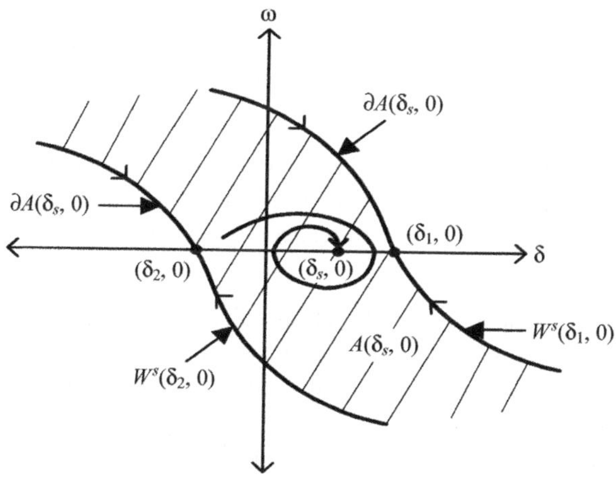

Figure 8.2 The position of the stable equilibrium point $(\delta_s, 0)$ along with its stability $A(\delta_s, 0)$ (the shaded area). The stability boundary $\partial A(\delta_s, 0)$ is composed of the stable manifold of the UEP $(\delta_1, 0)$ and the stable manifold of the UEP $(\delta_2, 0)$.

where $K(\omega) = \frac{1}{2}M\omega^2$ and $U(\delta) = -P_m\delta - P_0 \cos\delta$. The potential energy function $U(.)$ as a function of δ is shown in Figure 8.1. We notice that function $U(\delta)$ reaches its local maximum at UEPs δ_1 and δ_2 of the following (reduced) system:

$$\dot{\delta} = -P_0 \sin\delta + P_m. \tag{8.4}$$

The system is two-dimensional (2-D). Hence, the stability region of $(\delta_s, 0)$, shown in Figure 8.2, is 2-D. Regarding the stability region $A(\delta_s, 0)$, we have the following:

1. The exact stability region $A(\delta_s, 0)$ is completely characterized by its stability boundary $\partial A(\delta_s, 0)$, which is composed of the stable manifold of the UEP $(\delta_1, 0)$ and the stable manifold of the UEP $(\delta_2, 0)$ (see Figure 8.2).
2. The intersection between the stability region $A(\delta_s, 0)$ and the angle space $\{(\delta, \omega): \delta = R, \omega = 0\}$ is $A_\delta := \{(\delta, \omega): \delta \in [\delta_2, \delta_1], \omega = 0\}$.
3. The boundary of this one-dimensional region, A_δ, is composed of two points, δ_1 and δ_2, where $(\delta_1, 0)$ and $(\delta_2, 0)$ are the UEPs on the stability boundary $\partial A(\delta_s, 0)$.

4. These two points, δ_1 and δ_2, are characterized as being the local maxima of the potential energy function $U(.)$.

The stability for this simple system can be directly assessed on the basis of the energy function $U(\delta)$, without the knowledge of the postfault trajectory. The methods presented in the next three sections can be employed to assess the system's stability:

8.3 CLOSEST UEP METHOD

The closest UEP method is the classical direct method developed in the late 1960s. This method uses the constant energy surface $\{(\delta, \omega): V(\delta, \omega) = U(\delta_1)\}$, passing through the closest UEP $(\delta_1, 0)$ to approximate the stability boundary $\partial A(\delta_s, 0)$. Regarding the simple system discussed in the previous section, we have the following observation:

- Of the exact stability boundary $\partial A(\delta_s, 0)$, the UEP $(\delta_1, 0)$ has the lowest energy function value among all the UEPs on the stability boundary $\partial A(\delta_s, 0)$. Hence, $(\delta_1, 0)$ is termed the closest UEP of $(\delta_s, 0)$ with respect to the energy function $U(\delta)$.

If a given state, say, $(\delta_{cl}, \omega_{cl})$, whose energy function value $V(\delta_{cl}, \omega_{cl})$ is less than $U(\delta_1)$, then the state $(\delta_{cl}, \omega_{cl})$ is classified to be lying inside the stability region of $(\delta_s, 0)$ (see Figure 8.3). Thus, one can assert without numerical integration that the resulting trajectory will converge to $(\delta_s, 0)$.

This closest UEP method, although simple, can give considerable conservative stability assessments, especially for those fault-on trajectories crossing the stability boundary $\partial A(\delta_s, 0)$ through $W^s(\delta_2, 0)$ (see Figure 8.4a). For example, the postfault trajectory starting from state P, which lies inside the stability region $A(\delta_s, 0)$, is classified to be unstable by the closest UEP method. This is incorrect because the resulting trajectory will converge to $(\delta_s, 0)$ and hence it is stable (see Figure 8.4b).

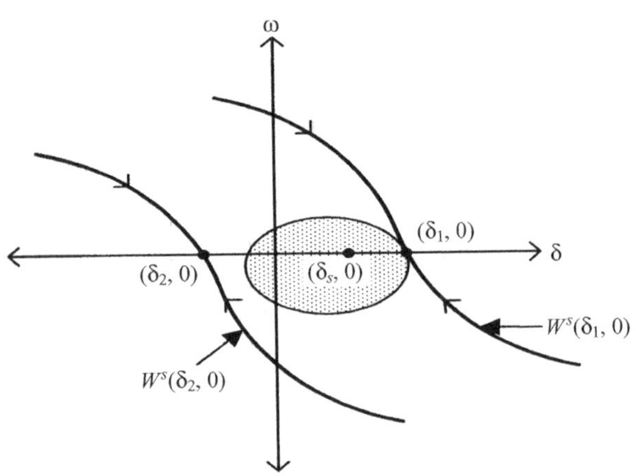

Figure 8.3 The closest UEP method uses the constant energy surface passing through the closest UEP $(\delta_1, 0)$ to approximate the (entire) stability boundary $\partial A(\delta_s, 0)$. The shaded area is the estimated stability region by the closest UEP method.

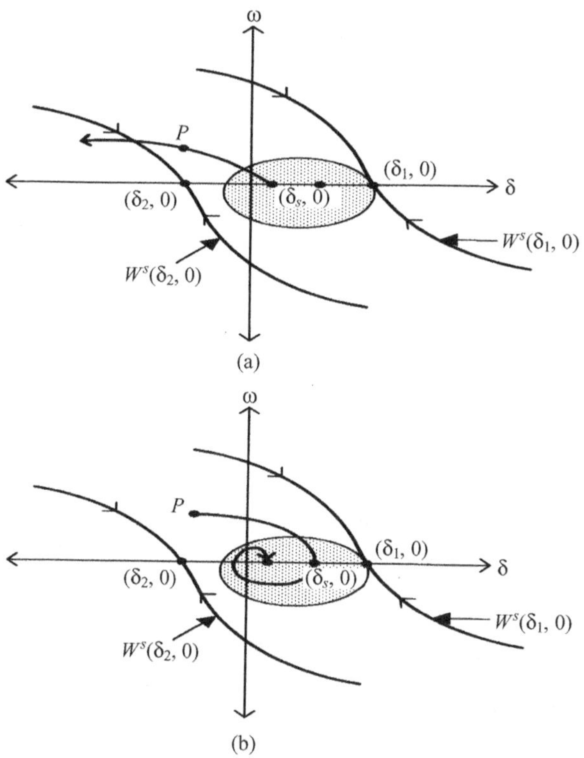

Figure 8.4 (a) The closest UEP method gives considerable conservative stability assessments for those fault-on trajectories crossing the stability boundary $\partial A(\delta_s, 0)$ through $W^s(\delta_2, 0)$. (b) The postfault trajectory starting from state P, which lies inside the stability region $A(\delta_s, 0)$, is classified to be unstable by the closest UEP method, while in fact the resulting trajectory will converge to $(\delta_s, 0)$ and hence it is stable.

The closest UEP method does provide an accurate approximation for the entire stability boundary. However, it does not provide an accurate approximation for the relevant stability boundary. This is due to the fact that the fault-on trajectory is not taken into account in the closest UEP method. The conservativeness of the closest UEP method is now well recognized. Because of this severe conservativeness, the closest UEP method is not popular as a direct method in practical applications. A rigorous analysis of the closest UEP method along with its theoretical foundation will be presented in Chapter 9.

8.4 CONTROLLING UEP METHOD

The controlling UEP method, developed in the 1980s, aims to reduce the conservativeness of the closest UEP method by taking the dependence of the fault-on trajectory into account. We illustrate this method using the simple system as an example. To assess the stability of the postfault trajectory whose corresponding fault-on trajectory $(\delta(t), \omega(t))$ moves towards δ_1, the controlling UEP method uses the constant energy surface passing through the UEP $(\delta_1, 0)$, which is $\{(\delta, \omega): E(\delta, \omega) = U(\delta_1)\}$ as the local approximation for the relevant stability boundary of the postfault system.

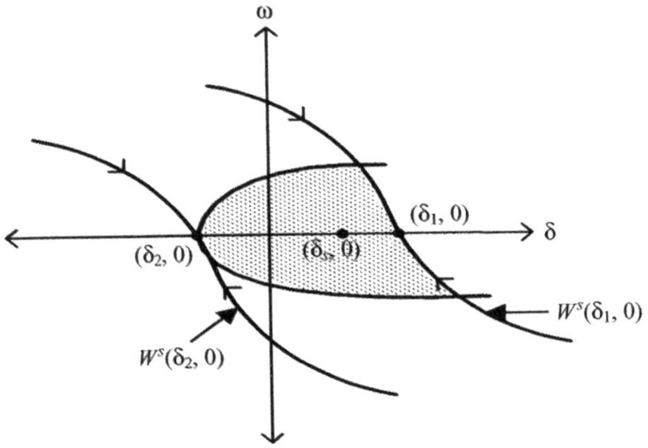

Figure 8.5 The controlling UEP method uses the constant energy surface passing through the controlling UEP to approximate the relevant stability boundary. The shaded area is the estimated stability region by the controlling UEP method. The controlling UEP method does not provide an approximation for the entire stability boundary. However, it provides an accurate approximation for the relevant stability boundary.

In the same manner, for those fault-on trajectories $(\delta(t), \omega(t))$ whose $\delta(t)$ component moves towards δ_2, the controlling UEP method uses the constant energy surface passing through the UEP $(\delta_2, 0)$, which is $\{(\delta, \omega): E(\delta, \omega) = U(\delta_2)\}$ as the local approximation for the relevant stability boundary (see Figure 8.5). Therefore, for each fault-on trajectory, there exists a unique corresponding UEP whose stable manifold constitutes the relevant stability boundary. This unique UEP is the controlling UEP.

The constant energy surface passing through the controlling UEP can be used to accurately approximate the relevant part of the stability boundary towards which the fault-on trajectory is heading. If the energy function value of a given state is less than that of the controlling UEP, then the state is classified as lying inside the stability region of $(\delta_s, 0)$ by the controlling UEP method. Thus, one can assert without numerical integration that the resulting trajectory will converge to $(\delta_s, 0)$.

The controlling UEP method uses the constant energy surface passing through the controlling UEP to approximate the relevant stability boundary. Apparently, the controlling UEP method does not provide an approximation for the entire stability boundary. However, it provides an accurate approximation for the relevant stability boundary. The controlling UEP method, although more complex than the closest UEP method, gives much more accurate and less conservative stability assessments than the closest UEP method. We illustrate this point on Figure 8.6. Suppose a fault is cleared when the fault-on trajectory reaches the state $(\bar{\delta}, \bar{\omega})$ in Figure 8.6. The postfault trajectory starting from the state $(\bar{\delta}, \bar{\omega})$, which lies inside the stability region $A(\delta_s, 0)$, will converge to the SEP $(\delta_s, 0)$. Hence, the postfault system is stable. This postfault trajectory is correctly classified as stable by the controlling UEP method, while it is classified to be unstable by the closest UEP method. Again, this

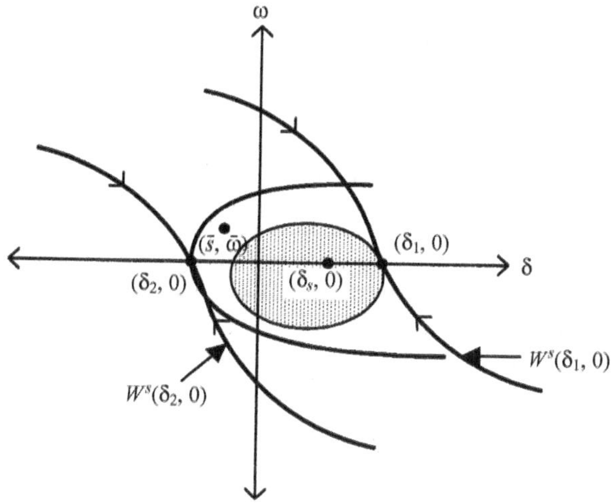

Figure 8.6 The postfault trajectory starting from the state $(\bar{\delta}, \bar{\omega})$, which lies inside the stability region $A(\delta_s, 0)$, is correctly classified as stable by the controlling UEP method, while it is classified to be unstable by the closest UEP method.

shows the conservativeness of the closest UEP method, which does not take into account the dependence of the fault-on trajectory.

The following three-step procedure is the basis of the controlling UEP method:

Step 1. Integrate the fault-on trajectory $(x(t),\dot{x}(t))$. Suppose the fault-on trajectory $(x(t),\dot{x}(t))$ intersects the stable manifold of a UEP lying on the stability boundary of the postfault SEP.

Step 2. Let the potential energy value of the UEP (i.e. the controlling UEP) be v. Then, the critical energy value for the fault-on trajectory $(x(t),\dot{x}(t))$ is v.

Step 3. Use the constant energy surface $\{(x,\dot{x}):V(x,\dot{x}) < v\}$ as a local approximation of the relevant stability boundary. If the energy value when the fault is cleared is less than the critical energy value v, then the postfault system will be stable; otherwise, it may be unstable.

A rigorous analysis of the controlling UEP method, along with its theoretical foundation, will be presented in Chapters 11 through 13.

8.5 PEBS METHOD

The main problem with the controlling UEP method is that it is generally very challenging to find the controlling UEP relative to a fault-on trajectory. The PEBS method attempts to find a local approximation of the relevant stability boundary without the need for computing the UEPs of the postfault system.

A heuristic explanation of the PEBS method using the one-machine infinite bus system is presented below:

1. From the potential energy function V_p (.), find the local maxima x_1 and x_2.

2. For those fault-on trajectories $(x(t),\dot{x}(t))$ whose $x(t)$ component moves towards x_1, the set, that is, the constant energy surface, $\{(x,\dot{x}):V(x,\dot{x}) = V_p(x_1)\}$ is chosen as the local approximation of the (relevant) stability boundary of $(\delta_s, 0)$. For those fault-on trajectories $(x(t),\dot{x}(t))$ whose $x(t)$ component moves towards x_2, the set $\{(x,\dot{x}):V(x,\dot{x}) = V_p(x_2)\}$ is chosen as the local approximation of the (relevant) stability boundary of $(\delta_s, 0)$.

It should be pointed out that for the one-machine infinite bus system, (1) the intersection between the stability boundary and the set $\dot{x} = 0$ is determined solely by the potential energy function $V_p(x)$, and (2) for this special case, the PEBS method behaves like the controlling UEP method.

Kakimoto et al. suggest that a "PEBS" be formed by joining these V_p maxima through the "ridge" of the potential energy level surface (Kakimoto et al., 1978). The PEBS constructed this way is clearly orthogonal to the equipotential curve $V_p(.)$. Furthermore, along the direction orthogonal to the PEBS, the potential energy $V_p(.)$ achieves a local maximum at the PEBS. The following two-step procedure was proposed as the basis of the PEBS method:

Step 1. Integrate the fault-on trajectory $(x(t),\dot{x}(t))$. Suppose $x(t)$ crosses PEBS at an intersection point whose potential energy value is v. The critical energy value for the fault-on trajectory $(x(t),\dot{x}(t))$ is v.

Step 2. Use the constant energy surface $\{(x,\dot{x}):V(x,\dot{x}) < v\}$ as a local approximation of the relevant stability boundary. If the energy value when the fault is cleared is less than the critical energy value v, then the postfault system will be stable; otherwise, it may be unstable.

The derivation of the PEBS method for the multimachine case is based on heuristic arguments. Theoretical justification for the PEBS procedure, except for the one-machine infinite bus case, is lacking. It is not clear, for the multimachine case, whether the PEBS is the intersection of the stability boundary of the multimachine system with the subspace $\{(x,\dot{x}):\dot{x} = 0\}$. Moreover, because of the heuristic way the PEBS is derived, it is difficult to check whether the method provides a good local approximation of the stability boundary or under what conditions the approximation is good. A rigorous analysis of the PEBS method, along with its theoretical foundation, will be presented in Chapter 10.

8.6 CONCLUDING REMARKS

The presented exposition on the one-machine infinite bus system outlines the key bases for the closest UEP method, the controlling UEP method, and the PEBS method. In addition, it has highlighted the differences between the direct stability assessments among these three methods. The above exposition also reveals the following three main steps needed for direct methods:

Step 1. Construct an energy function for the postfault system, say, $V(\delta, \omega)$.

Step 2. Compute the critical energy value V_{cr} for a given fault-on trajectory (say, based on the controlling UEP method).

Step 3. Compare the energy value of the state when the fault is cleared, say, $V(\delta_{cl}, \omega_{cl})$, with the critical energy value V_{cr}. If $V(\delta_{cl}, \omega_{cl}) < V_{cr}$, the postfault trajectory will be stable. Otherwise, it may be unstable.

Extending the above justification for direct methods to general multimachine power systems is nontrivial. This is partly due to the intrinsic differences between the nonlinear dynamics of 2-D systems and that of higher-dimensional nonlinear systems. Rigorous analysis and a theoretical foundation of direct methods will be presented in the following chapters. In particular, we will apply and extend the energy function theory to power system transient stability models to develop theoretical foundations for direct methods including the closest UEP method, the controlling UEP method, and the PEBS method. The focus of this book will be on the controlling UEP method since it is the best direct method.

In the context of direct methods for power system transient stability analysis, we point out one challenging task: given a point in the state space (say, the initial point of the postfault system), it is generally difficult to determine which connected component of a level set contains the point by simply comparing the energy at the given point and the energy of the level set. This is due to the fact that a level set usually contains several connected components and these components are not easy to differentiate based on an energy function value. For instance, the two states, x_1 and x_2, have the same energy function value. However, state x_1 lies outside the stability region, while state x_2 lies inside the stability region (see Figure 8.7 for the closest UEP method and Figure 8.8 for the controlling UEP method).

Fortunately, in the context of direct methods, this difficulty can be circumvented using direct methods since they compute the relevant pieces of information regarding (1) a prefault SEP, (2) a fault-on trajectory, and (3) a postfault SEP. These pieces of information are sufficient to identify the connected component of a level set that contains the initial point of the postfault system.

The above analysis also leads to the following fundamental assumption required for direct methods:

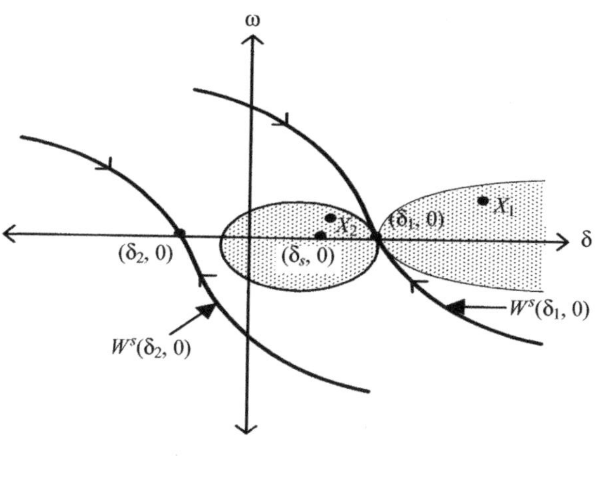

Figure 8.7 The closest UEP method uses the connected constant energy surface passing through the closest UEP $(\delta_1, 0)$ and containing the SEP to approximate the entire stability boundary $\partial A(\delta_s, 0)$. However, it is difficult to differentiate the two states highlighted solely on the basis of energy function values.

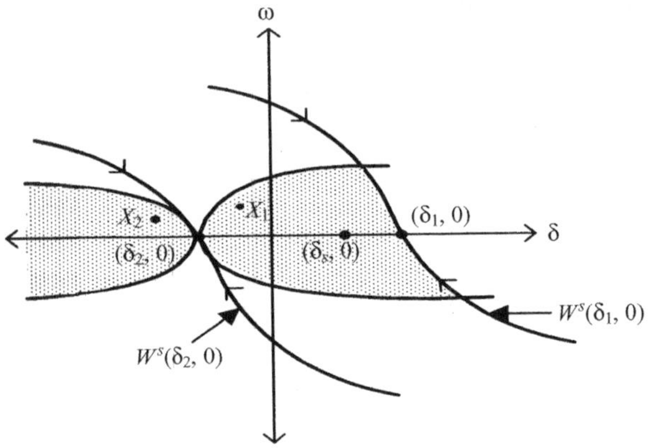

Figure 8.8 The controlling UEP method uses the constant energy surface passing through the controlling UEP and containing the SEP to approximate the relevant stability boundary. However, it is difficult to differentiate the highlighted two states solely on the basis of energy function values.

Fundamental Assumption: The prefault SEP lies inside the stability region of the postfault SEP.

This assumption is trivial if the prefault system equals the postfault system. Since the stability region is an open set of a certain size, this assumption is plausible for a great majority of contingencies. When this fundamental assumption is not satisfied for certain very extreme contingencies, then the time–domain simulation approach must be used for the stability analysis of these contingencies.

Chapter 9

Foundation of the Closest UEP Method

9.1 INTRODUCTION

The closest unstable equilibrium point (UEP) method is a classical tool for problems associated with power system dynamic security assessment (Chiang and Thorp, 1989a; Cook and Eskicioglu, 1983; El-Abiad and Nagappan, 1966). In this chapter, we study the closest UEP method from both theoretical and computational points of view. The robustness of the closest UEP with respect to changes in different parameters of a power system will be shown. This robustness property of the closest UEP heightens the practical applicability of the closest UEP method to power system dynamic security assessment (Lee and Chiang, 2004; Pavella et al., 2000; Qiu et al., 1989; Ribbens-Pavella and Evans, 1985).

An improved closest UEP method with a solid theoretical foundation is proposed. This improved closest UEP method explores the model structure to develop a reduced-state method to compute the closest UEP of a reduced-state model. It will be shown that the improved closest UEP method is optimal in the sense that the estimated stability region characterized by the corresponding energy function is the largest one within the entire stability region. Issues regarding the existence and uniqueness of the closest UEP are investigated. Topological and dynamic characterizations of the closest UEP will be derived.

9.2 A STRUCTURE-PRESERVING MODEL

The structure-preserving model of power systems will be used in this chapter. Consider a power system consisting of n_g generators and $(n_1 - n_g)$ load buses. Each generator is represented by an internal bus, a terminal bus, and an impedance between them. For simplicity, the transmission lines are assumed to be lossless and are represented by the Π-model. The voltage magnitude of the internal bus is a

Direct Methods for Stability Analysis of Electric Power Systems, by Hsiao-Dong Chiang

constant. The real power demand at both the terminal bus and the load bus has two components: a constant and an affine frequency component. The reactive power demand is a function of voltage magnitude at the bus.

Let $J_L := \{1, 2, \ldots, n_1 - n_g\}$ be load buses, $J_T := \{n_1 - n_g + 1, \ldots, n_1\}$ be generator terminal buses, $J_I := \{n_1 + 1, \ldots, n\}$ be generator internal buses, and $J := J_L \cup J_T \cup J_I$. Let V_i, θ_i denote the voltage magnitude and phase angle of bus i. We choose bus n as the reference bus and define the relative angles to be

$$\delta_i := \theta_i - \theta_n, i = 1, 2, \ldots, n-1.$$

The unknown state variables are relative angles $\delta := (\delta_1, \delta_2, \ldots, \delta_{n-1})^T$ and the voltage magnitudes, at load buses and terminal buses $V := (V_1, V_2, \ldots, V_{n_l})^T$.

At each generator internal bus, the real power balance is represented by the swing equation

$$\begin{aligned} m_i\dot{\omega}_i + d_i\omega_i + f_i(\delta, V) &= p_{m_i}, i \in J_I \\ f_i(\delta, V) &= V_iV_jB_{ij}\sin(\delta_i - \delta_j), j = i - n_g, \end{aligned} \tag{9.1}$$

where $\omega_i := \dot{\theta}_i$ generator speed with respect to the synchronous speed, $d_i > 0$ generator damping coefficient, $m_i > 0$ machine inertia constant, P_{m_i} is the mechanical power input, and $B_{ij}(i, j)$ is the element of the node susceptance matrix B. Note that $B_{ii} < 0$ and $B_{ij} > 0$ for $i \neq j$.

The real and reactive power balance at each generator terminal bus is

$$d_i\omega_i + f_i(\delta, \upsilon) = P_i, i \in J_T \text{ and} \tag{9.2}$$

$$g_i(\delta, \upsilon) = 0, i \in J_T, \tag{9.3}$$

where $\omega_i = \dot{\delta}_i$ frequency at bus i, $d_i > 0$ load frequency coefficient, P_i is the real power injection of bus i at synchronous frequency,

$$f_i(\delta, \upsilon) = \sum\nolimits_{j \in J - \{i\}} V_iV_jB_{ij}\sin(\delta_i - \delta_j), i \in J_T,$$

$$g_i(\delta, V) = -V_i^2B_{ii} - \sum\nolimits_{j \in J - \{i\}} V_iV_jB_{ij}\cos(\delta_i - \delta_j) - Q_i(V_i),$$

and $Q_i(V_i)$ is the reactive power demand at bus i.

At each load bus, the real and reactive power balance can be expressed as

$$d_i\omega_i + f_i(\delta, V) = P_i, i \in J_L \text{ and} \tag{9.4}$$

$$g_i(\delta, V) = 0, \tag{9.5}$$

where $\omega_i = \dot{\theta}_i$ and $d_i > 0$,

$$f_i(\delta, V) = \sum_{j \in J - \{i\}} V_iV_jB_{ij}\sin(\delta_i - \delta_j) \text{ and}$$

$$g_i(\delta, V) = -V_i^2B_{ii} - \sum_{j \in J - \{i\}} V_iV_jB_{ij}\cos(\delta_i - \delta_j) - Q_i(V_i).$$

Then the overall mathematical representation of the power systems becomes

$$\begin{aligned}\dot{\delta} &= T_2\omega_g - T_1 D_1^{-1} T_1^T [f(\delta, V) - P] \\ \dot{\omega}_g &= -M_g^{-1} D_g \omega_g - M_g^{-1} T_2^T [f(\delta, V) - P] \\ g(\delta, V) &= 0,\end{aligned} \tag{9.6}$$

where $P := (P_1, P_2, \ldots, P_{n-1})^T$, $\omega_g := (\omega_{n_1+1}, \ldots, \omega_n)^T$,

$f(\delta, V) := (f_1(\delta, V), \ldots, f_{n-1}(\delta, V))^T$,

$g(\delta, V) := (g_1(\delta, V), \ldots, g_{n_1}(\delta, V))^T$,

$Q(V) := (Q_{1(V_1)}, \ldots, Q_{n_1}(V_{n_1}))^T$, $D_1 := \text{diag}\{D_1, \ldots, D_{n_1}\}$, $D_g := \text{diag}\{D_{n_1+1}, \ldots, D_n\}$,

$$M_g := \text{diag}\{M_{n_1+1}, \ldots, M_n\}, T_1 = \begin{bmatrix} I_{n_1} \\ \cdots \\ 0 \end{bmatrix} \in R^{(n-1)\times n_1}, T_2 = \begin{bmatrix} 0 & \\ \cdots & e \\ I_{n_g-1} & \end{bmatrix} \in R^{(n-1)\times n_g},$$

I_{n_1} is the $n_1 \times n_1$ identity matrix, and $e := [-1, -1, \ldots, -1]^T \in R^{n-1}$. D_g, D_1, and M_g are positive diagonal matrices.

Let $x = (\delta, \omega_g)$. The Jacobian matrix of at any equilibrium point x is

$$J_x = \begin{bmatrix} -T_1 D_1^{-1} T_1^T \tilde{F}(x) & T_2 \\ -M_g^{-1} T_2^T \tilde{F}(x) & -M_g^{-1} D_g \end{bmatrix} \in R^{(2n-n_1-1)\times(2n-n_1-1)},$$

where

$$\tilde{F}(x) := \frac{\partial f}{\partial \delta} - \frac{\partial f}{\partial V}\left(\frac{\partial g(x)}{\partial V}\right)^{-1}\left(\frac{\partial g(x)}{\partial \delta}\right).$$

The power system stability model (Eq. 9.6), denoted by $d(M_g, D)$, is described by a set of differential and algebraic equations. This kind of system may not be well posed globally; namely, some trajectories may not be defined for positive time. To resolve this potential difficulty, we employ the singular-perturbation approach to view the system (Eq. 9.6) as being the degenerate system of the following singularly perturbed system:

$$\begin{aligned}\dot{\delta} &= T_2\omega_g - T_1 D_1^{-1} T_1^T [f(\delta, V) - P] \\ \dot{\omega}_g &= -M_g^{-1} D_g \omega_g - M_g^{-1} T_2^T [f(\delta, V) - P] \\ \varepsilon \dot{V} &= g(\delta, V).\end{aligned} \tag{9.7}$$

We assume the following:

Assumption (A1): The matrix $F(x)$ is nonsingular at any equilibrium point x of Equation 9.6. The following property of the equilibrium point of Equation 9.6 is easily derived from Assumption (A1).

Proposition 9.1: Hyperbolicity

The equilibrium points of the system (Eq. 9.6) are hyperbolic.

9.3 CLOSEST UEP

In this section, we study the closest UEP of system $d(M_g, D)$ and its related system, $d(I)$. The problem is concerned with the existence and uniqueness of the closest UEP of system $d(M_g, D)$. Dynamic and topological characterizations of the closest UEP will be derived. These characterizations are useful in the closest UEP method and in the development of a theoretical foundation of the method.

Suppose $(\delta^s, 0)$ is a stable equilibrium point (SEP) of system $d(M_g, D)$ and V^s is the corresponding voltage magnitude. Consider the following function:

$$V(\delta, \omega_g) = \frac{1}{2}\omega_g^T M_g \omega_g - \sum_{i=1}^{n}\sum_{k=1}^{n} V_i V_k B_{ik}\cos(\delta_i - \delta_k) + \sum_{j=1}^{n}\sum_{k=1}^{n} V_i^s V_k^s B_{ik}\cos(\delta_i^s - \delta_k^s) - P^T(\delta - \delta^s) + \sum_{i=1}^{n_1}\int_{V^s}^{V}\frac{Q_i(x_i)}{x_i}dx_i. \tag{9.8}$$

Differentiating the function $V(\delta, \omega_g)$ (Eq. 9.8) along a trajectory of Equation 9.6, it follows that

(i) $\dot{V}(\delta, \omega_g) = -\sum_{i=1}^{n_1} D_i\dot{\delta}_i^2 - \sum_{i=n_1+1}^{n} D_i\omega_i^2 \le 0$.

In addition, it can be shown that this function $V(.)$ in Equation 9.8 possesses the following properties:

(ii) If x is not an equilibrium point, then the set $\{t \in R: \dot{V}(\phi(x, t)) = 0\}$ has measure 0 in R, where $\phi(x, t)$ denotes the trajectory of system $d(M_g, D)$ with $\phi(x, 0) = x$.

(iii) $V(\phi(x, t))$ being bounded implies $\phi(x, t)$ is bounded.

Hence, the function $V(\delta, \omega_g)$ (Eq. 9.8) is an energy function for system $d(M_g, D)$ since it satisfies the three conditions (i)–(iii) of an *energy function*.

Because of the existence of the energy function (Eq. 9.8), the stability boundary of system $d(M_g, D)$ can be completely characterized. This complete characterization is obtained by applying the fundamental theorem of energy functions, Corollary 5.3, as follows.

Proposition 9.2: Stability Boundary

Let x_s be a SEP of system $d(M_g, D)$ and x_i, $i = 1, \ldots$ be the equilibrium points on the stability boundary of x_s, denoted by $\partial A(x_s)$. Then, $\partial A(x_s) \subseteq \cup_i W^s(x_i)$.

From the properties of the energy function, it can be derived that on the stable manifold $W^s(x_i)$ of an equilibrium point x_i, the point at which the energy function achieves the minimum is the equilibrium point x_i itself. Hence, the following property is derived by applying Theorem 5.11.

Proposition 9.3: Closest UEP

If x is a point with the minimum value of $V(.)$ on the stability boundary $\partial A(x_s)$ of a SEP of system $d(M_g, D)$, then x is an equilibrium point of system $d(M_g, D)$.

The concept of the closest UEP has been introduced and applied to the estimation of stability regions of power systems since the early 1970s (Prabhakara and El-Abiad, 1975; Ribbens-Pavella and Evans, 1985). Conventionally, the closest UEP of a SEP x_s with respect to the energy function $V(.)$ has been defined as the equilibrium point with the following property:

$$V(\hat{x}) = \min_{x \in E,\, V(x) > V(x_s)} V(x),$$

where E is the set of equilibrium points of $d(M_g, D)$. Working with this definition, the classical closest UEP method has been found to be very conservative in the estimation of the stability region.

A formal definition of the closest UEP is presented next.

Definition: An equilibrium point p is the *closest UEP* of a SEP x_s with respect to an energy function, $V(.)$, if $V(p) = \text{minimize}_{\,x \in A(x_s) \cap E} V(x)$.

This definition is different from the classical one in that the closest UEP is the UEP with the lowest value of $V(.)$ among all the UEPs on the stability boundary of the SEP x_s rather than among all the UEPs in the state space. For ease of exposition, sometimes we say an equilibrium point $\hat{x}$ is the closest UEP without referring to its corresponding SEP and energy function. With this definition of closest UEP, we now address the fundamental problems concerning the existence and uniqueness of the closest UEP for system $d(M_g, D)$ by applying Theorems 5.6 and 5.11. We provide another proof for Proposition 9.4.

Proposition 9.4: Existence and Uniqueness of Closest UEP

The closest UEP of the SEP x_s of system $d(M_g, D)$ with respect to the energy function $V(.)$ in Equation 9.8 exists. Moreover, this closest UEP is unique generically.

Proof: In order to show the existence of the closest UEP, it will be shown that (1) the point with the lowest value of function $V(.)$ over the stability boundary $\partial A(x_s)$ exists, and (2) the point with the lowest value of function $V(.)$ over the stability boundary $\partial A(x_s)$ is an equilibrium point. Part 1 is followed by the facts that (1) the function $V(.)$ is continuous; (2) the value of function $V(.)$ over the stability boundary $\partial A(x_s)$ is bounded below by $V(x_s)$; and (3) the stability boundary is a closed set (the stability region $A(x_s)$ is an open set, while the boundary of an open set is a close set). Part 2 follows from Propositions 9.2 and 9.3. Thus, the proof for the existence of a closest UEP is complete. To prove the closest UEP is generically unique, it is sufficient to show that for any two equilibrium points of system $d(M_g, D)$, say, x_1 and x_2, generically, $V(x_1) \neq V(x_2)$. It is constructive to see that the conditions to make $V(x_1) = V(x_2)$ are the existence of the solution of the following simultaneous equations:

$$V(x_1) = V(x_2)$$

$$\frac{\partial V}{\partial x_i}(x_1) = 0, 1 \le i \le 2n$$

$$\frac{\partial V}{\partial x_i}(x_2) = 0, 1 \le i \le 2n.$$

The difficulty is that there are $(4n + 1)$ equations in the $4n$ unknowns (x_1, x_2). This amounts to saying that, generically, no two equilibrium points have the same value of $V(.)$. A rigorous proof of the property, all critical values of a C^2 function $W(.)$ being distinct, is generic in $C^2(R^{2n}, R)$, as shown in Hirsch (1976) and in Munkres (1975). Thus, part 2 is true. This completes this proof.

We notice that the above theoretical result on the existence and uniqueness of the closest UEP holds for any energy function as long as it satisfies the required three conditions.

9.4 CHARACTERIZATION OF THE CLOSEST UEP

After having shown the existence and uniqueness of the closest UEP, we next address the problem of characterizing the closest UEP. The characterization of the closest UEP to be presented below is in terms of its unstable manifold. It is shown that the necessary condition for $\hat{x}$ being the closest UEP of x_s is that the unstable manifold of $\hat{x}$, denoted by $W^u(\hat{x})$, converges to x_s. This characterization of the closest UEP will play an important role in the algorithm of finding the closest UEP.

Theorem 9.5: Dynamic Characterization of the Closest UEP

If $\hat{x}$ is the closest UEP of x_s of system $d(M_g, D)$, then its unstable manifold $W^u(\hat{x})$ converges to x_s.

Proof: Since $\hat{x}$ is the closest UEP of x_s, then (1) $\hat{x}$ is an equilibrium point on the stability boundary $\partial A(x_s)$ and (2) $\hat{x}$ is the point with the lowest value of an energy function, $V(.)$, on the stability boundary $\partial A(x_s)$. According to the fundamental theorem of an equilibrium point lying on the stability boundary, a hyperbolic equilibrium point $\hat{x}$ is on the stability boundary $\partial A(x_s)$ if and only if the unstable manifold of $\hat{x}$ excluding $\hat{x}$ has a nonempty intersection with the closure of $A(x_s)$. It follows that $\{W^u(\hat{x}) - \hat{x}\} \cap \bar{A}(x_s) \neq \phi$. Next, it is shown that $\{W^u(\hat{x}) - \hat{x}\} \cap A(x_s) \neq \phi$. By contradiction, suppose $\{W^u(\hat{x}) - \hat{x}\} \cap A(x_s) = \phi$ (since $\{W^u(\hat{x}) - \hat{x}\} \cap \bar{A}(x_s) \neq \phi$; then this implies that $\{W^u(\hat{x}) - \hat{x}\} \cap \partial A(x_s) \neq \phi$. In accordance with Proposition 9.2, which states that every trajectory on the stability boundary $\partial A(x_s)$ converges to one of the equilibrium points on the stability boundary, it is deduced that $\{W^u(\hat{x}) - \hat{x}\}$ converges to an equilibrium point, say, $\hat{p} \in \partial A(x_s)$. However, the value of the energy function $V(.)$ along every trajectory of $d(M_g, D)$ is strictly decreasing; therefore, $V(\hat{x}) > V(\hat{p})$. This is contradictory to part 2, which states that $\hat{x}$ is the point with the lowest value of an energy function, $V(.)$, over the stability boundary $\partial A(x_s)$. Consequently, we conclude that $W^u(\hat{x})$ converges to x_s. Hence, $\{W^u(\hat{x}) - \hat{x}\} \cap A(x_s) \neq \phi$. This proof is completed.

The closest UEP lies on the stability boundary as well as on the quasi-stability boundary. Since the quasi-stability boundary is contained in the union of the closure of type-one equilibrium points, as shown in Theorem 4.11, it follows that the closest UEP must be a type-one equilibrium point, and hence we have the following topological characterization of closest UEP.

Theorem 9.6: Topological Characterization

If p is the closest UEP of x_s of system $d(M_g, D)$, then p is a type-one equilibrium point whose one-dimensional unstable manifold $W^u(p)$ converges to x_s.

Remarks:

1. Theorem 9.5 also holds for the reduced-state system $d(I)$. It is stated as follows: if $\bar{\delta}$ is the closest UEP of δ_s of system $d(I)$, then its unstable manifold $W^u\left(\hat{\bar{\delta}}\right)$ converges to δ_s.

2. Theorem 9.6 also holds for the reduced-state system $d(I)$. It is stated as follows: if $\bar{\delta}$ is the closest UEP of δ_s of system $d(I)$, then it is a type-one UEP.

9.5 CLOSEST UEP METHOD

The closest UEP method as well as its theoretical foundation will be presented in this section. In particular, it will also be shown that the closest UEP method is optimal in the sense that the estimated stability region characterized by the corresponding energy function in this method is the largest one within the entire stability region.

Closest UEP Method for the System $d(M_g, D)$

Step 1. Find all the type-one equilibrium points of $d(M_g, D)$.

Step 2. Order these type-one equilibrium points according to the values of their energy function, $V(.)$.

Step 3. Check whether or not these type-one equilibrium points, starting from the one with the lowest value of $V(.)$ but greater than $V(x_s)$, lie on the stability boundary $\partial A(x_s)$ (by checking if its unstable manifold of dimension one converges to the corresponding SEP $[x_s]$). The first point with the lowest energy function value lying on the stability boundary is the closest UEP, say, $\hat{x}$.

Step 4. The connected component of $\{x\colon V(x) < V(\hat{x})\}$ containing x_s is the estimated stability region of x_s.

This closest UEP method is different from the classical UEP method in one fundamental way: it only searches for those type-one UEPs on the stability boundary $\partial A(x_s)$ rather than for all types of UEPs in the state space. In other words, this closest UEP method considers not only the static behavior of the system, equilibrium points, but also the dynamic behavior of the system, the type-one equilibrium point on the stability boundary $\partial A(x_s)$, and its unstable manifold. In comparison, the classical closest UEP method only considers the static behavior of the system—the equilibrium points.

We next show that the closest UEP method is optimal in the sense that the estimated stability region characterized by the corresponding level set of the energy function is the largest one within the entire stability region. The theorem below is obtained via applying Theorem 5.13 to system $d(M_g, D)$.

Theorem 9.7: Optimal Estimation

Let $\hat{x}$ be the closest UEP of the SEP x_s of system $d(M_g, D)$ with respect to an energy function, $V(.)$. Let $S_c(r)$ denote the connected component of the set $\{x\colon V(x) < r\}$ containing x_s. Then, $S_c(r) \subset A(x_s)$ for $V(x_s) < r < V(\hat{x})$ and $S_c(r) \cap \partial A(x_s) \neq \phi$ for $r > V(\hat{x})$.

9.6 IMPROVED CLOSEST UEP METHOD

From a computational point of view, the closest UEP method can be very involved. In the following, we propose an improved closest UEP method that is computationally competitive with the classical closest UEP method. From a practical point of view, the improved closest UEP method yields the same estimated stability region as that of the closest UEP method (Chiang and Thorp, 1989a).

Our approach is to study another dynamical system, $d(I)$, which is a reduction of system $d(M_g, D)$. The motivation for investigating this reduced-state system will be clear later on. Indeed, it will be shown that there is a close relationship between the closest UEP of the reduced-state system $d(I)$ and the closest UEP of system $d(M_g, D)$. Furthermore, it will be shown that one is able to explore several properties of the closest UEP of system $d(M_g, D)$ via the reduced-state system $d(I)$.

Consider the following system $d(I)$, which is reduced from system $d(M_g, D)$ (we use the same notations as that in Eq. 9.6):

$$\dot{\delta} = -[f(\delta, V) - P]$$
$$0 = g(\delta, V). \tag{9.9}$$

The Jacobian matrix of system $d(I)$ at equilibrium point $\hat{\delta}$ is $\tilde{F}(x)$. Because of Assumption (A1) and the fact that $\tilde{F}(x)$ is symmetric, the equilibrium points of system $d(I)$ are hyperbolic.

Suppose δ^s is a SEP of system $d(I)$ and V^s is the corresponding voltage magnitude. Consider the following function:

$$V_p(\delta) = -\sum_{i=1}^{n}\sum_{k=1}^{n} V_i V_k B_{ik} \cos(\delta_i - \delta_k)$$
$$+ \sum_{j=1}^{n}\sum_{k=1}^{n} V_i^s V_k^s B_{ik} \cos(\delta_i^s - \delta_k^s) - P^T(\delta - \delta^s)$$
$$+ \sum_{i=1}^{n_l} \int_{V^s}^{V} \frac{Q_i(x)}{x} dx. \tag{9.10}$$

To show that $V_p(.)$ is an energy function for $d(I)$, we first differentiate $V_p(.)$ along the trajectory of Equation 9.9:

$$\dot{V}_p(\delta) = -\sum_{i=1}^{n} \dot{\delta}_i^2 \leq 0. \tag{9.11}$$

Second, it is observed that $\dot{V}_p(\delta) = 0$ if and only if δ is an equilibrium point of $d(I)$. Lastly, we will show that the function $V_p(.)$ also satisfies condition (iii) of the energy function. Hence, $V_p(.)$ is an energy function for system $d(I)$. Let $B_\varepsilon(\delta)$ be the ball with radius ε and center δ. Let $\hat{E}$ be the set of equilibrium points of $d(I)$. Choose two positive numbers, ε, η, such that the distance between two different balls is at least ε and $|f(\delta, V) - P| > \eta$ for all $x \notin \cup_{\delta \in \hat{E}} B_\varepsilon(\delta)$. Along the trajectory $\delta(t)$ of $d(I)$, we set

$$T_\varepsilon := \left\{t : \delta(t) \notin \bigcup_{\delta \in \hat{E}} B_\varepsilon(\delta)\right\}, \tag{9.12}$$

and let $L(T_\varepsilon)$ denote the Lebsegue measure of T_ε. From Equation 9.11, we notice that the function $V_p(.)$ is strictly decreasing along its trajectory; hence, it follows that

$$\int_{T_\varepsilon} \langle \dot{\delta}, \dot{\delta} \rangle dt \leq \int_0^\infty \langle \dot{\delta}, \dot{\delta} \rangle dt$$
$$= V_p(\delta(0)) - V_p(\delta(\infty)), \tag{9.13}$$

but

$$\int_{T_\varepsilon} \langle \dot{\delta}, \dot{\delta} \rangle dt \leq \eta^2 L(T_\varepsilon). \tag{9.14}$$

Combining Equations 9.13 and 9.14 gives

$$L(T_\varepsilon) \leq \frac{V_p(\delta(0)) - V_p(\delta(\infty))}{\eta^2}. \tag{9.15}$$

From Equation 9.15, we see that $L(T_\varepsilon)$ is bounded if $V_p(\delta(\infty))$ is bounded. Note that the vector field of $d(I)$, $-f(\delta, V) + P$, is bounded for all $\delta \in R^n$. These two imply that, after some finite distance, the trajectory $\delta(t)$ will remain in some ball $B_\varepsilon(\delta)$. Thus, we have shown that when $V_p(\delta(t))$ is bounded, $\delta(t)$ is also bounded. Hence, the function $V_p(.)$ also satisfies condition (iii) of the energy function.

With the existence of the energy function (Eq. 9.10) for system $d(I)$, we are in a position to characterize the stability boundary of system $d(I)$.

Proposition 9.8: Stability Boundary

Let δ_s be a SEP of system $d(I)$ and δ_i, $i = 1, \ldots, n$ be the equilibrium points on the stability boundary of δ_s, denoted by $\partial A(\delta_s)$. Then, $\partial A(\delta_s) \subseteq \cup_i W^s(\delta_i)$.

The following results concerning the existence and uniqueness of the closest UEP of system $d(I)$ come from the properties of energy functions, Theorems 5.6 and 5.11.

Proposition 9.9: Existence and Uniqueness of the Closest UEP

The closest UEP of the SEP δ_s of system $d(I)$ with respect to the energy function $V_p(.)$ in Equation 9.10 exists. Moreover, this closest UEP is unique generically.

The approach we take for developing the improved closest UEP method is to explore the relationship between the closest UEP of system $d(M_g, D)$ and the closest UEP of system $d(I)$. This approach is motivated by the fact that the value of the energy function $V_p(.)$ in Equation 9.10 at the equilibrium point (δ) of system $d(I)$ is equal to the value of energy function $V(.)$ in Equation 9.8 at the equilibrium point $(\delta, 0)$ of system $d(M_g, D)$. In particular, it will be shown that the closest UEP of the SEP $(\delta_s, 0)$ of system $d(M_g, D)$ with respect to the energy function $V(.)$ in (Eq. 9.8) corresponds to the closest UEP of the SEP (δ_s) of system $d(I)$ with respect to the energy function $V_p(.)$ in Equation 9.10. This result leads to the development of an improved closest UEP method.

We begin with developing a static relationship between the equilibrium points of system $d(M_g, D)$ and that of system $d(I)$.

Proposition 9.10: Static Relationship

1. δ_s is a SEP of system $d(I)$ if and only if $(\delta_s, 0)$ is a SEP of system $d(M_g, D)$.
2. δ_u is a type-k equilibrium point of system $d(I)$ if and only if $(\delta_u, 0)$ is a type-k equilibrium point of system $d(M_g, D)$.

Proof: We shall prove this result by applying the inertia theorem (Yee and Spading, 1997), which states that if H is a nonsingular Hermitian matrix and A has no eigenvalues on the imaginary axis, then $AH + HA^* \geq 0$ implies that the matrices A and H have the same number of eigenvalues with strictly positive and negative real parts.

The Jacobian matrix of $d(I)$ at δ_s is $F(\delta_s)$ and the Jacobian matrix of $d(M_g, D)$ at $(\delta_s, 0)$ is

$$J(\delta_s,0)=\begin{bmatrix} -T_1D_1^{-1}T_1^TF(\delta_s) & T_2 \\ -M_g^{-1}T_2^TF(\delta_s) & -M_g^{-1}D_g \end{bmatrix}.$$

We choose matrix H to be the symmetric matrix:

$$H=\begin{bmatrix} -F(\delta_s)^{-1} & 0 \\ 0 & -M_g^{-1} \end{bmatrix}.$$

It follows that

$$J(\delta_s,0)H+HJ(\delta_s,0)^T=2\begin{bmatrix} T_1D_1^{-1}T_1^T & 0 \\ 0 & M_g^{-1}D_gM_g^{-1} \end{bmatrix}\geq 0.$$

Applying the inertia theorem, it follows that the matrix $J(\delta_s, 0)$ has the same number of eigenvalues with a positive real part as the matrix $F(\delta_s)^{-1}$ does. Since $F(\delta_s)^{-1}$ and $F(\delta_s)$ have the same number of eigenvalues with positive (and negative) real parts, the matrices $J(\delta_s, 0)$ and $F(\delta_s)$ have the same number of eigenvalues with a positive real part, and we obtain the required result. This completes the proof.

The closest UEP method can be applied to system $d(I)$ as stated in Proposition 9.11. Its proof is similar in procedure to that of Theorem 5.13 and hence is omitted.

Proposition 9.11: Closest UEP Method

Let $\hat{\delta}$ be the closest UEP of the SEP δ_s of system $d(I)$ with respect to an energy function, $V_p(.)$. Let $S_p(r)$ denote the connected component of the set $\{\delta: V(\delta) < r\}$ containing δ_s. Then, $S_p(r) \subset A(\delta_s)$ for $V(\delta_s)<r<V(\hat{\delta})$ and $S_p(r) \cap \partial A(x_s) \neq \phi$ for $r>V(\hat{\delta})$.

Next, we establish a dynamic relationship between the closest UEP of system $d(M_g, D)$ and the closest UEP of the reduced-state system $d(I)$, on which the improved closest UEP method is based.

Theorem 9.12: Relationship between the Closest UEP

$(\hat{\delta},0)$ is the closest UEP of the SEP $(\delta_s, 0)$ of system $d(M_g, D)$ with respect to the energy function $V(.)$ in Equation 9.8 if and only if $\hat{\delta}$ is the closest UEP of the SEP δ_s of system $d(I)$ with respect to the energy function $V_p(.)$ in Equation 9.10.

Proof: See Appendix 9A.

Theorem 9.12 suggests that, in order to find the closest UEP with respect to the function $V(.)$ of system $d(M_g, D)$, one only needs to find the closest UEP of system $d(I)$, thus dramatically reducing the required computational effort.

We next present an improved closest UEP method for estimating the stability region $A(x_s)$ of system $d(M_g, D)$.

Improved Closest UEP Method for the System $d(M_g, D)$

Step 1. Find all the type-one equilibrium points of $d(I)$.

Step 2. Order these type-one equilibrium points according to the values of their energy function, $V_p(.)$.

Step 3. Check whether or not these type-one equilibrium points, starting from the one with the lowest value of $V_p(.)$ greater than $V_p(\delta_s)$, lie on the stability boundary $\partial A(\delta_s)$ by checking if its unstable manifold of dimension one converges to the corresponding SEP δ_s. The first lowest energy function that lies on the stability boundary is the closest UEP of system $d(I)$, say, $\bar{\delta}$.

Step 4. The connected component of $\{(\delta,0):V(\delta,0)<V_p(\bar{\delta})\}$ containing $(\delta_s, 0)$ is the estimated stability region of $(\delta_s, 0)$ of system $d(M_g, D)$.

9.7 ROBUSTNESS OF THE CLOSEST UEP

The problem concerning the robustness of the closest UEP with respect to changes in different parameters of the system $d(M_g, D)$ is studied in this section. In particular, we consider the changes in (1) the real power injection (say, due to the change in the load demand), (2) the network topology (e.g., due to different fault locations), and (3) the matrices M_g and D (perhaps, due to the error in modeling the system). It is noted that the changes in (1) and (2) do result in position changes of equilibrium points, whereas the changes in (3) do not alter the positions of equilibrium points.

Consider the case where the real power injection of the system (Eq. 9.6) is changed, say, due to a change in load demand, into another real power injection, $\bar{P}$, where $\bar{P} := (\bar{P}_1, \bar{P}_2, \ldots, \bar{P}_{n-1})^T$. Under this situation, the new system is described by

$$\begin{aligned}\dot{\delta} &= T_2\omega_g - T_1D_1^{-1}T_1^T\left[f(\delta,V)-\bar{P}\right]\\ \dot{\omega}_g &= -M_g^{-1}D_g\omega_g - M_g^{-1}T_2^T\left[f(\delta,V)-\bar{P}\right]\\ 0 &= g(\delta,V).\end{aligned} \tag{9.16}$$

One would like to know whether the information on the closest UEP of the original system (Eq. 9.6) is useful for the stability assessment of this new system (Eq. 9.16). This is to be clarified in the following. Note that the reduced-state system of Equation 9.16 is the following:

$$\dot{\delta} = -\left[f(\delta,V)-\bar{P}\right]. \tag{9.17}$$

It can be shown that the following function, $\bar{V}(.)$, is an energy function for the system (Eq. 9.16).

$$
\begin{aligned}
V(\delta, \omega_g) &= \frac{1}{2}\omega_g^T M_g \omega_g - \sum_{i=1}^{n}\sum_{k=1}^{n} V_i V_k B_{ik} \cos(\delta_i - \delta_k) \\
&+ \sum_{i=1}^{n}\sum_{k=1}^{n} V_i^s V_k^s B_{ik} \cos(\delta_i^s - \delta_k^s) - P^T(\delta - \delta^s) \\
&+ \sum_{i=1}^{n_1} \int_{V^s}^{V} \frac{Q_i(x)}{x} dx \\
&:= \bar{V}_k(\omega_g) + \bar{V}_p(\delta),
\end{aligned}
\tag{9.18}
$$

where $\bar{V}_k(\omega_g) = (1/2)\omega_g^T M_g \omega_g$.

Next, we explore a relationship between the closest UEP of the original system (Eq. 9.6) and the closest UEP of the new system (Eq. 9.16).

Theorem 9.13: Relationship between the Original System and the New System

Let $\hat{x}$ be the closest UEP of a SEP, x_s, of the system (Eq. 9.6) with respect to the energy function V in Equation 9.8. Then, for any given neighborhood $\hat{u}$ of $\hat{x}$, there exists a number $r > 0$ such that if the system (Eq. 9.16) satisfies $|P - \bar{P}| < r$, then the unique equilibrium point $p \in \hat{u}$ of the system (Eq. 9.16) is the closest UEP of the SEP $\hat{x}_s \in S_c(V(\hat{x}))$ of the system (Eq. 9.16) with respect to the energy function $\bar{V}(.)$ in Equation 9.18.

Proof: See Appendix 9B.

Theorem 9.13 suggests that, for a small disturbance in the real power injections P of the system (Eq. 9.6), the stability assessment of a new system (Eq. 9.16) can be carried out as follows:

Step 1. Solve the following nonlinear algebraic equation using $\hat{\delta}$ as starting point, where $\hat{x} := (\hat{\delta}, 0)$ is the closest UEP of SEP $(\delta_s, 0)$ of the system (Eq. 9.6), $[f(\delta, V) - \bar{P}] = 0$.

Let $\bar{\delta}$ be the solution.

Step 2. The connected component of $\{x : \bar{V}(x) < \bar{V}_p(\bar{\delta})\}$ containing $(\delta_s, 0)$ is an estimated stability region of $(\bar{\delta}_s, 0)$, where $(\bar{\delta}_s, 0) \in S_c(\ddot{V}(x))$ is a SEP of the new system.

The case to be considered next is a small disturbance occurring in the power system network of Equation 9.6. Suppose that the system (Eq. 9.6) is subjected to a line switch between bus l and bus m due to a fault, which is reflected in the network flow function $f(.)$ and $g(.)$, we say the function $f(.)$ is changed into $\bar{f}(.)$ and the function $g(.)$ is changed into $\bar{g}(.)$. Under the assumption that all other parameters are not changed, the new system is described by

$$\begin{aligned}\dot{\delta} &= T_2\omega_g - T_1 D_1^{-1} T_1^T [f(\delta, V) - P] \\ \dot{\omega}_g &= -M_g^{-1} D_g \omega_g - M_g^{-1} T_2^T [f(\delta, V) - P] \\ 0 &= \bar{g}(\delta, V).\end{aligned} \tag{9.19}$$

The following function, $V_1(.)$, is an energy function for the system (Eq. 9.19):

$$\begin{aligned}V_1(\delta, \omega_g) = \frac{1}{2}\omega_g^T M_g \omega_g - \sum_{i=1}^{n}\sum_{k=1}^{n} V_i V_k \bar{B}_{ik} \cos(\delta_i - \delta_k) \\ + \sum_{j=1}^{n}\sum_{k=1}^{n} V_i^s V_k^s \bar{B}_{ik} \cos(\delta_i^s - \delta_k^s) - P^T(\delta - \delta^s) \\ + \sum_{i=1}^{n_1 - n_g} \int_{V^s}^{V} \frac{Q_i(x)}{x} dx.\end{aligned} \tag{9.20}$$

Note that $B_{jk} = \bar{B}_{jk}$ for $j, k \in n$, except that, say, $j = l$, $k = m$, where the transmission line between bus l and bus m is disconnected due to line switch and $\bar{B}_{lm} = 0$.

We summarize the result concerning the relation between the closest UEP of the original system (Eq. 9.6) and the closest UEP of the new system (Eq. 9.19) in the following proposition, the proof of which is similar to that of Theorem 9.13 and is therefore omitted.

Proposition 9.14: Robustness to Network Topology

Let $\hat{x}$ be the closest UEP of a SEP, x_s, of the system (Eq. 9.6) with respect to the energy function $V(.)$ in Equation 9.8. Then, for any given neighborhood u_1 of $\hat{x}$, there exists a number $r_1 > 0$ such that if the system (Eq. 9.19) satisfies $|f(.) - f(.)| < r_1$, then the unique equilibrium point $p \in u_1$ of the system (Eq. 9.19) is the closest UEP of the SEP $\bar{x}_s \in S_c(V(\hat{x}))$ of the system (Eq. 9.19) with respect to the energy function $V_1(.)$ in Equation 9.20.

We consider another case where, due to the error in modeling the system matrices M_g and D or due to the changes of parameter settings in the control devices of generators, the machine inertia matrix and the damping matrix of the system (Eq. 9.6) are changed into anther machine matrix $\hat{M}_g$ and damping matrix $\hat{D}$. The new system denoted by $d(\hat{M}, \hat{D})$ is described by

$$\begin{aligned}\dot{\delta} &= T_2\omega_g - T_1 \hat{D}_1^{-1} T_1^T [f(\delta, V) - P] \\ \dot{\omega}_g &= -\hat{M}_g^{-1} \hat{D}_g \omega_g - \hat{M}_g^{-1} T_2^T [f(\delta, V) - P] \\ 0 &= g(\delta, V).\end{aligned} \tag{9.21}$$

It can be easily seen that the positions of the equilibrium points of system $d(M_g, D)$ are independent of matrices M_g and D. In addition, it can be deduced that the type of equilibrium points of system $d(M_g, D)$ is independent of matrices M_g and D as long as they are positively definite. In the following, we will show that the closest UEP of $d(M_g, D)$ is also independent of the matrices M_g and D as long as they are positively definite.

It can be shown that the following function is an energy function of Equation 9.21:

$$\hat{V}(\delta, \omega_g) = \frac{1}{2}\omega_g^T \hat{M}_g \omega_g - \sum_{i=1}^{n}\sum_{k=1}^{n} V_i V_k B_{ik} \cos(\delta_i - \delta_k)$$
$$+ \sum_{i=1}^{n}\sum_{k=1}^{n} V_i^s V_k^s B_{ik} \cos(\delta_i^s - \delta_k^s) - P^T(\delta - \delta^s)$$
$$+ \sum_{i=1}^{n_1 - n_g} \int_{V^s}^{V} \frac{Q_i(x)}{x} dx. \tag{9.22}$$

With this energy function, we are in a position to establish a relation between the closest UEP of system $d(M_g, D)$ and the closest UEP of the new system, $d(\hat{M}, \hat{D})$. Theorem 9.15 below asserts that the closest UEP of system $d(M_g, D)$ is independent of inertia matrix M_g and damping matrix D.

Theorem 9.15: Invariant Property of the Closest UEP

Let $\hat{x}$ be the closest UEP of a SEP, x_s, of system $d(M_g, D)$ with respect to the energy function $V(.)$ in (Eq. 9.8). If the system $d(M_g, D)$ is changed into $d(\hat{M}, \hat{D})$, then

1. $\hat{x}$ is on the stability boundary $\partial A(x_s)$ of $d(\hat{M}, \hat{D})$ and
2. $\hat{x}$ is the closest UEP of the SEP x_s of $d(\hat{M}, \hat{D})$ with respect to $\hat{V}(.)$ in Equation 9.22.

Proof: From Theorem 9.12, we notice that $\hat{x} := (\hat{\delta}, 0)$ is the closest UEP of the SEP $(\delta_s, 0)$ of system $d(M_g, D)$ with respect to the energy function $V(.)$ in Equation 9.8 if and only if $\hat{\delta}$ is the closest UEP of the SEP δ_s of system $d(I)$ with respect to the energy function $V_p(.)$ in Equation 9.10. Applying Theorem 9.12 again shows that $\hat{\delta}$ is the closest UEP of the SEP δ_s of system $d(I)$ with respect to the energy function $V_p(.)$ in Equation 9.10 if and only if $\hat{x} := (\hat{\delta}, 0)$ is the closest UEP of the SEP $(\delta_s, 0)$ of system $d(\hat{M}, \hat{D})$ with respect to the energy function $\hat{V}(.)$ in Equation 9.22. Combining these two facts, we conclude this result.

In Theorem 9.15, we have illustrated another application of energy functions in the understanding of the dynamic behavior of power systems. By appropriate selection of an energy function, we are able to show that a particular equilibrium point (i.e., the closest UEP) on the stability boundary of $d(M_g, D)$ remains on the stability boundary of the new system, $d(\hat{M}, \hat{D})$, under large variations of both the machine inertia matrix and the damping matrix. By Theorem 9.5, the unstable manifold of this particular equilibrium point always converges to the SEP x_s even if the system is under large changes of both matrices.

9.8 NUMERICAL STUDIES

In order to illustrate the improved closest UEP method in a simple context, we consider the following examples, which nearly represent a three-machine system with machine number 3 as the reference machine. Here, the classical model without losses is used.

Consider the following system:

$$\begin{aligned}
\dot{\delta}_1 &= \omega_1 \\
\dot{\omega}_1 &= -\sin\delta_1 - 0.5\sin(\delta_1 - \delta_2) - 0.4\delta_1 \\
\dot{\delta}_2 &= \omega_2 \\
\dot{\omega}_2 &= -0.5\sin\delta_2 - 0.5\sin(\delta_2 - \delta_1) - 0.5\delta_2 + 0.05.
\end{aligned}$$

The following function is an energy function for this system:

$$V(\delta_1, \delta_2, \omega_1, \omega_2) = \omega_1^2 + \omega_2^2 - 2\cos\delta_1 - \cos\delta_2 - \cos(\delta_1 - \delta_2) - 0.1\delta_2. \quad (9.23)$$

The point $x^s = (\delta_1^s, \omega_1^s, \delta_2^s, \omega_2^s) = (0.02001, 0, 0.06003, 0)$ is a SEP whose stability boundary we are interested in. Applying the closest UEP method to approximate the stability boundary $\partial A(x_s)$, we have

Step 1. There are three type-one equilibrium points within the region $\{(\delta_1, \delta_2): \delta_1^s - \pi < \delta_1 < \delta_1^s + \pi, \delta_2^s - \pi < \delta_2 < \delta_2^s + \pi\}$, as summarized in Table 9.1.

Step 2. The type-one equilibrium point (0.03333, 0, 3.10823, 0) is the closest UEP because its unstable manifold converges to the SEP (0.02001, 0, 0.06003, 0) and is the one with the lowest energy function value among all UEPs on the stability boundary.

Step 3. The constant energy surface of $V(.)$, with the level value –0.31249 containing the SEP x^s, approximates the entire stability boundary as shown in Figure 9.1. Curve A in this figure is the intersection between the exact stability boundary and the angle space $\{\delta_1, \delta_2: \delta_1 \in R, \delta_2 \in R\}$. Curve B is the intersection between the approximated stability boundary by the closest UEP method and the angle space.

Table 9.1 The Coordinates of Three Type-One Equilibrium Points and Their Energy Function Values

Type-one equilibrium point	δ_1	ω_1	δ_2	ω_2	$V(.)$
1	0.03333	0	3.10823	0	–0.31249
2	–2.69489	0	1.58620	0	2.07859
3	–3.03807	0	0.31170	0	1.98472

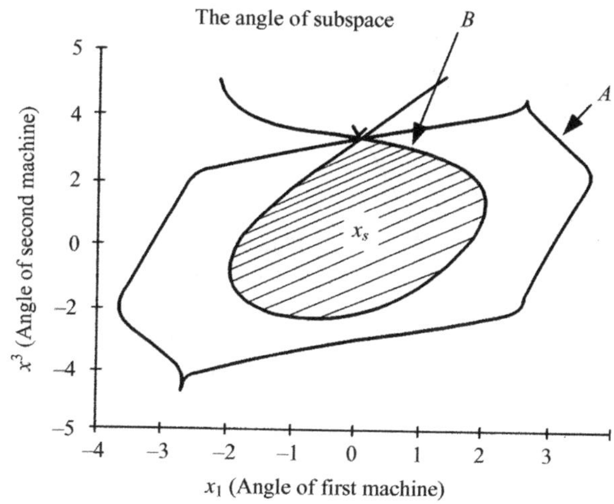

Figure 9.1 Curve *A* is the intersection between the exact stability boundary and the angle space. Curve *B* is the intersection between the approximated stability boundary by the closest UEP method and the angle space.

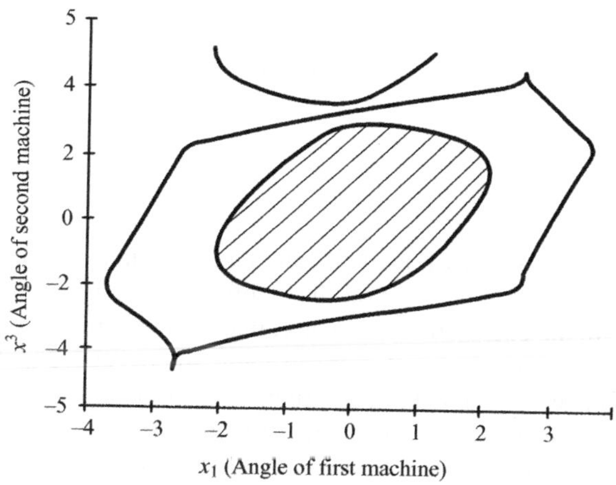

Figure 9.2 The approximated stability boundary by the level surface with a level value lower than the critical value, −0.31249, gives more conservative estimation results than the one given by the closest UEP method.

To show the optimality of the method, the approximated stability boundary by the level surfaces with different level values is shown in Figure 9.2 (level value = −0.2) and in Figure 9.3 (level value = −0.4). It can be seen from these figures that the approximated stability boundary by the level surface with a level value lower than −0.31249 gives a more conservative estimation than the closest UEP method, whereas the approximated stability boundary by the level surface with a level value higher than −0.31249 gives an inaccurate estimation because it contains points that are outside the exact stability boundary.

It is interesting to note that if the following scheme suggested in Prabhakara and El-Abiad (1975) is applied to this system,

$$\text{critical value} = \min\{V(\delta_{s1}, \pi - \delta_{s2}, 0, 0), V(\pi - \delta_{s1}, \delta_{s2}, 0, 0)\}, \qquad (9.24)$$

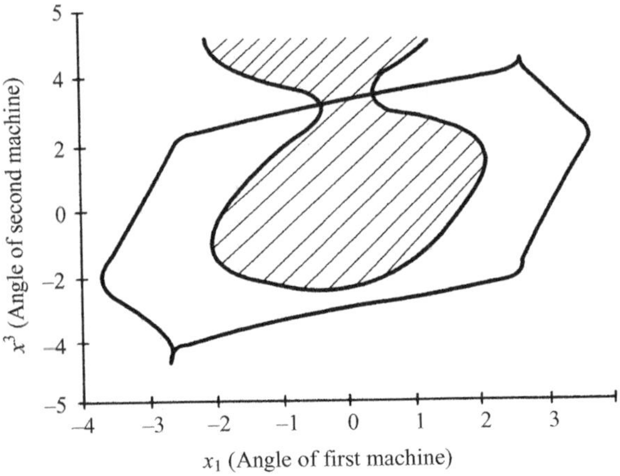

Figure 9.3 The approximated stability boundary by the level surface with a level value higher than the critical value, −0.31249, gives an inaccurate estimation because it contains points that are outside the exact stability boundary.

then the critical value of $V(.)$ would be −0.31276. This value is smaller than that obtained by the improved method, thus giving a very conservative estimate of the stability boundary.

9.9 CONCLUSIONS

The classical as well as improved closest UEP methods are studied in this chapter. The closest UEP of a SEP is defined as the UEP on the stability boundary of the SEP with the smallest distance in the energy function sense, that is, the one with the lowest energy function value. From a theoretical point of view, the existence and uniqueness of the closest UEP have been shown. In addition, it has been shown that the closest UEP method is optimal in the sense that the estimated stability region characterized by the corresponding energy function is the largest one within the entire stability region. A topological and dynamic characterization of the closest UEP has been derived. These characterizations have been incorporated in the improved closest UEP method. The closest UEP has been shown to be independent of both the machine inertia matrix and the damping matrix in power system models.

From a computational point of view, an improved closest UEP method has been proposed based on theoretical results. This improved closest UEP method uses a reduced model approach and is computationally efficient. Furthermore, it has been shown that the closest UEP is continuous (in the mathematical sense) in the real power injection and network flow function. This robustness property of the closest UEP heightens the practical applicability of the closest UEP method in power system dynamic security assessments.

It has been demonstrated that a structural exploration of the energy function can lead to a deep understanding of power system dynamic behaviors. It has also been shown by the use of an appropriate energy function that the closest UEP of

$d(M_g, D)$ remains on the stability boundary of the new system, $d(\hat{M}, \hat{D})$, which is obtained under large variations in both the machine inertia matrix and the damping matrix of $d(M_g, D)$. Moreover, the unstable manifold of the closest UEP always converges to the SEP x_s during the process of parameter variations in both the machine inertia matrix and the damping matrix.

Chapter 10

Foundations of the Potential Energy Boundary Surface Method

10.1 INTRODUCTION

The potential energy boundary surface (PEBS) method is a fast direct method intended to circumvent the problem of determining the controlling unstable equilibrium point (controlling UEP, or CUEP). For a given fault-on trajectory, the PEBS method uses a rather fast method to find a local approximation of the relevant stability boundary of the original system model and to perform direct transient stability analysis. The PEBS method was first proposed by Kakimoto et al. (1978, 1984) and was later pursued and extended by several researchers (e.g., see Athay et al., 1979; Padiyar and Ghosh, 1989; Sauer et al., 1989).

Unlike the controlling UEP method, the PEBS method cannot consistently give conservative stability assessments. It gives either overestimated or underestimated (i.e., conservative) stability assessments for transient stability analysis. Hence, one critical issue with the PEBS method is whether it provides a good local approximation of the relevant stability boundary (i.e., the stability boundary that the fault-on trajectory is heading towards) or under what conditions it does.

The derivation of the PEBS method was based on heuristic arguments. Theoretical justifications for the PEBS method, except for the one-machine infinite bus, are lacking. In this chapter, we present the theoretical foundation for the PEBS method. Sufficient conditions under which the PEBS method provides a good local approximation of the relevant stability boundary are derived. Most of the proofs in this chapter are based on the material in Chiang et al. (1988).This theoretical foundation also paves the development of the BCU method for direct transient stability analysis. This development will be described in later chapters.

Direct Methods for Stability Analysis of Electric Power Systems, by Hsiao-Dong Chiang

10.2 PROCEDURE OF THE PEBS METHOD

The PEBS method was originally proposed for the classical transient stability model, which was derived under the assumption that the transfer conductance of the reduced network after eliminating all load buses is zero. In this model, the dynamics of the ith generator can be represented by the equation

$$\dot{\delta}_i = \omega_i$$
$$M_i\dot{\omega}_i = P_i - D_i\omega_i - \sum_{\substack{j=1 \\ j\neq i}}^{n+1} V_i V_j B_{ij} \sin(\delta_i - \delta_j), \tag{10.1}$$

where the voltage at node $i + 1$ serves as the reference; that is, $\delta_{i+1}: = 0$. The following function is a well-known energy function. It is a summation of a kinetic function and a potential function, $VP(\delta)$:

$$V(\delta,\omega) = \frac{1}{2}\sum_{i=1}^{n} M_i\omega_i^2 - \sum_{i=1}^{n} P_i(\delta_i - \delta_i^s) - \sum_{i=1}^{n}\sum_{j=i+1}^{n+1} V_i V_j B_{ij}\left\{\cos(\delta_i - \delta_j) - \cos(\delta_i^s - \delta_j^s)\right\}$$
$$= \frac{1}{2}\sum_{i=1}^{n} M_i\omega_i^2 + V_P(\delta). \tag{10.2}$$

The above energy function is a summation of the potential energy and the kinetic energy, $V(\delta(t), \omega(t)) = V_P(\delta(t)) + V_K(\omega(t))$. The set of Equation 10.1 can be written in matrix form:

$$\dot{\delta} = \omega$$
$$M\dot{\omega} = P - D\omega - f(\delta, V) \tag{10.3}$$

or

$$\dot{\delta} = \omega$$
$$M\dot{\omega} = -D\omega - \frac{\partial V_P(\delta)}{\partial \delta}, \tag{10.4}$$

where $\delta, \omega \in R^n$ and M, D are positive diagonal matrices.

The PEBS method can be described as a two-step procedure:

Step 1. Integrate the fault-on trajectory $(\delta(t), \omega(t))$ until it crosses the PEBS at a point, termed exit point, which is characterized by the first maximum potential energy along the fault-on trajectory. Let the value of potential energy at the exit point be $V_p(\delta_{exit})$.

Step 2. Use the constant energy surface $\{(\delta, \omega): V(\delta, \omega) \leq V_p(\delta_{exit})\}$ as a local approximation of the stability boundary of the postfault dynamical system.

If the fault-on trajectory is cleared before the constant energy surface, then it will be stable; otherwise, it may be unstable.

It has been well recognized in several numerical studies that the PEBS method cannot consistently give conservative stability assessments. Hence, the PEBS method cannot provide a good local approximation of the relevant stability boundary. One practical issue with the PEBS method is under what conditions it provides a good local approximation. We next present a detailed analysis and a theoretical foundation for the PEBS method.

The following analysis is based on a suggestion by Varaiya et al to view the PEBS as the stability boundary of the following reduced-state artificial model:

$$\dot{\delta} = \frac{\partial V_P(\delta)}{\partial \delta}, \tag{10.5}$$

which is a gradient system (Varaiya et al., 1985). We note the following relations between the original model (Eq. 10.4) and the reduced-state artificial model (Eq. 10.5):

- The potential energy $V_P(\delta)$ of the original model (Eq. 10.4) is an energy function of the artificial model (Eq. 10.5).
- $(\delta_s, 0)$ is an equilibrium point of the original model (Eq. 10.4) if and only if (δ_s) is an equilibrium point of the artificial model (Eq. 10.5).
- The energy function value at an equilibrium point of the original model (Eq. 10.4) is the same as that of the corresponding equilibrium point of the artificial model (Eq. 10.5); in other words, at the equilibrium points, $V(\delta_s, 0) = V_P(\delta_s)$.

We next analyze the relationship between the original model (Eq. 10.4) and the reduced-state artificial model (Eq. 10.5), in particular, the relationship between the stability region of the original model (Eq. 10.4) and the stability region of the reduced-state artificial model (Eq. 10.5). The motivation behind this approach is that the transient stability problem is closely related to the stability region of the original postfault system. This relationship will be explored to provide the theoretical foundation of the PEBS method.

10.3 ORIGINAL MODEL AND ARTIFICIAL MODEL

The study of the relation between the stability boundaries of the original model (Eq. 10.4) and the reduced-state artificial model (Eq. 10.5) is quite challenging. We employ the following three steps to derive the relation:

Step 1. Determine the relation between the stability boundaries of the following two systems:

$$\dot{\delta} = -D^{-1}\frac{\partial V_P(\delta)}{\partial \delta} \tag{10.6}$$

and

$$\dot{\delta} = -\frac{\partial V_P(\delta)}{\partial \delta}, \tag{10.7}$$

where D^{-1}, the inverse of matrix D, is also a positive diagonal matrix.

Step 2. Determine the relation between the stability boundaries of the following two systems:

$$\begin{aligned} \dot{\delta} &= \omega \\ M\dot{\omega} &= -D\omega - \frac{\partial V_P(\delta)}{\partial \delta} \end{aligned} \tag{10.8}$$

and

$$\begin{aligned} \dot{\delta} &= \omega \\ \varepsilon I\dot{\omega} &= -D\omega - \frac{\partial V_P(\delta)}{\partial \delta}. \end{aligned} \tag{10.9}$$

Step 3. Determine the relation between the stability boundaries of the following two systems:

$$\begin{aligned} \dot{\delta} &= \omega \\ \varepsilon I\dot{\omega} &= -D\omega - \frac{\partial V_P(\delta)}{\partial \delta} \end{aligned} \tag{10.10}$$

and

$$\dot{\delta} = -D^{-1}\frac{\partial V_P(\delta)}{\partial \delta}. \tag{10.11}$$

With the above three steps, the relation between the stability boundaries of the original model (Eq. 10.4) (i.e., Eq. 10.8) and the reduced-state artificial model (Eq. 10.5) (i.e., Eq. 10.7) is established by connecting the relation between the stability boundaries of Equations 10.8 and 10.9 in Step 2, the relation between the stability boundaries of Equations 10.10 and 10.11 in Step 3, and the relation between the stability boundaries of Equations 10.6 and 10.7 in Step 1 (see Figure 10.1).

Instead of studying the relation in Step 1, we study the relation between the stability boundaries of the following generalized gradient systems:

$$\dot{\delta} = -D_1 f(\delta) \tag{10.12}$$

and

$$\dot{\delta} = -D_2 f(\delta), \tag{10.13}$$

where D_1 and D_2 are positive diagonal matrices and $f(\delta)$ is a gradient vector.

Instead of studying the relation in Step 2, we study the relation between the stability boundaries of the following systems:

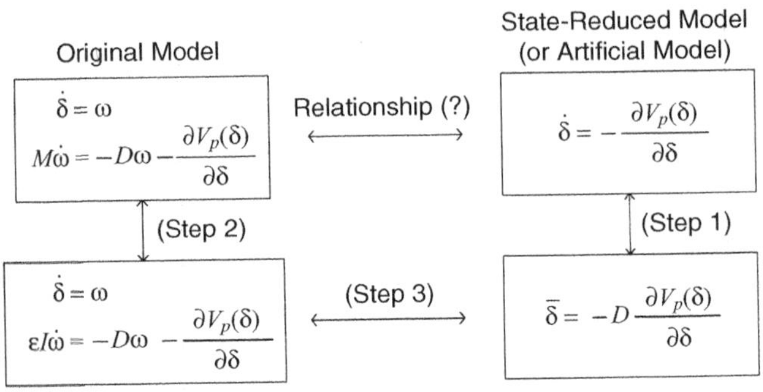

Figure 10.1 A procedure for establishing the relationship between the stability region of the original model (Eq. 10.4) and the stability region of the reduced-state artificial model (Eq. 10.5).

$$\begin{aligned} \dot{\delta} &= \omega \\ M_1\dot{\omega} &= -D\omega - \frac{\partial V_P(\delta)}{\partial \delta} \end{aligned} \tag{10.14}$$

and

$$\begin{aligned} \dot{\delta} &= \omega \\ M_2\dot{\omega} &= -D\omega - \frac{\partial V_P(\delta)}{\partial \delta}, \end{aligned} \tag{10.15}$$

where M_1 and M_2 are positive diagonal matrices and $f(\delta)$ is a gradient vector.

We employ a similar approach to establish the relation in Steps 1 and 2. In this approach, we first derive a complete characterization of the stability boundaries of these two systems and then conduct a qualitative analysis of the changes of the stability boundaries when the vector fields of these two systems vary. In Step 3, we apply the singular perturbation technique to establish the relation of stability boundaries of different dimensions.

We point out that the above procedure can be extended to the power system model with the transfer conductance G_{ij}, which is small as compared with B_{ij} of the reduced network. In this model, the dynamics of the ith generator is then represented by the following equation:

$$\begin{aligned} \dot{\delta}_i &= \omega_i \\ M_i\dot{\omega}_i &= P_i - D_i\omega_i - \sum_{\substack{j=1 \\ j\neq i}}^{n+1} V_iV_jB_{ij}\sin(\delta_i - \delta_j) - \sum_{\substack{j=1 \\ j\neq i}}^{n+1} V_iV_jG_{ij}\cos(\delta_i - \delta_j), \end{aligned} \tag{10.16}$$

or in compact form,

$$\begin{aligned} \dot{\delta} &= \omega \\ M\dot{\omega} &= -D\omega - \frac{\partial V_P(\delta)}{\partial \delta} - \varepsilon g(\delta). \end{aligned} \tag{10.17}$$

The associated reduced-state system of the above original model will then be the following:

$$\dot{\delta} = -\frac{\partial V_P(\delta)}{\partial \delta} - \varepsilon g(\delta). \tag{10.18}$$

To establish the relation between the stability boundaries of the original model (Eq. 10.17) and the reduced-state artificial model (Eq. 10.18), we add the following two steps in addition to the three steps used to establish the relation between the stability boundaries of Equations 10.4 and 10.5:

Step 4. Determine the relation between the stability boundaries of the following two systems:

$$\dot{\delta} = \omega$$

$$M\dot{\omega} = -D\omega - \frac{\partial V_P(\delta)}{\partial \delta} - \varepsilon g(\delta) \text{ and}$$

$$\dot{\delta} = \omega$$

$$M\dot{\omega} = -D\omega - \frac{\partial V_P(\delta)}{\partial \delta}.$$

Step 5. Determine the relation between the stability boundaries of the following two systems:

$$\dot{\delta} = -\frac{\partial V_P(\delta)}{\partial \delta} - \varepsilon g(\delta) \text{ and}$$

$$\dot{\delta} = -\frac{\partial V_P(\delta)}{\partial \delta}.$$

With loss of generality, we show how to establish the relation between the stability boundaries of Equations 10.4 and 10.5.

10.4 GENERALIZED GRADIENT SYSTEMS

We define a generalized gradient system, $d(D)$, via the following differential equation:

$$d(D): \dot{x} + Df(x) = 0, \tag{10.19}$$

where D is a positive diagonal matrix and f: $R^n \rightarrow R^n$ is a bounded gradient vector field. Zero is assumed to be a regular value of $f(x)$; that is, the Jacobian matrix of $f(x)$ at the equilibrium point x is assumed to be nonsingular. The following notations will be used. Set E, defined as E: = {x: $f(x) = 0$}, is the set of equilibrium points. The set $B_r(x_s)$ (respectively, $B_r^o(x_s)$) denotes the closed (respectively, open) ball centered at x_s with radius r. The vector $f(x)$ in Equation 10.19 is assumed to satisfy the following conditions: there exist $\varepsilon > 0$ and $\delta > 0$, such that

$$|B_\varepsilon(x_i) - B_\varepsilon(x_j)| > \varepsilon, \text{ for all } x_i, x_j \in E, \text{ and} \tag{10.20}$$

$$|f(x)| > \delta, \text{ for } x \notin \bigcup_{\bar{x} \in E} B_\varepsilon(\bar{x}). \tag{10.21}$$

Let $\partial A_D(x_s)$ denote the stability boundary of a stable equilibrium point, x_s, of system $d(D)$. (Sometimes, we write $\partial A(D)$ instead of $\partial A_D(x_s)$ when it is clear from the context.) Let $W_D^s(x_i)$ denote the stable manifold of the equilibrium point x_i of system $d(D)$. Also, $\phi D(p, t)$ denotes the trajectory of system $d(D)$ starting from p at $t = 0$.

Remarks:

1. It can be shown that if $f(x)$ is a periodic vector field, then the above conditions (Eqs. 10.20 and 10.21) are satisfied. Power system transient stability models satisfy these two conditions.
2. If there exists a $\delta_1 > 0$ and $\delta_2 > 0$ such that $|f(x)| > \delta_2$ for $x \notin B_{\delta_1}(0)$, then the conditions (Eqs. 10.20 and 10.21) are satisfied (Chiang and Chu, 1996).

In this section, we will analyze and characterize the stability boundaries of the generalized gradient system (Eq. 10.19). Furthermore, the relationship between the stability boundaries $\partial A_{D_1}(x_s)$ and $\partial A_{D_2}(x_s)$ of two generalized gradient systems will be investigated. The relationship between the equilibrium points of two generalized gradient systems of the form (Eq. 10.19) are summarized in Theorem 10.1. Before stating the theorem, we state one definition. The *inertia* of an $n \times n$ matrix A is defined as $In(A) = (n_s(A), n_u(A), n_c(A))$, where $n_s(A)$, $n_u(A)$, and $n_c(A)$ are, respectively, the number of eigenvalues of A with positive, negative, and zero real parts, respectively. We say two equilibrium points have the same inertia if the Jacobian of their corresponding vector fields has the same inertia.

Theorem 10.1: Same Equilibrium Points with Same Inertia

All generalized gradient systems of the form (Eq. 10.19) have the same equilibrium points and the same inertia. Moreover, all the equilibrium points are hyperbolic.

Proof: $d(D_1)$ and $d(D_2)$ have the same equilibrium points since they are defined by $f(x) = 0$. Let λ_x denote the eigenvalue of the Jacobian matrix

$$D_1 J_x = D_1 \left[\frac{\partial f(x)}{\partial x}\right]_x$$

at the equilibrium point x, and let υ be the associated eigenvector; that is,

$$D_1 J_x \upsilon = \lambda_x \upsilon. \tag{10.22}$$

Since D_1 is a positive diagonal matrix,

$$D_1^{1/2}\left(D_1^{1/2} J_x - \lambda_x D_1^{-1/2}\right)\upsilon = 0 \tag{10.23}$$

or

$$D_1^{1/2}\left(D_1^{1/2}J_xD_1^{1/2}-\lambda_xI\right)D_1^{-1/2}\upsilon=0. \tag{10.24}$$

Because matrices $D_1^{1/2}$ and $D_1^{-1/2}$ are nonsingular, the eigenvalues of $D_1^{1/2}J_xD_1^{1/2}$ and D_1J_x are the same. Moreover, $D_1^{1/2}J_xD_1^{1/2}$ is congruent to J_x. From Sylvester's inertia theorem, the congruence transformation preserves the inertia of matrix. Similarly, it follows that the eigenvalues of $D_2^{1/2}J_xD_2^{1/2}$ and D_2J_x are the same, and $D_2^{1/2}J_xD_2^{1/2}$ is congruent to J_x. The proof is completed.

Remark:

Let D be a positive diagonal matrix and J be a symmetrical matrix, which is nonsingular. Theorem 10.1 also reveals that the eigenvalues of matrix DJ will not change its inertia and will remain to be real if we perturb the matrix D into another positive diagonal matrix, $\bar{D}$.

The following two assumptions, which are generic properties, are needed in the following analysis.

Assumption (Cl): Transverse Intersection. The intersections of the stable and unstable manifolds of the equilibrium points on the stability boundary satisfy the transversality condition.

Assumption (C2): Finite Number of Equilibrium Points on the Stability Boundary. The number of equilibrium points on the stability boundary is finite.

Theorem 10.2 below asserts that if Assumption (Cl) is satisfied, then the stability boundary of Equation 10.19 is the union of the stable manifolds of the equilibrium points on the stability boundary.

Theorem 10.2: Characterization of the Stability Boundary of *d*(*D*)

Let x_s be a stable equilibrium point of $d(D)$ and x_i, $i = 1, 2, \ldots$ be the equilibrium points lying on $\partial A_D(x_s)$. If Assumption (Cl) is satisfied, then

$$\partial A_D(x_s)=\bigcup_{x_i\in E\cap\partial A_D(x_s)} W_D^s(x_i). \tag{10.25}$$

Proof: We apply Theorem 3.12 to the generalized gradient system (Eq. 10.19). Assumptions (A1) and (A2) of Theorem 3.12 are satisfied. We only need to show that Assumption (A3) of Theorem 3.12 is also satisfied.

Since $f(x)$ is a gradient vector, let $f(x) = \nabla V(x)$ (gradient of $V(x)$); V: $R^n \to R$. The derivative of $V(x)$ along the trajectory of $d(D)$ is

$$\dot{V}(x)=\langle f(x),-Df(x)\rangle\leq -d|f(x)|^2\leq 0,$$

where $d = \min\{D^1, D^2 \ldots, D^n\}$, D: $= \mathrm{diag}\{D^1, D^2 \ldots, D^n\}$. It therefore follows from Theorem 5.2 that Assumption (A3) of Theorem 3.12 is also satisfied. Thus, the proof is complete.

In the remainder of this section, we study the equilibrium points on the stability boundary of the generalized gradient system (Eq. 10.19). We first show that the equilibrium points on the stability boundary are invariant under certain "small" perturbations of the vector field. We then show that this invariance property also holds under certain large perturbations of the vector field.

Theorem 10.3: Local Invariant Property

If $d(D)$ satisfies Assumptions (Cl) and (C2), then there exists a positive number ε_D such that whenever $\|\bar{D} - D\| < \varepsilon_D$, every equilibrium point on the stability boundary of $d(D)$, $\partial A(D)$ is also on the stability boundary of $d(\bar{D})$, $\partial A(\bar{D})$. In other words, these two stability boundaries, $\partial A(D)$ and $\partial A(\bar{D})$, contain the same equilibrium points.

Proof: See Proof of Theorem 10.3 in the Appendix.

Theorem 10.3 is concerned with an invariance property of equilibrium points lying on the stability boundary in a "local" sense. It asserts that if the dynamical system $d(D)$ satisfies Assumptions (Cl) and (C2), then there exists a "neighborhood" of D such that the following is true: for every system $d(\bar{D})$ with $\bar{D}$ in this neighborhood, the stability boundary of $d(\bar{D})$ contains the equilibrium points on the stability boundary of system $d(D)$. Theorem 10.4 below generalizes this local property to a global property and shows that the stability boundaries of two generalized gradient systems, $d(D_1)$ and $d(D_2)$, contain the same set of equilibrium points under certain conditions. Furthermore, there are no other types of ω-limit set on the stability boundaries.

Theorem 10.4: Global Invariant Property

Let $D_\lambda = \lambda D_1 + (1 - \lambda)D_2$, $\lambda \in [0, 1]$. If $d(D_\lambda)$ satisfies Assumptions (Cl) and (C2) for all $\lambda \in [0, 1]$, then the following results hold:

1. The stability boundaries of $d(D_1)$ and $d(D_2)$ have the same set of equilibrium points; in other words, the stability boundaries of two generalized gradient systems, $d(D_1)$ and $d(D_2)$, contain the same set of equilibrium points, say, $(x_1, x_1, \ldots, x_N)$.
2. The stability boundary of $d(D_1)$ is the union of the stable manifolds of these equilibrium point of $d(D_1)$; that is, $\partial A(D_1) = \cup_{i=1}^{N} W^s_{D(1)}(x_i)$.
3. The stability boundary of $d(D_2)$ is the union of the stable manifolds of the same set of equilibrium points, with respect to the dynamics $d(D_2)$; that is, $\partial A(D_2) = \cup_{i=1}^{N} W^s_{D(2)}(x_i)$.

Proof: See Proof of Theorem 10.4 in the Appendix.

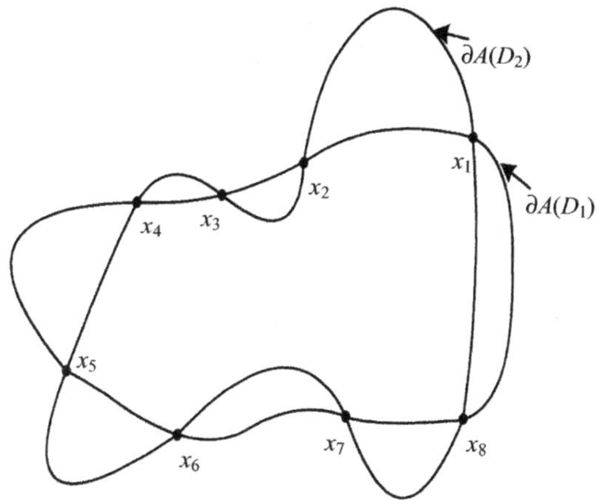

Figure 10.2 Under the conditions stated in Theorem 10.4, the stability boundaries of two generalized gradient systems, $d(D_1)$ and $d(D_2)$, contain the same set of equilibrium points. Moreover, $\partial A(D_1) = \cup_{i=1}^{N} W_{D(1)}^{s}(x_i)$. $\partial A(D_2) = \cup_{i=1}^{N} W_{D(2)}^{s}(x_i)$.

Theorem 10.4 asserts that the relationship between the stability boundaries of two generalized gradient systems is close in the sense that the stability boundaries have the same ω-limit sets while they are composed of the stable manifolds of these ω-limit sets, as depicted in Figure 10.2.

10.5 A CLASS OF SECOND-ORDER DYNAMICAL SYSTEMS

In this section, we consider a nonlinear dynamical system described by the following second-order vector differential equation:

$$M\ddot{x} + D\dot{x} + f(x) = 0, \tag{10.26}$$

where M and D are positive diagonal matrices and f: $R^n \to R^n$ is a bounded gradient vector with a bounded Jacobian. We assume that zero is a regular value of $f(x)$. We also assume that $f(x)$ satisfies the conditions (Eqs. 10.20 and 10.21).

The stability boundaries of the second-order dynamical system (Eq. 10.26) are considered in this section. The analysis of the second-order dynamical systems parallels that of the generalized gradient systems.

Let us rewrite Equation 10.26 into the state-space form

$$\begin{cases} \dot{x} = y \\ M\dot{y} = -Dy - f(x) \end{cases} \tag{10.27}$$

and refer to it as the system $d(M, D)$. Note that all equilibrium points of Equation 10.27 are of the form $\{(x, 0): f(x) = 0, x \in R^n, 0 \in R^n\}$. Let $A_{(M, D)}(x_s, 0)$ denote the stability region of system $d(M, D)$ with respect to the stable equilibrium point $(x_s, 0)$, and let $\partial A_{(M, D)}(x_s, 0)$ denote its stability boundary. We sometimes use the notation $A(M, D)$ $\partial A(M, D)$ instead of $A_{(M, D)}(x_s, 0)$, $\partial A_{(M, D)}(x_s, 0)$ when the meaning is clear from the context. The notation $W^s_{(M,D)}(x, 0)$ denotes the stable manifold of the equilibrium point $(x, 0)$ of system $d(M, D)$.

First, we consider the relationship between equilibrium points of two second-order dynamical systems of the form (Eq. 10.27).

Theorem 10.5: Static Relationship

All dynamical systems of the form (Eq. 10.27) have the same equilibrium points and the same inertia. Moreover, all the equilibrium points are hyperbolic.

It follows from Theorem 10.5 that

- $(x_s, 0)$ is a stable equilibrium point of $d(M_1, D_1)$ if and only if $(x_s, 0)$ is a stable equilibrium point of $d(M_2, D_2)$ and
- $(x, 0)$ is a type-k equilibrium point of $d(M_1, D_1)$ if and only if $(x, 0)$ is a type-k equilibrium point of $d(M_2, D_2)$.

Next, we give a complete characterization of the stability boundary for the second-order dynamical system $d(M, D)$. Theorem 10.6 below states that, under Assumption (Cl), the stability boundary of $d(M, D)$ is the union of the stable manifolds of the equilibrium points on its stability boundary.

Theorem 10.6: Characterization of the Stability Boundary

Let $(x_s, 0)$ denote a stable equilibrium point of $d(M, D)$ and $(x_i, 0)$, $i = 1, 2, \ldots$ be the equilibrium points on $\partial A_{(M, D)}(x_s, 0)$. If Assumption (Cl) is satisfied, then

$$\partial A_{(M,D)}(x_s, 0) = \bigcup_i W^s_{(M,D)}(x_i, 0).$$

Proof: We apply Theorem 3.15 to the second-order dynamical systems of the form (Eq. 10.27). The Assumptions (A1) and (A2) of Theorem 3.15 are satisfied because of Theorem 10.1 and Assumption (Cl). It has been shown in the previous chapter that the second-order dynamical system satisfies Assumption (A3). Hence, the three assumptions of Theorem 3.15 are satisfied and the proof is completed.

We next show the invariance property of the equilibrium points on the stability boundary of the second-order system $d(M, D)$, as we did in the previous section for the generalized gradient system $d(D)$. First, let us define the ε-ball neighborhood of $d(M, D)$.

Definition: ε-Ball Neighborhood

1. We say the dynamical system $d(M_n, D_n)$ is *in the ε-ball neighborhood of* $d(M, D)$ if the following holds:

$$\left\| \begin{bmatrix} 0 & I \\ -M_n^{-1} & -M_n^{-1}D_n \end{bmatrix} - \begin{bmatrix} 0 & I \\ -M^{-1} & -M^{-1}D \end{bmatrix} \right\| < \varepsilon.$$

2. We say the dynamical system $d(M_n, D_n)$ is *on the ε-ball neighborhood of* $d(M, D)$ if the following holds:

$$\left\| \begin{bmatrix} 0 & I \\ -M_n^{-1} & -M_n^{-1}D_n \end{bmatrix} - \begin{bmatrix} 0 & I \\ -M^{-1} & -M^{-1}D \end{bmatrix} \right\| = \varepsilon.$$

One challenge in the global analysis of the dynamical system $d(M, D)$ is that the vector field of the second-order system (Eq. 10.27) is unbounded. The unboundedness of the vector field is due to the variable y in Equation 10.27. However, it is discovered that there is an invariant set for the y-component as shown in Theorem 10.3. This fact is very useful for the global analysis of system $d(M, D)$.

Let $D = \text{diag}\{D^1, D^2, \ldots, D^n\}$, $d = \min\{D^1, D^2, \ldots, D^n\}$. Recall that $f(x)$ is bounded, say, $|f(x)| < k$. We define a set $Q(D) := \{(x, y): x \in R^n, |y| < k/d\}$.

Theorem 10.7: Positive Invariant Set *Q*(*D*)

Let $(x(t), y(t))$ denote the trajectory of system $d(M, D)$ starting from (x_0, y_0) with $|y_0| < \frac{k}{d}$; then, the entire trajectory satisfies the condition that its component, $y(t)$, is bounded; in particular, the set $Q(D)$ is a positive invariant set and the following inequality holds:

$$|y(t)| < \frac{k}{d} \text{ for all } t \geq 0.$$

Proof: From Equation 10.27, we have

$$\dot{y} = -M^{-1}D\dot{x} - M^{-1}f(x).$$

Thus, it holds for all $i \in \{1, 2, \ldots, n\}$ that

$$\dot{y}_i(t) < 0, \quad \text{if } y_i(t) > \frac{k}{d}, \text{ and}$$

$$\dot{y}_i(t) < 0, \quad \text{if } y_i(t) < -\frac{k}{d}.$$

Therefore, for any trajectory $(x(t), y(t))$ of system $d(M, D)$ starting from (x_0, y_0) with $|y_0| < k/d$, we have $|y_i(t)| < k/d$. Hence, the set $Q(D)$ is a positive invariant set for the system $d(M, D)$. The proof is completed.

In the rest of this section, we will examine the invariance property of the equilibrium points lying on the stability boundary $\partial A(M, D)$. In particular, we will show that the equilibrium points on the stability boundary $\partial A(M, D)$ are invariant under certain small perturbations of the system $d(M, D)$. We will further show that this invariance property holds under certain large perturbations of the system $d(M, D)$.

Theorem 10.8: Local Invariant Property

If $d(M, D)$ satisfies Assumptions (Cl) and (C2), then there exists a positive number, $\varepsilon(M, D)$, such that every equilibrium point on $\partial A(M, D)$ is also on $\partial A(\bar{M}, D)$ if the dynamical system $d(\bar{M}, D)$ is in the $\varepsilon(M, D)$-ball neighborhood of $d(M, D)$.

Proof: See Proof of Theorem 10.8 in the Appendix.

Theorem 10.9: Global Invariant Property

Let $M_\lambda = \lambda M_1 + (1 - \lambda)M_2, \lambda \in [0, 1]$. If $d(M_\lambda, D)$ satisfies Assumptions (Cl) and (C2), then these two sets, $\partial A(M_1, D)$ and $\partial A(M_2, D)$, contain the same equilibrium points.

Proof: See Proof of Theorem 10.9 in the Appendix.

The results of this section suggest that we may view the relationship between the stability regions of two second-order dynamical systems as similar to that depicted in Figure 10.2. (replace $\partial A(D_1)$ by $\partial A(M_1, D)$ and $\partial A(D_2)$ by $\partial A(M_2, D)$). Under the conditions stated in Theorem 10.5, the stability boundaries of two second-order dynamical systems, $d(M_1, D)$ and $d(M_2, D)$, contain the same set of equilibrium points, $(x_1, 0), (x_2, 0), \ldots, (x_N, 0)$, on the stability boundary. The stability boundary of $d(M_1, D)$ is the union of the stable manifolds of these equilibrium points with respect to the dynamics of $d(M_1, D)$ (i.e., $\partial A(M_1, D) = \cup_{i=1}^{N} W^s_{(M_1, D)}(x_i, 0)$). The stability boundary of $d(M_2, D)$ is the union of the stable manifolds of the same set of equilibrium points with respect to the dynamics of $d(M_2, D)$ (i.e., $\partial A(M_2, D) = \cup_{i=1}^{N} W^s_{(M_2, D)}(x_i, 0)$).

10.6 RELATION BETWEEN THE ORIGINAL MODEL AND THE ARTIFICIAL MODEL

The purpose of this section is to establish the static relationship between the equilibrium points of the generalized gradient system $d(\bar{D})$ and the second-order dynamical system $d(M, D)$. In addition, the dynamic relationship between the stability boundary of the generalized gradient system $d(\bar{D})$ and that of the second-order dynamical system $d(M, D)$ will be established. We shall express the Jacobian matrix of system $d(\bar{D})$ at equilibrium point x as $J(\bar{D})|_{(x)}$. A similar notation is used for $J(M, D)|_{(x, 0)}$. In this section E denotes the set of the equilibrium points of system $d(D)$, and $\bar{E}$ denotes the set of the equilibrium points of system $d(M, D)$.

Theorem 10.10: Static Relationship

(i) (x) is an equilibrium point of $d(\bar{D})$ if and only if $(x, 0)$ is an equilibrium point of $d(M, D)$.

(ii) $n_s(J(d(\bar{D}))|_{(x)}) = n_s(J(d(M,D))|_{(x,0)}), n_c(J(d(\bar{D}))|_{(x)}) = n_c(J(d(M,D))|_{(x,0)}) = 0.$

Proof: It is obvious that $\bar{x}$ is an equilibrium point of $d(\bar{D})$ if and only if $f(\bar{x}) = 0$. From Equation 10.27, it follows that $(\bar{x}, 0)$ is an equilibrium point of $d(M, D)$ if and only if $f(\bar{x}) = 0$. This proves the first part of this theorem. From Theorem 10.1, we have

$$In\left(\begin{bmatrix} I & 0 \\ 0 & M^{-1} \end{bmatrix}\begin{bmatrix} 0 & I \\ -J(d(I))|_{(x)} & -D \end{bmatrix}\right) = In\left(\begin{bmatrix} I & 0 \\ 0 & I \end{bmatrix}\begin{bmatrix} 0 & I \\ -J(d(I))|_{(x)} & -I \end{bmatrix}\right). \tag{10.28}$$

Without loss of generality, we may consider the dynamical system $d(I, I)$ instead of $d(M, D)$. For convenience, we drop the subscript x or $(x, 0)$. Let λ be an eigenvalue of $J(I, I)$ and let (x_1, y_1) be the corresponding eigenvector, where $x_1 \in R^n$, $y_1 \in R^n$; that is,

$$y_1 = \lambda x_1$$
$$-J(d(I))x_1 - y_1 = \lambda y_1.$$

It follows that

$$-J(d(I))x_1 = (\lambda + \lambda^2)x_1, \tag{10.29}$$

which means x_1 is also an eigenvector of $-J(d(I))$.

Now, let u be the eigenvalue of $-J(d(I))$ whose corresponding eigenvector is x_1. Then,

$$u = \lambda + \lambda 2 \tag{10.30}$$

since u is a real number. Let λ_1, λ_2 be the solution of Equation 10.30. It follows that

$$\lambda_1 + \lambda_2 = -1$$
$$\lambda_1 \times \lambda_2 = -u.$$

Consider then the following four situations:

(i) If $u > 0$, then λ_1 and λ_2 are all real but have opposite signs.

(ii) If $-1/4 \leq u < 0$, then λ_1 and λ_2 are negative numbers.

(iii) If $u < -1/4$, then λ_1 and λ_2 are complex numbers but with negative real parts.

(iv) If λ_1 or λ_2 equals 0, then $u = 0$, which is a contradiction.

Conditions (i)–(iii) imply $n_s(-J(d(I))) = n_s(J(d(M, D)))$. Conditions (i)–(iv) imply $0 = n_c(-J(d(I))) = n_c(J(d(M, D)))$.

By applying Theorem 10.1, it follows that $J(d(\bar{D}))$ and $J(d(I))$ have the same equilibrium points and the same respective inertia. Thus, this part is also true. Hence, the proof is completed.

The proof of Theorem 10.10 shows that X_s is a stable equilibrium point of $d(\bar{D})$ if and only if $(X_s, 0)$ is a stable equilibrium point of $d(M, D)$. Hence, it makes sense to establish the relationship between the stability boundary of X_s and the stability boundary of $(X_s, 0)$.

Next, we establish the relationship between the stability boundaries of $d(\bar{D})$ and $d(\varepsilon I, D)$, where ε is a small positive number, $\bar{D}$ is the inverse matrix of D, and I is the identity matrix. The stability region of $d(\bar{D})$ is n-dimensional, while the stability region of $d(\varepsilon I, D)$ is $2n$-dimensional. For the purpose of comparison, the dynamical system $d(\bar{D})$ is viewed as sitting in the product space $R^n \times 0$, where $0 \in R^n$.

The following theorem is established by means of the singular perturbation technique. In singular perturbation terminology, the dynamical system $d(\bar{D})$, called the *degenerate system*, is obtained by setting $\varepsilon = 0$ in the perturbed system $d(\varepsilon I, D)$.

Theorem 10.11: Dynamic Relationship

Let $A(\varepsilon I, D)$ denote the stability region of a stable equilibrium point $(X_s, 0)$ of $d(\varepsilon I, D)$, and let $A(\bar{D})$ denote the stability region X_s of $d(\bar{D})$. Then, for small ε, $(X_i, 0) \in \{\partial A(\varepsilon I, D) \cap \bar{E}\}$ if and only if $X_i \in \{\partial A(\bar{D}) \cap E\}$.

Proof: Consider the general form of perturbed systems:

$$\begin{aligned} \dot{x} &= f(x, y, \varepsilon) \quad x(0) = x_0 \quad x \in R^n \\ \varepsilon\dot{y} &= g(x, y, \varepsilon) \quad y(0) = y_0 \quad y \in R^m \end{aligned} \tag{10.31}$$

if they satisfy the following conditions:

(a) The Jacobian matrix $[\partial g/\partial y]_{\varepsilon=0}$ is nonsingular for all $x, y \in U$, where U is the region under study.

(b) The eigenvalues of $[\partial g/\partial y]_{\varepsilon=0}$ all have negative real parts.

We then can invoke the singular perturbation technique to study the behavior of perturbed systems.

When condition (a) is satisfied, the implicit function theorem ensures that there is a smooth function $\bar{y} = \bar{y}(\bar{x})$ such that $g(\bar{x}, \bar{y}(\bar{x}), 0) = 0$. The degenerate system can be rewritten in a more compact form:

$$\dot{\bar{x}} = f(\bar{x}, \bar{y}(\bar{x}), 0) \quad \bar{x}(0) = x_0. \tag{10.32}$$

Condition (b) ensures that the trajectories of the perturbed system (Eq. 10.31) have limiting trajectories as $\varepsilon \to 0$.

Consider the perturbed system $d(\varepsilon I, D)$:

$$\begin{aligned} \dot{x} &= y \\ \varepsilon\dot{y} &= -Dy - f(x). \end{aligned} \tag{10.33}$$

Obviously, this system, $d(\varepsilon I, D)$, satisfies the above two conditions. The degenerate system, which is obtained by setting $\varepsilon = 0$, is

$$\begin{aligned} \dot{\bar{x}}(t) &= -D^{-1}f(\bar{x}) \quad \bar{x}(0) = x_0 \\ \bar{y}(t) &= -D^{-1}f(\bar{x}(t)) \quad \bar{x}(0) = x_0. \end{aligned} \tag{10.34}$$

The "boundary layer" system of Equation 10.31 can be obtained as follows:

(i) introducing the timescale variable $\tau = t/\varepsilon$ and the fast variables $\tilde{x}(\tau)$ and $\tilde{y}(\tau)$ to obtain the following:

$$\begin{aligned} x(t) &= \bar{x}(t) + \tilde{x}(\tau) \\ y(t) &= \bar{y}(t) + \tilde{y}(\tau), \text{ and} \end{aligned} \tag{10.35}$$

(ii) substituting Equation 10.35 into Equation 10.33 and setting $\varepsilon = 0$, we have the boundary layer system of Equation 10.33:

$$\frac{d\tilde{x}}{d\tau} = 0, \quad \text{with } \tilde{x}(0) = 0, \text{ and} \tag{10.36}$$

$$\frac{d\tilde{y}}{dx} = -D(\tilde{y}(\tau)), \quad \text{with } \tilde{y}(0) = y(0) + D^{-1}f(x_0). \tag{10.37}$$

It is obvious from Equation 10.37 that $\tilde{y}(\tau)$ converges to the equilibrium point $\tilde{y} = 0$ from any initial state $\tilde{y} = 0$. Consequently, the following two timescale approximations are valid (Hoppensteadt, 1974):

$$\begin{aligned} x(t) &= \tilde{x}(t) + O(\varepsilon) \\ y(t) &= \bar{y}(t) + \tilde{y}(\tau) + O(\varepsilon). \end{aligned} \tag{10.38}$$

Then, from Equations 10.34, 10.37, and 10.38, we have

$$y(t) = -D^{-1}f(\bar{x}(t)) + e^{-D\tau}(y_0 + D^{-1}f(x_0)) + O(\varepsilon). \tag{10.39}$$

Note that Equations 10.38 and 10.39 indicate that the trajectory $(x_\varepsilon(t), y_\varepsilon(t))$ of $d(\varepsilon I, D)$ starting from $(x_0, 0)$ will first quickly approach the equilibrium manifold $y = D^{-1}f(x)$ due to the fast variables $e^{-D\tau}(y_0 + D^{-1}f(x_0))$, and then will converge to the degenerated trajectory $(\bar{x}(t), -D^{-1}f(\bar{x}(t)))$.

Next, we make use of Equation 10.38 to prove this theorem. By contradiction, suppose that there exists a point

$$x_j \in \partial A(\bar{D}) \cap E, \text{but } (x_j, 0) \notin \partial A(\varepsilon I, D) \cap \bar{E}$$

for any small number ε. Since there exists an energy function,

$$V(x, y) = \frac{1}{2}\langle y, \varepsilon I y\rangle + \int_0^x \langle f(x_1), \mathrm{d}x_1\rangle$$

for the perturbed system $d(\varepsilon I, D)$, such that

$$\dot{V}(x, y) = -\langle y, Dy\rangle \le 0.$$

It follows from Theorem 5.1 that every trajectory of $d(\varepsilon I, D)$ converges to one equilibrium point or goes to infinity. Now, because $(x_j, 0) \notin \partial A(\varepsilon I, D) \cap \bar{E}$, by Theorem 5.1, there exists a neighborhood u of $(x_j, 0)$ such that every trajectory $(x_\varepsilon(t), y_\varepsilon(t))$ starting from $(x_0, y_0) \in u$ approaches equilibrium point $(\hat{x}, 0) \neq (x_s, 0)$ or goes to infinity. In particular, $x_\varepsilon(t) \to \hat{x} \neq x_s$ (or $|x_\varepsilon(t) \to \infty$) for any small number $\varepsilon > 0$. Applying Equation 10.38, we have $\bar{x}(t) \to (\hat{x} - O(\varepsilon))$ (or $|\bar{x}(t)| \to \infty$), where $\bar{x}(t)$ is the trajectory of $d(\bar{D})$ starting from a neighborhood $\bar{u}$ of x_j. However, by Theorem 5.1, this contradicts the assumption that $x_j \in \partial A(\bar{D}) \cap E$. Thus, $x_j \in \partial A(\bar{D}) \cap E$ implies that $(x_j, 0) \in \partial A(\varepsilon I, D) \cap \bar{E}$ for the small number ε. Similar arguments go in the reverse direction. Therefore, this proof is completed.

10.7 ANALYSIS OF THE PEBS METHOD

The PEBS method will be analyzed in this section using the analytical results derived in previous sections. A sufficient condition under which the PEBS method works will be derived. We consider a classical power system model consisting of n generators. Let the loads be modeled as constant impedances. Under the assumption that the transfer conductances are zero, the original system model is described as Equation 10.1.

The equations describing the prefault, fault-on, and postfault systems all have the same form as Equation 10.1 except that the Y_{ij} are different due to the change in network topology. Let (δ^s, ω^s) be a stable equilibrium point of Equation 10.1. The original system model (Eq. 10.1) as shown in Equation 10.4 can be expressed as

$$\dot{\delta}_i = \omega_i$$

$$M_i\dot{\omega}_i = -D_i\omega_i - \frac{\partial V_p(\delta)}{\partial \delta_i}, i = 1, 2, \ldots, n. \tag{10.40}$$

We shall assume that zero is a regular value of $\partial V_p(\delta)/\partial \delta$, which is a generic property. Note that (i) $M_i > 0$, $D_i > 0$; (ii) $-\partial V_p()/\partial \delta$ is a bounded vector field with bounded Jacobian; and (iii) $-\partial V_p(\delta)/\partial \delta$ satisfies the conditions (Eqs. 10.20 and 10.21) since it is a periodic vector in δ.

The original system model (Eq. 10.40) is therefore of the same form as the dynamical system (Eq. 10.27) with $(x, y) = (\delta, \omega)$ We shall use the notation $d_p(M, D)$ to denote the system (Eq. 10.40) and the notation $\partial V_p(M, D)$ to denote the stability boundary of a stable equilibrium point of $d_p(M, D)$.

Differentiating the function $V(\delta, \omega)$ along the trajectories of the system model (Eq. 10.40) gives

$$\dot{V}(\delta, \omega) = \frac{\partial V}{\partial \delta}\dot{\delta} + \frac{\partial V}{\partial \omega}\dot{\omega} = \sum_{i=1}^{n} D_i \omega_i^2 \leq 0. \tag{10.41}$$

The function $V(\delta, \omega)$ is therefore an energy function for the system $d_p(M, D)$. Since all of the equilibrium points of the system (Eq. 10.40) lie on the subspace

$$\{(\delta, \omega) | \delta \in R^n, \omega = 0\},$$

the energy function $V(\delta, \omega)$ at the equilibrium point (δ_e, ω_e) is of the following form:

$$V(\delta_e, \omega_e) = V_p(\delta_e).$$

This observation motivates us to study the stability region in the δ-subspace $\{(\delta, \omega)$: $\omega = 0\}$ instead of in the whole state space. In other words, instead of studying the original system (Eq. 10.40), we consider the following reduced-state system, which is a gradient system:

$$\dot{\delta} = \frac{\partial V_p(\delta)}{\partial \delta}. \tag{10.42}$$

We define a dynamical system,

$$d_p(D): \dot{\delta} + D\frac{\partial V_p(\delta)}{\partial \delta} = 0,$$

where D is a positive diagonal matrix. The system $d_p(D)$ becomes the gradient system (Eq. 10.42) when the matrix D is an identity matrix. Note that $d_p(D)$ is of the same form as the generalized gradient system $d(D)$ defined in Equation 10.19. Let $\partial A(\delta_s)$ denote the stability boundary of a stable equilibrium point δ_s of the system (Eq. 10.42) (or $d_p(I)$).

Next, we give a rigorous definition of PEBS.

Definition: PEBS
The stability boundary $\partial A(\delta_s)$ of the reduced-state system (Eq. 10.42) is the PEBS associated with the original system model (Eq. 10.40).

We next show that the PEBS defined above is consistent with the geometrical construction procedure suggested by Kakimoto et al. (1978).

Theorem 10.12: Geometry of PEBS

The PEBS intersects the level surface $\{\delta: V_p(\delta) = c\}$ orthogonally.

Proof: By Theorem 10.2, we have

$$\partial A(\delta_s) = \bigcup_i W^s(\delta_i),$$

where δ_i is the equilibrium point on the stability boundary $\partial A(\delta_s)$.

Since the gradient of $V_p(\delta)$ is a vector perpendicular to the level surface $\{\delta: V_p(\delta) = c\}$ in the increasing direction, it follows that the vector field at every regular point of the system (Eq. 10.42) is perpendicular to the level surface $\{\delta: V_p(\delta) = c\}$. Thus, Theorem 10.12 holds.

Remarks:

1. Theorem 10.12 implies that in the direction orthogonal to the PEBS, the potential energy function $V_p(\cdot)$ achieves a local maximum at the PEBS. Kakimoto et al. (1978) suggest an approximating scheme to find the PEBS on the basis of this theorem. This scheme identifies those points at which the directional derivative of V_p along the direction of the fault-on trajectory achieves a local maximum as the PEBS. Athay et al. (1984) go even further in suggesting a working "definition" of PEBS: they consider that all rays emanating from x_s and the local maxima of $V_p(\cdot)$ along these rays are joined together as the "PEBS."
2. The intuition behind the use of the potential energy maxima in direct stability analysis is that they represent the maximum energy (along the fault-on trajectory) that can be converted into potential energy. Hence, if the fault is cleared before reaching this point, all the kinetic energy can be converted into potential energy before the fault-on trajectory exits the stability region. One problem with this intuition is that the PEBS is the stability boundary of the reduced-state system, not the stability boundary of the original system. Hence, the fact that the projected fault-on trajectory exits the PEBS does not necessarily guarantee that the fault-on trajectory exits the stability boundary of the original system.

The PEBS method provides a local approximation of the relevant stability boundary of the original system (Eq. 10.40). We will elaborate on this point later. The application of the PEBS as a local approximation of the relevant stability boundary for a given fault-on trajectory proceeds in three steps:

Step 1. From the fault-on trajectory $(\delta(t), \omega(t))$ and the corresponding projected fault-on trajectory $\delta(t)$, detect the point δ^* at which the projected trajectory $\delta(t)$ intersects with the PEBS. Let the value of $V_p(\cdot)$ at δ^* be a critical energy.

Step 2. Use the connected constant energy surface of the set $\{(\delta, \omega): V(\delta, \omega) = V_p(\delta^*)\}$ containing the stable equilibrium point as the local approximation of the relevant stability boundary $\partial A(\delta_s, 0)$ for the original fault-on trajectory $(\delta(t), \omega(t))$.

Step 3. If the fault is cleared before reaching the PEBS, then the postfault trajectory is determined to be stable; otherwise, it is unstable.

Next, we evaluate how well the local approximation of the stability boundary that the PEBS method provides represents the relevant stability boundary of the original system. For this purpose, we compare the stability boundary of the original system, $d(M, D)$, and the stability boundary of the gradient system, $d(I)$

(i.e., PEBS). We first examine the relationship between the equilibrium points on the stability boundaries of $d(D)$ and that of $d(M, D)$.

The following result, which is a corollary of Theorem 10.12, reveals the relationship between the equilibrium points of the reduced-state system $d(D)$ and those of the original system $d(M, D)$.

Theorem 10.13: Static Relationship between *d*(*D*) and *d*(*M*, *D*)

[1] (δ) is an equilibrium point of $d(D)$ if and only if $(\delta, 0)$ is an equilibrium point of $d(M, D)$.

[2] (δ^s) is a stable equilibrium point of $d(D)$ if and only if $(\delta^s, 0)$ is a stable equilibrium point of $d(M, D)$.

[3] (δ) is a type-k equilibrium point of $d(D)$ if and only if $(\delta, 0)$ is a type-k equilibrium point of $d(M, D)$; that is,

$$n_s\left(J(d(D))|_{(\delta)}\right) = n_s\left(J(d(M, D))|_{(\delta,0)}\right)$$

$$n_c\left(J(d(D))|_{(\delta)}\right) = n_c\left(J(d(M, D))|_{(\delta,0)}\right) = 0.$$

The dynamic relationship between the stability boundary of the original system (Eq. 10.40) and PEBS can be obtained using the results developed in the previous sections. This dynamic relationship is summarized in Theorem 10.14 below. Let $M\lambda = \lambda M + (1 - \lambda)\varepsilon I$, where ε is a small positive number, and $\lambda \in [0, 1]$. Let $\bar{D}_\lambda = \lambda \bar{D} + (1 - \lambda) I$, where $\bar{D}$ is the inverse matrix of D, $\lambda \in [0, 1]$.

Theorem 10.14: Dynamic Relationship

If the dynamical systems $d(M_\lambda, D)$ and $d(\bar{D}_\lambda)$ satisfy Assumptions (Cl) and (C2), then

[1] the equilibrium points (δ_i) on the PEBS correspond to the equilibrium points $(\delta_i, 0)$ on $\partial A_p(M, D)$, and

[2] the stability boundary $\partial A_p(M, D)$ and the PEBS are completely characterized as the following:

$$\partial A(M, D) = \bigcup_{(\delta_i, 0) \in \partial A_p(M, D)} W^s_{(M, D)}(\delta_i, 0), \ \text{PEBS} = \bigcup_i W^s_{(I)}(\delta_i).$$

Proof: Part [1] follows from Theorems 10.4, 10.9, and 10.12. Part [2] follows from Theorem 10.4, Theorem 10.9, and part [1]. Hence, this theorem follows.

Theorem 10.14 establishes the dynamic relationship between the PEBS and the stability boundary of the original system. This dynamic relationship is important in the analysis of the PEBS method in approximating the relevant stability boundary of the original system. This theorem asserts that, if the one-parameter transversality condition is satisfied, then the equilibrium points on the PEBS correspond to the equilibrium points on the stability boundary of the original system (Eq. 10.40). Moreover, the PEBS is the union of the stable manifolds of the equilibrium points, say, (δ_i), $i = 1, 2, \ldots, n$, and the stability boundary of the original system is the union of the stable manifolds of the equilibrium points $(\delta_i, 0)$, $i = 1, 2, \ldots, n$ of the original system (Eq. 10.40). Recall that the intersection point between the fault-on trajectory and the stability boundary (of the postfault system) is called the *exit point* of the original model. An exit point of the original model must lie on the stable manifold of the controlling UEP. What we mean by the relevant stability boundary for a given fault-on trajectory is the stable manifold on which the exit point of the original model lies.

Since the PEBS is the union of stable manifolds of all the UEPs lying on the PEBS, a stable manifold in the PEBS, say, $W^s_{(I)}(\delta_i)$, and the intersection between the subspace $\{(\delta, \omega): \omega = 0\}$ and the stable manifold of the corresponding equilibrium point $W^s_{(M,D)}(\delta_i, 0)$ of the original system (Eq. 10.40) are different (see Figure 10.3). These two objects, that is, a stable manifold in the PEBS and the intersection, are indeed different because of the above theorem and the property that the stable manifold depends continuously on the underlying vector field. Therefore, unlike the one-machine infinite bus case, the PEBS is different from the intersection of the stability boundary of the original model (Eq. 10.40) with the subspace $\{(\delta, \omega): \omega = 0\}$

We are now in a position to evaluate the accuracy of the PEBS method, particularly the conditions under which the PEBS method gives a good approximation of the relevant stability boundary of the original system (Eq. 10.40) for a given fault-on trajectory. Given a sustained fault-on trajectory, $(\delta(t), \omega(t))$, the corresponding projected fault-on trajectory, $\delta(t)$, is the projection of the fault-on trajectory $(\delta(t), \omega(t))$ onto the δ-space of the reduced-state system (Eq. 10.42) (see Figure 10.4).

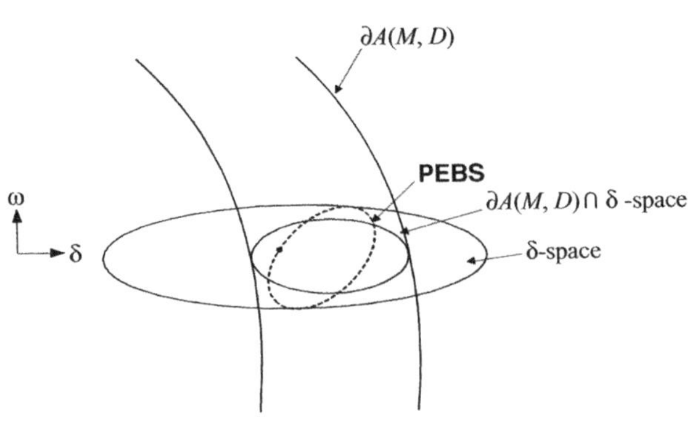

Figure 10.3 Illustration of the relationship between the PEBS and the stability boundary of the original system. Note that the PEBS is different from the intersection between the stability boundary of the original system and the subspace $\{(\delta, \omega): \omega = 0\}$.

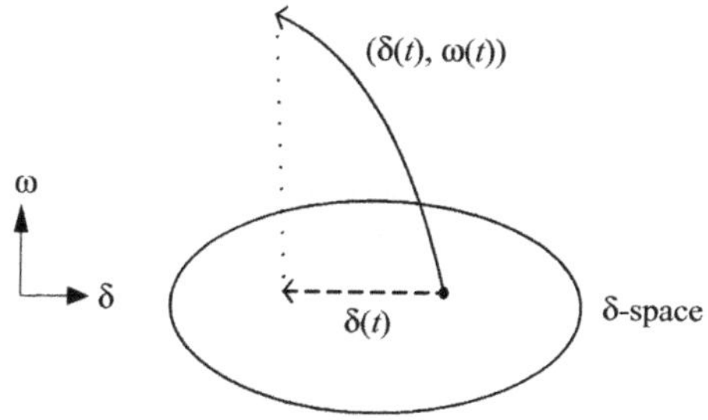

Figure 10.4 The fault-on trajectory (δ(t), ω(t)) of the original model and the projected fault-on trajectory δ(t) on the reduced-state model.

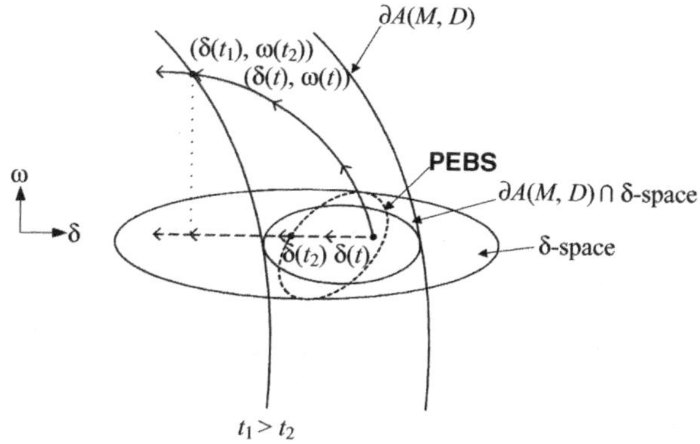

Figure 10.5 $(\delta(t_1), \omega(t_1))$ is the exit point of the fault-on trajectory $(\delta(t), \omega(t))$, which lies on the stability boundary of the original system. $\delta(t_2)$ is the exit point of the projected fault-on trajectory $\delta(t)$, which is the intersection between the PEBS and the projected fault-on trajectory $\delta(t)$. The projection of $(\delta(t_1), \omega(t_1))$ onto the subspace $\{(\delta, \omega): \omega = 0\}$ is usually different from $\delta(t_2)$.

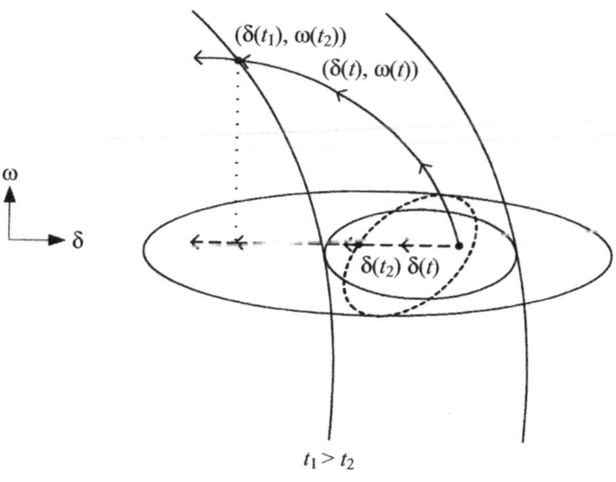

Figure 10.6 Illustration of the relationship between the PEBS and the stability boundary of the original system.

Let $(\delta(t_1), \omega(t_1))$ denote the exit point of the fault-on trajectory $(\delta(t), \omega(t))$, which is the intersection between the stability boundary of the original system and the fault-on trajectory $(\delta(t), \omega(t))$. Let $\delta(t_2)$ denote the exit point of the projected fault-on trajectory $\delta(t)$, which is the intersection between the PEBS and the projected fault-on trajectory $\delta(t)$ (see Figures 10.5 and 10.6). These two time instants, t_1 and t_2, are generally not equal.

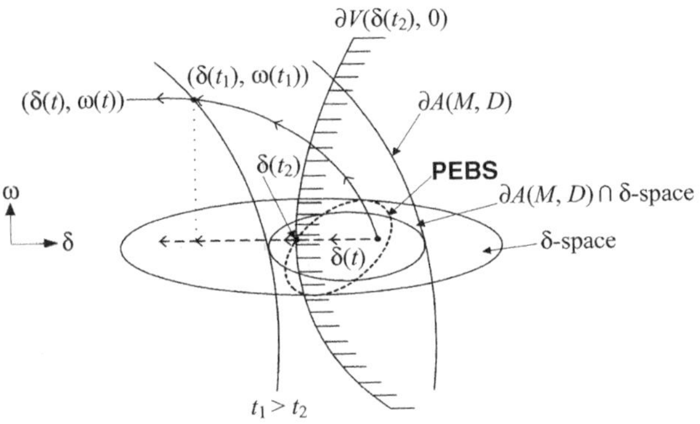

Figure 10.7 The PEBS method uses the connected constant energy surface passing through point $(\delta(t_2), 0)$; that is, (δ, ω): $V(\delta, \omega) = V_p(\delta(t_2))$ containing the stable equilibrium point $(\delta_s, 0)$ as a local approximation for the relevant stability boundary $\partial A(\delta_s, 0)$ of the original fault-on trajectory $(\delta(t), \omega(t))$.

We next give an analytical explanation of the PEBS method. Let $\partial V(x)$ denote the connected constant energy surface of $V(\cdot)$ containing the stable equilibrium point $(\delta_s, 0)$ passing through the point $x = (\delta, \omega)$; that is, $\partial V(x)$: = the connected component of the set $\{(\delta, \omega): V(\delta, \omega) = V(x)\}$ containing the stable equilibrium point $(\delta_s, 0)$. The PEBS method proceeds as follows: use the connected constant energy surface passing through the point $(\delta(t_2), 0)$; that s, (δ, ω): $V(\delta, \omega) = V_p(\delta(t_2))$ containing the stable equilibrium point $(\delta_s, 0)$ as a local approximation for the relevant stability boundary $\partial A(\delta_s, 0)$ (see Figure 10.7). Here, the state vector $\delta(t_2)$ of the exit point lies on the stability boundary of the reduced-state model.

To examine how well the PEBS method is in approximating the relevant (postfault) stability boundary, we compare how well the constant energy surface $\partial V(\delta(t_2), 0)$ is in approximating the relevant (postfault) stability boundary. To this end, we will develop analytical results to examine this approximation. We use the notation $V_c(x)$ to denote the connected component of the set $\{(\delta, \omega): V(\delta, \omega) < V(x)\}$ containing the stable equilibrium point $(\delta_s, 0)$.

Theorem 10.15

Let $(\delta, 0)$ be an equilibrium point on the stability boundary $\partial A(\delta_s, 0)$ of the system (Eq. 10.40). Then,

[1] The connected constant energy surface $\partial V(\hat{\delta}, 0)$ intersects with the stable manifold $W^s(\hat{\delta}, 0)$ only at point $(\hat{\delta}, 0)$; moreover, the set $V_c(\hat{\delta}, 0)$ has an empty intersection with the stable manifold $W^s(\hat{\delta}, 0)$.

[2] Suppose $(\bar{\delta}, \bar{\omega}) \in W^s(\hat{\delta}, 0)$, $(\bar{\delta}, \bar{\omega}) \neq (\hat{\delta}, 0)$. Then, the set $V_c(\bar{\delta}, \bar{\omega})$ has a nonempty intersection with the stable manifold $W^s(\hat{\delta}, 0)$.

Proof: Since the function $V(.)$ is decreasing along the trajectories of the original system (Eq. 10.40), the point with the minimum value of $V(.)$ in the stable manifold $W^s(\hat{\delta}, 0)$ must be the equilibrium point $(\hat{\delta}, 0)$ itself. Hence, part [1] follows. Because $(\delta_1, \omega_1) \in W^s(\hat{\delta}, 0)$ and $(\bar{\delta}, \bar{\omega}) \neq (\hat{\delta}, 0)$, it follows that $V(\bar{\delta}, \bar{\omega}) > V(\hat{\delta}, 0)$. The set $V_c(\hat{\delta}, 0)$ therefore contains points that belong to the stable manifold $W^s(\hat{\delta}, 0)$, Furthermore, $V(\bar{\delta}, \bar{\omega})$ is a regular value of the function $V(.)$. The preimage theorem states that if y is a regular value of f: $X \to Y$, then the preimage $f^{-1}(y)$ is a submanifold of X and $\dim f^{-1}(y) = \dim X - \dim Y$ (Guillemin and Pollack, 1974, p. 21). The dimension of $\partial V(\bar{\delta}, \bar{\omega})$ is therefore $2n - 1$. Now, because the dimension of the stable manifold $W^s(\hat{\delta}, 0)$ is $2n - k$, according to the intersection theorem, the dimension of the intersection is $2n - k - 1$, and part [2] is true (Guillemin and Pollack, 1974, p. 30).

We next apply Theorem 10.15 to examine the PEBS method. Part [1] of Theorem 10.4 asserts that for any fault-on trajectory $x_f(t)$ starting from a point p with $p \in A(\delta_s, 0)$ and $V(p) < V(\hat{\delta}, 0)$, if the exit point of this trajectory $x_f(t)$ lies on the set $W^s(\hat{\delta}, 0)$, then the fault-on trajectory $x_f(t)$ must pass through the connected constant energy surface $\partial V(\hat{\delta}, 0)$ before it passes through the stable manifold $W^s(\hat{\delta}, 0)$ (thus leaving the stability boundary $\partial A(\delta_s, 0)$). This suggests that the connected constant energy surface $\partial V(\hat{\delta}, 0)$ can be used to approximate part of the stability boundary $\partial A(\delta_s, 0)$ (i.e., the stable manifold $W^s(\hat{\delta}, 0)$ part) for the fault-on trajectory $x_f(t)$. In fact, this is the essence of the controlling UEP method (see Figure 10.8).

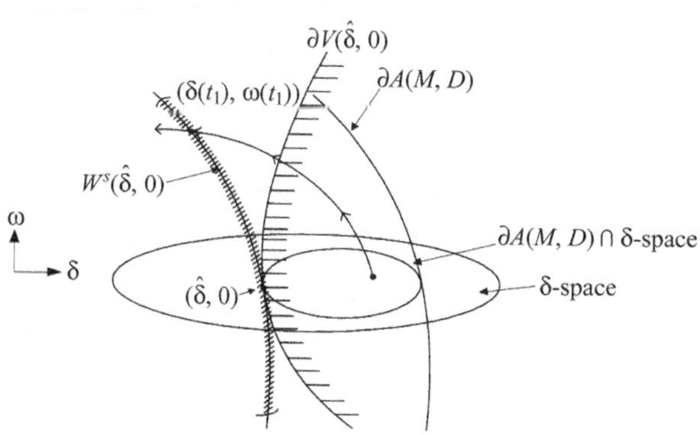

Figure 10.8 The connected constant energy surface $\partial V(\hat{\delta}, 0)$ can be used to approximate part of the stability boundary $\partial A(\delta_s, 0)$ (i.e., the stable manifold $W^s(\hat{\delta}, 0)$ part) for the fault-on trajectory $x_f(t)$ if the exit point of this trajectory $x_f(t)$ lies on the set $W^s(\hat{\delta}, 0)$. Under this situation, the fault-on trajectory $x_f(t)$ must pass through the connected constant energy surface $\partial V(\hat{\delta}, 0)$ before it passes through the stable manifold $W^s(\hat{\delta}, 0)$.

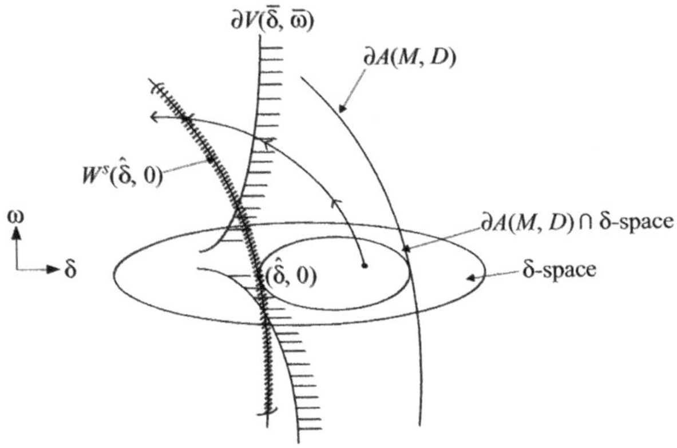

Figure 10.9 The fault-on trajectory $x_f(t)$ may pass through the connected constant energy surface $\partial V(\bar{\delta}, \bar{\omega})$ after it passes through the stable manifold $W^s(\hat{\delta}, 0)$. It is therefore incorrect to approximate the relevant part of the stability boundary (i.e., the stable manifold $W^s(\hat{\delta}, 0)$) by using the connected constant energy surface $\partial V(\bar{\delta}, \bar{\omega})$.

Part [2] of Theorem 10.15 asserts that the fault-on trajectory $x_f(t)$ may pass through the connected constant energy surface $\partial V(\bar{\delta}, \bar{\omega})$ after it passes through the stable manifold $W^s(\hat{\delta}, 0)$. It is therefore incorrect to approximate the relevant part of stability boundary (i.e., the stable manifold $W^s(\hat{\delta}, 0)$) by using the connected constant energy surface $\partial V(\bar{\delta}, \bar{\omega})$ (see Figure 10.9). It is clear that part of the connected constant energy surface $\partial V(\bar{\delta}, \bar{\omega})$ may lie outside the stability boundary $\partial A(\delta_s, 0)$. Hence, using the constant energy surface passing through the exit point of the reduced-state system, the PEBS method gives either overestimated stability assessments or underestimated stability assessments. Moreover, if the equilibrium point $(\hat{\delta}, 0)$ is of type-k, then the dimension of the set, which is the intersection between the connected constant energy surface $\partial V(\bar{\delta}, \bar{\omega})$ and the stable manifold $W^s(\hat{\delta}, 0)$, is $2n - k - 1$.

In summary, unlike the controlling UEP method, the PEBS method cannot consistently give conservative stability assessments. It gives either overestimated or underestimated (i.e., conservative) stability assessments for power system transient stability analysis. To elaborate on this point, we use the notation $\delta(t_2)$ to denote the exit point of the projected fault-on trajectory on the PEBS and the notation $(\hat{\delta}, 0)$ to denote the exit point of the fault-on trajectory. The essence of the PEBS method is to use the connected constant energy surface passing through the state $(\delta(t_2), 0)$; that is, $\{(\delta, \omega): V(\delta, \omega) = V_p(\delta(t_2))\}$ containing the stable equilibrium point $(\delta_s, 0)$ as a local approximation for the relevant stability boundary of the original fault-on trajectory $(\delta(t), \omega(t))$. The PEBS method gives conservative stability assessments if the fault-on trajectory $x_f(t)$ passes through the surface $\partial V(\delta(t_2), 0)$ before it passes through the stable manifold $W^s(\hat{\delta}, 0)$. Under this scenario, the PEBS method still provides a good approximation of the relevant local stability boundary for the fault-on trajectory $x_f(t)$ (see Figure 10.10).

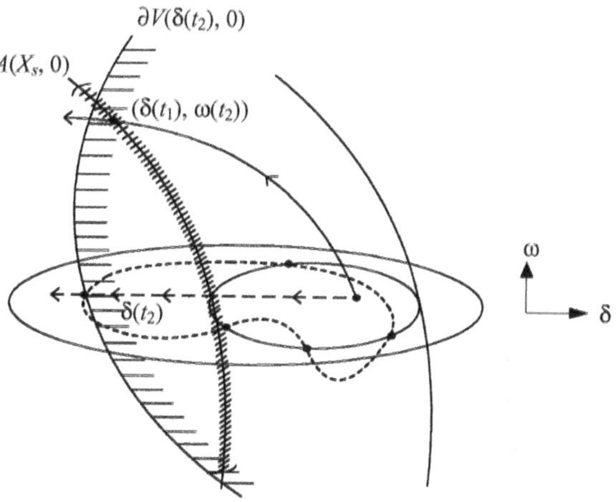

Figure 10.10 (1) The connected constant energy surface $\partial V(\bar{\delta},0)$ can be used to approximate part of the stability boundary $\partial A(\delta_s, 0)$ (i.e., the stable manifold $W^s(\bar{\delta},0)$ part). (2) It is inappropriate to approximate the relevant part of stability boundary (i.e., the stable manifold $W^s(\hat{\delta},0)$ part) by the connected constant energy surface $\partial V(\delta_1, \omega_1)$. Part of the connected constant energy surface $\partial V(\delta_1, \omega_1)$ clearly lies outside the stability boundary $\partial A(\delta_s, 0)$.

On the other hand, the PEBS method fails in the sense that it classifies unstable contingencies as stable if the fault-on trajectory $x_f(t)$ passes through the surface $\partial V(\delta(t_2), 0)$ after it passes through the stable manifold $W^s(\hat{\delta},0)$. Note that in this case, the fault-on trajectory, $x_f(t)$, after passing through the stable manifold, $W^s(\hat{\delta},0)$, becomes unstable for the postfault system but is still classified as stable by the PEBS method as long as this fault-on trajectory does not pass through the surface $\partial V(\delta_1, \omega_1)$ (see Figure 10.11). Therefore, based on the above analytical results, it follows that if the following condition is satisfied, then the PEBS method gives a good approximation of the relevant local stability boundary of the original system for the given fault-on trajectory $(\delta(t), \omega(t))$.

Sufficient Condition: The fault-on trajectory $(\delta(t), \omega(t))$ passes through the constant energy surface $\partial V(\delta(t_2), 0)$ before it passes through the stable manifold $\partial V(\delta(t_1), \omega(t_1))$.

Remarks:

1. The above sufficient condition is equivalent to the condition that $V(\delta(t_1), \omega(t_1)) > V(\delta(t_2), 0)$; consequently, the PEBS method gives conservative stability assessments in the following sense:
   - It may classify stable contingencies as unstable ones.
   - It classifies unstable contingencies as unstable ones.
2. If the sufficient condition is satisfied and $V(\delta(t_1), \omega(t_1)) > V(\delta(t_2), 0) > V(CUEP)$, then the PEBS method gives conservative but accurate stability assessments. Under this condition, the PEBS method gives better performance than the controlling UEP method in the following sense: the ratio of

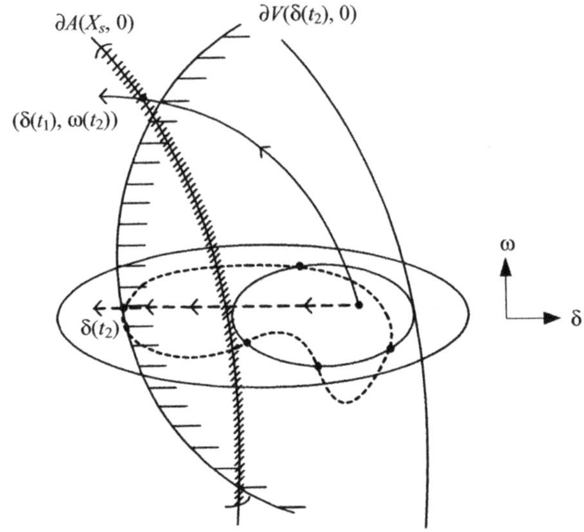

Figure 10.11 (a) The fault-on trajectory $x_f(t)$ passes through the surface $\partial V(\delta_1, \omega_1)$ before it passes through the stable manifold $W^s(\hat{\delta}, 0)$. In this case, the PEBS method still provides good approximation to the relevant local stability boundary for this fault-on trajectory $x_f(t)$. (b) The fault-on trajectory $x_f(t)$ passes through the surface $\partial V(\delta_1, \omega_1)$ after it passes through the stable manifold $W^s(\hat{\delta}, 0)$. In this case, the PEBS $W^s(\hat{\delta}, 0)$ method may fail to approximate the relevant local stability boundary for the fault-on trajectory $x_f(t)$.

classifying stable contingencies into unstable ones made by the PEBS method will be less than that of the controlling UEP method.

3. If the sufficient condition is satisfied and $V(\delta(t_1), \omega(t_1)) > V(CUEP) > V(\delta(t_2), 0)$, then the PEBS method gives more conservative stability assessments than the controlling UEP method. The ratio of classifying stable contingencies into unstable ones made by the PEBS method will be larger than the controlling UEP method.
4. If the sufficient condition is not satisfied, that is, $V(\delta(t_2), 0) > V(\delta(t_1), \omega(t_1))$, then the PEBS method gives overestimated stability assessments in the following sense:
   - It may classify unstable contingencies as stable ones.
   - It may classify unstable contingencies as unstable ones.
   - It classifies stable contingencies as stable ones.
5. The condition $\dot{V}(\delta(t_2), 0) > V(CUEP)$ is satisfied under the following sufficient condition: the exit point $\delta(t_2)$ lies on the stable manifold of $(\hat{\delta})$ implies that the exit point $(\delta(t_1), \omega(t_1))$ of the original model lies on the stable manifold of $(\hat{\delta}, 0)$. This sufficient condition contains the following two subconditions:

(ia) The equilibrium point $(\hat{\delta})$ is on the PEBS if and only if the equilibrium point $(\hat{\delta}, 0)$ is on the stability boundary of the original system (Eq. 10.40).

(ib) The exit point $\delta(t_2)$ is on the stable manifold $W^s_{(I)}(\hat{\delta})$ if and only if $(\delta(t_1), \omega(t_1))$ is on the stable manifold $W^s_{(M,D)}(\hat{\delta}, 0)$.

In Theorem 10.2, we have laid down the foundation for subcondition (ia) to be true.

6. In applying the PEBS method, it is not practical to check whether condition (ib) above is satisfied or not. This is mainly because finding the exit point $(\delta(t_1), \omega(t_1))$ of the original model for the fault-on trajectory $(\delta(t), \omega(t))$ is computationally demanding. It usually takes five to six iterations of time–domain simulation to identify the exit point of the original trajectory.

10.8 CONCLUDING REMARKS

The PEBS method proposed by Kakimoto et al. was based on heuristic arguments. In this chapter, we have presented a detailed analysis of the PEBS method and have developed a theoretical foundation for this method. Several sufficient conditions under which the PEBS method works for direct stability analysis have been derived. Specifically, sufficient conditions under which the PEBS method provides a good local approximation of the relevant stability boundary have been derived. The following static and dynamic relationships between the original system and the reduced-state system have also been derived:

Static Relationship

- The potential energy $V_P(\delta)$ of the original model (Eq. 10.40) is an energy function of the reduced-state model (Eq. 10.42).
- $(\delta_s, 0)$ is a stable equilibrium point of the original model (Eq. 10.40) if and only if (δ_s) is a stable equilibrium point of the reduced-state model (Eq. 10.42).
- $(\delta_s, 0)$ is a type-k equilibrium point of the original model (Eq. 10.40) if and only if (δ_s) is a type-k equilibrium point of the reduced-state model (Eq. 10.42).
- The energy function value at an equilibrium point of the original model (Eq. 10.40) is the same as that of the corresponding equilibrium point of the reduced-state model (Eq. 10.42), in other words, at the equilibrium points $V(\delta_s, 0) = V_P(\delta_s)$.

Dynamic Relationship

- The stability boundary of the original model (Eq. 10.40) equals the union of the stable manifold of the equilibrium points on the stability boundary.

- The stability boundary of the reduced-state model (Eq. 10.42) equals the union of the stable manifold of the equilibrium points on the stability boundary.
- Under certain conditions, the equilibrium points $(\delta_i, 0)$ on the stability boundary of $(\delta_s, 0)$ of the original model (Eq. 10.40) coincide with the equilibrium points (δ_i) on the stability boundary of (δ_s) of the reduced-state model (Eq. 10.42).

This comprehensive analysis is general and the approach employed in this chapter has other applications as will be shown in later chapters. This theoretical foundation also paves the development of the BCU method for direct transient stability analysis. This development will be described in Chapters 15 and 17.

Chapter 11

Controlling UEP Method: Theory

11.1 INTRODUCTION

Several methods are available for determining critical energy values for direct stability analysis. The classical closest unstable equilibrium point (UEP) method has been found to yield unduly conservative results when applied to power system transient stability analysis. The potential energy boundary surface (PEBS) method gives fairly fast stability assessments, but it can be inaccurate (i.e., giving overestimated and unduly underestimated stability assessments). A desirable method for determining the critical energy value is one that can provide the most accurate approximation of the relevant stability boundary towards which the fault-on trajectory is heading. This is the spirit of the controlling UEP method (Chiang, 1991; Chiang et al., 1987).

It is now well recognized that among several methods for determining the critical energy value, the controlling UEP method is the most viable for direct stability analysis of practical power systems. The controlling UEP method uses the constant energy surface passing through the controlling UEP to approximate the relevant part of the stability boundary towards which the fault-on trajectory is heading. If, when the fault is cleared, the system state lies inside the energy surface passing through the controlling UEP, then the postfault system is stable (i.e., the postfault trajectory will settle down to a stable operating point); otherwise, the postfault system may be unstable. The success of the controlling UEP method, however, hinges upon its ability to find the correct controlling UEP.

This chapter presents a rigorous introduction to the controlling UEP method and its theoretical foundations. In particular, we discuss the concept of a controlling UEP, present the controlling UEP method, develop a theoretical basis for the controlling UEP method, and derive dynamic and geometric characterizations of the controlling UEP. These characterizations are useful for the development of solution methodologies for computing controlling UEPs.

Direct Methods for Stability Analysis of Electric Power Systems, by Hsiao-Dong Chiang

11.2 THE CONTROLLING UEP

The concept of the controlling UEP can be traced back to the mid-1970s. Prabhakara and El-Abiad (1975) argue that the controlling UEP is the UEP that is closest to the fault-on trajectory. Athay et al. (1979) suggest that the controlling UEP is the UEP "in the direction" of the fault-on trajectory. Another viewpoint of the controlling UEP rests on physical arguments. Ribbens-Pavella and Lemal (1976) relate the controlling UEP to the machine or groups of machines that first go out of synchronism if the fault is sustained. Fouad and Vittal (1991) associate the controlling UEP with the "mode of instability" of machines. These concepts have been explored to develop numerical methods for computing the controlling UEP. These methods, however, suffer from several problems: (1) the UEPs obtained by these methods are not the controlling UEPs; (2) these methods give both overestimated and very conservative stability assessments; and (3) these methods are heuristic and have no theoretical foundations.

Before we further explain the controlling UEP method, the following concept of an exit point is useful (see Figure 11.1). Let X_s^{pre} be a prefault stable equilibrium point (SEP) and $X_f(t)$ be the corresponding fault-on trajectory. Let $\partial A(X_s)$ denote the stability region of the postfault SEP X_s. Then, the exit point is a relationship between the fault-on trajectory and the postfault system.

Definition: Exit Point
The point at which a sustained fault-on trajectory intersects with the stability boundary of the postfault SEP is called the *exit point* of the fault-on trajectory (relative to the postfault system). In addition, the fault-on trajectory exits the stability region after the exit point.

Another equivalent definition of exit point is as follows.

Definition: Exit Point
The point at which a (sustained) fault-on trajectory intersects with the quasi-stability boundary of the postfault SEP is called the *exit point* of the fault-on trajectory (relative to the postfault system).

We next present a formal definition of the controlling UEP

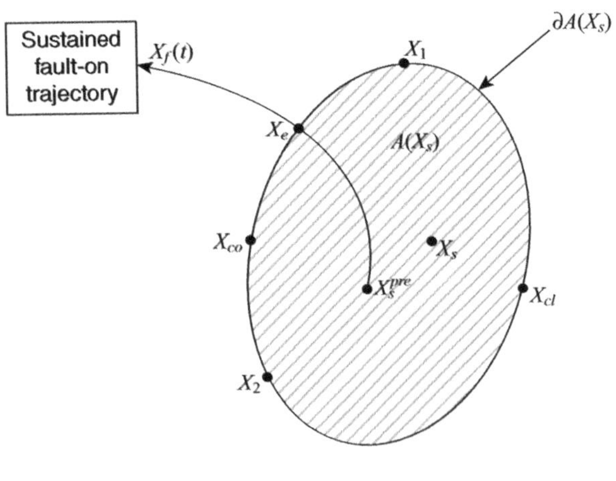

Figure 11.1 The sustained fault-on trajectory $x_f(t)$ moves towards the stability boundary $\partial A(x_s)$ and intersects it at the exit point, x_e. The exit point lies on the stable manifold of the controlling UEP.

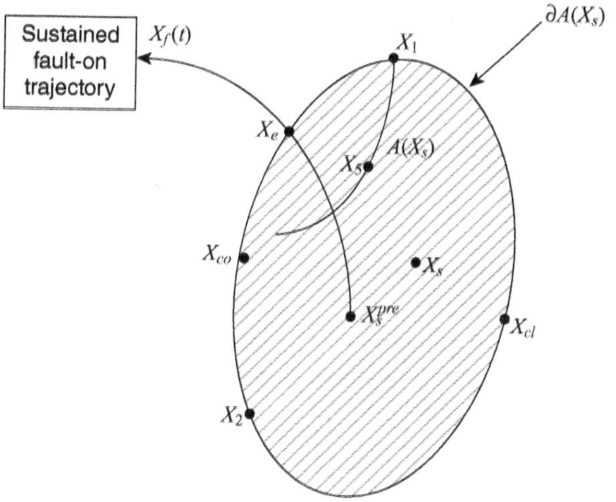

Figure 11.2 There are five UEPs (X_{co}, X_1, X_2, X_{cl}, and X_5) lying on the stability boundary, while there are four UEPs (X_{co}, X_1, X_2, and X_{cl}) lying on the quasi-stability boundary. By definition, the controlling UEP must lie on the quasi-stability boundary instead of just on the stability boundary so that the fault-on trajectory leaves the stability region after the exit point. The UEP x_5 is not the controlling UEP because it does not lie on the quasi-stability boundary.

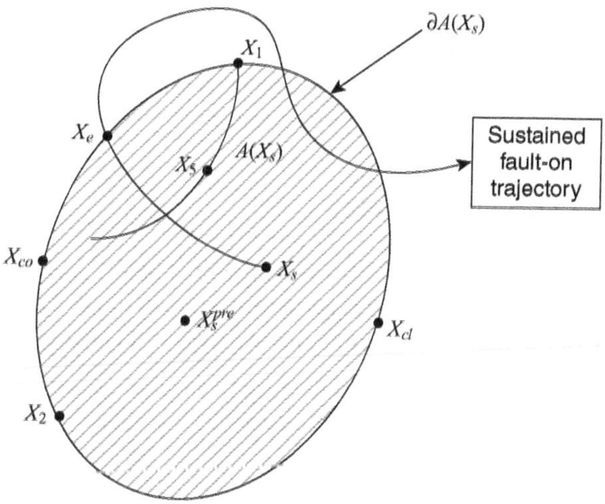

Figure 11.3 The fault-on trajectory may intersect the quasi-stability boundary several times; by definition, the controlling UEP is always the first UEP whose stable manifold intersects the fault-on trajectory $X_f(t)$ at the exit point.

Definition: Controlling UEP

The *controlling UEP* of a fault-on trajectory, $X_f(t)$, is the UEP whose stable manifold contains the exit point of $X_f(t)$ (i.e., the controlling UEP is the first UEP whose stable manifold is hit by the fault-on trajectory $X_f(t)$ at the exit point).

The controlling UEP must lie on the quasi-stability boundary, instead of just on the stability boundary. The fault-on trajectory leaves the stability region after the exit point. Hence, this definition of the controlling UEP does not apply to the UEP that lies inside the interior of the quasi-stability region as shown in Figure 11.2. In addition, from a mathematical viewpoint, the fault-on trajectory may have several intersections with the quasi-stability boundary; the controlling UEP is the first UEP whose stable manifold intersects the fault-on trajectory $X_f(t)$ at the exit point (see Figure 11.3). The issue of existence and uniqueness of the controlling UEP defined above will be investigated in the next section.

11.3 EXISTENCE AND UNIQUENESS

The issues of existence and uniqueness of the controlling UEP with respect to a fault-on trajectory and a postfault system are addressed in this section. We first present a complete characterization of the qausi-stability boundary, which is useful in explaining the controlling UEP.

Theorem 11.1: Postfault Stability Boundary

If a general postfault system with a SEP X_s has an energy function $V(\cdot)$: $R^n \to R$, then the quasi-stability boundary $\partial A_p(x_s)$ is contained in the union of the stable manifolds of the UEPs on the boundary; that is,

$$\partial A_p(x_s) \subseteq \bigcup_{x_i \in \{E \cap \partial A(x_s)\}} W^s(x_i).$$

Since the quasi-stability boundary of the SEP of a postfault system is contained in the set, which is the union of the stable manifolds of the UEPs on the boundary, the exit point must lie on the stable manifold of a UEP on the quasi-stability boundary. This UEP is the controlling UEP relative to the fault-on trajectory.

Theorem 11.2: Existence and Uniqueness

Given a prefault SEP, a fault-on system, and a postfault system with a SEP X_s, which has an energy function $V(\cdot)$: $R^n \to R$, let the stability region of X_s contain the prefault SEP. Then, the controlling UEP of the fault-on trajectory always exists and is unique.

Proof: The existence of the exit point of a fault-on trajectory is ensured as long as the energy function value increases along the fault-on trajectory. This proof is built on the following facts.

Fact A: A sustained fault-on trajectory must exit the stability boundary of a postfault system.

Fact B: The exit point of the fault-on trajectory must lie on the stable manifold of a UEP on the stability boundary of the postfault system.

Fact A is a consequence of the following two conditions: (1) the fundamental assumption of direct methods that the prefault SEP lies inside the stability region of the postfault SEP and (2) the energy value increases along a fault-on trajectory. Fact B is a consequence of Theorem 11.1. Combining Facts A and B, we complete this proof.

From an engineering viewpoint, the quasi-stability region is practically the stability region. Like the stability region and the stability boundary, the quasi-stability region is an open, invariant set that is diffeomorphic to $\Re^n$, and the quasi-stability boundary is a closed invariant set of dimension $n-1$. The key difference in the characterization of the UEP lying on the stability boundary and that on the quasi-stability boundary is described by Theorem 4.8. Hence, strictly speaking, the controlling UEP lies on the quasi-stability boundary; however, from a practical viewpoint, it can be stated that the controlling UEP lies on the stability boundary of a postfault SEP. Hence, in the rest of this book, the controlling UEP is defined as the first UEP whose stable manifold intersects the fault-on trajectory $X_f(t)$. This proof is completed.

11.4 THE CONTROLLING UEP METHOD

We now formalize the controlling UEP method. We start with an example. Let us observe Figure 11.4 depicting a fault-on trajectory, $X_f(t)$, which moves towards the stability boundary $\partial A(x_s)$ and intersects it at the exit point x_e. The critical clearing time (CCT) is the time difference between the prefault SEP and the exit point. If the fault is cleared before the fault-on trajectory reaches the exit point, say, at $x(t_{cl})$, then the fault clearing point must lie inside the stability region of the postfault SEP. Hence, the postfault trajectory starting from the fault clearing point must converge to the postfault SEP x_s and the postfault system is stable (see Figure 11.4).

Now the issue is determining whether or not the fault clearing point $x(t_{cl})$ lies inside the stability region of the postfault SEP without performing a time–domain simulation or without knowledge of exit point. The controlling UEP method addresses this issue as follows. Instead of computing the exit point, the controlling UEP method approximates the relevant stability boundary by the constant energy surface passing through the controlling UEP. The method replaces the task of detecting the exit point with the task of detecting the intersection between the fault-on trajectory

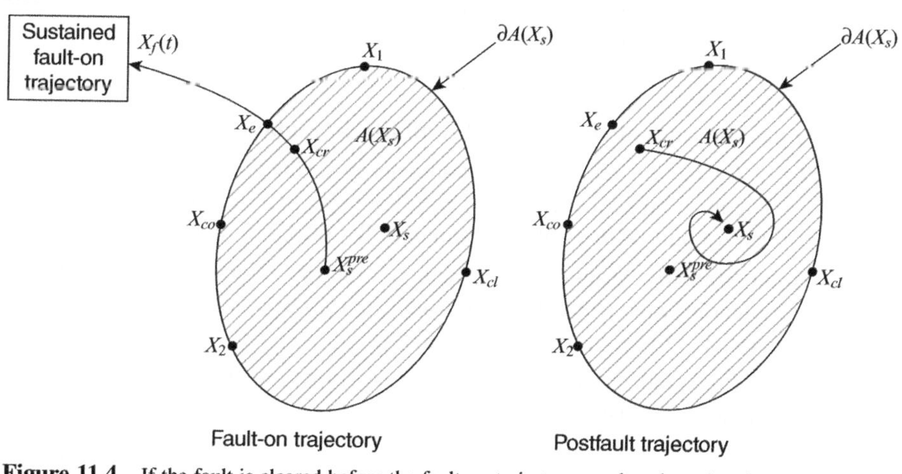

Figure 11.4 If the fault is cleared before the fault-on trajectory reaches the exit point, say, at $x(t_{cl})$, then the fault clearing point must lie inside the stability region of the postfault SEP. Hence, the postfault trajectory starting from the fault clearing point must converge to the postfault SEP x_s.

and the constant energy surface passing through the controlling UEP. The latter task only involves a comparison between the energy at the initial state of the post-fault trajectory and the energy at the controlling UEP and is much easier than the former task.

The controlling UEP method for the direct stability analysis of large-scale power systems proceeds as follows:

1. **Determination of the Critical Energy**

   Step 1.1. Find the controlling UEP, X_{co}, for a given fault-on trajectory, $X_f(_t)$.

   Step 1.2. The critical energy, v_{cr}, is the value of the energy function $V(\cdot)$ at the controlling UEP; that is,

$$v_{cr} = V(X_{co}). \tag{11.1}$$

2. **Direct Stability Assessment**

   Step 2.1. Calculate the value of the energy function $V(\cdot)$ at the time of fault clearance (say, at time t_{cl}):

$$vf = V(Xf(tcl)). \tag{11.2}$$

   Step 2.2. If $v_f < v_{cr}$, then the postfault system is stable. Otherwise, it may be unstable.

In fact, this method is computationally equivalent to the following method, and it is from this viewpoint that we build a framework to analyze the controlling UEP method.

Another Viewpoint of the Controlling UEP Method

1. **Determination of the Critical Energy**

   Step 1.1. Find the controlling UEP, X_{co}, for a given fault-on trajectory, $X_f(t)$.

   Step 1.2. The critical energy, v_{cr}, is the value of the energy function, $V(\cdot)$, at the controlling UEP; that is,

$$v_{cr} = V(X_{co}). \tag{11.3}$$

2. **Approximation of the Relevant Stability Boundary**

   Step 2.1. Use the connected constant energy surface of $V(\cdot)$ passing through the controlling UEP X_{co} and containing the SEP X_s to approximate the relevant part of the stability boundary for the fault-on trajectory $X_f(t)$.

3. **Direct Stability Assessment.** Check whether the fault-on trajectory at the fault clearing time (t_{cl}) is located inside the stability boundary characterized in Step 2.1. This is carried out as follows:

   Step 3.1. Calculate the value of the energy function $V(\cdot)$ at the time of fault clearance (t_{cl}) using the fault-on trajectory

$$v_f = V(X_f(t_{cl})). \tag{11.4}$$

   Step 3.2. If $v_f < v_{cr}$, then the point $X_{f(cl)}$ is located inside the stability boundary and the postfault system is stable. Otherwise, it may be unstable.

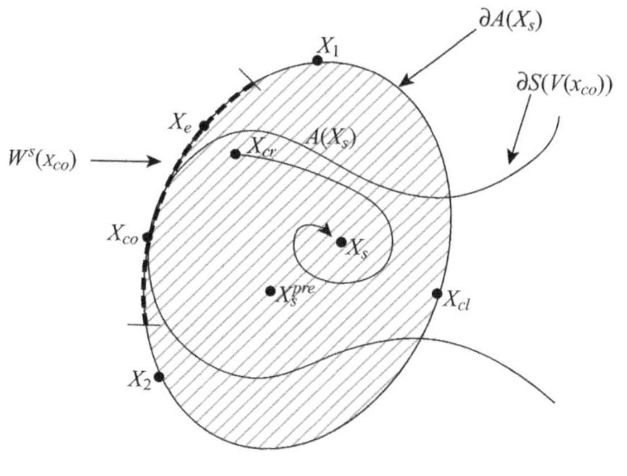

Figure 11.5 The controlling UEP method approximates the relevant stability boundary, the stable manifold of the controlling UEP $W^S(X_{co})$, by the constant energy surface, $\partial S(V(X_{co}))$, passing through the controlling UEP.

The above viewpoint reveals that the controlling UEP method yields an approximation of the relevant part of the stability boundary of the postfault system to which the fault-on trajectory is heading. It uses the connected constant energy surface passing through the controlling UEP to approximate the relevant part of the stability boundary.

To ensure the conservative nature of the controlling UEP method, it is important that the fault-on trajectory passes through the constant energy surface containing the controlling UEP before it exits the stability region at the exit point. If the fault is cleared before the fault-on trajectory reaches the constant energy surface containing the controlling UEP, then the postfault trajectory will converge to the SEP and will become stable (see Figure 11.5). To check when the fault-on trajectory passes through the constant energy surface containing the controlling UEP, one can monitor the value of $V(x_f(t))$ along the fault-on trajectory $x_f(t)$ and can identify when $V(x_f(t)) = V(x_{co})$ (see Figure 11.5).

The above analysis suggests that the energy value at the controlling UEP be used as the critical energy with respect to an energy function for the fault-on trajectory and that the constant energy surface containing the controlling UEP be used to approximate the relevant part of the stability boundary. This is the essence of the controlling UEP method. We will present a theoretical analysis of the controlling UEP method in the next section.

11.5 ANALYSIS OF THE CONTROLLING UEP METHOD

The controlling UEP method asserts that the energy value at the controlling UEP be used as the critical energy for the fault-on trajectory $X_f(t)$ to directly assess the stability of the postfault system. Theorem 11.3 below gives a rigorous theoretical justification for the controlling UEP method for the direct stability analysis of postfault systems by comparing the energy value of the state vector at which the fault is cleared with the energy value at the controlling UEP.

We define the following two components:

1. $S(r) \triangleq$ is the connected component of the energy level set $\{X \in R^n: V(X) < r\}$ containing X_s, and
2. $\partial S(r)$: $\triangleq$ is the (topological) boundary of $S(r)$, which is the constant energy surface with energy value r.

Theorem 11.3: Controlling UEP Method for Approximating the Relevant Stability Boundary

Consider a general postfault system that has an energy function $V(\cdot): R^n \to R$. Let X_{co} be an equilibrium point on the stability boundary $\partial A(X_s)$ of a SEP X_s of this system. Then, the following results hold:

1. The connected constant energy surface $\partial S(V(X_{co}))$ intersects with the stable manifold $W^s(X_{co})$ only at point X_{co}; moreover, the set $S(V(X_{co}))$ has an empty intersection with the stable manifold $W^s(X_{co})$.
2. Any connected path starting from a point $P \in \{S(V(X_{co})) \cap A(X_s)\}$ and intersecting with $W^s(X_{co})$ must also intersect with $\partial S(V(X_{co}))$.

We elaborate on the above fundamental theorem. Result 1 implies that $\partial S(V(X_{co})) \cap W^s(X_{co}) = X_{co}$ and $S(V(X_{co})) \cap W^s(X_{co}) = \phi$ (see Figure 11.5). Results 1 and 2 of Theorem 11.3 assert that for any fault-on trajectory $X_f(t)$ starting from a point $X_s^{pre} \in A(X_s)$ and $V\left(x_s^{pre}\right) < V\left(x_{co}\right)$, if the exit point of the fault-on trajectory $X_f(t)$ lies on the stable manifold of X_{co}, then the fault-on trajectory $X_f(t)$ must pass through the connected constant energy surface $\partial S(V(X_{co}))$ before it passes through the stable manifold of X_{co} (thus exiting the stability boundary $\partial A(X_s)$). Therefore, the connected constant energy surface $S(V(X_{co}))$ is adequate for approximating the relevant part of the stability boundary. Theorem 11.3 also confirms the slightly conservative nature of the controlling UEP method in direct stability assessment. The only scenario in which the controlling UEP method gives conservative stability assessments is the situation where the fault is cleared, when the fault-on trajectory $X_f(t)$ lies between the connected constant energy surface $\partial S(V(X_{co}))$ and the stable manifold of X_{co} (see Figure 11.6).

Using the energy value at a different UEP instead of the controlling UEP as the critical energy can give an erroneous stability assessment. We show in Theorem 11.4 below that among all the UEPs on the stability boundary, the controlling UEP is the one that gives the most accurate critical energy. Indeed, there are usually multiple UEPs lying on the stability boundary. As such, it is inappropriate to choose the energy values at other UEPs instead of the controlling UEP as the critical energy.

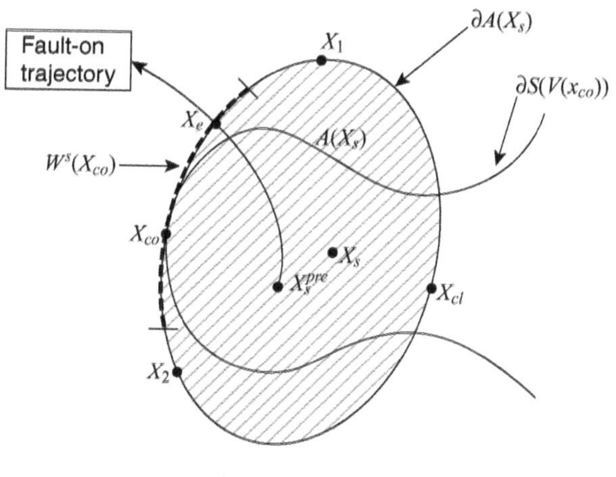

Figure 11.6 The only scenario in which the controlling UEP method gives conservative stability assessments is the situation where the fault is cleared when the fault-on trajectory $X_f(t)$ lies between the connected constant energy surface $\partial S(V(X_{co}))$ and the stable manifold of X_{co}, which is highlighted in the figure.

Theorem 11.4: Controlling UEP Method for Accurate Critical Energy

Consider a general postfault system that has an energy function $V(\cdot): R^n \to R$. Let X_{co} be an equilibrium point on the stability boundary $\partial A(X_s)$ of a SEP X_s of this system.

Then, the following results hold:

1. If X^u is a UEP and $V(X^u) > V(X_{co})$, then $S(V(X^u)) \cap W^s(X_{co}) \neq \phi$.
2. If X^u is a UEP and $V(X^u) < V(X_{co})$, then $S(V(X^u)) \cap W^s(X_{co}) = \phi$.
3. If X is a state vector on the stability boundary but is not the closest UEP, then the constant energy surface passing through X has a nonempty intersection with the complement of the closure of the stability region; that is, $\partial S(V(\hat{X})) \cap (\bar{A}(X_s))^c \neq \phi$.

Results 1 and 3 of Theorem 11.4 assert that the following two situations may occur:

Case 1: The energy level set $S(V(X_1))$ contains only part of the stable manifold $W^s(X_{co})$ (see Figure 11.7).

Case 2: The energy level set $S(V(X_1))$ contains the entire stable manifold $W^s(X_{co})$ (see Figure 11.8).

In Case 1, the fault-on trajectory $X_f(t)$ may pass through the connected constant energy surface $\partial S(V(X_1))$ before it passes through the stable manifold $W^s(X_{co})$ (see Figure 11.9). In this situation, incorrect use of X_1 as the controlling UEP still gives an accurate stability assessment. Alternatively, the fault-on trajectory $X_f(t)$ may pass through the connected constant energy surface $\partial S(V(X_1))$ after it passes through the stable manifold $W^S(X_{co})$ (see Figure 11.10). In this situation, using the energy value

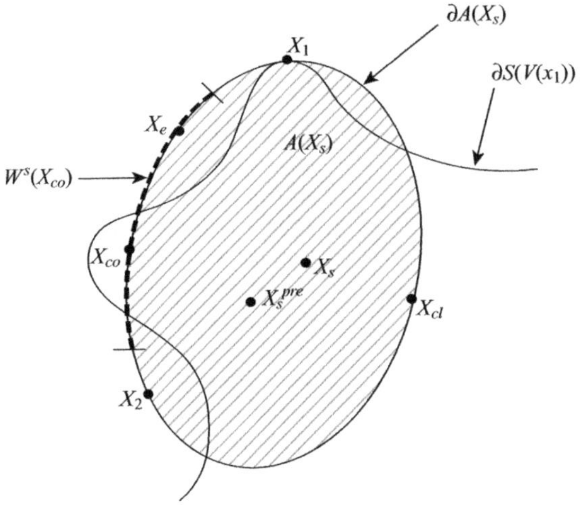

Figure 11.7 The constant energy surface passes through the UEP X_1 and the corresponding energy level set contains part of the stable manifold $W^s(X_{co})$.

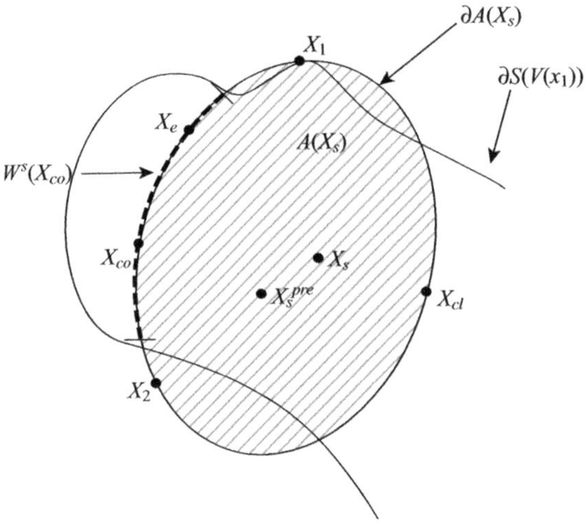

Figure 11.8 The constant energy surface passes through the UEP X_1 and the corresponding energy level set contains the entire stable manifold $W^s(X_{co})$.

at X_1 as the critical energy gives an inaccurate stability assessment. This is an incorrect stability assessment.

In Case 2, the fault-on trajectory $X_f(t)$ passes through the connected constant energy surface $\partial S(V(X_1))$ after it passes through the stable manifold $W^S(X_{co})$ (see Figure 11.11). Under this situation, using the energy value at X_1 as the critical energy can give inaccurate stability assessments. In particular, it can classify the postfault trajectory to be stable when in fact it is unstable.

Results 2 and 3 of Theorem 11.4 assert that the set $S(V(X_1))$ has an empty intersection with the stable manifold $W^s(X_{co})$. Under this situation, the fault-on trajectory $X_f(t)$ always passes through the connected constant energy surface $\partial S(V(X_1))$ first before it passes through the connected constant energy surface $\partial S(V(X_{co}))$. Thus,

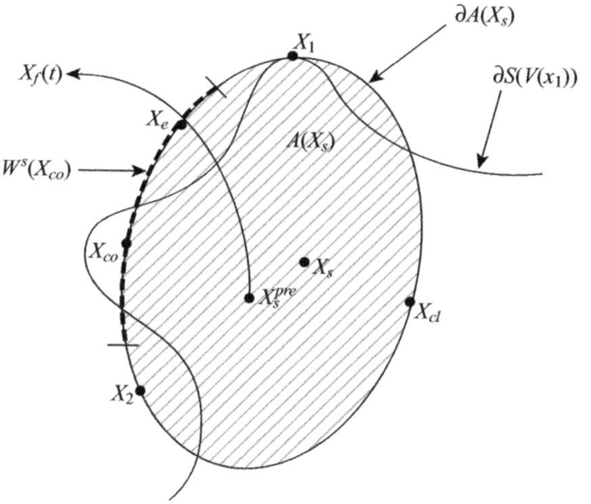

Figure 11.9 The energy level set $S(V(X_1))$ contains only part of the stable manifold $W^s(X_{co})$. The fault-on trajectory $X_f(t)$ passes through the connected constant energy surface $\partial S(V(X_1))$ before it passes through the stable manifold $W^s(X_{co})$. In this situation, incorrect use of the energy function value of X_1 as the critical energy still gives a correct stability assessment.

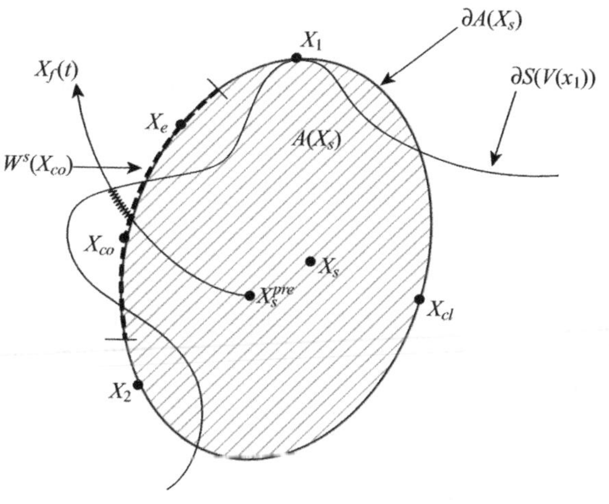

Figure 11.10 The set $S(V(X_1))$ contains only part of the stable manifold $W^s(X_{co})$. The fault-on trajectory $X_f(t)$ may pass through the connected constant energy surface $\partial S(V(X_1))$ after it passes through the stability boundary. Under this situation, using the energy value at X_1 as the critical energy can give inaccurate stability assessments. It classifies an unstable contingency as a stable one when the fault is cleared in the highlighted segment of the fault-on trajectory.

using the energy value at X_1 as the critical energy value always gives more conservative stability assessments than using that of the controlling UEP, X_{co}.

From the above analysis, it is clear that for a given fault-on trajectory, $X_f(t)$, if the exit point of this fault-on trajectory $X_f(t)$ lies on the stable manifold of X_{co}, then using the energy value at a UEP other than X_{co} can give incorrect or rather conservative stability assessments: too conservative stability assessments (e.g., classify stable trajectories as unstable) or too optimistic stability assessments (e.g., classify unstable trajectories as stable). Hence, it is essential to identify the correct controlling UEP for direct stability analysis.

Historically, direct methods have been said to be only applicable to first-swing stability analysis. We point out that once the initial point of the postfault system lies

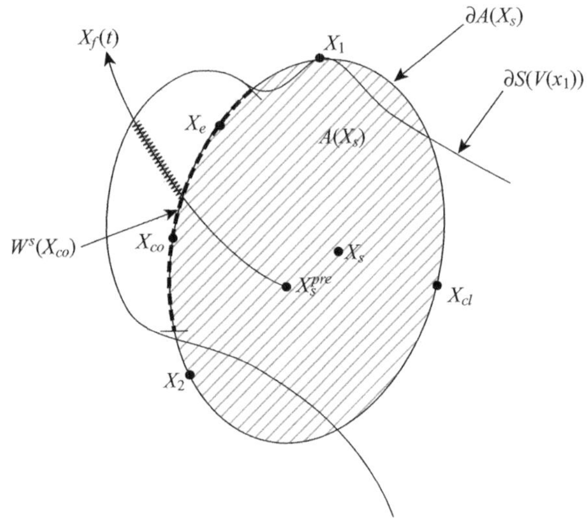

Figure 11.11 The energy level set $S(V(X_1))$ contains the whole stable manifold $W^s(X_{co})$. In this situation, the fault-on trajectory $X_f(t)$ always passes through the connected constant energy surface $\partial S(V(X_1))$ after it passes through the stability boundary. In this situation, using the energy value at X_1 as the critical energy gives inaccurate stability assessments. It classifies an unstable contingency as a stable one when the fault is cleared in the highlighted segment of the fault-on trajectory.

inside the stability region $A(x_s)$, the postfault trajectory will converge to X_s after single or multiple swings. Theorems 11.3 and11.4 collectively assert that the controlling UEP method can directly detect both first-swing and multiswing stability or instability.

11.6 NUMERICAL EXAMPLES

In order to illustrate the concept of the controlling UEP, we use the following simple numerical example, which closely represents a three-machine system, with machine number 3 as the reference machine:

$$\begin{aligned}
\dot{\delta}_1 &= \omega_1 \\
\dot{\omega}_1 &= -\sin\delta_1 - 0.5\sin(\delta_1 - \delta_2) - 0.4\omega_1 \\
\dot{\delta}_2 &= \omega_2 \\
\dot{\omega}_2 &= -0.5\sin\delta_2 - 0.5\sin(\delta_2 - \delta_1) - 0.5\omega_2 + 0.05.
\end{aligned} \tag{11.5}$$

It is easy to show that the following function is an energy function for this system:

$$V(\delta_1, \delta_2, \omega_1, \omega_2) = \omega_1^2 + \omega_2^2 - 2\cos\delta_1 - \cos\delta_2 - \cos(\delta_1 - \delta_2) - 0.1\delta_2. \tag{11.6}$$

The point $X_s = (\delta_1^s, \omega_1^s, \delta_2^s, \omega_2^s) = (0.02001, 0, 0.06003, 0)$ is a SEP of the above postfault system. There are six type-two equilibrium points (see Table 11.1) and six type-one equilibrium points (see Table 11.2) lying on the stability boundary of X_s. The unstable manifold of each of these equilibrium points converges to the SEP, X_s.

The stability boundary, $\partial A(X_s)$, is contained in the set, which is the union of the stable manifolds of these six type-one equilibria and of these six type-two equilibria.

Table 11.1 Coordinates of Type-Two Equilibrium Points Lying on the Stability Boundary of X_s

Type-two equilibrium point	δ_1	ω_1	δ_2	ω_1
1	3.60829	0	1.58620	0
2	2.61926	0	4.25636	0
3	−2.67489	0	1.58620	0
4	−3.66392	0	−2.02684	0
5	−2.67489	0	−4.69699	0
6	2.61926	0	−2.02684	0

Table 11.2 Coordinates of Six Type-One Equilibrium Points Lying on the Stability Boundary of X_s

Type-one equilibrium point	δ_1	ω_1	δ_2	ω_1
1	3.24512	0	0.31170	0
2	3.04037	0	3.24387	0
3	0.03333	0	3.10823	0
4	−3.03807	0	0.3117	0
5	−3.24282	0	−3.03931	0
6	0.03333	0	−3.17496	0

Figure 11.12 depicts the intersection between the stability boundary $\partial A(X_s)$ and the angle space, $\{(\delta_1, \delta_2): \delta_1 \in \pi, \delta_2 \in R\}$. For illustrational purposes, we assume that the fault-on trajectory, $X_f(t)$, shown in Figure 11.13, is due to the fault.[1] The fault-on trajectory, $X_f(t)$, starting from the prefault SEP x_s, first intersects with the constant energy surface passing through the type-one equilibrium point, $X_{co} = (0.03333, 0, 3.10823, 0)$, and then intersects with the stability boundary, $\partial A(X_s)$, by passing through the stable manifold of the type-one equilibrium point, X_{co}, at the exit point, X_e. Among the stable manifolds of the six type-one UEPs on the stability boundary, the fault-on trajectory $X_f(t)$ only passes through the stable manifold of the type-one equilibrium point, X_{co}. Hence, X_{co} is the controlling UEP relative to the fault-on trajectory $X_f(t)$. The time duration between the prefault SEP X_s^{pre} and the exit point X_e along the fault-on trajectory $X_f(t)$ is the so-called CCT. If the fault is cleared before the fault-on trajectory reaches the exit point, then the postfault trajectory will converge to X_s. Note that this relationship is in agreement with the analytical results developed in Theorem 11.3.

Given a fault-on trajectory $X_f(t)$ with a preset fault clearing time, let X_{cl} be the state vector when the fault is cleared, which is also the initial point of the postfault trajectory. Let $W^s(X_{co})$ be the stable manifold of the corresponding controlling UEP

[1]The fault-on trajectory $X_f(t)$, in practice, has nonzero components of ω.

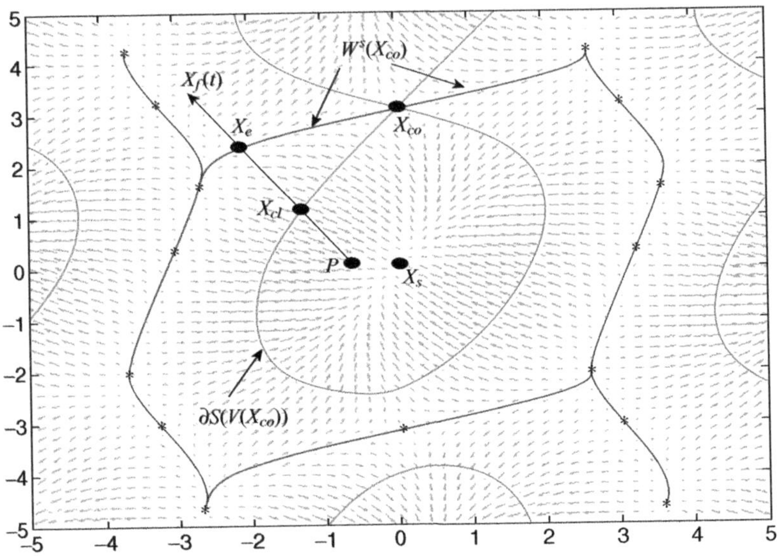

Figure 11.12 The intersection between the stability boundary $\partial A(X_s)$ and the angle subspace. The stability boundary, $\partial A(X_s)$, is contained in the set, which is the union of the stable manifolds of these six type-one equilibrium points and six type-two equilibrium points. The constant energy surface passing through one UEP is drawn. The vector field at many points is plotted in the figure, which can serve to confirm the location of the exact stability boundary.

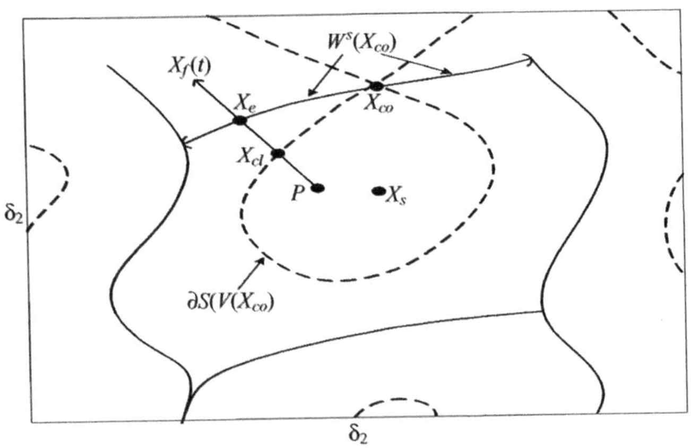

Figure 11.13 The fault-on trajectory, $X_f(t)$, starting from the prefault SEP x_s, first intersects the constant energy surface passing through the type-one equilibrium point $X_{co} = (0.03333, 0, 3.10823, 0)$ and then intersects the stability boundary, $\partial A(X_s)$, by passing through the stable manifold of the type-one equilibrium point, X_{co}, at the exit point, X_e. The controlling UEP relative to the fault-on trajectory is hence the type-one equilibrium point $X_{co} = (0.03333, 0, 3.10823, 0)$.

X_{co}. The task of determining whether or not the segment of the fault-on trajectory, $X_f(t)$, between X_s^{pre} and X_{cl} lies inside the stability region $A(X_s)$ based on the stable manifold $W^S(X_{co})$ is numerically very difficult, if not impossible. The main reason for this difficulty is due to the lack of an explicit expression for stable manifolds. However, the task becomes relatively easy if an energy function instead of the stable manifold $W^s(X_{co})$ is given.

To elaborate on this point, let us examine Figure 11.13, which shows the relationship between the stable manifold $W^s(X_{co})$ and the constant energy surface passing through X_{co}. In order for the fault-on trajectory $X_f(t)$ to pass through the constant energy surface, the point X_{cl} must have an energy value greater than the energy value at the controlling UEP X_{co}; that is, $V(X_{cl}) > V(X_{co})$. Hence, the task of determining whether the segment of the fault-on trajectory $X_f(t)$ between X_s^{pre} and X_{cl} lies inside the stability region $A(X_s)$ is boiled down to the task of comparing two scalars: $V(X_{cl})$ and $V(X_{co})$. If $V(X_{cl}) \leq V(X_{co})$, then the controlling UEP method asserts that X_{cl} lies inside the stability region. Indeed, the segment of the fault-on trajectory $X_f(t)$ between X_s^{pre} and X_{cl} lies inside the stability region $A(X_s)$. In this case, the postfault trajectory starting from X_{cl} will converge to the SEP X_s.

On the other hand, if $V(X_{cl}) > V(X_{co})$, the controlling UEP method asserts that X_{cl} may lie outside the stability region; however, the segment from X_s^{pre} to X_{cl} may not lie entirely inside the stability region $A(X_s)$. Two cases are possible. In the first case, X_{cl} lies outside the stability region $A(X_s)$, in which we have $V(X_{co}) < V(X_e) < V(X_{cl})$, and the postfault trajectory starting from X_{cl} will not converge to x_s. In the second case, X_{cl} still lies inside the stability region $A(X_s)$, in which we have $V(X_{co}) < V(X_{cl}) < V(X_e)$, and the postfault trajectory starting from X_{cl} will still converge to X_s. The second case reveals the slightly conservative nature of the controlling UEP method in assessing the stability property of the postfault trajectory.

11.7 DYNAMIC AND GEOMETRIC CHARACTERIZATIONS

In this section, dynamic and geometric characterizations of the controlling UEP will be derived. These characterizations will lead to several advances in computing and identifying the controlling UEP. The notion and the complete characterization of the quasi-stability region will be employed to show this property.

We next present a dynamic characterization of the controlling UEP.

Theorem 11.5: Dynamic Characterization of Controlling UEP

Given a prefault SEP, a fault-on trajectory, and a postfault system with a SEP, X_s, let the stability region of X_s contain the prefault SEP. If the postfault system admits an energy function, then the controlling UEP, say, X_{co}, of the fault-on trajectory always exists. If the transversality condition is satisfied at X_{co}, then the unstable manifold of X_{co} converges to the SEP X_s; that is, $W^u(X_{co}) \cap A(X_s) \neq \phi$. Moreover, its unstable manifold intersects the complement of the closure of the stability region; that is, $W^u(X_{co}) \cap (\bar{A}(X_s))^c \neq \phi$.

Proof: This theorem follows from the definition of a controlling UEP and the following fact, which is derived from Theorem 4.8:

Fact 11.6

Let $A_p(x_s)$ and $A(x_s)$ be the quasi-stability region and the stability region of SEP x_s of a general nonlinear system, respectively. Let $\sigma \neq x_s$ be a hyperbolic equilibrium point. If Assumptions (A1)–(A3) are satisfied, then

1. $\sigma \in \partial A_p$ if and only if $W^u(\sigma) \cap A \neq \emptyset$ and $W^u(\sigma) \cap (\bar{A})^c \neq \emptyset$.
2. $\sigma \in \partial A$ if and only if $W^u(\sigma) \cap A \neq \emptyset$.

Hence, the unstable manifold of the controlling UEP converges to the SEP and also intersects the complement of the closure of the stability region. And this theorem follows.

We next present a geometric characterization of the controlling UEP. It has been shown that the quasi-stability boundary is the union of stable manifolds of all the critical elements on the quasi-stability boundary. In addition, this characterization can be sharpened by the following theorem, which asserts that the quasi-stability boundary is the union of the closure of the stable manifolds of all the type-one critical elements on the quasi-stability boundary. By applying the characterization of the quasi-stability boundary to a controlling UEP, it follows that the controlling UEP is a type-one UEP as described below:

Theorem 11.7: Geometric Characterization

Given a prefault SEP, a fault-on trajectory, and a postfault system with a SEP X_s, let the stability region of X_s contain the prefault SEP. If the postfault system admits an energy function, then the controlling UEP, say, X_{co}, of the fault-on trajectory always exists. If the transversality condition is satisfied at every equilibrium point on the stability boundary of X_s, then X_{co} is a type-one UEP.

Proof: This theorem follows from the definition of controlling UEP and the following theorem, which is derived from Theorem 4.11.

Theorem

Let $A_p(x_s)$ be the quasi-stability region of x_s of a general nonlinear system, which satisfies Assumptions (A1)–(A3). Let σ_i, $i = 1, 2, \ldots$ be the type-one equilibrium points lying on the quasi-stability boundary $\partial A_p(x_s)$ of the SEP x_s. Then,

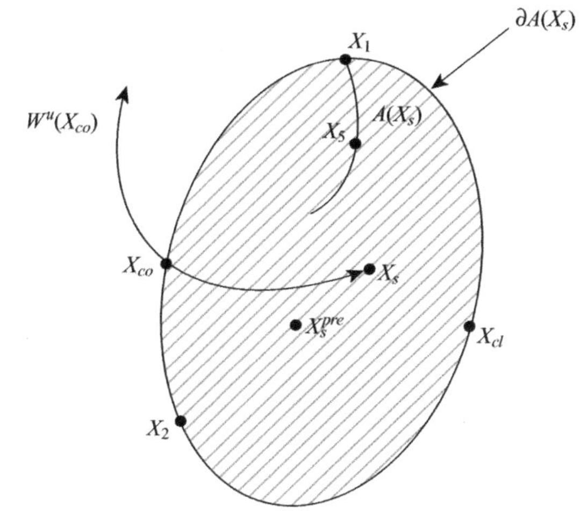

Figure 11.14 The controlling UEP is a type-one UEP lying on the stability boundary, and its one-dimensional unstable manifold always converges to the postfault SEP and also intersects with regions that are "outside the stability region."

$$\partial A_p(x_s) \bigcup_{\sigma_i \in \partial A_p(x_s)} \overline{W^s(\sigma_i)}. \tag{11.7}$$

Hence, the controlling UEP must be of type one. And this theorem follows.

Hence, Theorems 11.5 and 11.7 collectively assert that the controlling UEP is a type-one UEP lying on the stability boundary and that the one-dimensional unstable manifold of the controlling UEP converges to the postfault SEP and also intersects the regions that are "outside the stability region" (see Figure 11.14).

11.8 CONCLUDING REMARKS

In the past, the Lyapunov function theory was employed by several researchers to explain the conservativeness of their proposed direct methods in determining critical energy. This explanation, however, is not convincing because the Lyapunov function was not developed for direct stability assessments. Conservativeness comes from other reasons such as the fact that method itself is independent of fault-on trajectories.

The classical closest UEP method has been found to yield unduly conservative results when applied to power system transient stability analysis. We have shown that the approximation of the stability boundary of the postfault system by the closest UEP method is independent of the fault-on trajectory. Thus, the closest UEP method usually yields very conservative results for transient stability analysis.

This chapter has presented a rigorous introduction to the concept of the controlling UEP and has shown its existence and uniqueness. A controlling UEP method, along with its theoretical foundations, has been developed. Some dynamic and geometric characterizations of the controlling UEP have also been derived. These characterizations are useful for the development of solution methodologies for computing controlling UEPs.

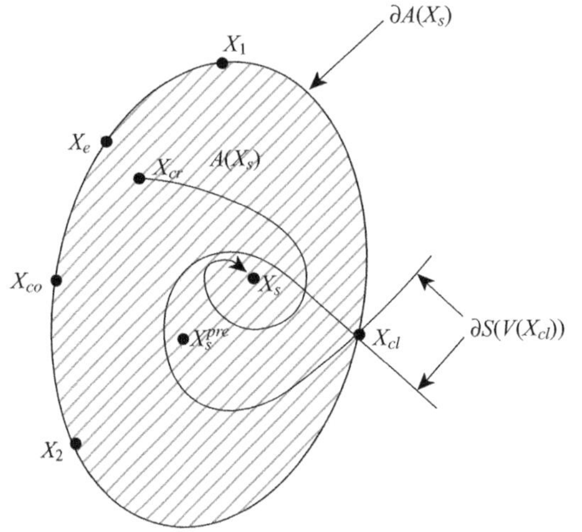

Figure 11.15 A postfault trajectory is stable when the fault is cleared between the constant energy surface passing through the closest UEP and the exit point, yet it is assessed as unstable by the closest UEP method.

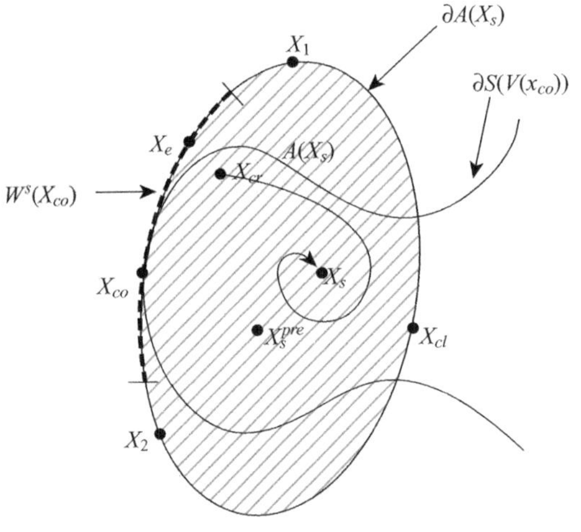

Figure 11.16 A postfault trajectory is stable when the fault is cleared between the constant energy surface passing through the controlling UEP and the exit point, yet it is assessed as unstable by the controlling UEP method. This is the only scenario where the controlling UEP method gives conservative estimations.

The controlling UEP method takes into account the fault-on trajectory, while the closest UEP method does not. The closest UEP method aims to approximate the entire stability boundary of a postfault system in an optimal way, while the controlling UEP method seeks to approximate the relevant stability boundary of a postfault system in an optimal way. It is clear that the critical energy value according to the controlling UEP method is always no less than the critical energy value given by the closest UEP method. Hence, the controlling UEP method does not suffer from the unduly conservative nature of the closest UEP method. When a fault-on trajectory is cleared after it hits the constant energy surface passing through the closest UEP and before the exit point of the original model, the postfault trajectory will be assessed as unstable by the closest UEP method, while it is correctly assessed as stable by the controlling UEP method (see Figures 11.15 and 11.16).

The only scenario for which the controlling UEP method gives a conservative stability assessment is the following: when the fault is cleared after the fault-on trajectory hits the constant energy surface passing through the controlling UEP and before it reaches the exit point. Indeed, this is the only scenario where the controlling UEP method classifies a stable contingency as unstable; otherwise, the controlling UEP method classifies an unstable contingency as unstable and a stable one as stable.

Chapter 12

Controlling UEP Method: Computations

12.1 INTRODUCTION

It is now well recognized that the controlling unstable equilibrium point (controlling UEP) method is the most viable method for direct stability analysis of practical power systems. The success of applying the controlling UEP method hinges upon the ability to find the correct controlling UEP. However, the task of finding the controlling UEP of a given fault for general power system transient stability models is very challenging.

From a theoretical viewpoint, the existence and uniqueness of the controlling UEP with respect to a fault-on trajectory are assured. From a computational viewpoint, computing the controlling UEP is very challenging as summarized into seven challenges in the following section. These computational challenges should be taken into account during the development of numerical methods for computing the controlling UEP. These challenges collectively serve to explain why previous methods proposed in the literature fail to compute the controlling UEP. We note that since the controlling UEP is independent of the energy function used in the direct stability assessment, the tasks of constructing an energy function and of computing the controlling UEP are not interrelational.

In this chapter, we focus on the computational aspects of the controlling UEP method. Specifically, the following will be described:

- The computational challenges in computing the controlling UEP are described. A total of seven challenges are highlighted.
- The task of computing stable equilibrium points (SEPs) and UEPs is formulated as solving a set of constrained nonlinear algebraic equations.
- Numerical techniques in solving this set of constrained nonlinear algebraic equations are presented.

Direct Methods for Stability Analysis of Electric Power Systems, by Hsiao Dong Chiang

- The issue of convergence regions of equilibrium points is raised, in which the computational challenges stemming from this issue are highlighted.
- Two conceptual benchmark methods for computing the controlling UEPs of general transient stability models are presented and illustrated on a simple transient stability model.

12.2 COMPUTATIONAL CHALLENGES

The task of finding the controlling UEP is very difficult due to the following computational challenges:

Challenge I: The controlling UEP is a particular UEP embedded in a large-degree state space.

Challenge II: The controlling UEP is the first UEP whose stable manifold has a nonempty intersection with the fault-on trajectory at the exit point.

Challenge III: The task of computing the exit point is very involved; it usually requires an iterative time–domain approach.

Challenge IV: The task of computing the controlling UEP requires solving a large set of constrained nonlinear algebraic equations.

Challenge V: A good initial guess for computing the controlling UEP is difficult to provide.

Challenge VI: The size of the convergence region of a controlling UEP with respect to a numerical method can be very small and irregular.

Challenge VII: Starting from the exit point as an initial guess, a numerical method such as the Newton method may not converge to the controlling UEP.

The above seven challenges make the task of developing robust numerical methods for computing the controlling UEP extremely difficult. Challenges I–III raise the doubt of any method attempting to directly compute the controlling UEP of transient stability models. These challenges collectively serve to explain why previous methods proposed in the literature fail to compute the controlling UEP. It is because these methods attempt to directly compute the controlling UEP, which, as observed in Challenges I–III, is extremely difficult if not impossible without using the iterative time–domain method.

The theoretical aspect of Challenges I–III is described by the controlling UEP and the theoretical foundation of the controlling UEP method. Regarding Challenge VI, it is known that, with respect to a selected numerical method, each equilibrium point has its own convergence region (i.e., the region from which the sequence generated by the numerical method starting from a point in the region will converge to the equilibrium point). It has been observed and theoretically investigated by several researchers that, under the Newton method, the size of the convergence region of the UEP can be much smaller than that of the SEP (Thorp and Naqavi, 1989). In addition, the convergence region of either a SEP or a UEP with respect to the Newton method is a *fractal* (Thorp and Naqavi, 1989), meaning the structure of the convergence region cannot be described by typical geometrical objects such as lines, surfaces, and solids. Irregular shape (no smooth boundaries) and self-similarity (each tiny piece we observe is similar to the form of the entire shape) are characteristics of fractals.

Due to the small size and the irregular, fractal-like shape of the convergence region of the UEP, the task of computing the controlling UEP is very challenging. If an initial guess is not sufficiently close to the controlling UEP, then the resulting sequence, generated by, say, Newton's method, will diverge or converge to an equilibrium point of the following possibilities:

Case 1: Converge to the postfault SEP

Case 2: Converge to another SEP

Case 3: Converge to a UEP on the stability boundary

Case 4: Converge to a UEP outside the stability boundary.

Of the above cases, Case 1 is easy to detect, while the other three cases are not. The resulting stability assessments will be incorrect if a method encounters one of Cases 2, 3, or 4. Unfortunately, finding an initial guess sufficiently close to the controlling UEP, such that the initial guess lies within the convergence region of the controlling UEP, is a difficult task.

In the literature, a number of methods based on physical reasonings attempt to find the controlling UEP (or the relevant UEP). These methods are classified as the following: the machine acceleration method (Pavella and Murthy, 1994; Ribbens-Pavella and Lemal, 1976), the system separation method (Fouad and Stanton, 1981), and the mode of disturbance (MOD) method (Fouad and Vittal, 1991). These three methods, unfortunately, may not find the controlling UEP because the initial point provided by these three methods may lie far away from the controlling UEP or may lie outside the convergence region of the controlling UEP. From a computational viewpoint, both the machine acceleration method and the MOD method provide an initial guess on the basis of a prediction of the "mode" of generators. Furthermore, this initial guess does not depend on the loading conditions. As the loading conditions change, the mode changes, while the location of the controlling UEP always changes. As the loading conditions change, the convergence region also changes. But these methods do not take into account these changes.

Another important factor that is overlooked by the methods proposed in the literature is that the controlling UEP is the first UEP whose stable manifold has a nonempty intersection with the fault-on trajectory. This fact must be incorporated into a numerical method for computing the controlling UEP. Without the incorporation of this fact, a numerical method is unlikely to be consistent in finding the correct controlling UEP. This fact has been incorporated into the BCU method to be described in later chapters.

12.3 CONSTRAINED NONLINEAR EQUATIONS FOR EQUILIBRIUM POINTS

Consider a power system stability model described generically by the following set of equations:

$$\begin{aligned} \dot{x} &= f(x, y) \\ 0 &= g(x, y). \end{aligned} \tag{12.1}$$

The task of computing the UEP and the SEP of a (postfault) transient stability model involves solving the following algebraic equation with different initial guesses (i.e., starting points for the intended numerical method):

$$\begin{aligned} 0 &= f(x, y) \\ 0 &= g(x, y). \end{aligned} \tag{12.2}$$

For an illustrational purpose, the following dynamic equations are used for transient stability analysis.

12.3.1 Generator Motion Dynamic Equations

For generator i, the motion dynamic equations in center of inertia (COI) coordinates are expressed as follows (Fouad and Vittal, 1991; La Salle and Lefschetz, 1961; Sauer and Pai, 1998):

$$\begin{aligned} \frac{d\delta_i}{dt} &= \omega_i \\ \frac{d\omega}{dt} &= P_{nii} - P_{ei} - \frac{M_i}{M_T} P_{COI}, \end{aligned} \tag{12.3}$$

where δ_i is the internal angle of generator i, ω_i is the angular velocity of generator i, and P_{ei} and P_{mi} are the electrical power output and mechanical power input at generator i, respectively. M_i is the machine inertia of generator i, M_T is the total machine inertia of the system, and P_{COI} is the power of the COI. The P_{COI} is computed as follows:

$$P_{COI} = \sum_i (P_{mi} - P_{ei}). \tag{12.4}$$

12.3.2 Generator Electric Dynamic Equations

The electric dynamic equations of generator i can be expressed as follows:

$$\frac{dx_{ei}}{dt} = f_e(x, y), \tag{12.5}$$

where x_{ei} can include E'_{qi}, E'_{di}, E''_{qi}, E''_{di}, and other electrical variables of generator i. (x, y) represents the relevant state variables. Depending on the generator type and the level of detail in the study, Equation 12.5 can have different expressions.

12.3.3 Excitation System and Power System Stabilizer (PSS)

The excitation system and the PSS of generator i can be expressed as follows:

$$\frac{dx_{exi}}{dt} = f_{ex}(x, y), \tag{12.6}$$

where x_{exi} represents all the state variables in the excitation system and in the PSS of generator i. There are several types of excitation systems and PSSs. The mathematical expression of Equation 12.6 is different for different types of excitation systems and PSSs.

12.3.4 Network Equations

The network equations are composed of real power network equations and reactive power network equations:

$$\begin{aligned} f_P(x, y) &= 0 \\ f_Q(x, y) &= 0. \end{aligned} \tag{12.7}$$

12.3.5 Equilibrium Equations

An equilibrium point (SEP or UEP) is the state vector at which the right-hand sides of the dynamic equations (Eqs. 12.3, 12.5, and 12.6) are zero and satisfy the network equation (Eq. 12.7). More specifically, a SEP or a UEP must satisfy the following equations:

$$\begin{aligned} P_{mi} - P_{ei} - \frac{M_i}{M_T} P_{COI} &= 0 \\ f_e(x, y) &= 0 \\ f_{ex}(x, y) &= 0 \\ f_P(x, y) &= 0 \\ f_Q(x, y) &= 0. \end{aligned} \tag{12.8}$$

Remarks:

1. If the classical generator model is considered (i.e., both the salient effect and the excitation system are not modeled), the set of equilibrium equations (Eq.

12.8) is reduced to the network equation (Eq. 12.7), and the equilibrium equations for transient stability analysis with the classical generator model are the following:

$$P_{mi} - P_{ei} - \frac{M_i}{M_T} P_{COI} = 0$$
$$f_P(x, y) = 0 \tag{12.9}$$
$$f_Q(x, y) = 0.$$

2. If the excitation system is considered, then the effect of electrical dynamic and excitation systems modeled as the one gain, one time constant, and one limiter are represented as follows (regardless of which detailed generator model is used) (Fouad et al., 1989):

$$E_{fdi} = V_{qi} + I_{di} X_{di}$$
$$E_{fdi} = \begin{cases} E_{efmax,i} & \text{if } K_{exi}(V_{refi} - V_i) > E_{efmax,i} \\ E_{efmin,i} & \text{if } K_{exi}(V_{refi} - V_i) < E_{efmin,i}, \\ K_{exi}(V_{refi} - V_i) & \text{otherwise} \end{cases} \tag{12.10}$$

where V_{qi} is the q-axis component of the terminal voltage of generator i, I_{di} is the d-axis component of the current of generator i, K_{exi} is the total gain of the excitation system and the PSS of generator i, and $E_{fdmax,i}$ and $Ef_{dmin,i}$ are the upper and lower limits of the excitation system of generator i. Hence, the equilibrium equations become a set of constrained nonlinear algebraic equations.

12.4 NUMERICAL TECHNIQUES FOR COMPUTING EQUILIBRIUM POINTS

Two equilibrium points are required in the controlling UEP method: one is the postfault SEP and the other is the controlling UEP. The controlling UEP is a type-one equilibrium point lying on the stability boundary of the postfault SEP, and its stable manifold contains the exit point of the fault-on trajectory. The numerical techniques for solving these two equilibrium points are basically the same; the only difference is the initial points needed for a numerical method to solve the set of nonlinear algebraic equations (Eq. 12.8) with the constraint (Eq. 12.10). For computing the postfault SEP, the prefault SEP is usually used as the initial point. For computing the controlling UEP, the great challenge is providing a good initial point such that Challenge II is met.

We consider the case where all the generators are represented by a two-axis model equipped with simplified one-gain, one-time constant, and one-limiter exciters. The constrained nonlinear algebraic equations to be solved are arranged as follows:

$$
\left.\begin{aligned}
&V_i\sum_{j=1}^{n+m}V_j(G_{ij}\cos\theta_{ij}+B_{ij}\sin\theta_{ij})+P_{Li}-P_{Gi}=0\\
&V_i\sum_{j=1}^{n+m}V_j(G_{ij}\sin\theta_{ij}+B_{ij}\cos\theta_{ij})+Q_{Li}-Q_{Gi}=0
\end{aligned}\right\}\quad \begin{array}{c}\text{(Network equations)}\\ i=1,\ldots,n+m\end{array}
$$

$$
\left.\begin{aligned}
E_{fdi}-E'_{qi}+(x_{di}-x'_{qi})I_{di}&=0\\
E'_{di}+(x_{qi}-x'_{qi})I_{qi}&=0\\
P_{mi}-P_{ei}-\frac{M_i}{M_T}P_{COI}&=0\\
-E'_{fdi}+K_{ei}(V_{refi}-V_i)&=0
\end{aligned}\right\}\quad \begin{array}{c}\text{(Generator equations)}\\ i=1,\ldots,n\end{array}
\qquad (12.11)
$$

$$
E_{fdi}=\begin{cases}E_{fdimax} & \text{if } E'_{fdi}\geq E_{fdmin}\\ E'_{fdi} & \text{if } E_{fdimin}<E'_{fdi}<E_{fdimax}\\ E_{fdmin} & \text{if } E_{fidmax}\geq E'_{fdi}\end{cases}\quad \text{(Excitor limiter).}\qquad (12.12)
$$

We explain a partitioned Newton method to solve the above set of constrained nonlinear equations. In this method, a generator terminal bus is chosen as a slack bus. The solution procedure is described below:

Step 1. From the initial point, compute P_{COI}, E'_{qi}, E'_{di}, and E_{fdi} and allocate P_{COI} among the generators (note that P_{mi} is fixed while P_{ei} is unknown):

$$
P_{ei}=P_{mi}-\frac{M_i}{M_T}P_{COI}\quad i=1,\ldots,n.
$$

Step 2. Convert the two-axis generator model into the classical model, and treat the generator internal buses as PV buses where the specified P_i is P_{ei} and the specified V_i is the equivalent internal bus voltage E'.

Step 3. Solve the power flow network equations using the Newton method.

Step 4. Check the slack bus's mismatch equation to examine whether the extended power flow equations have converged or not. If not (meaning that the real power mismatch is large), reallocate the mismatch among the n generators by

$$
P_{ei}=P_{mi}-\frac{M_i}{M_T}(P_{COI}+P_{Slack})\quad i=1,\ldots,n.
$$

Step 5. Repeat Steps 3–4 until the real power mismatch at the slack bus is sufficiently small.

Step 6. Check the equations associated with generator internal buses. If the mismatch at each internal bus is also sufficiently small, stop the computation; otherwise, go to Step 1.

The above partitioned Newton method is applicable for computing both the controlling UEP and the postfault SEP. The only difference is the initial points provided for solving the set of constrained nonlinear algebraic equations. As discussed before, the prefault SEP is usually used as the initial point for postfault SEP computation. In the controlling UEP computation, the great challenge is providing a good initial point such that Challenge II is met.

We point out that after the postfault SEP and controlling UEP have been obtained, care should be taken to modify the rotor angles and voltage angles to satisfy the COI reference frame in order to compute an energy function value, in particular, the critical energy value. Note that the computation of the energy margin is performed in the COI reference frame.

12.5 CONVERGENCE REGIONS OF EQUILIBRIUM POINTS

The Newton method and its variations are widely used for solving a system of nonlinear equations:

$$F(x)=0, \tag{12.13}$$

where F: $\Re^n \rightarrow \Re^n$ because of their computational efficiency. These methods can achieve superlinear convergence if the initial starting point is within a certain neighborhood. However, this type of method fails if the initial starting point is not sufficiently close to the solution or if singular points are encountered. In other words, these methods are locally convergent and cannot provide guarantees for obtaining solutions.

The convergent region of a solution of Equation 12.13 under the Newton method is finite, irregular, and can be fractal. This study of convergence regions is motivated by the need to determine an initial guess, which converges to the solution of interest using the Newton method. In addition, it may improve the reliability of the Newton method in obtaining solutions. The knowledge of convergence regions also improves the efficiency of numerical methods by avoiding a revisit of the same solution from different initial points.

We next describe a geometric definition of a convergence region.

Definition: Convergence Region
The convergence region $C_F(\tilde{x})$ of a solution $\tilde{x}$ of Equation 12.13 using the numerical method is the set of all initial points, where the numerical method with initial point converges to the solution.

Each equilibrium point of a nonlinear dynamical system has its own convergence region under a numerical method, regardless of its stability (Lee and Chiang, 2001). Hence, a SEP has a convergence region and a UEP also has its own convergence region. The convergence region of an equilibrium point under a numerical method is likely to be different from the convergence region of the equilibrium point under another numerical method. An illustration of the difference of convergence regions of the controlling UEP under two different numerical methods is shown in Figures 12.1 and 12.2.

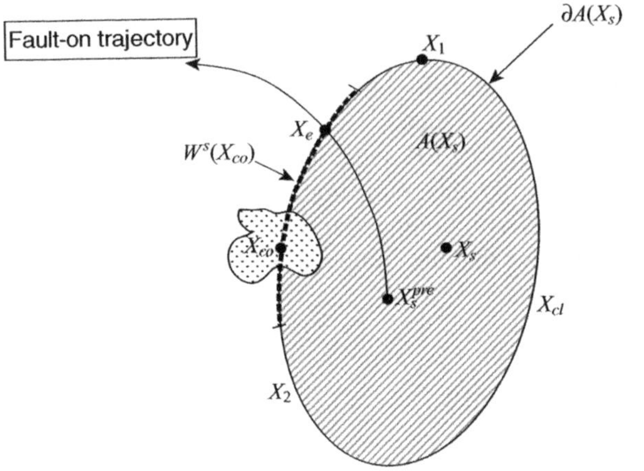

Figure 12.1 Due to the small size and the irregular, fractal-like shape of the convergence region of UEPs under the Newton method (see the dotted area), computing the controlling UEP is very challenging. If an initial guess is not sufficiently close to the controlling UEP, then the resulting sequence generated by the Newton method will diverge or converge to another equilibrium point.

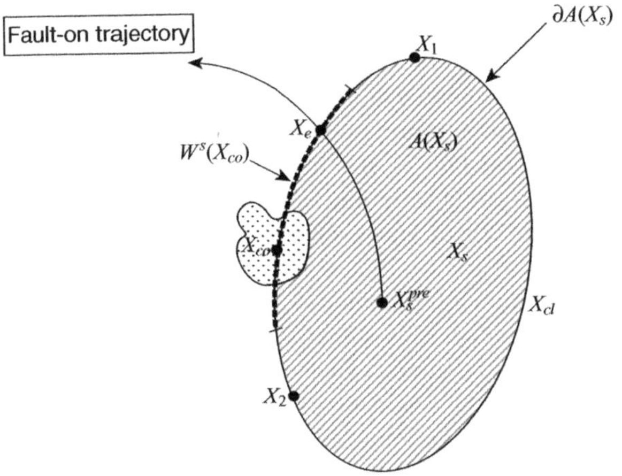

Figure 12.2 The convergence regions of a controlling UEP under different numerical methods are different.

It has been found that the size of the convergence region of a UEP under the Newton method is relatively small, while the size of the convergence region of a SEP under the Newton method is relatively large. For example, convergence regions of a small-size power system stability model using high-speed computer simulation were derived by Thorp and Naqavi (1989). It was numerically shown that the convergence region of a SEP as well as a UEP has a fractal boundary. It was further shown by numerical simulations that the fractal features of a convergence region persist under different loading conditions and under the rectangular coordinates expressing the power flow equations. It is worth noting that the fractal feature of the convergence region is retained under a quasi-Newton method with a line search (Thorp and Naqavi, 1989). Similar fractal domains of convergence regions using the Newton method have also been observed in other engineering areas.

This fractal property of convergence region using the Newton method does not depend on the size of the study system (Thorp and Naqavi, 1989). Hence, convergence regions of equilibrium points of large-scale power systems are expected to have a fractal boundary. One key implication of a fractal boundary is that a small change in the initial condition will produce convergence to different equilibrium solutions.

The observation of and the experience with the ease of computing SEPs compared to computing UEPs may be linked to the size of the convergence region and the ease of the initial guess for SEPs. Since the size of the convergence region of a UEP is usually smaller than that of a SEP under the Newton method, and since the boundary of the convergence region is a fractal, Newton-based equilibrium point solvers can behave in unexpected ways when the required initial guess is far away from normal operating points. These observations lead to the following:

- The task of solving for UEPs is more difficult than that of solving for SEPs because of the difference between convergence regions and the difficulty in providing the required initial guess.
- The size of the controlling UEP convergence region makes a stringent requirement on the initial guess: it must be sufficiently close to the desired UEP. If the initial guess is not close to the controlling UEP, then it may converge to other UEPs and SEPs. For example, if the initial guess is the exit point, then it may converge to another UEP. If the initial guess is close to the exit point but is distant from the controlling UEP, then it may converge to another UEP.

12.6 CONCEPTUAL METHODS FOR COMPUTING THE CONTROLLING UEP

The controlling UEP is a particular UEP embedded in its state space. Furthermore, the controlling UEP is the first UEP whose stable manifold has a nonempty intersection with the fault-on trajectory at the exit point. These two facts make the task of verifying whether or not a computed UEP is the controlling UEP very challenging.

We present in this section two conceptual methods for computing controlling UEPs. The term "conceptual" implies that the computational requirements of these methods are too intensive to be practical. However, these two methods can be used as a benchmark method for computing (exact) controlling UEPs. Of these two methods, one method is based on the geometric and dynamic characterizations of the controlling UEP, while the other method is based on the computation of the exit point and on the fact that the stable manifold of the controlling UEP contains the exit point.

Characterization-Based Method for Computing the Controlling UEP

Step 1. Compute all the type-one equilibrium points of the postfault system in a certain subregion containing the postfault SEP.

Step 2. Identify those type-one equilibrium points whose unstable manifolds, a one-dimensional curve, converge to the postfault SEP.

Step 3. Derive an approximation of the stable manifolds of those type-one equilibrium points identified in Step 2.

Step 4. Integrate the fault-on trajectory and identify the first type-one equilibrium point among those identified in Step 2 such that the fault-on trajectory passes through its stable manifold. This type-one equilibrium point is the controlling UEP of the fault-on trajectory.

The major computational effort needed in the above characterization-based method is mostly for the computation involved in Steps 1 and 3. Since the exact series representation of the stable manifold may not be computationally feasible, two schemes have been proposed to approximate stable manifolds, namely, the hyperplane scheme and the hypersurface scheme. The hyperplane scheme proposed in Yee and Spading (1997) is a first-order approximation of the stable manifold, while the hypersurface method described in Cook and Eskicioglu (1983) is a quadratic approximation. Yet, even though the quadratic approximation scheme is generally more accurate than the first-order approximation scheme, these two schemes all suffer from accuracy problems.

We discuss a time–domain-based method that can compute the exact controlling UEP. This method is slow in nature because it contains an iterative time–domain-based procedure for computing the exit point of the original model. This method is intended as a benchmark method for computing the exact controlling UEPs. We note that there is no method available that can reliably compute the exact controlling UEP of every contingency. In addition, the exact controlling UEP computed by the time–domain-based method can serve as a bench mark to check the correctness of the controlling UEP computed by other methods.

We next present a time–domain-based method for finding the controlling UEP relative to a contingency via the exit point. This method is applicable to general power system transient stability models and does not require the existence of an energy function for the postfault power system. This method only requires the existence of the controlling UEP relative to the contingency under study. One sufficient condition to ensure the existence of the controlling UEP is that the underlying power system model admits an energy function, or at least a local energy function, whose domain covers the stability region of interest.

A Time–Domain-Based Controlling UEP Method

Step 1. Compute the exit point relative to a fault-on trajectory using a time–domain-based method.

Step 2. Use the exit point as an initial condition and integrate the postfault system until the norm of the end point of the simulated postfault trajectory is smaller than a threshold value.

Step 3. Use the end point as an initial guess and solve for an equilibrium point of the postfault system. Let the solution be x_{co}. The exact controlling UEP with respect to the fault-on trajectory is x_{co}.

Remarks:

1. Step 2 requires an effective numerical method to ensure that the simulated postfault trajectory moves along the stability boundary and stays close to it.
2. From a computational viewpoint, it is very difficult to compute the exact exit point because the simulated trajectory is composed of a sequence of points along the entire trajectory. Hence, Step 1 of the above method usually gives a point close to the exit point. If the point computed by Step 1 is not sufficiently close to the exact exit point, then a divergence problem might occur at Step 3. If this situation arises, it is recommended that the step size T involved in Step 1 be reduced. This recommendation is based on the general property that the system trajectory depends continuously on initial conditions for general nonlinear dynamical systems.
3. A time–domain-based method required in Step 1 is presented below.

Time–Domain Method for Computing the Exit Point

Step 1. Integrate the postfault system starting from a point on the fault-on trajectory, say, $x_f(t)$. If the resulting postfault trajectory does not converge to the postfault SEP, then the point, say, $x_f(k_1T)$, lies outside the stability region; stop and go to Step 2; otherwise, repeat the process of integrating the postfault system starting from the next point (in the forward-time sense) along the fault-on trajectory until the first point along the fault-on trajectory whose corresponding postfault trajectory does not converge to the postfault SEP. Let the first point be denoted as $x_f(kT)$ and its previous point on the fault-on trajectory be denoted as $x_f((k-1)T)$. Go to Step 3.

Step 2. Integrate the postfault system starting from $x_f((k_1-1)T)$. If the resulting postfault trajectory does not converge to the postfault SEP, then set $k_1 = k_1 - 1$ and repeat Step 2; otherwise, set $k_2 = k_1$ and go to Step 3.

Step 3. The exit point, denoted as x_{ex}, lies between the point $x_f(k_1T)$ and the point $x_f(k_2T)$. Apply the golden bisection method to these two points to obtain an accurate exit point.

12.7 NUMERICAL STUDIES

In order to illustrate the controlling UEP and the controlling UEP method presented in the previous section in a simple context, we consider the following equation, which nearly represents a three-machine system in an absolute angle coordinate (see Figure 12.3). Here, the system model considered is a network-reduction model:

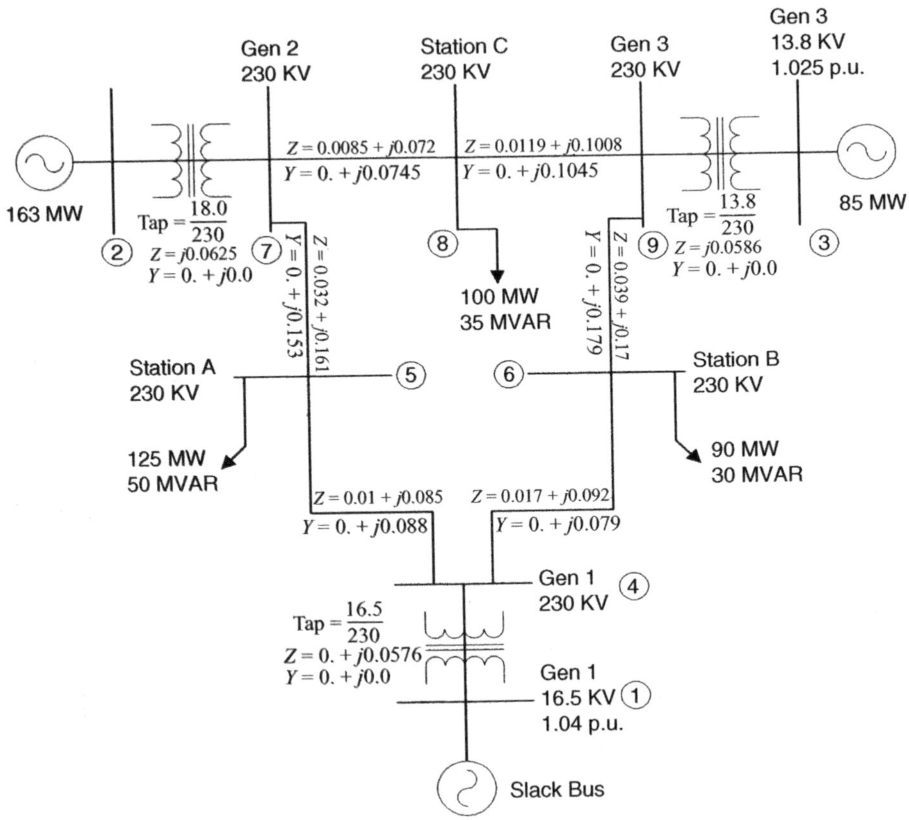

Figure 12.3 A three-machine, nine-bus system; the value of Y is half the line charging.

$$
\begin{aligned}
\dot{\delta}_1 &= \omega_1 \\
\dot{\delta}_2 &= \omega_2 \\
\dot{\delta}_3 &= \omega_3 \\
m_1\dot{\omega}_1 &= -d_1\omega_1 + P_{m_1} - P_{e_1}(\delta_1, \delta_2, \delta_3) \\
m_2\dot{\omega}_2 &= -d_2\omega_2 + P_{m_2} - P_{e_2}(\delta_1, \delta_2, \delta_3) \\
m_3\dot{\omega}_3 &= -d_3\omega_3 + P_{m_3} - P_{e_3}(\delta_1, \delta_2, \delta_3),
\end{aligned}
\tag{12.14}
$$

where

$$
P_{e_i}(\delta_1, \delta_2, \delta_3) = \sum_{j=1, j\neq 1}^{3} E_i E_j \left(B_{ij}\sin(\delta_i - \delta_j) + G_{ij}\cos(\delta_i - \delta_j)\right),
$$

$P_{m1} = 0.8980$, $P_{m2} = 1.3432$, $P_{m3} = 0.9419$, $E_1 = 1.1083$, $E_2 = 1.1071$, and $E_3 = 1.0606$. The prefault power flow solution is summarized in Table 12.1. The system admittance matrix of the prefault system is summarized in Figure 12.4.

Table 12.1 The Prefault Power Flow Solution of the Three-Machine, Nine-Bus System

Bus number	Bus type	Voltage (p.u.)	P_G (p.u.)	Q_G (p.u.)	$-P_L$ (p.u.)	$-Q_L$ (p.u.)
1	Swing	1.04	0.716	0.27	—	—
2	P-V	$1.025 \angle 9.3°$	1.63	0.067	—	—
3	P-V	$1.025 \angle 4.7°$	0.85	−0.109	—	—
4	P-Q	$1.026 \angle -2.2°$	—	—	—	—
5	P-Q	$0.996 \angle -4.0°$	—	—	1.25	0.5
6	P-Q	$1.013 \angle -3.7°$	—	—	0.9	0.3
7	P-Q	$1.026 \angle 3.7°$	—	—	—	—
8	P-Q	$1.016 \angle 0.7°$	—	—	1.00	0.35
9	P-Q	$1.032 \angle 2.0°$	—	—	—	—

$$
\begin{bmatrix}
-j17.361 & 0 & 0 & j17.361 & 0 & 0 & 0 & 0 & 0 \\
0 & -j16 & 0 & 0 & 0 & 0 & j16 & 0 & 0 \\
0 & 0 & -j17.065 & 0 & 0 & 0 & 0 & 0 & j17.065 \\
j17.361 & 0 & 0 & \begin{matrix}3.307\\-j39.309\end{matrix} & \begin{matrix}-1.365\\+j11.604\end{matrix} & \begin{matrix}-1.942\\+j10.511\end{matrix} & 0 & 0 & 0 \\
0 & 0 & 0 & \begin{matrix}-1.365\\+j11.604\end{matrix} & \begin{matrix}2.553\\-j17.338\end{matrix} & 0 & \begin{matrix}-1.188\\+j5.975\end{matrix} & 0 & 0 \\
0 & 0 & 0 & \begin{matrix}-1.942\\+j10.511\end{matrix} & 0 & \begin{matrix}3.224\\-j15.841\end{matrix} & 0 & 0 & \begin{matrix}-1.282\\+j5.588\end{matrix} \\
0 & j16 & 0 & 0 & \begin{matrix}-1.188\\+j5.975\end{matrix} & 0 & \begin{matrix}-2.805\\-j35.4460\end{matrix} & \begin{matrix}-1.617\\+j13.689\end{matrix} & 0 \\
0 & 0 & 0 & 0 & 0 & 0 & \begin{matrix}-1.617\\+j13.689\end{matrix} & \begin{matrix}-2.437\\-j32.1540\end{matrix} & \begin{matrix}-1.617\\+j13.689\end{matrix} \\
0 & 0 & j17.065 & 0 & 0 & \begin{matrix}-1.282\\+j5.588\end{matrix} & 0 & \begin{matrix}-1.617\\+j13.689\end{matrix} & \begin{matrix}-2.437\\-j32.1540\end{matrix}
\end{bmatrix}
$$

Figure 12.4 System admittance matrix of the base case system.

We consider a normal loading condition with a uniform damping factor $d_i/m_i = 0.1$ with $[d_1, d_2, d_3] = [0.0125, 0.0034, 0.0016]$. The coordinate of the prefault SEP is [−0.0482, 0.1252, 0.1124]. The postfault SEP relative to each contingency is computed using the method described in Section 12.3. The contingency list, the associate fault, and the postfault SEP are summarized in Table 12.2.

The controlling UEP relative to each contingency is computed using the time–domain-based controlling UEP method and is summarized in Table 12.3.

To illustrate the computational procedure involved in the controlling UEP method, we numerically simulate the following dynamic objects for several contingencies:

- prefault SEP and postfault SEP,
- UEPs on the stability boundary of the postfault system,

Table 12.2 The Contingency List, the Associate Fault, and the Postfault SEP

Contingency number	Faulted bus	Fault clearing type	Description		Postfault SEP (δ_1, δ_2, δ_3)
			From bus	To bus	
1	7	Line trip	7	5	(−0.1204, 0.3394, 0.2239)
2	7	Line trip	8	7	(−0.0655, 0.2430, −0.0024)
3	5	Line trip	7	5	(−0.1204, 0.3394, 0.2239)
4	5	Line trip	5	4	(−0.0231, 0.0467, 0.0817)
5	4	Line trip	4	6	(−0.0319, 0.0949, 0.0492)
6	4	Line trip	5	4	(−0.0231, 0.0467, 0.0817)
7	9	Line trip	6	9	(−0.0967, 0.2180, 0.2958)
8	9	Line trip	9	8	(−0.0462, 0.0728, 0.2082)
9	8	Line trip	9	8	(−0.0462, 0.0728, 0.2082)
10	8	Line trip	8	7	(−0.0655, 0.2430, −0.0024)
11	6	Line trip	4	6	(−0.0319, 0.0949, 0.0492)
12	6	Line trip	6	9	(−0.0967, 0.2180, 0.2958)

Table 12.3 The Controlling UEP Relative to Each Contingency

Contingency number	Controlling UEP (δ_1, δ_2, δ_3)
1	(−0.7589, 1.9528, 1.8079)
2	(−0.5424, 2.1802, −0.3755)
3	(−0.7589, 1.9528, 1.8079)
4	(−0.8364, 2.0797, 2.1466)
5	(−0.8256, 2.0830, 2.0549)
6	(−0.8364, 2.0797, 2.1466)
7	(−0.7576, 1.8583, 1.9986)
8	(−0.2910, −0.1011, 2.5008)
9	(−0.2910, −0.1011, 2.5008)
10	(−0.3495, 0.0745, 2.5864)
11	(−0.8256, 2.0830, 2.0549)
12	(−0.7576, 1.8583, 1.9986)

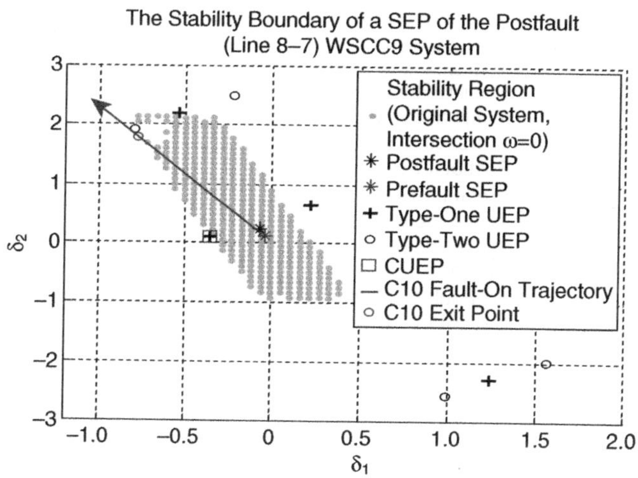

Figure 12.5 For Contingency 10, the stability region of the postfault system on the angle plane is highlighted. The fault-on trajectory intersects the stable manifold of the controlling UEP at the exit point, as shown in the figure. Contrary to the claim made in the literature, the controlling UEP is not the UEP closest to the fault-on trajectory.

- the stability boundary of a postfault system on the machine angle space, and
- fault-on trajectory.

These dynamic objects are simulated due to the considerations that the fault-on trajectory starting from the prefault SEP intersects the stability boundary of the postfault system at the exit point. The exit point lies on the stable manifold of the controlling UEP. The relevant stability boundary towards which the fault-on trajectory heads is the stable manifold of the controlling UEP.

For Contingency 10, a fault occurs near Bus 8, and the line between Buses 7 and 8 is tripped. The exact stability region of the postfault system on the angle plane is highlighted (see Figure 12.5). The fault-on trajectory intersects the stable manifold of the controlling UEP, a type-one UEP, highlighted in the figure. There are two type-one UEPs and one type-two UEP lying on the stability boundary. The stable manifold of one type-one UEP, the controlling UEP, contains the exit point, which is the intersection between the stability boundary and the fault-on trajectory. This exit point is at some distance from the controlling UEP, while it is very close to a type-two UEP lying on the stability boundary. However, it is incorrect to use this type-two UEP as the controlling UEP.

Contrary to the claim made in the literature, this numerical example clearly shows that the controlling UEP is not the UEP closest to the fault-on trajectory. The controlling UEP is the UEP whose stable manifold contains the exit point.

For Contingency 2, a fault occurs near Bus 7, and the line between Buses 7 and 8 is tripped. The stability region of the postfault system on the angle plane is highlighted (see Figure 12.6). The fault-on trajectory intersects the stable manifold of

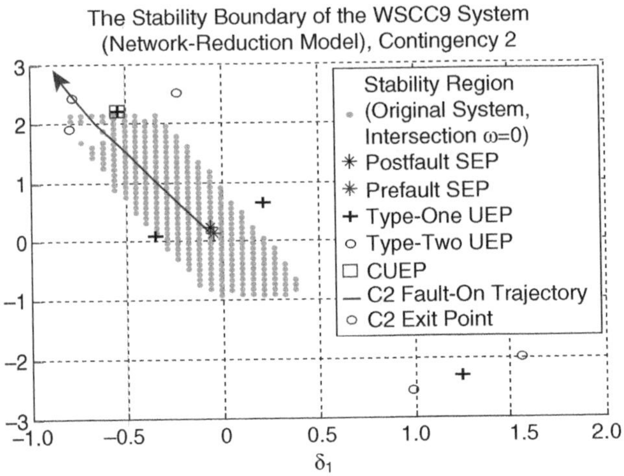

Figure 12.6 For Contingency 2, the stability region of the postfault system on the angle plane is highlighted. The fault-on trajectory intersects the stable manifold of the controlling UEP, a type-one UEP, at the exit point, as shown in the figure.

the controlling UEP at the exit point, as shown in the figure. There are two type-one UEPs and one type-two UEP lying on the stability boundary. The stable manifold of one type-one UEP, the controlling UEP, contains the exit point, which is the intersection between the stability boundary and the fault-on trajectory. This exit point is close to the controlling UEP, while it is at some distance from the other type-one UEP. It is interesting to note that the exit point is also close to a type-two UEP.

For Contingency 7, a fault occurs near Bus 9, and the line between Buses 6 and 9 is tripped. The stability region of the postfault system on the angle plane is highlighted (see Figure 12.7). The fault-on trajectory intersects the stable manifold of the controlling UEP, a type-one UEP, at the exit point, as shown in the figure. There is only one type-one UEP lying on the stability boundary. Hence, the stability boundary of the postfault system is unbounded in the direction of the machine speed.

For Contingency 12, a fault occurs near Bus 6, and the line between Buses 6 and 9 is tripped. The stability region of the postfault system on the angle plane is highlighted (see Figure 12.8). The fault-on trajectory intersects the stable manifold of the controlling UEP at the exit point, as shown in the figure. There is only one type-one UEP lying on the stability boundary. Hence, the stability boundary of the postfault system is unbounded (in the direction of the machine speed). The exit point is very close to the controlling UEP.

12.8 CONCLUDING REMARKS

The task of computing the controlling UEPs is very difficult due to the seven computational challenges presented in this chapter. These challenges serve to explain why previous methods proposed in the literature fail to compute the controlling UEP. The main reason for their failure is that these methods attempt to directly compute the controlling UEP of the power system stability model, which, as observed in Challenges I–III, is difficult if not impossible without using the iterative time–domain method. Another important factor that has been overlooked by the previously proposed methods is the fact that the controlling UEP is the first UEP whose stable manifold has a nonempty intersection with the fault-on trajectory. This fact must be incorporated into a numerical method for computing the controlling UEP.

Several methods for computing the controlling UEP, such as the machine acceleration method (Pavella and Murthy, 1994; Ribbens-Pavella and Lemal, 1976), the system separation method (Fouad and Stanton, 1981), and the MOD method (Fouad and Vittal, 1991), all share some common properties: (1) these methods provide initial guesses for computing the controlling UEP based on either heuristics or physical arguments, and (2) these methods are based on the original power system model. Challenges I–III question the correctness of any method attempting to directly compute the controlling UEP of a power system stability model. The only known method that can compute the controlling UEP of a power system stability model is the iterative time–domain-based method presented in this chapter.

The ability to compute the controlling UEP is vital to direct stability analysis. It may prove fruitful to develop a tailored method for finding controlling UEPs by exploiting special properties as well as some physical and mathematical insights into the underlying power system model. We will present such a systematic method, called the BCU method, along this line for computing controlling UEPs of power system transient stability models. The BCU method does not attempt to directly compute the controlling UEP of a power system stability model, termed the original model; instead, it computes the controlling UEP of a reduced-state model and relates the computed controlling UEP to the controlling UEP of the original model. The BCU method, its theoretical foundation, as well as its numerical implementation will be presented in great detail in the following chapters.

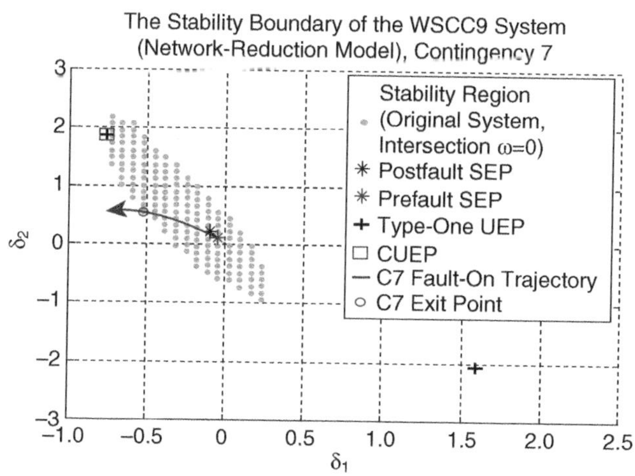

Figure 12.7 For Contingency 7, the stability region of the postfault system on the angle plane is highlighted. There is only one type-one UEP lying on the stability boundary; hence, the stability region is unbounded.

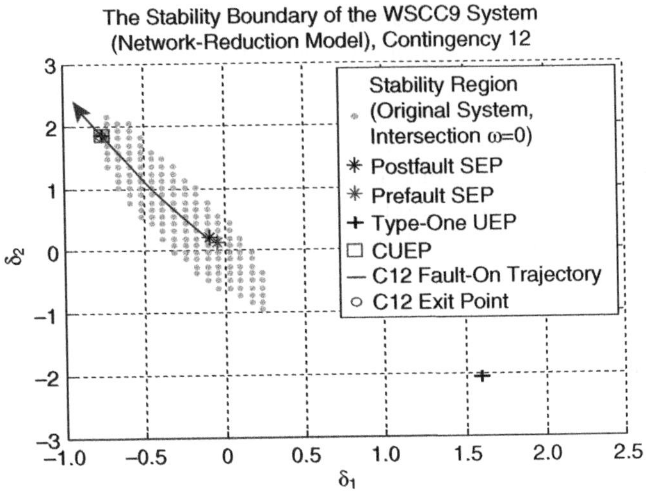

Figure 12.8 For Contingency 12, the stability region of the postfault system on the angle plane is highlighted. The fault-on trajectory intersects the stable manifold of the controlling UEP at the exit point, which is very close to the controlling UEP.

Chapter 13

Foundations of Controlling UEP Methods for Network-Preserving Transient Stability Models

13.1 INTRODUCTION

Many methods for direct stability analysis have been developed for the network-reduction models, which are described by a set of ordinary differential equations (ODEs). These types of classical models were common during the early developmental stage of direct methods. Network-preserving (i.e., structure-preserving) transient stability models were first proposed in the 1980s to improve some shortcomings of classical models. Network-preserving models are described by a set of differential and algebraic equations (DAEs). There are several advantages to using network-preserving power system models, as compared to the network-reduction models, for direct stability analysis:

1. From a modeling viewpoint, it allows for the development of more realistic representations of power system components such as load behavior and generator models.
2. From a computational viewpoint, it allows for the use of the sparse matrix technique for solving nonlinear algebraic equations involved in direct methods.
3. From the energy function viewpoint, the transfer conductance of the models is significantly smaller than that of the network-reduction models. The resulting numerical energy functions are "close" to (exact) energy functions.

Direct Methods for Stability Analysis of Electric Power Systems, by Hsiao-Dong Chiang

Nonlinear behaviors of DAE are very complex. For instance, the stable manifold of an equilibrium point of DAEs must lie on the algebraic submanifold of the DAEs. The stability regions of DAE systems must also lie on the algebraic submanifold, and these structures are complex. Singularity surfaces also exist in the algebraic submanifolds, making system trajectories near singularity surfaces difficult to analyze (Chen and Aihara, 2001; Chua and Deng, 1989; Praprost and Loparo, 1996). In addition, jump behaviors occur in DAE systems (Sastry and Desoer, 1981; Sastry and Varaiya, 1980). These jump behaviors are very complex and are only partially understood.

In the context of direct stability analysis, DAE formulation poses great analytical and computational challenges. For instance, the fault-on trajectory will not intersect with the stability boundary of its DAE postfault system because the algebraic submanifold of the fault-on system and the algebraic submanifold of the postfault system are different. For DAE formulation, jump behaviors consist of external and internal jump behaviors (Chiang et al., 1995; Zou et al., 2003). This internal jump results from the existence of singular surfaces located in the algebraic submanifold, while the external jump is due to the difference between the algebraic submanifold of the fault-on system and that of the postfault system. There are indeed some fundamental differences between DAE and ODE formulations.

The singular perturbation approach has been proposed to bridge the gap between the ODE and DAE systems. For instance, it was suggested that the DAE system be transformed into a system including a set of boundary layer equations to ensure that the fault-on trajectory intersects the stability boundary of the postfault boundary layer equations. Hence, the controlling unstable equilibrium point (UEP) for the DAE model can be defined via the controlling UEP of the associated singularly perturbed model, which is an ODE model.

This chapter presents a controlling UEP method for network-preserving transient stability models. In addition, theoretical foundations of controlling UEP methods for network-preserving transient stability models will be presented. The controlling UEP of a network-preserving transient stability model is defined using the controlling UEP of the associated singularly perturbed model. The controlling UEP method for DAE systems will be numerically illustrated on how the controlling UEP of a network-preserving transient stability model is computed and how the constant energy surface passing through the controlling UEP approximates the relevant stability boundary. Readers are advised to refer to an extended controlling UEP method for the direct stability analysis of network-preserving models (Zou et al., 2003). The extended controlling UEP method examines both external jump behaviors and the constrained stability boundaries of network-preserving transient stability models. A theoretical foundation for the extended controlling UEP method is also developed (Zou et al., 2003).

13.2 SYSTEM MODELS

Network-preserving transient stability models are mathematically described in the following DAE form:

$$\begin{aligned} \dot{x} &= f(x, y) \\ 0 &= g(x, y), \end{aligned} \tag{13.1}$$

where $x \in R^n$ and $y \in R^m$ are the corresponding dynamic and static variables of the systems, respectively. Regarding this DAE form, the transient stability problem is formulated in the following mathematical form. In the prefault regime, the system model is represented as

$$\begin{aligned} \dot{x} &= f_{pre}(x, y) \\ 0 &= g_{pre}(x, y) \end{aligned} \quad t \in \left[0, t_0^-\right]. \tag{13.2}$$

The system undergoes an event disturbance at time t_0, which results in a structural change in the network topology. Suppose the fault duration is confined to the time interval $\left[t_0^+, t_{cl}^-\right]$. During this interval, the system is governed by the fault-on dynamics,

$$\begin{aligned} \dot{x} &= f_{pre}(x, y) \\ 0 &= g_{pre}(x, y) \end{aligned} \quad t \in \left[t_0^+, t_{cl}^-\right]. \tag{13.3}$$

After the fault is clear at time t_{cl}, the system, termed the postfault system, is henceforth governed by postfault dynamics described by

$$\begin{aligned} \dot{x} &= f(x, y), \quad t_{cl} \le t \le \infty \\ 0 &= g(x, y). \end{aligned} \tag{13.4}$$

The network configuration may or may not be the same as the prefault configuration in the postfault system. We will use the notation $z(t_{cl}) = (x(t_{cl}), y(t_{cl}))$ to denote the fault-on state at switching time t_{cl}. The DAE system (Eq. 13.1) can be interpreted as an implicitly dynamic system defined on the algebraic manifold L:

$$L = \{(x, y) : g(x, y) = 0\}.$$

All the system states such as the equilibrium points, stable and unstable manifolds, and stability regions must lie in the above algebraic manifold (or termed constrained manifold). On the algebraic manifold L, a complex set, termed *singular surfaces*, defined as follows:

$$S = \left\{(x, y) : (x, y) \in L, \Delta(x, y) = \det \frac{\partial g}{\partial y}(x, y) = 0\right\}, \tag{13.5}$$

may exist. Singular surfaces, when they exist, decompose the algebraic manifold L into several disjoint components. If the points on a component are so that the corresponding Jacobian matrix $\frac{\partial g}{\partial y}(x, y)$ has all the eigenvalues with negative real parts, then the component is stable; otherwise, it is an unstable component.

Complicated dynamic behaviors of DAE systems occur in the vicinity of the singular surface because its vector field is unbounded on the singular surface.

Trajectories move very fast near singular surfaces. Once the trajectory reaches the singular surface, an unsmooth jump behavior (i.e., an internal jump) will occur and force the trajectory to reach another component of the algebraic manifold.

If the Jacobian matrix $\frac{\partial g}{\partial y}(x, y)$ is nonsingular, that is, the system is on the regular part of the DAE, then by the implicit function theorem, the system equations (Equation 13.1) are locally equivalent to the following equations:

$$\begin{aligned} \dot{x} &= f(x, y) \\ \dot{y} &= -\left(\frac{\partial g}{\partial y}(x, y)\right)^{-1} \frac{\partial g}{\partial x}(x, y) f(x, y). \end{aligned} \tag{13.6}$$

The existence and uniqueness of solutions of DAE in a neighborhood N can be guaranteed provided that functions f and g are smooth and the Jacobian matrix $\frac{\partial g}{\partial y}(x, y)$ has a constant full rank (Hill and Mareels, 1990). We say (x, y) is an equilibrium point of the system (Eq. 13.1) if $f(x, y) = 0$ and $g(x, y) = 0$. A regular equilibrium point is a type-k equilibrium point of Equation 13.1 if (x, y) is a type-k equilibrium point of Equation 13.6. The stability of an equilibrium point of DAE systems can be analyzed by using a local energy function as shown in the following.

Lemma 13.1: Stability of an Equilibrium Point

Let $(\bar{x}, \bar{y})$ be an equilibrium point of the system (Eq. 13.1) and N be a small neighborhood of $(\bar{x}, \bar{y})$ in L. If there exists a smooth positive definite function $V: N \to R$, such that

$$\dot{V} = \frac{\partial}{\partial x} V(x, y) f(x, y) - \frac{\partial}{\partial x} V(x, y) \left(\frac{\partial}{\partial x} g(x, y)\right)^{-1} \frac{\partial}{\partial x} V(x, y) f(x, y) \leq 0,$$

then the equilibrium point $(\bar{x}, \bar{y})$ is stable.

It is generally difficult to extend the above local results into global results since, once the trajectory intersects the singular surface S, some elements of the vector field of the DAE system become unbounded.

13.3 STABILITY REGIONS

The singular surface S decomposes the algebraic manifold L into several disjoint components, Γ_i. If all the points on some Γ_i are such that the Jacobian matrix $\frac{\partial}{\partial x} g(x, y)$ has eigenvalues with negative real parts, then Γ_i is a stable component; otherwise, it is an unstable component. Let Γ_s be a stable component of L and $\Phi_t(x, y)$ be a trajectory of the DAE system (Eq. 13.1) starting from (x, y). The stability region of a given stable equilibrium point (SEP) (x_s, y_s) of a DAE system (Eq. 13.1) is defined as

$$A(x_s, y_s) = \left\{(x, y) \in \Gamma_s : \lim_{t\to\infty} \Phi_t(x, y) = (x_s, y_s)\right\}. \tag{13.7}$$

Here we restrict the stability region to lying on the stable component Γ_s and exclude other stable components from which trajectories pass through a singular surface and converge to the SEP (x_s, y_s). Similarly, the stable manifold and the unstable manifold of an equilibrium point, $(\bar{x}, \bar{y})$, on Γ_s are defined as follows:

$$\begin{aligned} W^s(\bar{x}, \bar{y}) &= \left\{(x, y) \in \Gamma_s : \lim_{t\to\infty} \Phi_t(x, y) = (\bar{x}, \bar{y})\right\} \\ W^u(\bar{x}, \bar{y}) &= \left\{(x, y) \in \Gamma_s : \lim_{t\to-\infty} \Phi_t(x, y) = (\bar{x}, \bar{y})\right\}. \end{aligned} \tag{13.8}$$

Characterizations of the stability boundary $\partial A(x_s, y_s)$ of DAE systems have recently been developed. It has been shown that under certain conditions, the stability boundary $\partial A(x_s, y_s)$ consists of two parts: the first part is the stable manifolds of the equilibrium points on the stability boundary, while the second part contains points whose trajectories reach singular surfaces (Chiang and Fekih-Ahmed, 1992; Fekih-Ahmed, 1991). The second part can be further delineated as a union of the stable manifolds of pseudoequilibrium points and semisingular points on the stability boundary and parts of the singular surface (Venkatasubramanian et al., 1991, 1995a).

13.4 SINGULAR PERTURBATION APPROACH

The singular perturbation approach treats the set of algebraic equations describing a DAE system as a limit of the fast dynamics: $\varepsilon\dot{y} = g(x, y)$. In other words, as ε approaches zero, the fast dynamics will approach the algebraic manifold. Therefore, for the DAE system (Eq. 13.1), we can define an associated singularly perturbed system:

$$\begin{aligned} \dot{x} &= f(x, y) \\ \varepsilon\dot{y} &= g(x, y), \end{aligned} \tag{13.9}$$

where ε is a sufficiently small positive number. If f and g are both smooth functions and are bounded for all $(x, y) \in R^{n+m}$, then the vector field is globally well-defined. The state variables of the system (Eq. 13.1) have very different rates of dynamics, and they can be separated into two distinct time scales: slow variable x and fast variable y. A DAE system and its corresponding singularly perturbed system share several similar dynamic properties. The following results show an invariant topological relationship among equilibrium points between a DAE system and its associated singularly perturbed system.

Theorem 13.1: Topological Property

If an equilibrium point, say, $(\bar{x}, \bar{y})$, of the system (Eq. 13.1) lies on one stable component, Γ_s, of the constraint manifold, then there exists an $\varepsilon > 0$ such that for all $\varepsilon \in (0, \varepsilon)$, it follows that

1. $(\bar{x}, \bar{y})$ is a hyperbolic equilibrium point of the DAE system (Eq. 13.1) if and only if $(\bar{x}, \bar{y})$ is a hyperbolic equilibrium point of the singularly perturbed system (Eq. 13.9); moreover,
2. $(\bar{x}, \bar{y})$ is a type-k equilibrium point of the DAE system (Eq. 13.1) if and only if $(\bar{x}, \bar{y})$ is a type-k equilibrium point of the singularly perturbed system (Eq. 13.9).

The above result shows that the type of the equilibrium point of the DAE system (Eq. 13.1) is the same as the type of the corresponding equilibrium point of the singularly perturbed system (Eq. 13.9), provided ε is sufficiently small. Theorem 13.2 below extends this local result to a global result and shows that the stability boundaries of these two systems contain the same set of equilibrium points on stable components.

Theorem 13.2: (Chiang and Fekih-Ahmed, 1992; Fekih-Ahmed, 1991; Venkatasubramanian et al., 1991)

Let (x_s, y_s) and (x_u, y_u) be the SEP and UEP of the DAE system (Eq. 13.1) on the stable component Γ_s, respectively. Suppose that, for each $\varepsilon > 0$, the associated singularly perturbed system (Eq. 13.9) has an energy function and its equilibrium points are isolated. Then there exists an $\varepsilon > 0$ such that for all $\varepsilon \in (0, \varepsilon)$, (x_u, y_u) lies on the stability boundary $\partial A_0(x_s, y_s)$ of the DAE system (Eq. 13.1) if and only (x_u, y_u) lies on the stability boundary $\partial A\varepsilon(x_s, y_s)$ of the singularly perturbed system (Eq. 13.9).

The above theorem provides a theoretical basis for direct stability analysis of a network-preserving system via the stability analysis of its associated singularly perturbed system, which is an ODE system. This enables the controlling UEP method developed for network-reduction transient stability models to be extendable to network-preserving transient stability models.

Direct stability analysis of network-preserving models via the singular perturbation approach to extend the controlling UEP method for DAE systems also provides advantages in the following aspects:

1. **Energy Function.** An energy function for a singularly perturbed network-preserving model, which is an ODE system, can be easily obtained.
2. **Computing the Controlling UEP.** Since the final state of a fault-on trajectory will not lie on the algebraic manifold L of its postfault system, the

fault-on trajectory will not hit the stability boundary of its postfault DAE system. Instead, the fault-on trajectory will hit the stability boundary of the corresponding singularly perturbed postfault system. Hence, the exit point of a fault-on trajectory must lie on the stable manifold of the controlling UEP of its singularly perturbed postfault system.

Note that trajectories of the singularly perturbed system (Eq. 13.9) will not be confined to the algebraic manifold L and they are not exactly the same as that of the original DAE system (Eq. 13.1). However, trajectories generated by the singularly perturbed system are still valid approximations to those of the DAE system. A theoretical justification to ensure that the difference of solution trajectories between the original DAE (Eq. 13.1) and the singularly perturbed system (Eq. 13.9) is uniformly bounded by the order of $O(\varepsilon)$ is provided by Tikhonov's theorem over the infinite time interval (Hoppensteadt, 1974; Khalil, 2002; Sastry, 1999).

13.5 ENERGY FUNCTIONS FOR NETWORK-PRESERVING MODELS

The energy function theory, the controlling UEP method, and its theoretical foundation presented in the previous chapters are applicable to transient stability models described by ODEs. We next consider the extension of the energy function theory developed for ODE systems to DAE systems via the singular perturbation approach.

A majority of existing lossy network-preserving models can be rewritten as a set of general DAEs of the following compact form (Chu and Chiang, 2005):

$$
\begin{aligned}
0 &= -\frac{\partial U}{\partial u}(u, w, x, y) + g_1(u, w, x, y) \\
0 &= -\frac{\partial U}{\partial w}(u, w, x, y) + g_2(u, w, x, y) \\
T\dot{x} &= -\frac{\partial U}{\partial x}(u, w, x, y) + g_3(u, w, x, y) \\
\dot{y} &= z \\
M\dot{z} &= -Dz - \frac{\partial U}{\partial y}(u, w, x, y) + g_4(u, w, x, y),
\end{aligned}
\tag{13.10}
$$

where $u \in \Re^k$ and $w \in \Re^l$ are instantaneous variables, while $x \in \Re^m$, $y \in \Re^n$, and $z \in \Re^n$ are state variables. T is a positive definite matrix, and M and D are diagonal positive definite matrices. Here, differential equations describe generator and/or load dynamics, while algebraic equations express the power flow equations at each bus. $g_1(u, w, x, y)$, $g_2(u, w, x, y)$, $g_3(u, w, x, y)$, and $g_4(u, w, x, y)$ are vectors representing the effects of the transfer conductance in the network Y-bus matrix. With the aid of the singularly perturbed systems, the compact representation of the lossy network-preserving model becomes

$$
\begin{aligned}
\varepsilon_1 \dot{u} &= -\frac{\partial U}{\partial u}(u, w, x, y) + g_1(u, w, x, y) \\
\varepsilon_2 \dot{u} &= -\frac{\partial U}{\partial w}(u, w, x, y) + g_2(u, w, x, y) \\
T\dot{x} &= -\frac{\partial U}{\partial x}(u, w, x, y) + g_3(u, w, x, y) \\
\dot{y} &= z \\
M\dot{z} &= -Dz - \frac{\partial U}{\partial y}(u, w, x, y) + g_4(u, w, x, y),
\end{aligned} \tag{13.11}
$$

where ε_1 and ε_2 are sufficiently small positive numbers. The singular perturbation version of the network-preserving model (Eq. 13.11) belongs to the compact representation of existing network-reduction models. By employing the techniques developed for the network-reduction power system models, one can construct analytical energy functions for the class of lossless network-preserving models (without the transfer conductance) described as follows. For the compact representation of the singularly perturbed network-preserving power system model (Eq. 13.11) without the transfer conductance, we consider the following function W: $R^{k+l+2n+m} \rightarrow R$:

$$
W(u, w, x, y, z) = K(z) + U(u, w, x, y) = \frac{1}{2} z^T M z + U(u, w, x, y). \tag{13.12}
$$

If, along every nontrivial trajectory $(u(t), w(t), x(t), y(t), z(t))$ of the system (Eq. 13.11) with a bounded value of $W(u, w, x, y, z)$, the vector $(u(t), w(t), x(t))$ is also bounded for $t \in R^+$, then $W(u, w, x, y, z)$ is an energy function for the system (Eq. 13.11). As described in Chapter 6, the function (Eq. 13.12) is an analytical energy function for several existing network-preserving models without losses (i.e., lossless models).

Regarding general lossy network-preserving transient stability models, it can be shown that an analytical expression of energy functions does not exist for these models. Consequently, numerical energy functions must be resorted. A numerical network-preserving energy function, $W_{num}(u, w, x, y)$, can be constructed by combining an analytic energy function, $W_{ana}(u, w, x, y, z) = K(z) + U(u, w, x, y)$, and a path-dependent potential energy, $U_{path}(u, w, x, y)$, using the several schemes described in Chapter 7.

13.6 CONTROLLING UEP FOR DAE SYSTEMS

The concept of controlling UEP is well-defined for network-reduction models described by ODEs. One key difficulty in applying the concept of the controlling UEP to network-preserving models is that the final state of a fault-on trajectory will not lie on the algebraic manifold of its postfault DAE system. Instead, the fault-on trajectory will hit the stability boundary of the corresponding singularly perturbed postfault system.

Hence, the controlling UEP of a network-preserving model can be defined via the controlling UEP of the associated singularly perturbed model. To this end, we first study the controlling UEP of the singularly perturbed model (Eq. 13.11).

Definition: Controlling UEP for a Singularly Perturbed Model
The controlling UEP of the singularly perturbed model (Eq. 13.11) for a fixed, small $\varepsilon > 0$, with respect to a fault-on trajectory, is the UEP on the stability boundary $\partial A_\varepsilon(u_s, w_s, x_s, y_s, 0)$ of the singularly perturbed system (Eq. 13.11) whose stable manifold contains the exit point of the fault-on trajectory.

This definition is based on the fact that the exit point must lie on the stable manifold of some UEP on the stability boundary of the postfault singularly perturbed model (Eq. 13.11). By employing characterizations of the stability boundary of the postfault singularly perturbed system (Eq. 13.11), one can prove the existence and uniqueness of the controlling UEP of the singularly perturbed model (Eq. 13.11).

Theorem 13.3: Existence and Uniqueness of the Controlling UEP

If the postfault singularly perturbed system (Eq. 13.11) satisfies the following assumptions:

(A1) all the equilibriums are hyperbolic, and

(A2) there exists an energy function, for a sufficiently small $\varepsilon > 0$,

then the controlling UEP, with respect to a fault-on trajectory of the singularly perturbed system (Eq. 13.11), for each fixed small $\varepsilon > 0$, exists and is unique.

Having shown the existence and uniqueness of a controlling UEP for the singularly perturbed system (Eq. 13.11), it is desirable to establish a relationship between the controlling UEP of the singularly perturbed system (Eq. 13.5) and the controlling UEP of the DAE system (Eq. 13.1). In other words, it is desirable to study the behavior of the controlling UEP for the singularly perturbed system (Eq. 13.5) when $\varepsilon \to 0$. This requirement leads to the following definition of uniform controlling UEP.

Definition: Uniform Controlling UEP
Let $(u_{co}^{\varepsilon}, w_{co}^{\varepsilon}, x_{co}^{\varepsilon}, y_{co}^{\varepsilon}, z_{co}^{\varepsilon})$ be the controlling UEP of the singularly perturbed postfault system (Eq. 13.11) with respect to fault-on trajectory $(u^{\varepsilon}(t), w^{\varepsilon}(t), x^{\varepsilon}(t), y^{\varepsilon}(t), z^{\varepsilon}(t))$. Consider the map $\varepsilon \to \left(u_{co}^{\varepsilon}, w_{co}^{\varepsilon}, x_{co}^{\varepsilon}, y_{co}^{\varepsilon}, z_{co}^{\varepsilon}\right)$. If there exists $\varepsilon^* > 0$ such that the map is constant for all $\varepsilon \in (0, \varepsilon^*)$, then $\left(u_{co}^{0}, w_{co}^{0}, x_{co}^{0}, y_{co}^{0}, z_{co}^{0}\right) = \left(u_{co}^{\varepsilon}, w_{co}^{\varepsilon}, x_{co}^{\varepsilon}, y_{co}^{\varepsilon}, z_{co}^{\varepsilon}\right)$ is a *uniform controlling UEP* with respect to the fault-on trajectory $(u^{\varepsilon}(t), w^{\varepsilon}(t), x^{\varepsilon}(t), y^{\varepsilon}(t), z^{\varepsilon}(t))$ for all $\varepsilon \in (0, \varepsilon^*)$.

We next present analytical results establishing the relationship between the controlling UEP of the singularly perturbed system (Eq. 13.5) and the controlling UEP of the DAE system (Eq. 13.1).

Theorem 13.4: Controlling UEP of the DAE System and Uniform Controlling UEP

Consider the structure-preserving transient stability model (Eq. 13.10). Let $(u_s, w_s, x_s, y_s, 0)$ be a SEP of the postfault system on a stable constraint component and let $(u^0_{co}, w^0_{co}, x^0_{co}, y^0_{co}, z^0_{co})$ be the corresponding controlling UEP of the DAE system (Eq. 13.10) with respect to the DAE postfault trajectory $(u^0(t), w^0(t), x^0(t), y^0(t), z^0(t))$ on the stable constrained manifold. If the postfault singularly perturbed system (Eq. 13.11) satisfies the following assumptions:

(A1) all the equilibriums are hyperbolic, and

(A2) there exists an energy function for a sufficiently small $\varepsilon > 0$,

then the following results hold:

1. Given $\eta > 0$, there exists $\varepsilon^* > 0$ such that the exit point of the singularly perturbed fault-on trajectory is η-close to the exit point of the DAE fault-on trajectory for all $\varepsilon \in (0, \varepsilon^*)$.
2. The equilibrium point $(u^0_{co}, w^0_{co}, x^0_{co}, y^0_{co}, z^0_{co})$ is generically a uniform controlling UEP of the postfault singularly perturbed system (Eq. 13.11) with respect to the fault-on trajectory $(u^\varepsilon(t), w^\varepsilon(t), x^\varepsilon(t), y^\varepsilon(t), z^\varepsilon(t))$ for all $\varepsilon \in (0, \varepsilon^*)$.

Theorem 13.4 guarantees that, under the stated conditions, the controlling UEP of the DAE system is the uniform controlling UEP of the postfault singularly perturbed system (Eq. 13.11). Hence, this theorem ensures that the controlling UEP of the DAE system (Eq. 13.10) can be calculated via the controlling UEP of the singularly perturbed system (Eq. 13.11). Furthermore, it can be shown that the asymptotic behavior of the postfault trajectories of the DAE system is the same as that of the postfault trajectory of the singularly perturbed system, provided the initial conditions of these trajectories lie inside the stability region of the postfault singularly perturbed system (Eq. 13.11).

13.7 CONTROLLING UEP METHOD FOR DAE SYSTEMS

We are now in a position to present a controlling UEP method for the direct stability analysis of the generic network-preserving transient stability model (Eq. 13.10). Given a fault-on network-preserving trajectory $(u_f(t), w_f(t), x_f(t), y_f(t), z_f(t))$ and a postfault network-preserving power system model and an energy function, $W(u, w, x, y, z)$, for the postfault DAE power system model (Eq. 13.10), the controlling UEP method for direct stability analysis of the generic network-preserving transient stability model (Eq. 13.10) proceeds as follows:

A Controlling UEP Method for DAE Models

Step 1. Determination of the critical energy

Step 1.1. Find the controlling UEP (u_{co}, w_{co}, x_{co}, y_{co}, 0) relative to the fault-on DAE trajectory ($u_f(t)$, $w_f(t)$, $x_f(t)$, $y_f(t)$, $z_f(t)$).

Step 1.2. The critical energy value W_{cr} is the value of energy function $W(u, w, x, y, z)$ at the controlling UEP; in other words,

$$W_{cr} = W\left(u_{co}, w_{co}, x_{co}, y_{co}, 0\right).$$

Step 2. Direct determination of the stability of the postfault trajectory

Step 2.1. Calculate the state vector at the time of fault clearance (say, t_{cl}) using the fault-on DAE trajectory, ($u_f(t_{cl})$, $w_f(t_{cl})$, $x_f(t_{cl})$, $y_f(t_{cl})$, $z_f(t_{cl})$).

Step 2.2. Compute the initial state of the postfault trajectory, that is, ($u(t_{cl}^+)$, $w(t_{cl}^+)$, $x(t_{cl}^+)$, $y(t_{cl}^+)$, $z(t_{cl}^+)$), using the postfault DAE system and the state vector ($u_f(t_{cl})$, $w_f(t_{cl})$, $x_f(t_{cl})$, $y_f(t_{cl})$, $z_f(t_{cl})$).

Step 2.3. Compute the energy function value of the initial condition of the postfault trajectory, that is,

$$W\left(u\left(t_{cl}^+\right), w\left(t_{cl}^+\right), x\left(t_{cl}^+\right), y\left(t_{cl}^+\right), z\left(t_{cl}^+\right)\right).$$

Step 2.4. If $W\left(u\left(t_{cl}^+\right), w\left(t_{cl}^+\right), x\left(t_{cl}^+\right), y\left(t_{cl}^+\right), z\left(t_{cl}^+\right)\right) < W_{cr}$, then the postfault trajectory of the network-preserving model is stable. Otherwise, it may be unstable.

We next present a fundamental theorem for the controlling UEP method for network-preserving transient stability models.

Theorem 13.5: Fundamental Theorem

Consider a singularly perturbed network-preserving model described by the system (Eq. 13.10), which has an energy function $W(u, w, x, y, z)$. Let p_s be the postfault SEP of the model and let $\hat{p}$ be an equilibrium point on the stability boundary $\partial A(p_s)$. Let $r > W(p_s)$ and

- $S(r) \triangleq$ the connected component of the set $\{x \in R^n: W(x) < r\}$ containing p_s, and
- $\partial S(r) \triangleq$ the connected component of the set $\{x \in R^n: W(x) = r\}$ containing p_s.

Then,

[1] The connected constant energy surface $\partial S(V(\hat{p}))$ intersects the stable manifold $W^s(\hat{p})$ only at the equilibrium point $\hat{p}$; moreover, the set $S(V(\hat{p}))$ has an empty intersection with the stable manifold $W^s(\hat{p})$. In other words, $\partial S(V(\hat{p})) \cap W^s W(\hat{p}) = \hat{p}$ and $S(V(\hat{p})) \cap W^s(\hat{p}) = \phi$.

[2] $S(V(p_u)) \cap W^s(\hat{p}) \neq \phi$ if p_u is an equilibrium point and $V(p_u) > V(\hat{p})$.

[3] $S(V(p_u)) \cap W^s(\hat{p}) = \phi$ if p_u is an equilibrium point and $V(\hat{p}) > V(p_u)$.

[4] If $\hat{p}$ is not the closest UEP, then $\partial S(V(\hat{p})) \cap (\bar{A}(p_s)) \neq \phi$.

[5] Any connected path starting from $P \in \{S(V(\hat{p})) \cap (A(p_s)\}$ and passing through $W^s(\hat{p})$ must hit $\partial S(V(\hat{p}))$ first before the path hits $W^s(\hat{p})$.

The above theorem is similar to the fundamental theorem for the controlling UEP method for network-reduction transient stability models. This theorem asserts that one can use the constant energy surface passing the controlling UEP of the singularly perturbed model to approximate the relevant stability boundary for the fault-on trajectory of the network-preserving model. Once the initial condition of the postfault trajectory lies inside the constant energy surface, the ensuing postfault trajectory of the network-preserving model will converge to the SEP of the postfault DAE system. Thus, one can perform a direct stability analysis of a network-preserving transient stability model via an energy function and the controlling UEP method for the DAE model (Eq. 13.10). In addition, the stability assessments will be on the conservative side.

13.8 NUMERICAL STUDIES

In order to illustrate the controlling UEP method for the DAE system presented in the previous section in a simple context, we consider the following system, which nearly represents a three-machine nine-bus system in an absolute angle coordinate. This nine-bus system was studied in the previous chapter. The system model considered, however, is the network-preserving model of the following DAEs:

$$\dot{\delta}_1 = \omega_1$$

$$\dot{\delta}_2 = \omega_2$$

$$\dot{\delta}_3 = \omega_3$$

$$m_1\dot{\omega}_1 = -d_1\omega_1 + P_{m1} - P_{G1}(\delta_1, \theta_1, V_1)$$

$$m_2\dot{\omega}_2 = -d_2\omega_2 + P_{m2} - P_{G2}(\delta_2, \theta_2, V_2)$$

$$m_3\dot{\omega}_3 = -d_3\omega_3 + P_{m3} - P_{G3}(\delta_3, \theta_3, V_3)$$

$$0 = \sum_{k=1}^{9} V_1V_k(G_{1k}\cos(\theta_1 - \theta_k) + B_{1k}\sin(\theta_1 - \theta_k)) - P_{G_1}(\delta_1, \theta_1, V_1)$$

$$0 = \sum_{k=1}^{9} V_2V_k(G_{2k}\cos(\theta_2 - \theta_k) + B_{2k}\sin(\theta_2 - \theta_k)) - P_{G_2}(\delta_2, \theta_2, V_2)$$

$$0 = \sum_{k=1}^{9} V_3V_k(G_{3k}\cos(\theta_3 - \theta_k) + B_{3k}\sin(\theta_3 - \theta_k)) - P_{G_3}(\delta_3, \theta_3, V_3)$$

$$0 = \sum_{k=1}^{9} V_4V_k(G_{4k}\cos(\theta_4 - \theta_k) + B_{4k}\sin(\theta_4 - \theta_k))$$

$$\cdots$$
$$\cdots$$
$$\cdots$$

$$0=\sum_{k=1}^{9} V_9V_k(G_{9k}\cos(\theta_9-\theta_k)+B_{9k}\sin(\theta_9-\theta_k))$$

$$0=\sum_{k=1}^{9} V_1V_k(G_{1k}\sin(\theta_1-\theta_k)+B_{1k}\cos(\theta_1-\theta_k))-Q_{G_1}(\delta_1,\theta_1,V_1)$$

$$0=\sum_{k=1}^{9} V_2V_k(G_{2k}\sin(\theta_2-\theta_k)+B_{2k}\cos(\theta_2-\theta_k))-Q_{G_2}(\delta_2,\theta_2,V_2)$$

$$0=\sum_{k=1}^{9} V_3V_k(G_{3k}\sin(\theta_3-\theta_k)+B_{3k}\cos(\theta_3-\theta_k))-Q_{G_3}(\delta_3,\theta_3,V_3)$$

$$0=\sum_{k=1}^{9} V_4V_k(G_{4k}\sin(\theta_4-\theta_k)+B_{4k}\cos(\theta_4-\theta_k))$$

$$\cdots$$
$$\cdots$$
$$\cdots$$

$$0=\sum_{k=1}^{9} V_9V_k(G_{9k}\sin(\theta_9-\theta_k)+B_{9k}\cos(\theta_9-\theta_k)),$$

where $E_1 = 1.1083$, $E_2 = 1.1071$, $E_3 = 1.0606$, and

$$P_{G_i}(\delta,\theta,V)=\begin{cases}\dfrac{E'_{qi}V_i\sin(\delta_i-\theta_i)}{X'_{di}} & \text{if } i=1,\ldots,3\\ 0 & \text{if } i=4,\ldots,9\end{cases},$$

$$Q_{G_i}(\delta,\theta,V)=\begin{cases}-\dfrac{V_i^2}{X'_{di}}+\dfrac{E'_{qi}V_i\sin(\delta_i-\theta_i)}{X'_{di}} & \text{if } i=1,\ldots,3\\ 0 & \text{if } i=4,\ldots,9\end{cases}$$

$$P_{e_i}(\delta_1,\delta_2,\delta_3)=\sum_{j=1,j\neq 1}^{3} E_iE_j(B_{ij}\sin(\delta_i-\delta_j)+G_{ij}\cos(\delta_i-\delta_j)).$$

We consider a normal loading condition with a uniform damping factor $d_i/m_i = 0.1$ with $[d_1, d_2, d_3] = [0.0125, 0.0034, 0.0016]$. The contingency list and the associate fault are summarized in Table 13.1.

To illustrate the computational steps of the controlling UEP method, we numerically simulate the following dynamic information for two contingencies:

Table 13.1 The Contingency List and the Associate Fault

Contingency number	Faulted bus	Fault clearing type	Description	
			From bus	To bus
1	7	Line trip	7	5
2	7	Line trip	8	7
3	5	Line trip	7	5
4	5	Line trip	5	4
5	4	Line trip	4	6
6	4	Line trip	5	4
7	9	Line trip	6	9
8	9	Line trip	9	8
9	8	Line trip	9	8
10	8	Line trip	8	7
11	6	Line trip	4	6
12	6	Line trip	6	9

- prefault SEP and postfault SEP,
- UEPs on the stability boundary of the postfault system,
- the stability boundary of the postfault system on the machine angle space,
- the fault-on trajectory,
- the controlling UEP, and
- the exit point of the original model.

For Contingency 10, a fault occurs near Bus 8, and the line between Buses 7 and 8 is tripped. The exact stability region of the postfault system on the angle plane is highlighted (see Figure 13.1). The projected fault-on trajectory intersects the stable manifold of the controlling UEP, a type-one UEP, as highlighted in the figure. There are two type-one UEPs and one type-two UEP lying on the stability boundary. The stable manifold of one type-one UEP, the controlling UEP, contains the exit point (of the original model). This exit point is at some distance from the controlling UEP, while it is very close to a type-two UEP lying on the stability boundary.

It is interesting to note that, contrary to the claim made in the literature, the controlling UEP is not the UEP that is closest to the fault-on trajectory; for instance, the type-two UEP is closest to the fault-on trajectory, and it is incorrect to use this type-two UEP as the controlling UEP.

For Contingency 2, a fault occurs near Bus 7, and the line between Buses 7 and 8 is tripped. The exact stability region of the postfault system on the angle plane is highlighted (see Figure 13.2). The projected fault-on trajectory intersects the stable manifold of the controlling UEP, a type-one UEP, as highlighted in the figure. There are two type-one UEPs and one type-two UEP lying on the stability boundary. The

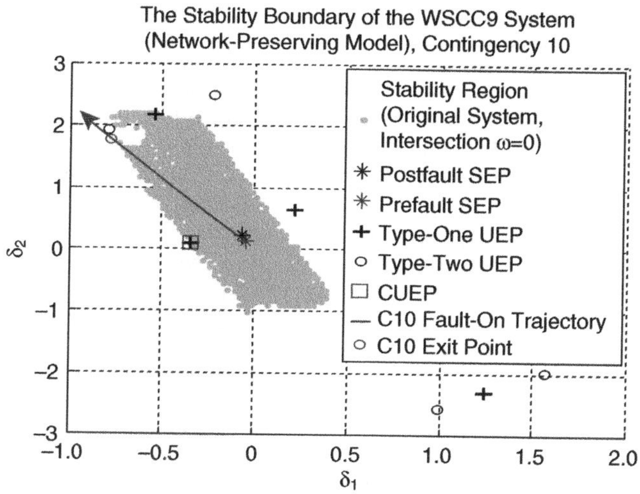

Figure 13.1 For Contingency 10, the exact stability region of the postfault system on the angle plane is highlighted. The projected fault-on trajectory intersects the stable manifold of the controlling UEP at the exit point (of the original model), which is at some distance from the controlling UEP, while it is close to a type-two UEP, which is not the controlling UEP.

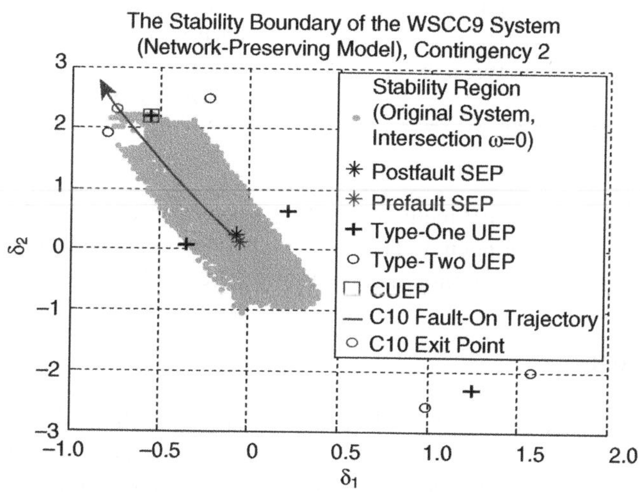

Figure 13.2 For Contingency 2, the exact stability region of the postfault system on the angle plane is highlighted. The projected fault-on trajectory intersects the stable manifold of the controlling UEP at the exit point. There are two type-one and one type-two UEPs.

stable manifold of one type-one UEP, the controlling UEP, contains the exit point (of the original model). This exit point is close to the controlling UEP, while it is at some distance from the other type-one UEP. On the other hand, the exit point is also close to a type-two equilibrium point.

Table 13.2 The Postfault SEP and the Controlling UEP Relative to Contingency 1

	$[\delta_1, \delta_2, \delta_3]$	$[V_1, V_2, \ldots, V_9]$	$[\theta_1, \theta_2, \ldots, \theta_9]$
Postfault SEP	[−0.1204, 0.3394, 0.2239]	[1.0781, 1.0845, 1.0669, 1.0511, 1.0066, 1.0402, 1.0802, 1.0657, 1.0740]	[−0.1584, 0.2091, 0.0756, −0.1963, −0.2816, −0.1621, 0.1396, 0.0589, 0.0284]
Controlling UEP	[−0.7589, 1.9528, 1.8079]	[0.8652, 0.9073, 0.6730, 0.6349, 0.6080, 0.3222, 0.8113, 0.6945, 0.5598]	[−0.7557, 1.8228, 1.5978, −0.7504, −0.8357, −0.3298, 1.7305, 1.6101, 1.4697]

Table 13.3 The Postfault SEP and the Controlling UEP Relative to Contingency 2

	$[\delta_1, \delta_2, \delta_3]$	$[V_1, V_2, \ldots, V_9]$	$[\theta_1, \theta_2, \ldots, \theta_9]$
Postfault SEP	[−0.0655, 0.2430, −0.0024]	[1.0930, 1.0940, 1.0538, 1.0801, 1.0547, 1.0615, 1.0947, 1.0235, 1.0568]	[−0.1048, 0.1129, −0.1534, −0.1430, −0.1361, −0.2089, 0.0443, −0.2823, −0.2022]
Controlling UEP	[−0.5424, 2.1802, −0.3755]	[0.9045, 0.6391, 0.9241, 0.7114, 0.3721, 0.7647, 0.4195, 0.8577, 0.8856]	[−0.5447, 1.9911, −0.5301, −0.5481, −0.3621, −0.6079, 1.7293, −0.6699, −0.5898]

We next apply the controlling UEP method for DAE systems to compute the controlling UEP relative to each contingency. The postfault SEP and the controlling UEP relative to each contingency are summarized in Tables 13.2–13.13.

High-Loading Condition

We next consider the nine bus at a high-loading condition at which loads are increased by 50%. The prefault SEP is [−0.0636, 0.1700, 0.1379]. The contingency list and the associate fault are the same as for the normal loading condition. The controlling UEPs relative to four contingencies at the high-loading condition are summarized in Tables 13.14–13.17.

13.9 CONCLUDING REMARKS

Structure-preserving transient stability models are described by a set of DAEs. The concept of controlling UEP for network-reduction transient stability models, described by ODEs, is not applicable to the structure-preserving transient stability models. In this chapter, we have defined the controlling UEP of a structure-preserving transient stability model using the controlling UEP of its associated singularly perturbed model, which is an ODE model.

A controlling UEP method for network-preserving transient stability models has been presented and illustrated on simple power system models. In addition, theoretical foundations of the controlling UEP methods for network-preserving transient stability models have been developed. The next issue, computing the controlling UEP of large-scale network-preserving transient stability modes, will be addressed in Chapter 16.

Table 13.4 The Postfault SEP and the Controlling UEP Relative to Contingency 3

	$[\delta_1, \delta_2, \delta_3]$	$[V_1, V_2, \ldots, V_9]$	$[\theta_1, \theta_2, \ldots, \theta_9]$
SEP	[−0.1204, 0.3394, 0.2239]	[1.0781, 1.0845, 1.0669, 1.0511, 1.0066, 1.0402, 1.0802, 1.0657, 1.0740]	[−0.1584, 0.2091, 0.0756, −0.1963, −0.2816, −0.1621, 0.1396, 0.0589, 0.0284]
Controlling UEP	[−0.7589, 1.9528, 1.8079]	[0.8652, 0.9073, 0.6730, 0.6349, 0.6080, 0.3222, 0.8113, 0.6945, 0.5598]	[−0.7557, 1.8228, 1.5978, −0.7504, −0.8357, −0.3298, 1.7305, 1.6101, 1.4697]

Table 13.5 The Postfault SEP and the Controlling UEP Relative to Contingency 4

	$[\delta_1, \delta_2, \delta_3]$	$[V_1, V_2, \ldots, V_9]$	$[\theta_1, \theta_2, \ldots, \theta_9]$
SEP	[−0.0231, 0.0467, 0.0817]	[1.1056, 1.0684, 1.0732, 1.1043, 0.9660, 1.0858, 1.0558, 1.0563, 1.0821]	[−0.0579, −0.0843, −0.0649, −0.0911, −0.3085, −0.1435, −0.1559, −0.1715, −0.1111]
Controlling UEP	[−0.8364, 2.0797, 2.1466]	[0.8654, 0.8759, 0.6241, 0.6353, 0.6997, 0.2712, 0.7648, 0.6470, 0.4981]	[−0.8372, 1.9429, 1.9175, −0.8386, 1.6871, −0.5897, 1.8398, 1.7801, 1.7605]

Table 13.6 The Postfault SEP and the Controlling UEP Relative to Contingency 5

	$[\delta_1, \delta_2, \delta_3]$	$[V_1, V_2, \ldots, V_9]$	$[\theta_1, \theta_2, \ldots, \theta_9]$
SEP	[−0.0319, 0.0949, 0.0492]	[1.0986, 1.0898, 1.0654, 1.0910, 1.0672, 1.0205, 1.0884, 1.0710, 1.0720]	[−0.0718, −0.0362, −0.1005, −0.1102, −0.1665, −0.2695, −0.1056, −0.1577, −0.1482]
Controlling UEP	[−0.8256, 2.0830, 2.0549]	[0.8458, 0.7324, 0.8280, 0.5973, 0.2403, 0.7232, 0.5512, 0.6318, 0.7597]	[−0.8133, 1.9274, 1.8882, −0.7917, −0.4936, 1.6920, 1.7643, 1.7541, 1.8133]

Table 13.7 The Postfault SEP and the Controlling UEP Relative to Contingency 6

	$[\delta_1, \delta_2, \delta_3]$	$[V_1, V_2, \ldots, V_9]$	$[\theta_1, \theta_2, \ldots, \theta_9]$
SEP	[−0.0231, 0.0467, 0.0817]	[1.1056, 1.0684, 1.0732, 1.1043, 0.9660, 1.0858, 1.0558, 1.0563, 1.0821]	[−0.0579, −0.0843, −0.0649, −0.0911, −0.3085, −0.1435, −0.1559, −0.1715, −0.1111]
Controlling UEP	[−0.8364, 2.0797, 2.1466]	[0.8654, 0.8759, 0.6241, 0.6353, 0.6997, 0.2712, 0.7648, 0.6470, 0.4981]	[−0.8372, 1.9429, 1.9175, −0.8386, 1.6871, −0.5897, 1.8398, 1.7801, 1.7605]

Table 13.8 The Postfault SEP and the Controlling UEP Relative to Contingency 7

	$[\delta_1, \delta_2, \delta_3]$	$[V_1, V_2, \ldots, V_9]$	$[\theta_1, \theta_2, \ldots, \theta_9]$
SEP	[−0.0967, 0.2180, 0.2958]	[1.0836, 1.0805, 1.0614, 1.0617, 1.0360, 1.0294, 1.0742, 1.0600, 1.0667]	[−0.1352, 0.0868, 0.1464, −0.1734, −0.1657, −0.2388, 0.0164, 0.0160, 0.0985]
Controlling UEP	[−0.7576, 1.8583, 1.9986]	[0.8530, 0.7528, 0.8390, 0.6116, 0.2914, 0.5929, 0.5806, 0.6515, 0.7739]	[−0.7406, 1.7099, 1.8361, −0.7115, −0.3019, −0.7769, 1.5622, 1.6264, 1.7644]

Table 13.9 The Postfault SEP and the Controlling UEP Relative to Contingency 8

	$[\delta_1, \delta_2, \delta_3]$	$[V_1, V_2, \ldots, V_9]$	$[\theta_1, \theta_2, \ldots, \theta_9]$
SEP	[−0.0462, 0.0728, 0.2082]	[1.0942, 1.0786, 1.0802, 1.0825, 1.0563, 1.0710, 1.0716, 1.0457, 1.0915]	[−0.0868, −0.0599, 0.0604, −0.1262, −0.1860, −0.1223, −0.1313, −0.1887, 0.0141]
Controlling UEP	[−0.2910, −0.1011, 2.5008]	[0.9326, 0.9938, 0.4550, 0.7666, 0.8091, 0.4447, 0.9423, 0.9196, 0.2993]	[−0.3080, −0.2298, 2.1701, −0.3312, −0.3764, −0.1971, −0.3085, −0.3659, 1.7890]

Table 13.10 The Postfault SEP and the Controlling UEP Relative to Contingency 9

	$[\delta_1, \delta_2, \delta_3]$	$[V_1, V_2, \ldots, V_9]$	$[\theta_1, \theta_2, \ldots, \theta_9]$
SEP	[−0.0462, 0.0728, 0.2082]	[1.0942, 1.0786, 1.0802, 1.0825, 1.0563, 1.0710, 1.0716, 1.0457, 1.0915]	[−0.0868, −0.0599, 0.0604, −0.1262, −0.1860, −0.1223, −0.1313, −0.1887, 0.0141]
Controlling UEP	[−0.2910, −0.1011, 2.5008]	[0.9326, 0.9938, 0.4550, 0.7666, 0.8091, 0.4447, 0.9423, 0.9196, 0.2993]	[−0.3080, −0.2298, 2.1701, −0.3312, −0.3764, −0.1971, −0.3085, −0.3659, 1.7890]

Table 13.11 The Postfault SEP and the Controlling UEP Relative to Contingency 10

	$[\delta_1, \delta_2, \delta_3]$	$[V_1, V_2, \ldots, V_9]$	$[\theta_1, \theta_2, \ldots, \theta_9]$
SEP	[−0.0655, 0.2430, −0.0024]	[1.0930, 1.0940, 1.0538, 1.0801, 1.0547, 1.0615, 1.0947, 1.0235, 1.0568]	[−0.1048, 0.1129, −0.1534, −0.1430, −0.1361, −0.2089, 0.0443, −0.2823, −0.2022]
Controlling UEP	[−0.3495, 0.0745, 2.5864]	[0.9318, 1.0088, 0.4284, 0.7647, 0.8074, 0.4451, 0.9633, 0.2576, 0.2660]	[−0.3387, −0.0384, 2.2596, −0.3239, −0.2914, −0.2163, −0.1060, 1.7530, 1.8332]

Table 13.12 The Postfault SEP and the Controlling UEP Relative to Contingency 11

	$[\delta_1, \delta_2, \delta_3]$	$[V_1, V_2, \ldots, V_9]$	$[\theta_1, \theta_2, \ldots, \theta_9]$
SEP	[−0.0319, 0.0949, 0.0492]	[1.0986, 1.0898, 1.0654, 1.0910, 1.0672, 1.0205, 1.0884, 1.0710, 1.0720]	[−0.0718, −0.0362, −0.1005, −0.1102, −0.1665, −0.2695, −0.1056, −0.1577, −0.1482]
Controlling UEP	[−0.8256, 2.0830, 2.0549]	[0.8458, 0.7324, 0.8280, 0.5973, 0.2403, 0.7232, 0.5512, 0.6318, 0.7597]	[−0.8133, 1.9274, 1.8882, −0.7917, −0.4936, 1.6920, 1.7643, 1.7541, 1.8133]

Table 13.13 The Postfault SEP and the Controlling UEP Relative to Contingency 12

	$[\delta_1, \delta_2, \delta_3]$	$[V_1, V_2, \ldots, V_9]$	$[\theta_1, \theta_2, \ldots, \theta_9]$
SEP	[−0.0967, 0.2180, 0.2958]	[1.0836, 1.0805, 1.0614, 1.0617, 1.0360, 1.0294, 1.0742, 1.0600, 1.0667]	[−0.1352, 0.0868, 0.1464, −0.1734, −0.1657, −0.2388, 0.0164, 0.0160, 0.0985]
Controlling UEP	[−0.7576, 1.8583, 1.9986]	[0.8530, 0.7528, 0.8390, 0.6116, 0.2914, 0.5929, 0.5806, 0.6515, 0.7739]	[−0.7406, 1.7099, 1.8361, −0.7115, −0.3019, −0.7769, 1.5622, 1.6264, 1.7644]

Table 13.14 The Postfault SEP and the Controlling UEP Relative to Contingency 1

	$[\delta_1, \delta_2, \delta_3]$	$[V_1, V_2, \ldots, V_9]$	$[\theta_1, \theta_2, \ldots, \theta_9]$
SEP	[−0.1739, 0.4922, 0.3192]	[1.0671, 1.0913, 1.0705, 1.0078, 0.9331, 0.9828, 1.0712, 1.0412, 1.0567]	[−0.2285, 0.3122, 0.1187, −0.2868, −0.4207, −0.2353, 0.2110, 0.0913, 0.0490]
Controlling UEP	[−0.6844, 1.7827, 1.5845]	[0.8965, 0.9498, 0.7514, 0.6722, 0.6224, 0.4140, 0.8555, 0.7441, 0.6429]	[−0.6746, 1.6131, 1.3316, −0.6591, −0.7930, −0.2173, 1.4935, 1.3341, 1.1872]

Table 13.15 The Postfault SEP and the Controlling UEP Relative to Contingency 2

	$[\delta_1, \delta_2, \delta_3]$	$[V_1, V_2, \ldots, V_9]$	$[\theta_1, \theta_2, \ldots, \theta_9]$
SEP	[−0.0943, 0.3587, −0.0217]	[1.0894, 1.0949, 1.0528, 1.0513, 1.0002, 1.0188, 1.0769, 0.9760, 1.0338]	[−0.1520, 0.1771, −0.2272, −0.2109, −0.2022, −0.3112, 0.0756, −0.4216, −0.3003]
Controlling UEP	[−0.4976, 1.9662, −0.2729]	[0.9290, 0.7051, 0.9479, 0.7354, 0.4184, 0.7714, 0.5049, 0.8451, 0.8952]	[−0.4957, 1.7315, −0.4753, −0.4929, −0.2463, −0.5861, 1.4493, −0.6797, −0.5585]

Table 13.16 The Postfault SEP and the Controlling UEP Relative to Contingency 3

	$[\delta_1, \delta_2, \delta_3]$	$[V_1, V_2, \ldots, V_9]$	$[\theta_1, \theta_2, \ldots, \theta_9]$
SEP	[−0.1739, 0.4922, 0.3192]	[1.0671, 1.0913, 1.0705, 1.0078, 0.9331, 0.9828, 1.0712, 1.0412, 1.0567]	[−0.2285, 0.3122, 0.1187, −0.2868, −0.4207, −0.2353, 0.2110, 0.0913, 0.0490]
Controlling UEP	[−0.6844, 1.7827, 1.5845]	[0.8965, 0.9498, 0.7514, 0.6722, 0.6224, 0.4140, 0.8555, 0.7441, 0.6429]	[−0.6746, 1.6131, 1.3316, −0.6591, −0.7930, −0.2173, 1.4935, 1.3341, 1.1872]

Table 13.17 The Postfault SEP and the Controlling UEP Relative to Contingency 4

	$[\delta_1, \delta_2, \delta_3]$	$[V_1, V_2, \ldots, V_9]$	$[\theta_1, \theta_2, \ldots, \theta_9]$
SEP	[−0.0408, 0.0966, 0.1152]	[1.1178, 1.0564, 1.0848, 1.1056, 0.8661, 1.0675, 1.0191, 1.0220, 1.0752]	[−0.0889, −0.0871, −0.0810, −0.1357, −0.4247, −0.2094, −0.1957, −0.2280, −0.1481]
Controlling UEP	[−0.7973, 1.9874, 2.0360]	[0.8931, 0.8847, 0.6615, 0.6658, 0.6450, 0.3217, 0.7590, 0.6501, 0.5307]	[−0.7821, 1.8081, 1.7507, −0.7577, 1.4365, −0.4324, 1.6655, 1.5754, 1.5533]

Chapter 14

Network-Reduction BCU Method and Its Theoretical Foundation

14.1 INTRODUCTION

The ability to compute the controlling unstable equilibrium point (UEP) is vital to direct stability analysis. However, the task of computing the exact controlling UEP relative to a given fault is very difficult. The computational challenges and complexities of computing the controlling UEP described in Chapter 12 serve to explain why the majority of existing methods fail. It has been fruitful to develop tailored methods for computing the controlling UEP by exploring special properties as well as some physical and mathematical insights of the underlying power system transient stability model. In this chapter, we will present such a method for computing the controlling UEP: the Boundary of stability region-based Controlling Unstable equilibrium point (BCU) method (Chiang, 1995, 1996; Chiang and Chu, 1995; Chiang et al., 1994).

The BCU method is a systematic method for computing the controlling UEP of large-scale power systems. It first explores the special structure of the underlying model so as to define an artificial, reduced-state model that captures all the equilibrium points on the stability boundary of the original model. It then computes the controlling UEP of the original model by computing the controlling UEP of the reduced-state model, which is much easier to compute than that of the original model. Given a power system stability model with certain properties, there exists a corresponding version of the BCU method. Other names have been given to the BCU method, such as the exit point method (Electric Power Research Institute [EPRI], 1995; Fouad and Vittal, 1991; Mokhtari et al., 1994) and the hybrid method (Pai, 1989).

Direct Methods for Stability Analysis of Electric Power Systems, by Hsiao-Dong Chiang

The BCU method has received worldwide attention and has been applied by researchers and engineers to several practical power systems. The applicability of the BCU method to online transient stability assessments was demonstrated by the Ontario Hydro Company (Kim, 1994) and by the Northern States Power Company (Ejebe et al., 1999; Mokhtari et al., 1994). A 3-year project, sponsored by EPRI, has also confirmed the practicability and reliability of the BCU method via extensive numerical evaluations (Rahimi, 1990; Rahimi et al., 1993). It should be noted that the BCU method is the only direct method adopted by EPRI in its latest version of DIRECT 4.0 (EPRI, 1995). In addition, the BCU method has also been extensively evaluated on large-scale power systems such as a 12,000-bus power system (Chiang et al., 2006; Tada et al., 2004, 2005). The BCU method has also been applied to fast derivation of power transfer limits (Tong et al., 1993) and has been applied to real power rescheduling to increase dynamic security (Kuo and Bose, 1995).

The BCU method and its improved variants have been installed in several utility companies around the world. This method has deep theoretical foundations and is currently the only computationally efficient, theory-based method for computing the controlling UEP. In this chapter, we present the BCU method for solving network-reduction power system stability models and develop its theoretical foundation. Most of the proofs presented in this chapter are taken from Chiang (1995) and Chiang and Chu (1995).

14.2 REDUCED-STATE SYSTEM

The fundamental ideas behind the BCU method are explained in the following. Given a power system transient stability model, the BCU method first explores special properties of the original model with the aim to define an artificial, reduced-state model, such that certain static and dynamic properties of the original model are captured by the reduced-state model. For illustrational purposes, we consider the following generic network-reduction model as the original transient stability model that encompasses the existing network-reduction models:

$$\begin{aligned} T\dot{x} &= -\frac{\partial U}{\partial x}(x, y) + g_1(x, y) \\ \dot{y} &= z \\ M\dot{z} &= -Dz - \frac{\partial U}{\partial y}(x, y) + g_2(x, y), \end{aligned} \tag{14.1}$$

where $x \in R^n$, $y \in R^n$, and $z \in R^n$ are state variables. T is a positive definite matrix and M and D are diagonal positive definite matrices. The vectors $g_1(x, y)$ and $g_2(x, y)$ represent the transfer conductances.

We then choose the following model as an artificial, reduced-state model associated with the original model:

$$
\begin{aligned}
T\dot{x} &= -\frac{\partial U}{\partial x}(x, y) + g_1(x, y) \\
\dot{y} &= -\frac{\partial U}{\partial x}(x, y) + g_2(x, y).
\end{aligned}
\tag{14.2}
$$

Note that the state space of the original model (Eq. 14.1) is (x, y, z). By eliminating the state variable z, we have the state space of the reduced-state model (Eq. 14.2), which is (x, y).

A reduced-state model associated with the original model must satisfy the following static and dynamic properties:

Static Properties

(S1) The locations of the equilibrium points of the reduced-state model correspond to the locations of the equilibrium points of the original model. For example, $(\bar{x}, \bar{y})$ is an equilibrium point of the reduced-state model (Eq. 14.2) if and only if $(\bar{x}, \bar{y}, 0)$ is an equilibrium point of the original model (Eq. 14.1), where $0 \in R^m$ and m is an appropriate positive integer.

(S2) The types of equilibrium points of the reduced-state model are the same as those of the original model. For example, (x_s, y_s) is a stable equilibrium point (SEP) of the reduced-state model (Eq. 14.2) if and only if $(x_s, y_s, 0)$ is a SEP of the original model (Eq. 14.1).

Dynamic Properties

(D1) There exists an energy function for the reduced-state model.

(D2) An equilibrium point, say, (x_i, y_i), is on the stability boundary $\partial A(x_s, y_s)$ of the reduced-state model if and only if the equilibrium point $(x_i, y_i, 0)$ is on the stability boundary $\partial A(x_s, y_s, 0)$ of the original model.

(D3) It is computationally feasible to efficiently detect when the projected fault-on trajectory $(x_f(t), y_f(t))$ of the fault-on trajectory $(x_f(t), y_f(t), z_f(t))$ intersects the stability boundary $\partial A(x_s, y_s)$ of the postfault reduced-state model without resorting to detailed time–domain simulation.

The dynamic property (D3) plays a pivotal role in the development of the BCU method for circumventing the difficulty of applying an iterative time–domain procedure to compute the exit point of the original model. The BCU method computes the controlling UEP of the reduced-state model by exploring the special structure of the stability boundary and the energy function of the reduced-state model. Given a power system stability model, there exists a corresponding reduced-state model and a corresponding version of the BCU method. The reduced-state model may not be unique.

14.3 ANALYTICAL RESULTS

We present in this section some analytical results for the generic network-reduction transient stability model (Eq. 14.1) and the associated reduced-state model

(Eq. 14.2). We will develop analytical results showing that, under certain conditions, the original model and the reduced-state model satisfy the static properties (S1) and (S2) as well as the dynamic properties (D1) and (D2). These properties will be explored to develop a network-reduction BCU method to compute the controlling UEP in the next section. Note that a type-zero equilibrium point is a SEP; a type-one equilibrium point is the one whose corresponding Jacobian has only one eigenvalue with a positive real part; a type-two equilibrium point is the one whose corresponding Jacobian has only two eigenvalues with a positive real part, and so on.

To show that the reduced-state model satisfies the static as well as dynamic properties stated earlier, we proceed to perform the following steps. In each step, we will determine the following static properties between the original and the reduced-state models:

- locations of the equilibrium points of these two models and
- types of equilibrium points of these two models.

We also determine in each step the following dynamic properties between these two models:

- UEPs lying on the stability boundary of each model and
- characterization of the stability boundary of each model.

Step 1. Determine the static as well as dynamic relationship between the reduced-state model and the following model:

$$\begin{aligned} T\dot{x} &= -\frac{\partial U}{\partial x}(x, y) \\ \dot{y} &= -\frac{\partial U}{\partial x}(x, y). \end{aligned} \tag{14.3}$$

Step 2. Determine the static as well as dynamic relationship between

$$\begin{aligned} T\dot{x} &= -\frac{\partial U}{\partial x}(x, y) \\ \dot{y} &= -\frac{\partial U}{\partial x}(x, y) \end{aligned}$$

and the following model:

$$\begin{aligned} T\dot{x} &= -\frac{\partial U}{\partial x}(x, y) \\ \dot{y} &= -\frac{\partial U}{\partial x}(x, y) \\ M\dot{z} &= -Dz. \end{aligned} \tag{14.4}$$

Step 3. Determine the static as well as dynamic relationship between

$$T\dot{x} = -\frac{\partial U}{\partial x}(x, y)$$

$$\dot{y} = -\frac{\partial U}{\partial x}(x, y)$$

$$M\dot{z} = -Dz$$

and the following one-parameter dynamic model $d(\lambda)$:

$$\begin{aligned} T\dot{x} &= -\frac{\partial}{\partial x}U(x, y) \\ \dot{y} &= (1-\lambda)z - \lambda\frac{\partial U}{\partial x}(x, y) \\ M\dot{z} &= -Dz + (\lambda - 1)\frac{\partial U}{\partial y}(x, y). \end{aligned} \tag{14.5}$$

Step 4. Determine the static as well as dynamic relationship between

$$T\dot{x} = -\frac{\partial U}{\partial x}(x, y)$$

$$\dot{y} = (1-\lambda)z - \lambda\frac{\partial U}{\partial x}(x, y)$$

$$M\dot{z} = -Dz + (\lambda - 1)\frac{\partial U}{\partial y}(x, y)$$

and the following model:

$$\begin{aligned} T\dot{x} &= -\frac{\partial U}{\partial x}(x, y) \\ \dot{y} &= z \\ M\dot{z} &= -Dz - \frac{\partial U}{\partial y}(x, y). \end{aligned} \tag{14.6}$$

Step 5. Determine the static as well as dynamic relationship between

$$T\dot{x} = -\frac{\partial U}{\partial x}(x, y)$$

$$\dot{y} = z$$

$$M\dot{z} = -Dz - \frac{\partial U}{\partial y}(x, y)$$

and the original model,

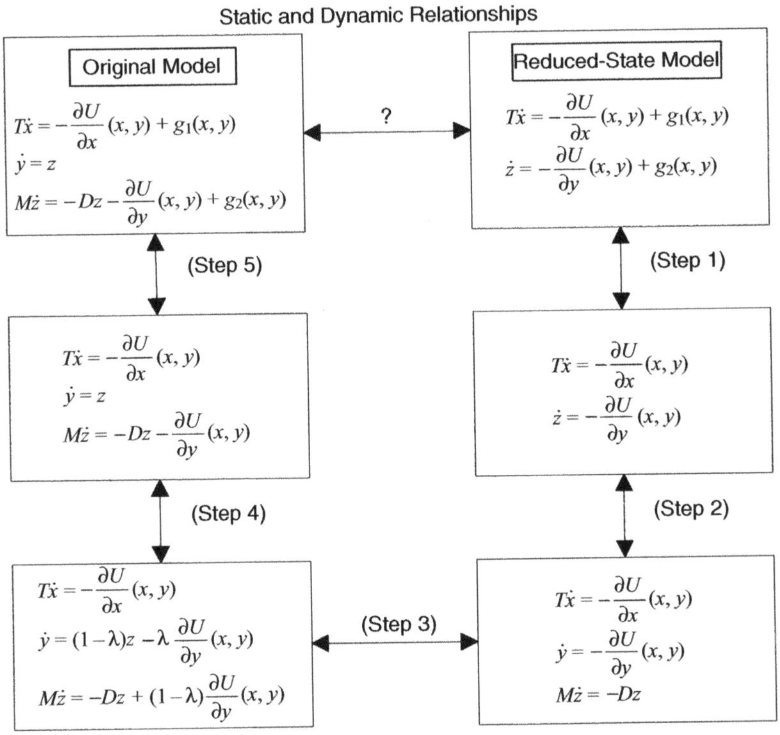

Figure 14.1 The five-step procedure showing the static properties (S1) and (S2) as well as the dynamic properties (D1) and (D2) between the original model (Eq. 14.1) and the reduced-state model (Eq. 14.2).

$$T\dot{x} = -\frac{\partial U}{\partial x}(x, y) + g_1(x, y)$$

$$\dot{y} = z$$

$$M\dot{z} = -Dz - \frac{\partial U}{\partial y}(x, y) + g_3(x, y).$$

The overall five-step procedure to show that the reduced-state model (Eq. 14.2) satisfies the static properties (S1) and (S2) as well as the dynamic properties (D1) and (D2) is summarized in Figure 14.1.

In the following, we will derive the properties stated in each step:

Steps 1 and 5. In these two steps, we shall show that if the transfer conductances are sufficiently small, then the following properties hold:

[(Static property)] $(\bar{x}, \bar{y})$ is a type-k equilibrium point of the reduced-state model (Eq. 14.2) if and only if $(\bar{x}, \bar{y})$ is a type-k equilibrium point of the model (Eq. 14.3).

[(Static property)] $(\bar{x}, \bar{y}, 0)$ is a type-k equilibrium point of the model (Eq. 14.6) if and only if $(\bar{x}, \bar{y}, 0)$ is a type-k equilibrium point of the original model (Eq. 14.1).

[(Dynamic property)] (x_i, y_i) is an equilibrium point on the stability boundary $\partial A(x_s, y_s)$ of the reduced-state model (Eq. 14.2) if and only if (x_i, y_i) is an equilibrium point on the stability boundary $\partial A(x_s, y_s)$ of the model (Eq. 14.3).

[(Dynamic property)] $(\bar{x}, \bar{y}, 0)$ is an equilibrium point on the stability boundary $\partial A(x_s, y_s, 0)$ of the model (Eq. 14.6) if and only if the equilibrium point $(x, y, 0)$ is on the stability boundary $\partial A(x_s, y_s, 0)$ of the original model (Eq. 14.1).

Step 2. In Step 2, the following properties will be shown:

[(Static property)] (x, y) is a type-k equilibrium point of the model (Eq. 14.3) if and only if $(\bar{x}, \bar{y}, 0)$ is a type-k equilibrium point of the model (Eq. 14.4), where $0 \in R^m$ and m is an appropriate positive integer.

[(Dynamic property)] (x_i, y_i) is an equilibrium point on the stability boundary $\partial A(x_s, y_s)$ of the model (Eq. 14.3) if and only if the equilibrium point $(xi, y_i, 0)$ is on the stability boundary $\partial A(x_s, y_s, 0)$ of the model (Eq. 14.4).

Steps 3 and 4. In these two steps, we shall show that under certain conditions, the following properties hold:

[(Static property)] $(\bar{x}, \bar{y}, 0)$ is a type-k equilibrium point of the model (Eq. 14.4) if and only if $(\bar{x}, \bar{y}, 0)$ is a type-k equilibrium point of the model (Eq 14.6), where $0 \in R^m$ and m is an appropriate positive integer.

[(Dynamic property)] $(xi, y_i, 0)$ is an equilibrium point on the stability boundary $\partial A(x_s, y_s, 0)$ of the model (Eq. 14.4), if and only if the equilibrium point $(x_i, y_i, 0)$ is on the stability boundary $\partial A(x_s, y_s, 0)$ of the model (Eq. 14.6).

Finally, the relationship between the original model (Eq. 14.1) and the reduced-state model (Eq. 14.2) will be established by combining the analytical results to be derived in Steps 1–5.

We next derive some relevant analytical results for Steps 1 through 5:

Steps 1 and 5. Mathematically speaking, we study in these two steps the "robustness" of both the types of equilibrium points and the equilibrium points on the stability under a small perturbation of the vector field. Here, we consider a C^1 (i.e., continuously differentiable) perturbation of the vector field. Let f: $M \to E$ be a C^1 vector field and M be an open set in a vector space, E. Let $v(M)$ be the set of all C^1 vector fields on M. If E has a norm, the C^1 vector field $h \in v(M)$ is the least upper bound of all the numbers:

$$\{|h(x)|, \|Dh(x)\|; x \in M\}.$$

With this topology, a neighborhood of $f \in v(M)$ is a subset of $v(M)$ that contains a set of the form $g \in v(M)$ and $|g - f|_1 < \varepsilon$ for some $\varepsilon > 0$ and some norm on M. We call g a $\varepsilon - C^1$ perturbation of the vector field f.

The property of persistence is a condition much weaker than the notion of structural stability. In the following theorem, it is shown that the property of persistence holds for "almost all" C^1 vector fields, while the stronger property of structural stability does not.

Theorem 14.1: Property of Persistence

Let f: $M \rightarrow E$ be a C^1 vector field and M be an open set in the vector space E. Let x_s be a SEP and $\hat{x}$ be a type-k hyperbolic equilibrium point on the stability boundary $\partial A(x_s)$ satisfying the transversality condition. Then, for any $\varepsilon > 0$, there exists a δ-neighborhood of $h \subset v(M)$ of f and a neighborhood $u \subset M$ of $\hat{x}$, such that for any $g \in h$,

1. there is a unique type-k hyperbolic equilibrium point $\hat{x}^g$ of the vector field g, such that $\hat{x}^g \in \mu$ moreover,
2. $\hat{x}^g$ is on the stability boundary $\partial A(x_s^g)$, where x_s^g is a SEP of the vector field g and $|x_s^g - x_s| < \varepsilon$

Proof: Since x is a type-k hyperbolic equilibrium point, there exists a δ_1-neighborhood of f, such that for any $g \in h$, x^g is also a type-k hyperbolic equilibrium point. For the second part, since $\hat{x}$ is on the stability boundary $\partial A(x_s^g)$, by Theorem 3.12, we have

$$W^u(\hat{x}) \cap A(x_s) \neq \varnothing.$$

In other words, $W^u(\hat{x})$ intersects $W^s(x_s)$ transversely. By the openness of the transversality condition, there exists a δ_2 neighborhood of f, such that $W^u(\hat{x}^g)$ intersects with $W^s(x_s^g)$ transversely for any vector field g in this neighborhood. It follows from Theorem 3.12 that $\hat{x}^g$ is on the stability boundary $\partial A(x_s^g)$. Taking $\delta = \min\{\delta_1, \delta_2\}$, we complete the proof.

We next generalize the above result to all the equilibrium points on a stability boundary. Let the equilibrium point of the vector field f be hyperbolic. Theorem 14.1 implies that for an equilibrium point, say, x_i, there exists a neighborhood h_i of f and a neighborhood u_i of x_i, such that any vector field $g \in h$ will have a unique critical element $x_i(g) \in u_i$. It can be shown by using the notion of filtration that the neighborhoods h_i and u_i can be shrunk so that $x_i(g)$ is the maximal invariant set of g entirely contained in u_i (Shub, 1987). Moreover, there exists a neighborhood h of f, such that there is no other nonwandering point outside the union of u_i. Next, suppose that the vector field f satisfies the transversality condition. By the open property of the transversality condition, it follows that, for a pair of critical elements x_i and x_j with x_j stable and $W^u(x_i) \cap W^s(x_j) \neq \varnothing$, there exists a neighborhood ε of f, such that for any vector field g in this neighborhood, the intersection between $W^u(x_i(g))$ and $W^s(x_j(g))$ is nonempty and transverse. Based on the above arguments, we have the following.

Theorem 14.2: Robustness of Equilibrium Points on the Stability Boundary

Let x_s be a SEP of the vector field f and let x_i, $i = 1, 2, \ldots, k$ be the hyperbolic equilibrium points on $\partial A(x_s)$ and the intersection between $W^u(x_i)$ and $W^s(x_j)$ satisfying the transversality condition and $\partial A(x_s) = \cup W^s(x_i)$. There exists a positive number ε such that, for any vector field g, which is a $\varepsilon - C^1$ perturbation of f, the equilibrium points $h(x_i)$, $i = 1, 2, \ldots, k$ are also on the stability boundary of $h(x_s\ (g))$ and $\partial A(x_s(g)) = \cup W^s(x_i(g))$, where $h(\cdot)$ is a one-to-one correspondence and $h(x_s\ (g))$ is a SEP of the vector field g.

Step 2. By observing the vector field of the model (Eq. 14.4), it is easy to derive the following:

- The dynamics of the state variables (x, y) of the model (Eq. 14.4) are completely decoupled from the dynamic of state variable (z).
- The dynamics of the state variables (x, y) of the model (Eq. 14.4) are exactly the same as the dynamics of the state variables (x, y) of the model (Eq. 14.3).

Hence, it follows that

[(Static property)] $(\bar{x}, \bar{y})$ is a type-k equilibrium point of the model (Eq. 14.3) if and only if $(\bar{x}, \bar{y}, 0)$ is a type-k equilibrium point of the model (Eq. 14.4), where $0 \in R^m$ and m is an appropriate positive integer.

[(Dynamic property)] (x_i, y_i) is an equilibrium point on the stability boundary $\partial A(x_s, y_s)$ of the model (Eq. 14.3) if and only if the equilibrium point $(x_i, y_i, 0)$ is on the stability boundary $\partial A(x_s, y_s, 0)$ of the model (Eq. 14.4).

Steps 3 and 4. The main analytical results to be derived are a static property between equilibrium points of the models (Eqs. 14.4 and 14.6) and a dynamic property between equilibrium points on the stability boundary of the model (Eq. 14.4) and those on the stability boundary of the model (Eq. 14.6).

Theorem 14.3: Static Property

If zero is a regular value of $\dfrac{\partial^2 U}{\partial x \partial y}(x_i, y_i)$, then $(\bar{x}, \bar{y}, 0)$ is a type-k equilibrium point of the model (Eq. 14.4) if and only if $(\bar{x}, \bar{y}, 0)$ is a type-k equilibrium point of the model (Eq. 14.6).

Proof: It is obvious that the model (Eq. 14.4) and the model (Eq. 14.6) have the same equilibrium points. The Jacobian matrix of the model (Eq. 14.4) is

$$J_4 = \begin{bmatrix} -T^{-1}\frac{\partial^2}{\partial x^2}U(x,y) & -T^{-1}\frac{\partial^2}{\partial x\partial y}U(x,y) & 0 \\ -\frac{\partial^2}{\partial x\partial y}U(x,y) & -\frac{\partial^2}{\partial y^2}U(x,y) & 0 \\ 0 & 0 & -M^{-1}D \end{bmatrix}.$$

The Jacobian matrix of the model (Eq. 14.6) is

$$J_6 = \begin{bmatrix} -T^{-1}\frac{\partial^2}{\partial x^2}U(x,y) & -T^{-1}\frac{\partial^2}{\partial x\partial y}U(x,y) & 0 \\ 0 & 0 & I \\ -M^{-1}\frac{\partial^2}{\partial x\partial y}U(x,y) & -M^{-1}\frac{\partial^2}{\partial y^2}U(x,y) & -M^{-1}D \end{bmatrix}.$$

By using Sylvester's inertia theorem, it follows that these two matrices have the same number of eigenvalues with negative real parts. This completes the proof.

Theorem 14.4: Dynamic Property

Let $(x_s, y_s, 0)$ be a SEP of the model (Eq. 14.4). If the intersection of the stable and unstable manifolds of the equilibrium points on the stability boundary of the following one-parameter dynamical system $d(\lambda)$:

$$T\dot{x} = -\frac{\partial}{\partial x}U(x,y)$$

$$\dot{y} = (1-\lambda)z - \lambda\frac{\partial U}{\partial x}(x,y)$$

$$M\dot{z} = -Dz + (\lambda-1)\frac{\partial U}{\partial y}(x,y)$$

satisfy the transversality condition for $\lambda \in [0, 1]$, then

1. $(x_i, y_i, 0)$ is an equilibrium point on the stability boundary $\partial A(x_s, y_s, 0)$ of the model (Eq. 14.4) if and only if the equilibrium point $(x_i, y_i, 0)$ is on the stability boundary $\partial A(x_s, y_s, 0)$ of the model (Eq. 14.6).
2. The stability boundary $\partial A(x_s, y_s, 0)$ of the model (Eq. 14.4) equals the union of the stable manifolds of the equilibrium points, say, $(x_i, y_i, 0)$, lying on $\partial A(x_s, y_s, 0)$, $i = 1, \ldots, n$. In addition, the stability boundary $\partial A(x_s, y_s, 0)$ of model (Eq. 14.6) equals the union of the stable manifolds of the equilibrium points $(x_i, y_i, 0)$ (i.e., the same equilibrium points), lying on $\partial A(x_s, y_s, 0)$, $i = 1, \ldots, n$.

Proof: This proof requires the following two theorems, which are essentially the same as Theorems 10.3 and 10.4. Here, we state the results without a detailed proof. We note that the model (Eq. 14.5) can be expressed as the following:

$$\begin{bmatrix} \dot{x} \\ \dot{y} \\ \dot{z} \end{bmatrix} = -[\bar{\Gamma}(\lambda)] \begin{bmatrix} \dfrac{\partial W}{\partial x} \\ \dfrac{\partial W}{\partial y} \\ \dfrac{\partial W}{\partial z} \end{bmatrix},$$

where $W(x, y, z) = U(x, y) + (1/2)z^T z$, which is an energy function.

Theorem 14.5: Local Invariant Property on the Stability Boundary

Suppose that the vector field $d(\lambda 1)$ satisfies the transversality condition and $(x_s, y_s, 0)$ is a SEP, then there exists an $\varepsilon > 0$, such that for all $\bar{\Gamma}(\lambda)$ with $\|\bar{\Gamma}(\lambda) - \bar{\Gamma}(\lambda_1)\| < \varepsilon$, the equilibrium point $(x_i, y_i, 0)$ lies on the stability boundary $\partial A(x_s, y_s, 0)$ of system $d(\lambda)$ if and only if the equilibrium point $(x_i, y_i, 0)$ lies on the stability boundary $\partial A(x_s, y_s, 0)$ of system $d(\lambda)$.

Theorem 14.6

Let X_s be a SEP of system $d(\lambda_1)$ and X_u be a UEP on the stability boundary $\partial A_{d(\lambda_1)}(X_s)$. Let $\varepsilon > 0$ be a sufficiently small number and $\bar{K} = \{k \in N\}$ be a collection of infinite positive integers, such that

(1) $\|\bar{\Gamma}(\lambda_k) - \bar{\Gamma}(\lambda_1)\| < \varepsilon$ and

(2) $\bar{\Gamma}(\lambda_k) \to \bar{\Gamma}(\lambda_{k^*})$ as $k \to k^*$ and $\|\bar{\Gamma}(\lambda_{k^*}) - \bar{\Gamma}(\lambda_1)\| = \varepsilon$.

Suppose that $X_u \in \partial A_{d(\lambda_k)}(X_s)$ for all $k \in \bar{K}$, then $X_u \in \partial A_{d(\lambda_{k^*})}(X_s)$.

We now are in a position to prove this theorem. Let E be the set of equilibrium points of the parameterized system (Eq. 14.5). From Theorem 14.5, it follows that there exists an $\bar{\varepsilon}_0$ and $\bar{\varepsilon}_0 > 0$ such that

$$\|\bar{\Gamma}(\lambda) - \bar{\Gamma}(0)\| < \bar{\varepsilon}_0 \Rightarrow E \cap \partial A_{d(0)} \subseteq E \cap \partial A_{d(\lambda)} \text{ and}$$

$$\|\bar{\Gamma}(\lambda) - \bar{\Gamma}(1)\| < \bar{\varepsilon}_0 \Rightarrow E \cap \partial A_{d(1)} \subseteq E \cap \partial A_{\Sigma(\lambda)}.$$

If $\bar{\varepsilon}_0 + \bar{\varepsilon}_1 > \|\bar{\Gamma}(0) - \bar{\Gamma}(1)\|$, then we complete the proof. Without loss of generality, suppose that $\bar{\varepsilon}_0 + \bar{\varepsilon}_1 \leq \|\bar{\Gamma}(0) - \bar{\Gamma}(1)\| = \bar{n}$. By Theorem 14.6, we conclude that

$$\|\bar{\Gamma}(\lambda_1) - \bar{\Gamma}(0)\| < \bar{\varepsilon}_0 \Rightarrow E \cap \partial A_{d(0)} \subseteq E \cap \partial A_{d(\lambda_1)} \text{ and}$$

$$\|\bar{\Gamma}(\lambda_2)-\bar{\Gamma}(\lambda_1)\| = \bar{\varepsilon}_1 \Rightarrow E \cap \partial A_{d(\lambda_1)} \subseteq E \cap \partial A_{\Sigma(\lambda_2)}.$$

Since $\bar{n}$ is finite and the convex hull $C_0(\bar{\Gamma}(0), \bar{\Gamma}(1))$ generated by $\bar{\Gamma}(0)$ and $\bar{\Gamma}(1)$ is compact, there exists a finite number of open covering. Thus, after repeating the above process a finite number of times, we can conclude that

$$E \cap \partial A_{d(0)} \subseteq E \cap \partial A_{d(1)} \text{ and}$$

$$E \cap \partial A_{d(1)} \subseteq E \cap \partial A_{d(0)}.$$

Combining the above two equations, we complete the proof.

14.4 STATIC AND DYNAMIC RELATIONSHIPS

The static as well as dynamic relationships between the original system (Eq. 14.1) and the reduced-state system (Eq. 14.2) will be established in this section. We first establish the static properties (S1) and (S2) and the dynamic property (D2) by combining the analytical results derived in Steps 1–5 of the previous section. These relationships are summarized as follows:

Theorem 14.7: Static Relationship

Consider the original system (Eq. 14.1) and the reduced-state system (Eq. 14.2). If zero is a regular value of $\frac{\partial^2 U}{\partial x \partial y}(x_i, y_i)$ for all the UEPs $(x_i, y_i, 0)$ on the stability boundary $\partial A(x_s, y_s, 0)$, then there exists a positive number, $\varepsilon > 0$, such that if the transfer conductance of the system (Eq. 14.1) satisfies $G_{ij} < \varepsilon$, $(\bar{x}, \bar{y})$ is a type-k equilibrium point of the system (Eq. 14.2) if and only if $(\bar{x}, \bar{y}, 0)$ is a type-k equilibrium point of the system (Eq. 14.1).

Theorem 14.7 asserts that, under the stated conditions, the static properties (S1) and (S2) between the original model (Eq. 14.1) and the reduced-state model (Eq. 14.2) hold. It not only establishes the static relationship but also justifies the efforts to establish the relationship between the stability region $A(x_s, y_s)$ of the reduced-state model (Eq. 14.2) and the stability region $A(x_s, y_s, 0)$ of the original model (Eq. 14.1).

Theorem 14.8: Dynamic Relationship

Let (x_s, y_s) be a SEP of the model (Eq. 14.2). If zero is a regular value of $\frac{\partial^2 U}{\partial x \partial y}(x_i, y_i)$ for all the UEP (x_i, y_i) on the stability boundary $\partial A(x_s, y_s)$, then there exists a positive number, $\varepsilon > 0$, such that if the transfer conductance of the model (Eq. 14.1) satisfies $G_{ij} < \varepsilon$ and the intersections of the stable and unstable manifolds of the equilibrium points on the stability boundary $\partial A(x_s, y_s, 0)$ of the parameterized model $d(\lambda)$ satisfy the transversality condition for $\lambda \in [0, 1]$, then

1. the equilibrium point $(x_i, y_i, 0)$ is on the stability boundary $\partial A(x_s, y_s, 0)$ of the original model (Eq. 14.1) if and only if the equilibrium point (x_i, y_i) is on the stability boundary $\partial A(x_s, y_s)$ of the reduced-state model (Eq. 14.2);
2. the stability boundary $\partial A(x_s, y_s, 0)$ of the original model (Eq. 14.1) is the union of the stable manifold of all the equilibrium points $(x_i, y_i, 0)$, $i = 1, 2, \ldots$ on the stability boundary $\partial A(x_s, y_s, 0)$; that is,

$$\partial A(x_s, y_s, 0) = \cup W^s(x_i, y_i, 0); \tag{14.7}$$

3. the stability boundary $\partial A(x_s, y_s)$ of the reduced-state model (Eq. 14.2) is the union of the stable manifold of all the equilibrium points (x_i, y_i), $i = 1, 2, \ldots,$ on the stability boundary $\partial A(x_s, y_s)$; that is,

$$\partial A(x_s, y_s) = \cup W^s(x_i, y_i). \tag{14.8}$$

Theorem 14.8 asserts that, under the stated condition, the dynamic property (D2) is satisfied. Furthermore, the stability boundaries of both the original model and the reduced-state model are completely characterized by Equations 14.7 and 14.8. As to the dynamic property (D1), it can be shown that, for any compact set S of the state space of the model (Eq. 14.2), there is a positive number α such that if the transfer conductance of the model satisfies $|G| < \alpha$, then there is an energy function defined on this compact set S.

14.5 DYNAMIC PROPERTY (D3)

We show in this section that the reduced-state model (Eq. 14.2) satisfies the dynamic property (D3). To this end, a numerical scheme for directly detecting the exit point, which is the intersection point between the projected fault-on trajectory $(xf(t), yf(t))$ and the stability boundary $\partial A(x_s, y_s)$ of the postfault reduced-state model, without resorting to detailed time–domain numerical simulation, is presented. Recall that a numerical energy function for the original model (Eq. 14.1) is in the following form, which is a summation of a numerical potential energy and a kinetic energy:

$$\begin{aligned} W_{num}(x, y, z) &= W_{ana}(x, y, z) + U_{path}(x, y) \\ &= \frac{1}{2} z^T M z + U(x, y) + U_{path}(x, y) \\ &= \frac{1}{2} z^T M z + U_{num}(x, y). \end{aligned} \tag{14.9}$$

The numerical scheme is described as follows:

Step 1. Along the fault-on trajectory ($x(t)$, $y(t)$, $z(t)$), detect the point (x^*, y^*) at which the projected fault-on trajectory ($x(t)$, $y(t)$) reaches the first local maximum (along the fault-on trajectory) of the numerical potential energy $U_{num}(x, y)$ in Equation 14.9.

Step 2. The point (x^*, y^*) is the exit point, which is an approximated point for the intersection point between the stability boundary of the (postfault) reduced-state model (Eq. 14.2) and the projected fault-on trajectory ($x(t)$, $y(t)$).

The above numerical scheme does not resort to detailed time–domain simulation of the postfault system. It has been found that this numerical scheme is very effective for a special class of nonlinear dynamical systems such as gradient systems and quasi-gradient systems. Another numerical scheme for detecting the exit point of the reduced-state stability boundary is the following:

Step 1. Along the fault-on trajectory ($x(t)$, $y(t)$, $z(t)$), compute the dot product of the following two vectors: (1) the fault-on generator speed vector and (2) the postfault power mismatch vector at each integration step. When the sign of the dot product changes from positive to negative, the exit point is detected.

We next present a numerical example to illustrate this numerical scheme to detect the stability boundary of a gradient system without resorting to a time–domain simulation. We consider the following gradient system:

$$\begin{aligned} x_1 &= f_1(x_1, x_2) = -2\sin(x_1) - \sin(x_1 - x_2) + 0.1 \\ x_2 &= f_2(x_1, x_2) = -2\sin(x_2) - \sin(x_2 - x_1) + 0.3. \end{aligned} \tag{14.10}$$

There are four type-one equilibrium points lying on the stability boundary of the SEP (0.10037, 0.20094). Since the system (Eq. 14.10) is a gradient system and

$$\frac{\partial f_1}{\partial x_2} = \frac{\partial f_2}{\partial x_1} = \cos(x_1 - x_2), \tag{14.11}$$

we employ the following procedure to construct an energy function:

$$g(x) = \frac{\partial V}{\partial x} = -f(x).$$

Hence,

$$\begin{aligned} \dot{V}(x_1, x_2) &= \left\langle \frac{\partial V}{\partial x}, f(x) \right\rangle \\ &= g^T(x) f(x) = -\|f(x)\|^2 < 0, \forall x \notin \{x : f(x) = 0\}. \end{aligned}$$

Consequently,

$$\begin{aligned} g_1(x) &= 2\sin(x_1) + \sin(x_1 - x_2) - 0.1 \\ g_2(x) &= 2\sin(x_2) + \sin(x_2 - x_1) - 0.3, \end{aligned}$$

and an energy function for the system (Eq. 14.10) is the following:

$$
\begin{aligned}
V(x_1, x_2) &= \int_0^x \langle g(x), dx \rangle \\
&= \int_0^{x_1} g_1(y_1, 0)dy_1 + \int_0^{x_2} g_2(x_1, y_2)dy_2 \\
&= [-3\cos(y_1) - 0.1y_1]_0^{x_1} + [-\cos(y_2) - \cos(y_2 - x_1) - 0.3y_2]_0^{x_2} \\
&= -2\cos(x_1) - \cos(x_2) - \cos(x_1 - x_2) - 0.1x_1 - 0.3x_2 + 4. \qquad (14.12)
\end{aligned}
$$

We next apply the above numerical scheme to identify the stability boundary of the gradient system (Eq. 14.10). Starting from the SEP, we construct a ray and detect the intersection point between the ray and the stability boundary. According to the numerical scheme, we calculate the energy function value along the ray. The first local maximum of the energy function along the ray occurs in the neighborhood of the intersection point between this ray and the stability boundary. We draw 36 rays out of the SEP (0.10037, 0.20094) and identify the corresponding intersection point between the stability boundary and each ray, and then connect the 36 intersection points to form the approximated stability boundary. In the numerical experiment, the angles of the 36 rays spread evenly within the range [0, 2π], and the step size along each ray is set as 0.0033. Numerical results show that 35 such local maxima have been found in the domain $\{(x_1, x_2) \| |x_1| \le 5.5, |x_1| \le 5.5\}$

According to Theorem 3.10, the exact stability boundary of the SEP x_s = (0.10037, 0.20094) is the union of the stable manifolds of the equilibrium points on the stability boundary

$$\partial A(x_s) = \bigcup_{x_1 \in \partial A(x_s)} W^s(x_i).$$

The exact stability boundary of the SEP x_s is represented by the red curve in Figure 14.2. The center point highlighted is the SEP x_s. The approximated stability

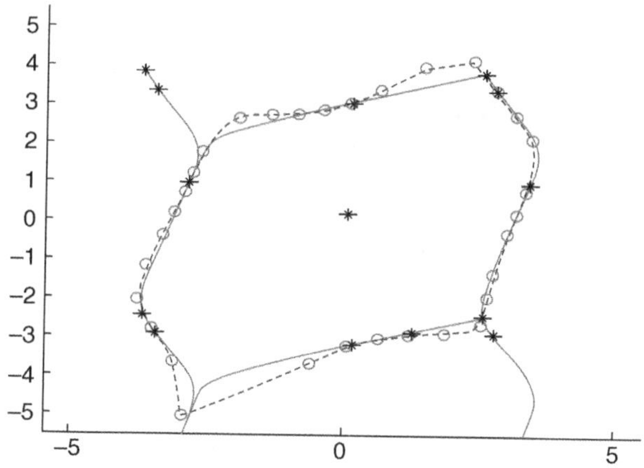

Figure 14.2 The approximated stability boundary is represented by the dashed lines with the circle representing the corresponding local maxima. The exact stability boundary is represented by the solid line. It can be seen that the approximated stability boundary by the proposed scheme is quite close to the exact stability boundary.

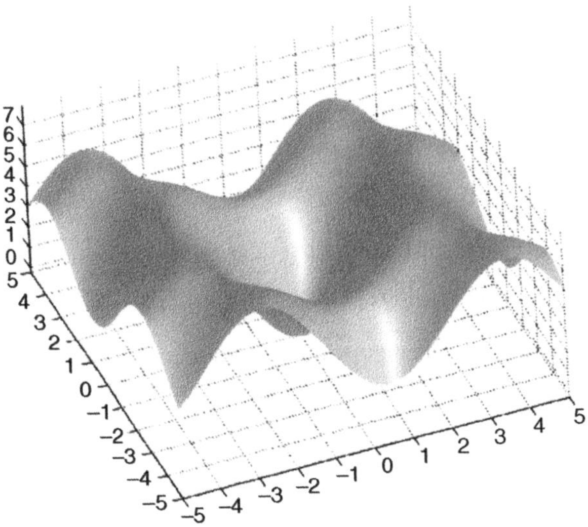

Figure 14.3 Contours of a constant energy function of the simple gradient system.

boundary is represented by the dashed lines in Figure 14.2, and the circles represent the corresponding local maxima.

A comparison between the exact and the approximated stability boundaries is thus visually obvious. It can be observed that the numerical scheme performs very well in the neighborhood of the type-one equilibrium points on the stability boundary, while the accuracy of the approximated stability boundary tends to be not so well in the neighborhood of type-two equilibrium points. For the convenience of comparison and observation, a surface plot of the energy function is shown in Figure 14.3.

In summary, the reduced-state model (Eq. 14.2) belongs to a quasi-gradient system, and the proposed numerical schemes presented in this section are effective for detecting exit points of quasi-gradient systems. Hence, the dynamic property (D3) is satisfied with the reduced-state model (Eq. 14.2).

14.6 A CONCEPTUAL NETWORK-REDUCTION BCU METHOD

The analytical results derived in Theorems 14.7 and 14.8 suggest a means of finding the controlling UEP of the original model by computing the controlling UEP of the reduced-state model. A theory-based method, called the BCU method for direct stability analysis of the generic network-reduction model, is presented next.

Network-Reduction BCU Method

Step 1. From the fault-on trajectory $(x(t), y(t), z(t))$, detect the exit point, say, (x^*, y^*), at which the projected trajectory $(x(t), y(t))$ exits the stability boundary of the reduced-state system. The projected fault-on trajectory $(x(t), y(t))$

is obtained from the original fault-on trajectory $(x(t), y(t), z(t))$ by removing the component $z(t)$.

Step 2. Use the point (x^*, y^*) as an initial condition and integrate the postfault reduced-state system to compute the controlling UEP of the reduced-state system, say, (x_{co}^*, y_{co}^*).

Step 3. The controlling UEP with respect to the fault-on trajectory is $(x_{co}^*, y_{co}^*, 0)$.

In essence, the BCU method computes the controlling UEP of the original system by computing the controlling UEP of the reduced-state system whose controlling UEP can be effectively computed. Steps 1 and 2 of the conceptual BCU method compute the controlling UEP of the reduced-state system (Eq. 14.2) and Step 3 relates the controlling UEP of the reduced-state system to the controlling UEP of the original system (Eq. 14.1). Note that the postfault reduced-state trajectory starting from the exit point (u^*, w^*, x^*, y^*), which is obtained at Step 2, will converge to an equilibrium point, which is the controlling UEP of the reduced-state system. The controlling UEP always exists and is unique, and the stable manifold of the controlling UEP of the reduced-state system $(u_{co}, w_{co}, x_{co}, y_{co})$ contains the exit point (u^*, w^*, x^*, y^*) (see Figure 14.4). Step 3 relates the controlling UEP of the reduced-state system, with respect to the projected fault-on trajectory, to the controlling UEP of the original system (see Figure 14.5).

Theorem 14.8 can provide a theoretical basis for Step 2 and a partial basis for Step 3 of the above conceptual BCU method. Indeed, the point (x^*, y^*) must lie on the stable manifold of the controlling UEP of the reduced-state system. Hence, numerically integrating the postfault reduced-state system from the point (x^*, y^*) yields a trajectory converging to the controlling UEP of the reduced-state system. The key step in the conceptual BCU method is Step 3. Hence, ensuring that $(x_{co}^*, y_{co}^*, 0)$ is the controlling UEP with respect to the fault-on trajectory of the original system is the major concern of the conceptual BCU method.

We next present a sufficient condition under which the BCU method finds the correct controlling UEP relative to a given fault: the exit point of the projected fault-on trajectory $x(t), y(t)$ is on the stable manifold $W^s(\hat{x}, \hat{y})$ if and only if the exit point of the fault-on trajectory $(x(t), y(t), z(t))$ is on the stable manifold $W^s(\hat{x}, \hat{y}, 0)$. Theorems 14.7 and 14.8 provide a theoretical basis for the sufficient condition to hold. Theorem 14.8 shows that under the one-parameter transversality condition, the equilibrium point $(\hat{x}, \hat{y})$ is on the stability boundary $\partial A(x, y)$ of the reduced-state system if and only if the equilibrium point $(\hat{x}, \hat{y}, 0)$ is on the stability boundary $\partial A(\hat{x}, \hat{y}, 0)$ of the original system.

There are several possible ways to numerically implement the conceptual BCU method for power system models. A numerical implementation of this method is presented in the next chapter.

14.7 CONCLUDING REMARKS

The ability to compute the controlling UEP is vital to direct stability analysis. However, the majority of the existing methods fail to compute the controlling UEP.

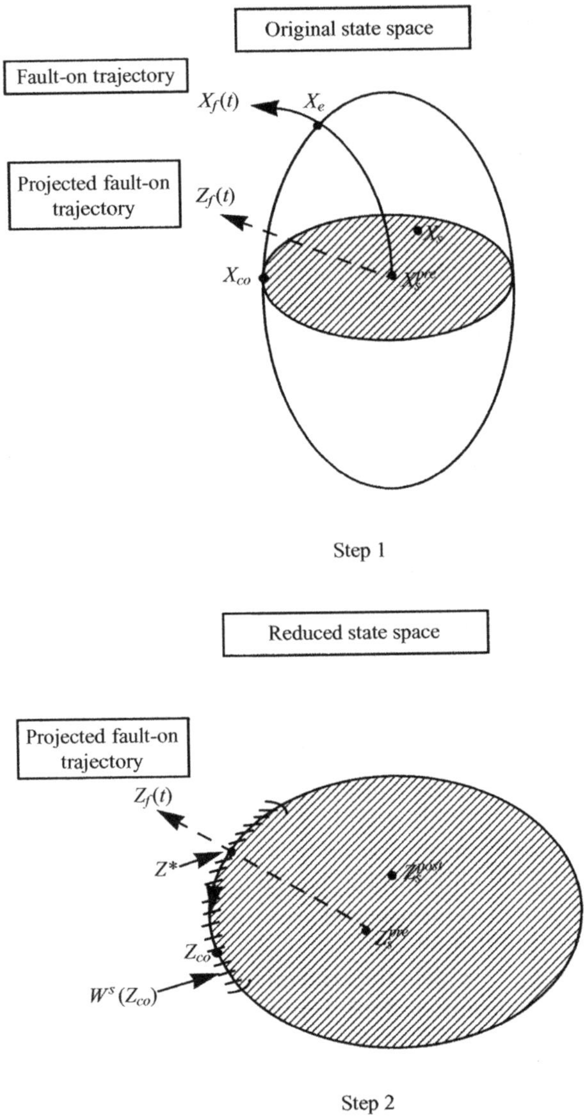

Figure 14.4 Steps 1 and 2 of the conceptual BCU method.

The key reason for this failure is that the task of directly computing the controlling UEP of the original transient stability model is very difficult, if not impossible. Another important factor overlooked by the methods proposed in the literature is that the controlling UEP is the first UEP whose stable manifold has a nonempty intersection with the fault-on trajectory. This fact must be incorporated into a numerical method for computing the controlling UEP. Without the incorporation of this fact, a numerical method is unlikely to be consistent in finding the correct controlling UEP. This fact has been incorporated into the BCU method presented in this

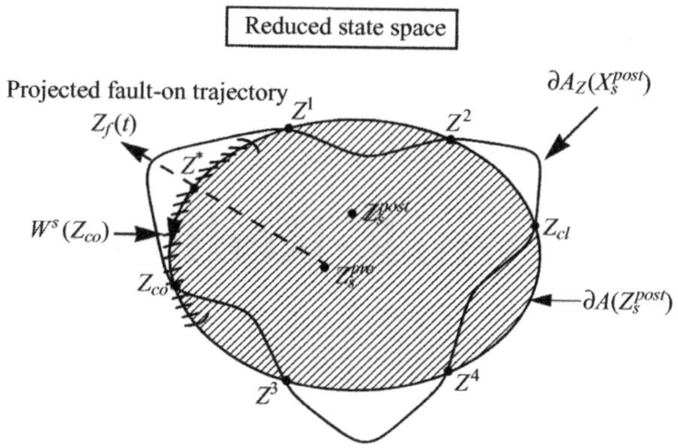

Another viewpoint of Step 2 of the conceptual BCU method, where $\partial A_Z(X_s^{post})$ is the intersection between the stability boundary $\partial A(X_s^{post})$ and the reduced state space and $\partial A(Z_s^{post})$ is the stability boundary of Z_s^{post} of the reduced-state system.

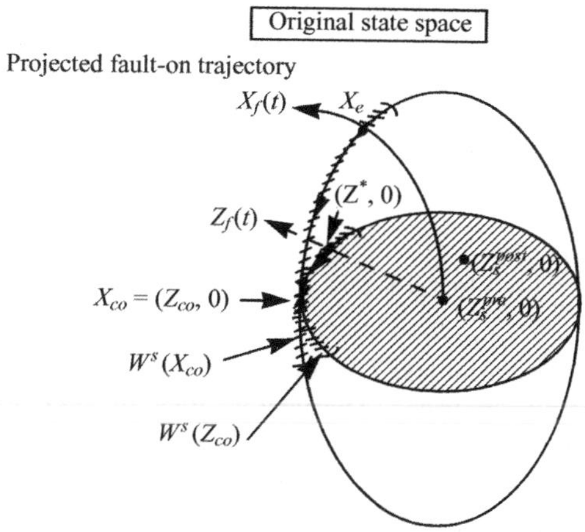

Step 3 of the conceptual BCU method

Figure 14.5 Step 3 of the conceptual BCU method.

chapter. The network-reduction BCU method computes the controlling UEP by exploiting special properties as well as some physical and mathematical insights of the underlying network-reduction transient stability model. A network-preserving BCU method will be presented in Chapter 16.

The BCU method is a systematic method for computing the controlling UEP of large-scale power systems. It computes the controlling UEP of the original model by computing the controlling UEP of the reduced-state model, which is much easier to compute than that of the original model. A sufficient condition under which the BCU method finds the correct controlling UEP has been derived. A comprehensive theoretical basis for the sufficient condition to hold has also been developed. Numerical network-reduction BCU methods for large-scale power systems will be presented in the next chapter.

Chapter 15

Numerical Network-Reduction BCU Method

15.1 INTRODUCTION

The fundamental ideas behind the conceptual network-reduction BCU method are explained in the following. We are given a generic network-reduction model described below, which encompasses the existing network-reduction models:

$$
\begin{aligned}
T\dot{x} &= -\frac{\partial U}{\partial x}(x, y) + g_1(x, y) \\
\dot{y} &= z \\
M\dot{z} &= -Dz - \frac{\partial U}{\partial y}(x, y) + g_2(x, y),
\end{aligned}
\tag{15.1}
$$

where $x \in R^n$, $y \in R^n$, and $z \in R^n$ are state variables. T is a positive definite matrix and M and D are diagonal positive definite matrices. The vectors $g_1(x, y)$, $g_2(x, y)$ represent the transfer conductances. The following artificial, reduced-state model associated with the original model has been shown to satisfy the static properties (S1) and (S2) as well as the dynamic properties (D1), (D2), and (D3):

$$
\begin{aligned}
T\dot{x} &= -\frac{\partial U}{\partial x}(x, y) + g_1(x, y) \\
\dot{y} &= -\frac{\partial U}{\partial y}(x, y) + g_2(x, y).
\end{aligned}
\tag{15.2}
$$

These properties have been explored to develop a conceptual network-reduction BCU method for computing the controlling unstable equilibrium point (controlling UEP). The reduced-state model is not unique. In the BCU method, a model can be a reduced-state model as long as it satisfies the static and dynamic properties.

Direct Methods for Stability Analysis of Electric Power Systems, by Hsiao-Dong Chiang

There are several possible ways to numerically implement the conceptual network-reduction BCU method. In this chapter, a numerical network-reduction BCU method is presented, explained in great detail, and applied, for illustrational purposes, to several test systems. We first present a numerical network-reduction BCU method. It should be noted that other numerical network-reduction BCU methods can also be developed.

A Numerical Network-Reduction BCU Method

Step 1. Construct a numerical energy function for the postfault system:

$$W_{num}(x, y, z) = \frac{1}{2} z^T M z + U_{num}(x, y).$$

Step 2. From the fault-on trajectory $(x_f(t), y_f(t), z_f(t))$, detect the exit point (x^*, y^*) at which the projected fault-on trajectory $(x_f(t), y_f(t))$ reaches the first local maximum along the fault-on trajectory of the numerical potential energy $U_{num}(x, y)$.

Step 3. Use the point (x^*, y^*) as an initial condition and integrate the postfault reduced-state system to find the minimum gradient point (MGP), which, along the postfault trajectory of the reduced-state system, has the first local minimum value of the norm of the vector field of the postfault reduced-state system, that is, the first local minimum of the following value:

$$\left\| -\frac{\partial}{\partial x} U(x, y) + g_1(x, y) \right\| + \left\| -\frac{\partial}{\partial x} U(x, y) + g_2(x, y) \right\|,$$

where $U(x, y)$, $g_1(x, y)$, and $g_2(x, y)$ are components of the vector field of the postfault reduced-state system, and the MGP is found at (x_o^*, y_o^*).

Step 4. Use the point (x_o^*, y_o^*) as the initial guess to solve the following set of nonlinear algebraic equations. Let the solution be (x_{co}^*, y_{co}^*):

$$\left\| -\frac{\partial}{\partial x} U(x, y) + g_1(x, y) \right\| + \left\| -\frac{\partial}{\partial x} U(x, y) + g_2(x, y) \right\| = 0.$$

Step 5. The controlling UEP with respect to the fault-on trajectory $(x_f(t), y_f(t), z_f(t))$ is $(x_{co}^*, y_{co}^*, 0)$.

Step 6. The critical energy, v_{cr}, is the value of the numerical energy function $W_{num}(\cdot)$ at the controlling UEP; that is, $v_{cr} = W_{num}(x_{co}^*, y_{co}^*, 0)$.

Step 7. Calculate the value of the numerical energy function $W_{num}(\cdot)$ at the time of fault clearance (t_{cl}) using the fault-on trajectory

$$v_f = W_{num}(x_f(t_{cl}), y_f(t_{cl}), z_f(t_{cl})).$$

Step 8. If $v_f < v_{cr}$, then the point $(x_f(t_{cl}), y_f(t_{cl}), z_f(t_{cl}))$ is located inside the stability boundary of the postfault system and the postfault trajectory is stable. Otherwise, it may be unstable.

Steps 1–5 are basically the computational steps for implementing the network-reduction method, while Steps 6–8 are basically the computational steps for implementing the controlling UEP method.

The exit point computed in Step 2 will not lie on the stability boundary of the reduced-state system; instead it will lie in the neighborhood of the stability boundary. Hence, the trajectory generated in Step 3 will not lie in the stable manifold of the controlling UEP of the reduced-state system. Rather, the trajectory will move off the stable manifold. One challenge of this step is generating a trajectory in Step 4 such that it moves along the stability boundary of the reduced-state system and the trajectory approaches the reduced-state controlling UEP (i.e., the controlling UEP of the reduced-state model relative to the projected fault-on trajectory). To this end, a stability-boundary-following procedure to numerically implement Step 4 will be presented. The postfault reduced-state model can be numerically stiff. In this situation, a stiff differential equation solver should be used to implement Step 3 of the numerical network-reduction BCU method.

A robust nonlinear algebraic solver is needed for computing the controlling UEP of the reduced-state model in Step 4. In this step, the computed MGP of Step 3 is usually close to the reduced-state controlling UEP.

15.2 COMPUTING EXIT POINTS

The exit point at which the projected sustained fault-on trajectory intersects the stability boundary of the postfault reduced-state system is computationally characterized by the first local maximum of the potential energy along the projected fault-on trajectory. If the exit point is not yet detected before the fault is cleared, then a sustained fault-on trajectory beyond the fault is cleared is simulated. Hence, the exit point (computed in Step 2) is computationally characterized by the first local maximum of the potential energy along the (sustained) projected fault-on trajectory.

Another scheme to detect the exit point is computing the dot product of the fault-on speed vector and postfault power mismatch vector at each integration step. As explained in the following, the exit point is detected when the sign of the dot product changes from positive to negative.

The derivative of the potential energy with respect to time, evaluated along the fault-on trajectory, can be calculated as

$$\frac{dV_{PE}}{dt} = -\sum_{i=1}^{n}(P_{mi} - P_{ei})\frac{d\delta_i}{dt}.$$

The right-hand side of the above equation is termed the *dot product*. If this dot product changes sign from positive to negative along the fault-on trajectory, it indicates that the point that has a local maximum of the potential energy along the (sustained) projected fault-on trajectory has just passed. The exit point can then be accurately computed by computing the point that has a zero dot product.

Due to the nature of simulating fault-on trajectories on a digital computer, a sequence of points along the trajectory is generated. The computed exit point cannot be positioned exactly as a point lying on the stability boundary of the reduced-state system. There are two possible ways of selecting the detected exit point along the simulated fault-on trajectory: one is the point along the simulated fault-on trajectory right before the sign of the dot product changes from positive to negative and another is the one after the sign changes. Either way is generally sufficient to compute an accurate exit point. However, for some cases where the accuracy of the computed exit point is of concern, the accuracy of computing the exit point can be improved via a linear interpolation method, described as follows.

Suppose that the value of the dot product along the projected sustained fault-on trajectory at time t_1 is d_1, $d_1 > 0$, and one more step integration to t_2 results in a negative dot product $d_2 < 0$. We apply a linear interpolation to the time interval $[t_1, t_2]$, which results in an interpolated time, $t_0 \in [t_1, t_2]$, and

$$t_0 = \frac{d_2 t_1 - d_1 t_2}{d_2 - d_1}.$$

The dot product at t_0 is then computed; say, for example, its value is d_0. If the absolute value of d_0 is sufficiently small, then t_0 is considered as the instant of PEBS crossing or the exit point time. If the absolute value of d_0 is not sufficiently small, two situations may arise: (i) $d_0 > 0$ or (ii) $d_0 < 0$. In the former case, one more linear interpolation is applied to the time interval $[t_0, t_2]$; in the latter case, one more linear interpolation is applied to the time interval $[t_1, t_0]$.

The above process is repeatedly executed until the absolute value of the dot product at the interpolated time is smaller than a specified tolerance. It usually takes three or four interpolations for the convergence tolerance to be satisfied. The procedure for accurate computation of the exit point by employing the linear interpolation method is implemented as follows:

Step 1. Integrate the fault-on trajectory until the dot product changes sign, say, between the interval $[t_1, t_2]$.

Step 2. Apply the linear interpolation to the interval $[t_1, t_2]$, which results in an intermediate time, t_0. Compute the exact dot product at t_0 for the postfault reduced-state system. If the value is smaller than a threshold value, then the exit point is detected and then stop; otherwise, go to Step 3.

Step 3. If the dot product is positive, then replace t_1 with t_0; otherwise, replace t_2 with t_0 and go to Step 2.

15.3 STABILITY-BOUNDARY-FOLLOWING PROCEDURE

The three important steps in the numerical BCU method are computations of the exit point, the MGP, and the controlling UEP. The role of MGP is used as an initial condition to search for the controlling UEP. The robustness of computing the controlling UEP depends on the quality of the MGP. A numerical inaccuracy in computing the exit point will probably cause a numerical difficulty in computing the MGP.

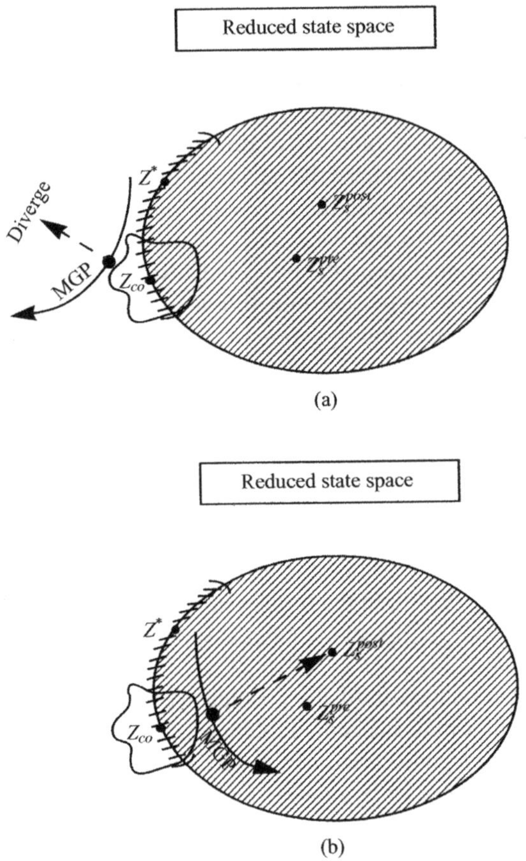

Figure 15.1 If the MGP does not lie inside the convergent region of the Newton method, then the sequence generated by the Newton method starting from the MGP will not converge to the controlling UEP The sequence may diverge as illustrated in (a) or it may converge to the stable equilibrium point as illustrated in (b).

Likewise, a numerical inaccuracy in computing the MGP will probably cause a numerical difficulty in computing the controlling UEP. It is well-known that if the MGP is sufficiently close to the controlling UEP, then the sequence generated by the Newton method starting from the MGP will converge to the controlling UEP. Otherwise, the sequence may converge to another equilibrium point or may diverge (see Figure 15.1).

If the exit point is not accurately computed and it lies in some significant distance from the controlling UEP, then the following numerical difficulties will occur:

- The MGP cannot be numerically found.
- The found MGP is not sufficiently close to the controlling UEP so that the nonlinear algebraic solver either diverges or converges to an incorrect controlling UEP.

To compute an adequate MGP to reliably compute the controlling UEP, the following stability-boundary-following procedure to guide the search process of an improved MGP is developed (see Figures 15.2–15.7):

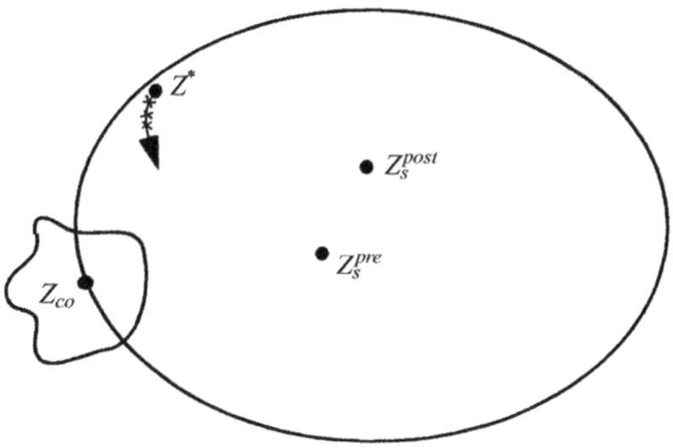

(a)

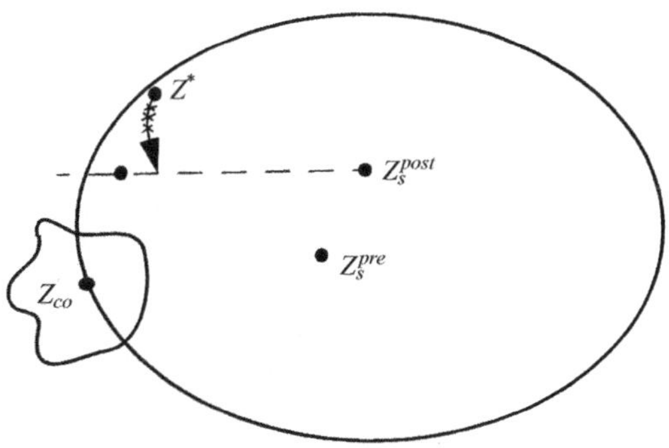

(b)

Figure 15.2 (a) Use the exit point, which, in this case, lies inside the stability region of the reduced-stated system, as an initial condition and integrate the postfault reduced-state system for a few time steps, say, five time steps. (b) Draw a ray connecting the current point on the trajectory and the SEP of the postfault reduced-state system. Replace the current point with the corrected exit point, which is the (first) local maximal point of potential energy along the ray starting from the SEP.

A Stability-Boundary-Following Procedure

Step 1. Integrate the postfault reduced-state model (Eq. 15.2) starting from the current point for a few time steps, say, four or five time-step points of the postfault reduced-state trajectory. In the first iteration, the current point is the exit point. The last point of the obtained four or five points is termed as the current point.

Step 2. Check whether the postfault reduced-state trajectory reaches a relative local minimum of the vector field of the reduced-state model along the trajectory. If yes, then the point with a local minimum of the vector field is the MGP, and then go to the next step; otherwise, go to Step 6.

Step 3. Apply a nonlinear algebraic solver, such as the Newton method, starting from the MGP to compute the reduced-state controlling UEP.

Step 4. Check the validity of the computed controlling UEP. Discard it and go to the next step if it is a postfault stable equilibrium point (SEP) or if the distance between the computed controlling UEP and the MGP is large;

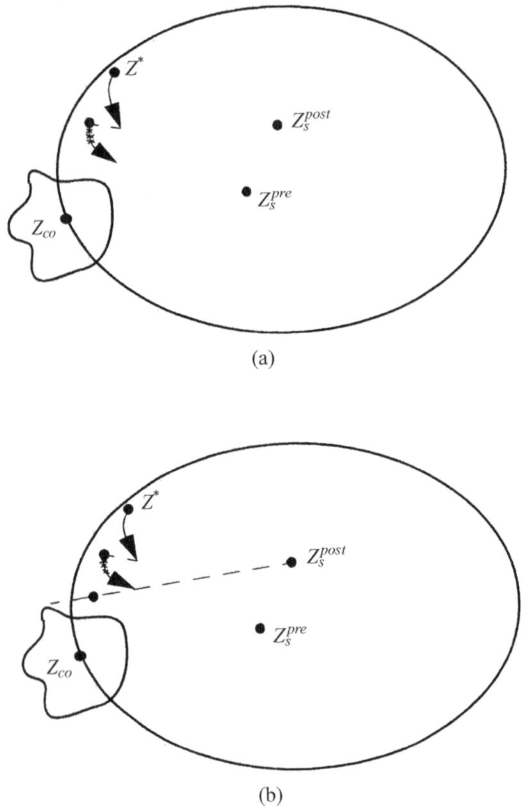

Figure 15.3 (a) Use the (corrected) exit point as an initial condition and integrate the postfault reduced-state system for a few time steps, say, five time steps. (b) Draw a ray connecting the current point on the trajectory and the SEP of the postfault reduced-state system. Replace the current point with the corrected exit point, which is the (first) local maximal point of potential energy along the ray starting from the SEP.

otherwise, return the computed controlling UEP as the reduced-state controlling UEP and stop the procedure.

Step 5. If the number of adjustments in the stability boundary following procedure is greater than a threshold, then stop the procedure and go back to the exit point detection procedure to compute an improved exit point and restart the stability-boundary-following procedure; otherwise, use the MGP as the current point and go to Step 1.

Step 6. Draw a ray connecting the current point on the postfault reduced-state trajectory with the SEP of the postfault reduced-state system.

Step 7. Start from the current point and move along the ray to find the point on the ray that has the first local maximum value of the potential energy function. Replace the current point by the point found in this step and go to Step 1.

The theoretical basis for the above stability-boundary-following procedure is the structure of the stability boundary of the reduced-state system. The stability boundary is composed of the stable manifolds of the UEPs lying on the stability

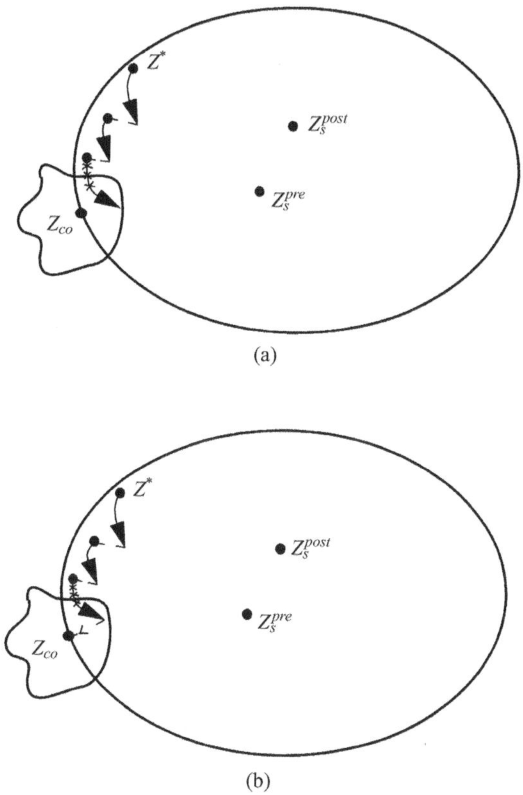

Figure 15.4 (a) Use the (corrected) exit point as an initial condition and integrate the postfault reduced-state system for a few time steps, say, five time steps. The end point of this trajectory lies inside the convergence region of the controlling UEP (with respect to the Newton method). Hence, the sequence generated by the Newton method starting from the end point will converge to the controlling UEP as shown in (b).

boundary. Steps 6 and 7 are based on the computational scheme developed to implement the exit point. Hence, these two steps serve to ensure the stability-boundary-following procedure moves along the stability boundary. Practical implementation of Step 7 can be executed by checking the zero crossing of the dot product between the power mismatch and the speed. The process of checking the zero crossing can be sped up by starting the search from the current point on the "trajectory" instead of starting from the SEP. The sign of the dot product at the current point determines the starting direction of the local maximal search. Since the controlling UEP is of type-one (the UEP whose corresponding Jacobian matrix contains only one eigenvalue with a positive real part), it has the nice feature of being a "relative SEP" within the stability boundary.

The stability-boundary-following procedure described above yields a search process close to the stability boundary. Replacing the current point generated in Step 1 of the procedure with the point having a local maximal energy function along the ray amounts to confining the MGP search process close to the stability boundary. A numerical implementation of the stability-boundary-following procedure may encounter oscillation around the controlling UEP when the norm of the sequence produced by the stability-boundary-following procedure reaches a small number.

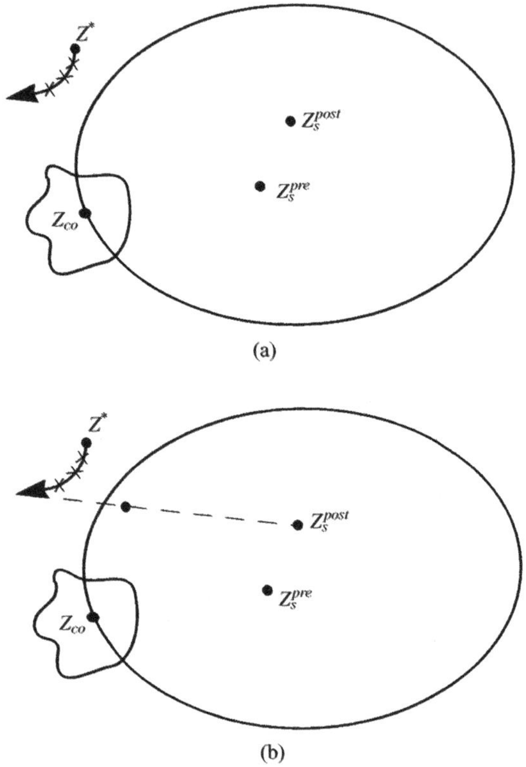

Figure 15.5 (a) Use the exit point, which, in this case, lies outside the stability region of the reduced-stated system, as an initial condition and integrate the postfault reduced-state system for a few time steps, say, five time steps. (b) Draw a ray connecting the current point on the trajectory and the SEP of the postfault reduced-state system. Replace the current point with the corrected exit point, which is the (first) local maximal point of potential energy along the ray starting from the SEP.

When the norm is smaller than a threshold value, we stop the procedure and apply a robust nonlinear algebraic equation solver such as the Newton method with the updated point as an initial guess to compute the exact controlling UEP. The convergence process to the controlling UEP is usually smooth without numerical difficulties. The closeness of the MGP to the controlling UEP contributes to the ease of convergence. Another similar procedure was described in Treinen et al. (1996).

15.4 A SAFEGUARD SCHEME

From a mathematical viewpoint, the MGP in the BCU method is not a well-defined state. Instead, it is a computational product in the BCU method. MGP is a good initial point for computing the controlling UEP. Both the coordinates and the energy function value of MGP are not used in BCU methods.

Some "ill-conditioned" phenomenon may occur during the stability-boundary-following procedure to find an MGP based on the guideline of the first local minimum of the norm of the vector field. The vector field of the reduced-state model is a gradient vector. This ill-conditioned situation arises when the norm is very small (close to zero in magnitude); the occurrence of a local minimum is usually caused

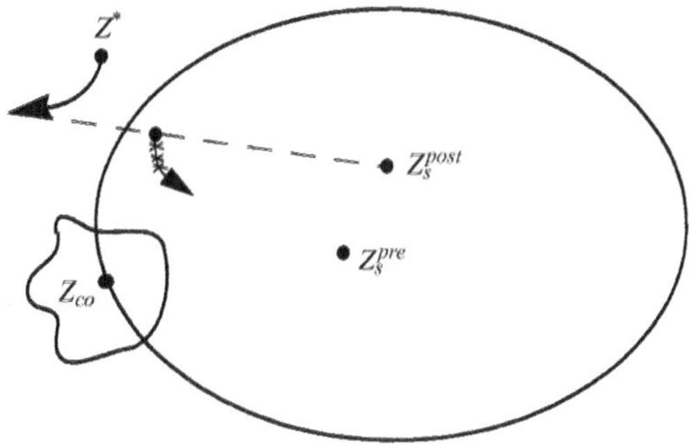

(a)

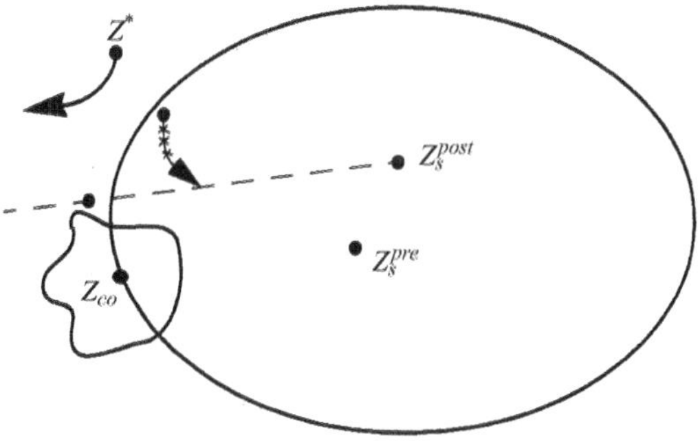

(b)

Figure 15.6 (a) Use the (corrected) exit point as an initial condition and integrate the postfault reduced-state system for a few time steps, say, five time steps. (b) Draw a ray connecting the current point on the trajectory and the SEP of the postfault reduced-state system. Replace the current point with the corrected exit point, which is the (first) local maximal point of potential energy along the ray starting from the SEP.

by the round-off error of the computation, which is not an indication of the MGP. A safeguard step against this false detection of MGP is to add a threshold checking of the variation in norm before checking the norm of the vector field.

For a numerical example, we use the nine-bus system with Contingency 7 to illustrate this phenomenon. During the stability-boundary-following procedure starting from the exit point [−0.5084, 0.5167, 2.8945], the first local minimum point occurs at [−0.5169, 0.5465, 2.8978], which, as seen from Figure 15.8, is not a true local minimum. As the MGP search continues, the second local minimum point occurs at [−0.7653, 1.6882, 2.4208], which, as seen from Figure 15.8, is a true (numerical) local minimum. Hence, this point is an MGP. A Newton method starting from this MGP converges to the controlling UEP, [−0.7576, 1.8583, 1.9986]. In this figure, the stability-boundary-following procedure is also numerically illustrated.

15.5 ILLUSTRATIVE EXAMPLES

In order to illustrate the numerical network-reduction BCU method in a simple context, we consider a simple three-machine system with a machine number 1 as the reference machine. The system model is described by the following equations:

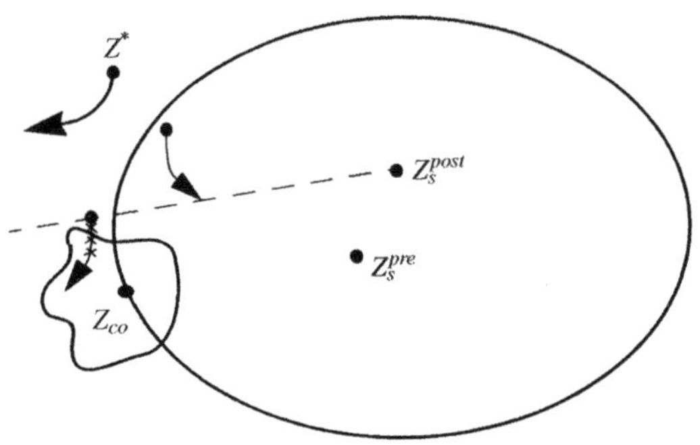

(a)

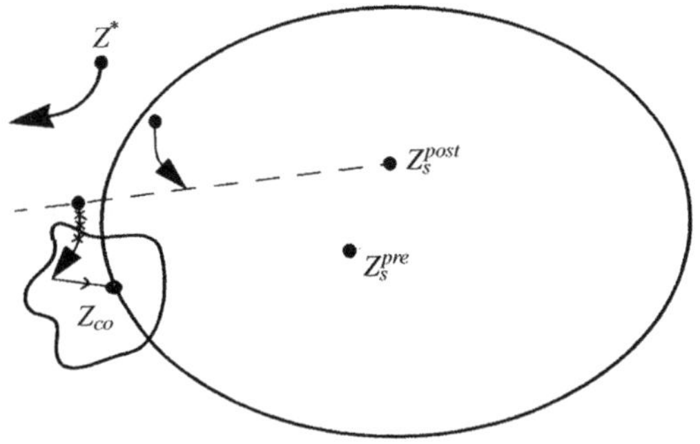

(b)

Figure 15.7 (a) Use the (corrected) exit point as an initial condition and integrate the postfault reduced-state system for a few time steps, say, five time steps. The end point of this trajectory lies inside the convergence region of the controlling UEP (with respect to the Newton method). Hence, the sequence generated by the Newton method starting from the end point will converge to the controlling UEP as shown in (b).

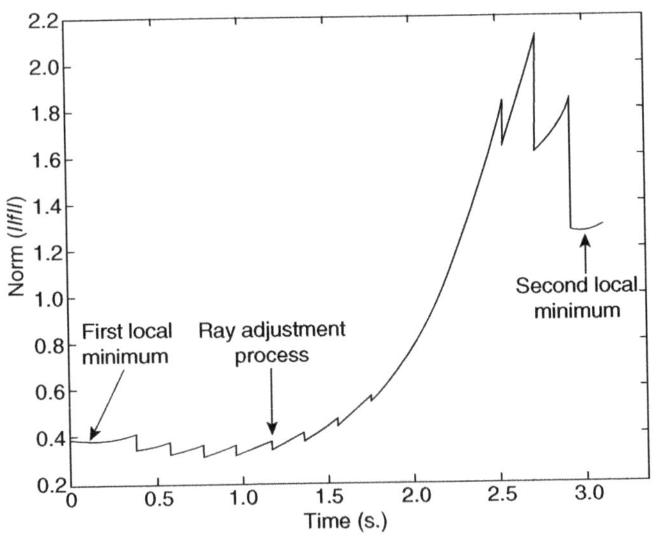

Figure 15.8 During the stability-boundary-following procedure, the first local minimum is not an MGP, while the second local minimum of the "stability-boundary-following trajectory" is an MGP. The vertical axis represents the norm of the state of the reduced-state system.

$$
\begin{aligned}
\dot{\delta}_1 &= \omega_1 \\
\dot{\delta}_2 &= \omega_2 \\
\dot{\delta}_3 &= \omega_3 \\
m_1\dot{\omega}_1 &= -d_1\omega_1 + P_{m_1} - P_{e_1}(\delta_1, \delta_2, \delta_3) \\
m_2\dot{\omega}_2 &= -d_2\omega_2 + P_{m_2} - P_{e_2}(\delta_1, \delta_2, \delta_3) \\
m_3\dot{\omega}_3 &= -d_3\omega_3 + P_{m_3} - P_{e_3}(\delta_1, \delta_2, \delta_3),
\end{aligned}
\tag{15.3}
$$

where

$$
P_{e_i}(\delta_1, \delta_2, \delta_3) = \sum_{j=1, j\neq 1}^{3} E_i E_j \left(B_{ij}\sin(\delta_i - \delta_j) + G_{ij}\cos(\delta_i - \delta_j)\right),
$$

$m_1 = 0.1254$, $m_2 = 0.034$, $m_3 = 0.016$, $P_{m1} = 0.8980$, $P_{m2} = 1.3432$, $P_{m3} = 0.9419$, $d_1 = 0.1254$, $d_2 = 0.0034$, $d_3 = 0.0016$, $E_1 = 1.1083$, $E_2 = 1.1071$, and $E_3 = 1.0606$.

The corresponding reduced-state model for the BCU method is

$$
\begin{aligned}
\dot{\delta}_1 &= P_{m_1} - P_{e_1}(\delta_1, \delta_2, \delta_3) \\
\dot{\delta}_2 &= P_{m_2} - P_{e_2}(\delta_1, \delta_2, \delta_3) \\
\dot{\delta}_3 &= P_{m_3} - P_{e_3}(\delta_1, \delta_2, \delta_3).
\end{aligned}
\tag{15.4}
$$

The application of the network-reduction BCU method to computing the exit point, MGP, and the controlling UEP of the reduced-state system is illustrated in this simple example. The three-machine system represented in the center of inertia (COI) angle coordinate is expressed as follows:

$$
\begin{aligned}
\dot{\tilde{\delta}}_1 &= \tilde{\omega}_1 \\
\dot{\tilde{\delta}}_2 &= \tilde{\omega}_2 \\
\dot{\tilde{\delta}}_3 &= \tilde{\omega}_3 \\
M_1\dot{\tilde{\omega}}_1 &= P_{m_1} - P_{e_1}(\tilde{\delta}) - \frac{M_1}{M_T}P_{COI} - D_1\omega_1 \\
M_2\dot{\tilde{\omega}}_2 &= P_{m_2} - P_{e_2}(\tilde{\delta}) - \frac{M_2}{M_T}P_{COI} - D_2\omega_2 \\
M_3\dot{\tilde{\omega}}_3 &= P_{m_3} - P_{e_3}(\tilde{\delta}) - \frac{M_3}{M_T}P_{COI} - D_3\omega_3,
\end{aligned}
\tag{15.5}
$$

where $\delta_0 = \frac{1}{M_T}\sum_{i=1}^{3} M_i\delta_i$, $\omega_0 = \frac{1}{M_T}\sum_{i=1}^{3} M_i\omega_i$, $M_T = \sum_{i=1}^{3} M_i$, $\tilde{\delta}_i = \delta_i - \delta_0$ for $i = 1, \ldots, 3$,

$\tilde{\omega}_i = \omega_i - \omega_0$ for $i = 1, \ldots, 3$, $\tilde{\theta}_i = \theta_i - \delta_0$ for $i = 1, \ldots, 3$,

$$P_{ei}(\tilde{\delta}) = \sum_{j=1}^{3} E'_{qi}E'_{qj}\left(G_{ij}\cos\left(\tilde{\delta}_i - \tilde{\delta}_j\right) + B_{ij}\sin\left(\tilde{\delta}_i - \tilde{\delta}_j\right)\right), \text{ and}$$

$$P_{COI} = \sum_{i=1}^{3} P_{m_i} - \sum_{i=1}^{3}\sum_{j=1}^{3} E'_{qi}E'_{qj}\left(G_{ij}\cos\left(\tilde{\delta}_i - \tilde{\delta}_j\right) + B_{ij}\sin\left(\tilde{\delta}_i - \tilde{\delta}_j\right)\right).$$

The system Jacobian matrix is as follows:

$$\left[\begin{array}{ccc|ccc} 0 & 0 & 0 & 1 & 0 & 0 \\ 0 & 0 & 0 & 0 & 1 & 0 \\ 0 & 0 & 0 & 0 & 0 & 1 \\ \hline & & & \frac{-D_1}{M_1} & 0 & 0 \\ & A^{COI} & & 0 & \frac{-D_2}{M_2} & 0 \\ & & & 0 & 0 & \frac{-D_3}{M_3} \end{array}\right]$$

where

$$a_{ij}^{COI} = \begin{cases} \left[-E_iE_j\left(G_{ij}\sin\left(\tilde{\delta}_i - \tilde{\delta}_j\right) - B_{ij}\cos\left(\tilde{\delta}_i - \tilde{\delta}_j\right)\right)\right]\cdot\dfrac{1}{M_i} - \dfrac{1}{M_T}\dfrac{\partial P_{COI}}{\partial \tilde{\delta}_i} & \text{if } i \neq j \\ \displaystyle\sum_{k=1}^{3} E_iE_k\left(G_{ik}\sin\left(\tilde{\delta}_i - \tilde{\delta}_k\right) + B_{ik}\cos\left(\tilde{\delta}_i - \tilde{\delta}_k\right)\right)\cdot\dfrac{1}{M_i} - \dfrac{1}{M_T}\dfrac{\partial P_{COI}}{\partial \tilde{\delta}_i} & \text{if } i = j \end{cases}$$

$$\frac{\partial P_{COI}}{\partial \tilde{\delta}_i} = 2\sum_{k\neq i} E_iE_k \sin\left(\tilde{\delta}_i - \tilde{\delta}_k\right).$$

The associated reduced-state model is

$$\begin{aligned} \dot{\tilde{\delta}}_1 &= P_{m_1} - P_{e_1}(\tilde{\delta}) - \frac{M_1}{M_T}P_{COI} \\ \dot{\tilde{\delta}}_2 &= P_{m_2} - P_{e_2}(\tilde{\delta}) - \frac{M_2}{M_T}P_{COI} \\ \dot{\tilde{\delta}}_3 &= P_{m_3} - P_{e_3}(\tilde{\delta}) - \frac{M_3}{M_T}P_{COI}. \end{aligned} \tag{15.6}$$

If we introduce new variables $\delta_{2,1} = \delta_2 - \delta_1$ and $\delta_{3,1} = \delta_3 - \delta_1$, then P_{e_i} $(\delta_1, \delta_2, \delta_3)$ can be expressed as a function of $\delta_{2,1}$ and $\delta_{3,1}$ as shown below:

$$P_{e_i}(\delta_{2,1},\delta_{3,1}) = \sum_{j=1,j\neq 1}^{3} E_iE_j(B_{ij}\sin(\delta_{i,1}-\delta_{j,1})+G_{ij}\cos(\delta_{i,1}-\delta_{j,1})).$$

The above network-reduction power system model can be transformed into the following:

$$\begin{aligned}
\dot{\delta}_{2,1} &= \omega_2 - \omega_1 \\
\dot{\delta}_{3,1} &= \omega_3 - \omega_1 \\
m_1\dot{\omega}_1 &= -d_1\omega_1 + P_{m_1} - P_{e_1}(\delta_{2,1},\delta_{3,1}) \\
m_2\dot{\omega}_2 &= -d_2\omega_2 + P_{m_2} - P_{e_2}(\delta_{2,1},\delta_{3,1}) \\
m_3\dot{\omega}_3 &= -d_3\omega_3 + P_{m_3} - P_{e_3}(\delta_{2,1},\delta_{3,1}).
\end{aligned} \tag{15.7}$$

The corresponding reduced-state system is

$$\begin{aligned}
\dot{\delta}_{2,1} &= P_{m_2} - P_{e_2}(\delta_{2,1},\delta_{3,1}) \\
\dot{\delta}_{3,1} &= P_{m_3} - P_{e_3}(\delta_{2,1},\delta_{3,1}).
\end{aligned} \tag{15.8}$$

We next present the numerical results of applying the numerical BCU method to Contingency 6, which is a fault that occurs near Bus 4, and the line between Buses 4 and 5 is tripped. Starting from the prefault SEP, $(\omega_1, \omega_2, \omega_3, \delta_1, \delta_2, \delta_3) = (0, 0, 0, -0.0482, 0.1252, 0.1124)$, the original fault-on trajectory is simulated. The intermediate results of the computed exit point are summarized in Table 15.1 and are displayed in Figure 15.9. It is observed that the dot product monotonically decreases from zero, which is at the SEP, to a local minimum and then monotonically increases to zero when the projected fault-on trajectory intersects the stability boundary of the reduced-state system at the exit point. Note that the dot product is negative when the time step is between 1 and 532, while the dot product is positive when the time step 533. The exit point hence can be approximated by the state vector of the time step at either 532 or 533.

After obtaining the exit point, $(\omega_1, \omega_2, \omega_3, \delta_1, \delta_2, \delta_3) = (-2.9278, 7.7395, 6.5385, -0.8327, 2.0543, 2.1723)$, the next step of the BCU method is to identify an MGP starting from the exit point. The intermediate results of the stability boundary following procedure to locate the MGP are summarized in Table 15.2 and are displayed in Figure 15.10. An MGP, $(\delta_1, \delta_2, \delta_3) = (-0.8327474193, 2.0542861779, 2.17233138)$, is computed via the stability-boundary-following procedure.

It is observed that, starting from the exit point, the norm of the computed state vector during the stability-boundary-following procedure monotonically decreases from a positive scalar until it reaches the MGP. Since the computed exit point is not precisely on the stability boundary and the numerical round-off error during the procedure is inevitable, the norm along the stability-boundary-following procedure will not decrease to zero. In addition, the norm of the computed state vector starts

Table 15.1 The State Vector and the Dot Product at Each Iteration during the Procedure

Iteration	δ_1	δ_2	δ_3	Dot product
1	−0.0482178	0.1252317347	0.1124204947	−0.0166233464
10	−0.0484520348	0.1256719991	0.1133240227	−0.0168545339
20	−0.0492385989	0.1271518699	0.1163549871	−0.0176657158
30	−0.0505863339	0.1296921547	0.1215385856	−0.0191797409
50	−0.0549632136	0.137981287	0.1382890802	−0.0251638405
70	−0.0615782079	0.1506184635	0.1633723149	−0.0372302798
90	−0.0704268774	0.1677217182	0.1965024533	−0.0588388746
110	−0.0815049891	0.1894487449	0.237310955	−0.0942420812
130	−0.0948086152	0.2159903883	0.2853611887	−0.1481712075
150	−0.1103342291	0.2475605049	0.3401707459	−0.2254730133
170	−0.1280786947	0.284393908	0.4012157244	−0.3306957017
190	−0.1480393709	0.3267354	0.4679548727	−0.4676617494
210	−0.1702140598	0.3748293155	0.5398514134	−0.6390393906
230	−0.1946009049	0.4289144394	0.6163830494	−0.8459066033
250	−0.2211983241	0.4892153515	0.6970598442	−1.0873349471
270	−0.250004768	0.5559315787	0.7814453859	−1.3599942812
290	−0.281018506	0.629233358	0.8691641168	−1.6577449526
310	−0.3142373933	0.7092548431	0.9599139478	−1.971244116
330	−0.3496590478	0.7960984758	1.0534583536	−2.2875377044
350	−0.3872802065	0.889818948	1.1496554197	−2.5897565795
370	−0.4270971339	0.9904332985	1.2484395799	−2.8568421154
390	−0.4691050968	1.0979084054	1.3498440844	−3.06351182
410	−0.5132986856	1.2121684831	1.4539875866	−3.180480125
430	−0.5596718492	1.333096534	1.5610713282	−3.1751479615
450	−0.6082177562	1.4605304777	1.6713862785	−3.0128950156
470	−0.6589296269	1.594284845	1.7852735452	−2.6591475404
490	−0.7117999082	1.7341293542	1.9031634431	−2.0822917871
510	−0.7668218287	1.8798338674	2.0254921196	−1.2576158807
530	−0.8239879682	2.0311260593	2.1527803286	−0.1719108308
531	−0.8269024287	2.0388322166	2.159284793	−0.1107527615
532	−0.8298222527	2.0465525653	2.1658012081	−0.048953087
533	−0.8327474193	2.0542861779	2.17233138	0.0134905258

to increase when the MGP is passed. It is observed from Table 15.2 that the norm of the computed state vector reaches its local minimum at the 38th iteration. An MGP whose coordinate,

$$[\delta_1, \delta_2, \delta_3] = [-0.8361126793, 2.0780524851, 2.1482285163],$$

is then found. The norm immediately increases at the 39th iteration.

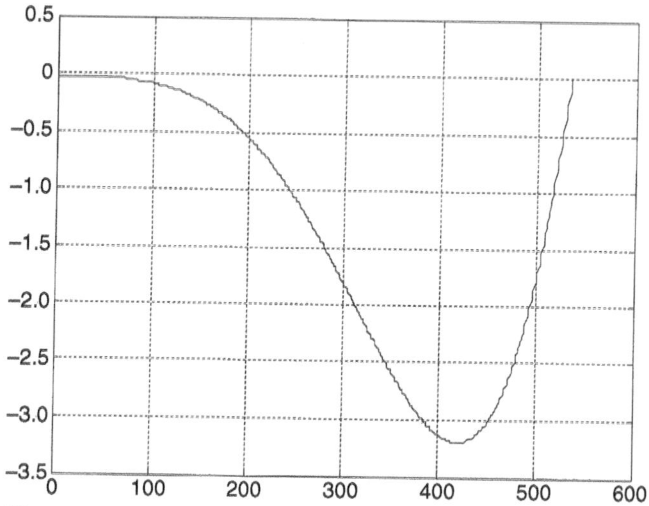

Figure 15.9 The dot product at each iteration during the procedure of computing the exit point.

Table 15.2 The State Vector and Its Norm during the Stability-Boundary-Following Procedure

Iteration	δ_1	δ_2	δ_3	Norm of the vector field
1	−0.8327474193	2.0542861779	2.17233138	0.0586610068
2	−0.832235193	2.0541649491	2.1685662088	0.0549363158
3	−0.8329685897	2.0570625194	2.1681652281	0.0507681272
4	−0.8329106923	2.0579172055	2.1658932397	0.0474005642
5	−0.833102442	2.0593221919	2.1644118605	0.0441331274
10	−0.8340972253	2.0655525606	2.1589774149	0.0306781155
15	−0.8346554728	2.0695451766	2.1548725071	0.0215092562
20	−0.8350300009	2.0722873221	2.1519835081	0.0152470273
25	−0.8353597421	2.0743582661	2.1501699001	0.0108223884
30	−0.8355930747	2.0758093797	2.1489170283	0.0077729197
35	−0.8357438527	2.076792088	2.1480117324	0.0057555441
36	−0.8357625985	2.0769366881	2.1478515035	0.0054698814
37	−0.8357792166	2.0770691849	2.1477002979	0.0052157099
38	−0.8363340225	2.0785399417	2.1489304533	0.004544846
39	−0.8358425257	2.0773775361	2.1475418862	0.0046087424

Starting from the MGP, the Newton method converges to the controlling UEP relative to the projected fault-on trajectory, $(\delta_1, \delta_2, \delta_3) = (-0.8363517729, 2.0796982602, 2.1466069919)$. Hence, the controlling UEP relative to the original fault-on trajectory is $(\omega_1, \omega_2, \omega_3, \delta_1, \delta_2, \delta_3) = (0, 0, 0, -0.8363517729, 2.0796982602, 2.1466069919)$.

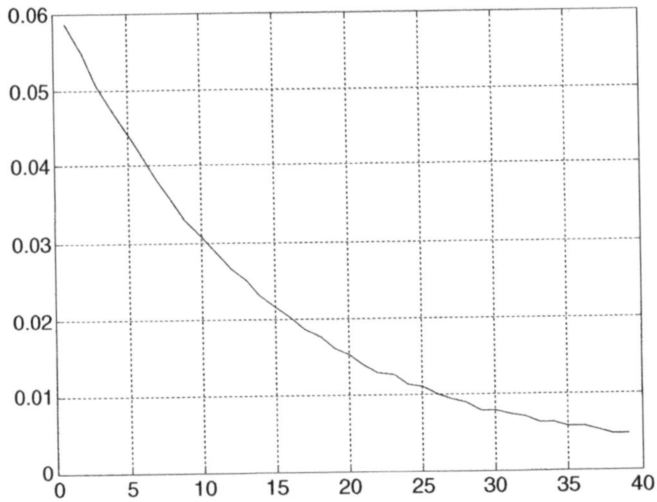

Figure 15.10 The norm of the state vector at each iteration during the stability-boundary-following procedure.

15.6 NUMERICAL ILLUSTRATIONS

The numerical network-reduction BCU method is applied to the simple three-machine test system (Eq. 15.3) with a list of 10 contingencies. The prefault SEP is [–0.0482, 0.1252, 0.1124]. The types of faults are three-phase faults with fault locations at both generator and load buses. Table 15.3 lists these 10 contingencies and the postfault SEP after each contingency. The second line of Table 15.3 states that a three-phase fault occurs at Bus 7, and the postfault system is the system with the transmission line between Buses 7 and 5 tripped due to the openings of circuit breakers at both ends of the line. The location of the SEP of the postfault system is [–0.1204, 0.3394, 0.2239].

For each contingency, the network-reduction BCU method computes the three important states: exit point, MGP, and controlling UEP. These three states of each contingency are summarized in Table 15.4 for the list of 10 contingencies. The second line of Table 15.4 states that, with respect to Contingency 1, the computed exit point, MGP, and controlling UEP by the BCU method is [–0.8387, 2.6561, 0.9391], [–0.7590, 1.9534, 1.8078], and [–0.7589, 1.9528, 1.8079], respectively.

The BCU method estimates the critical clearing time (CCT), t_{cl}, much faster than the time–domain numerical integration method. Table 15.5 presents the simulation results obtained by the network-reduction BCU method and by using the time–domain numerical integration method. The second line of Table 15.5 states that a three-phase fault occurs at Bus 4, and the postfault system is the system with the transmission line between Buses 4 and 5 tripped due to the openings of circuit breakers at both ends of the line. The location of the SEP of the postfault system is

Table 15.3 Contingency List and Postfault SEP after Each Contingency

Contingency number	Faulted bus	Fault clearing type	Description		Postfault system SEP $[\delta_1, \delta_2, \delta_3]$
			From bus	To bus	
1	7	Line trip	7	5	[−0.1204, 0.3394, 0.2239]
2	7	Line trip	8	7	[−0.0655, 0.2430, −0.0024]
3	5	Line trip	7	5	[−0.1204, 0.3394, 0.2239]
4	4	Line trip	4	6	[−0.0319, 0.0949, 0.0492]
5	9	Line trip	6	9	[−0.0967, 0.2180, 0.2958]
6	9	Line trip	9	8	[−0.0462, 0.0728, 0.2082]
7	8	Line trip	9	8	[−0.0462, 0.0728, 0.2082]
8	8	Line trip	8	7	[−0.0655, 0.2430, −0.0024]
9	6	Line trip	4	6	[−0.0319, 0.0949, 0.0492]
10	6	Line trip	6	9	[−0.0967, 0.2180, 0.2958]

Table 15.4 The Computed Exit Point, Minimum Gradient Point (MGP) and the Controlling UEP Computed by the Network-Reduction BCU Method for Each Contingency

Contingency number	Exit point $[\delta_1, \delta_2, \delta_3]$	MGP $[\delta_1, \delta_2, \delta_3]$	CUEP $[\delta_1, \delta_2, \delta_3]$
1	[−0.8387, 2.6561, 0.9391]	[−0.7590, 1.9534, 1.8078]	[−0.7589, 1.9528, 1.8079]
2	[−0.7694, 2.4080, 0.9224]	[−0.5425, 2.1803, −0.3754]	[−0.5424, 2.1802, −0.3755]
3	[−0.7786, 2.0854, 1.6808]	[−0.7591, 1.9535, 1.8078]	[−0.7589, 1.9528, 1.8079]
4	[−0.8298, 2.0461, 2.1667]	[−0.8259, 2.0824, 2.0586]	[−0.8256, 2.0830, 2.0549]
5	[−0.5084, 0.5167, 2.8945]	[−0.7653, 1.6882, 2.4208]	[−0.7576, 1.8583, 1.9986]
6	[−0.4686, 0.4964, 2.6252]	[−0.2911, −0.1010, 2.5009]	[−0.2910, −0.1011, 2.5008]
7	[−0.7754, 1.7590, 2.3495]	[−0.2913, −0.1002, 2.5008]	[−0.2910, −0.1011, 2.5008]
8	[−0.7731, 1.7540, 2.3421]	[−0.3495, 0.0746, 2.5864]	[−0.3495, 0.0745, 2.5864]
9	[−0.8298, 2.0159, 2.2310]	[−0.8258, 2.0824, 2.0582]	[−0.8256, 2.0830, 2.0549]
10	[−0.7585, 1.8366, 2.0523]	[−0.7576, 1.8574, 2.0007]	[−0.7576, 1.8583, 1.9986]

Table 15.5 Critical Clearing Time Estimated by the BCU Method and by the Time–Domain Simulation Approach, Respectively

Fault bus	Tripped line	CCT by BCU method (s)	CCT by time–domain method (s)
4	4–5	0.32	0.32
4	4–6	0.31	0.32
5	5–7	0.33	0.33
6	6–9	0.34	0.34
7	7–8	0.35	0.36
8	8–9	0.26	0.27

(15.44709, 9.98168). The CCT estimated by the BCU method is 0.32 s, while the CCT estimated by the time–domain simulation method is also 0.32 s.

To illustrate the computational procedure involved in the network-reduction BCU method, we numerically simulate the following objects related to the BCU method for two different contingencies:

- the prefault SEP and postfault SEP,
- UEPs on the stability boundary of the postfault reduced-state model,
- the stability boundary of the postfault reduced-state model,
- the projected fault-on trajectory, and
- the controlling UEP relative to the projected fault-on trajectory.

These objects are chosen because the projected fault-on trajectory intersects the stability boundary of the postfault reduced-state model at the exit point, which lies on the stable manifold of the reduced-state controlling UEP.

For Contingency 7, a fault occurs near Bus 9, and the line between Buses 6 and 9 is tripped. The exact stability boundary of the postfault reduced-state model is shown in Figure 15.11. There are two type-one UEPs and no type-two UEP lying on the stability boundary of the reduced-state system. It can be concluded that the stability region of the postfault reduced-state model is unbounded. The projected fault-on trajectory intersects the stable manifold of its controlling UEP, a type-one UEP at the exit point, as highlighted in the figure. We note that the controlling UEP of the reduced-state system corresponds to the controlling UEP of the original system.

For Contingency 12, a fault occurs near Bus 6, and the line between Buses 6 and 9 is tripped. The exact stability boundary of the postfault reduced-state model is shown in Figure 15.12. There are two type-one UEPs and no type-two UEP lying on the stability boundary of the reduced-state system. Hence, the stability region of the postfault reduced-state model is unbounded. The projected fault-on trajectory intersects the stable manifold of its controlling UEP, a type-one UEP at the exit point, as highlighted in the figure. The exit point is very close to the controlling UEP, making it unclear and separated from the type-one UEP in the figure.

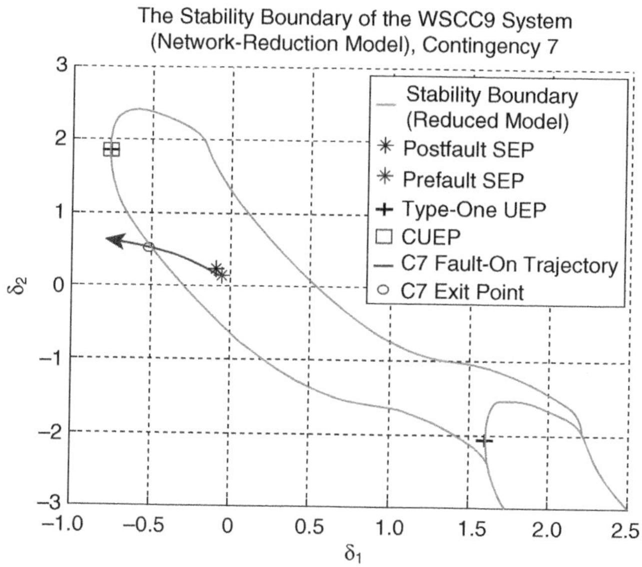

Figure 15.11 For Contingency 7, the prefault SEP and the postfault SEP are shown. The exact stability boundary of the postfault reduced-state model is highlighted. There are two type-one UEPs and no type-two UEPs on the stability boundary. The controlling UEP of the reduced-state model corresponds to the controlling UEP of the original model.

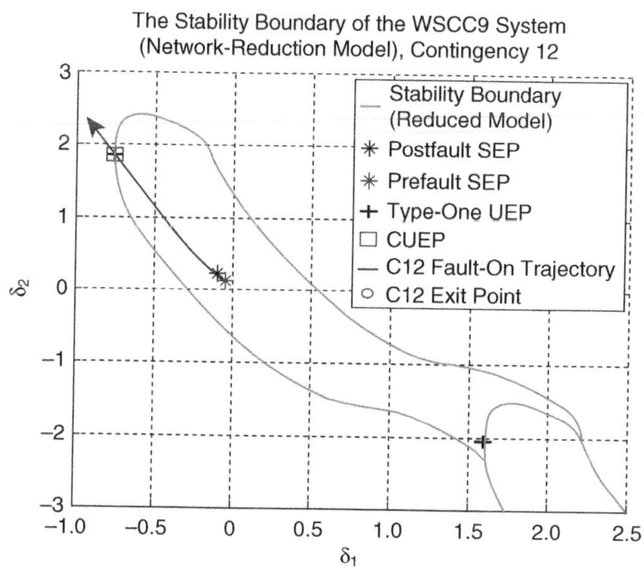

Figure 15.12 For Contingency 12, the exact stability region of the postfault system on the angle plane is highlighted. The projected fault-on trajectory intersects the stable manifold of the controlling UEP, a type-one UEP, as highlighted in the figure. The exit point is very close to the controlling UEP, making it unclear in the figure.

15.7 IEEE TEST SYSTEM

The application of the network-reduction BCU method to a 50-generator, 145-bus power system, a IEEE test system (IEEE Committee Report, 1992), is described in this section. The types of faults are three-phase faults with fault locations occurring at both generator and load buses. In this study, the test system is modeled by the network-reduction power system model with classical generators. Table 15.6 displays the estimated CCT of several faulted systems using two different methods: the time–domain simulation method and the BCU method. The results from the time–domain method are used as a bench mark.

Table 15.6 is explained as follows. The second line of Table 15.6 states that a three-phase fault occurs at Bus 7, and the postfault system is the system with the transmission line between Buses 6 and 7 tripped due to the openings of circuit breakers at both ends of the line. The CCT estimated by the BCU method is 0.115 s, while the CCT estimated by the time–domain simulation method is 0.126 s. It should be pointed out that, in these simulation results, the BCU method consistently gives slightly conservative results in estimating the CCTs. This is consistent with the

Table 15.6 Simulation Results of CCTs Estimated by the Network-Reduction BCU Method and by the Time–Domain-Based Method on the 50-Generator, 145-Bus IEEE Test System

Faulted bus	Tripped line	CCT by BCU method (s)	CCT by time–domain method (s)
7	7–6	0.115	0.126
6	6–12	0.170	0.197
12	12–14	0.170	0.197
102	63–102	0.195	0.201
97	97–66	0.260	0.260
67	67–66	0.175	0.278
98	98–72	0.205	0.206
96	96–73	0.242	0.244
108	108–75	0.270	0.278
109	109–73	0.310	0.320
82	82–75	0.300	0.312
91	91–75	0.295	0.297
100	100–72	0.315	0.324
103	103–59	0.290	0.296
89	89–59	0.265	0.272
90	90–92	0.265	0.277
33	33–1	0.480	0.517
95	95–138	0.110	0.123
135	135–138	0.130	0.136

analytical results of the controlling UEP method that the critical energy value based on the controlling UEP gives accurate and yet slightly conservative stability assessments, despite the fact that numerical energy functions were used in these simulations.

Tables 15.7 and 15.8 display the estimated CCTs of the IEEE test system base case with several faults by using six different methods. These six methods are the time–domain simulation method, the BCU method, the MOD method (Fouad and Vittal, 1991), the exit point method (Electric Power Research Institute, 1995), the hybrid method (Tang et al., 1994), and the DEEAC method (Xue et al., 1992). The results from the time–domain simulation method are used as a bench mark.

The second line of Table 15.7 and Table 15.8 states that a three-phase fault occurs at Bus 7 and that the postfault system is the system with the transmission line between Buses 6 and 7 tripped due to the opening of circuit breakers at both ends of the line. The CCT estimated using the BCU method is 0.102 s, while the exact CCT obtained by the time–domain simulation method is 0.108 s. The CCTs estimated by the MOD method, the exit point method, the hybrid method, and the mixed DEEAC method are 0.1125, 0.1125, 0.1075, and 0.108 s, respectively. The sign ** in Tables 15.7 and 15.8 means that the corresponding method does not converge to a UEP in that case. All the simulation results displayed in Tables 15.7 and 15.8, except for those using the BCU method, were taken from the CIGRE Task Force Report (1992) and from Pavella and Murthy (1994, pp. 203, 235). The results by using the BCU method were taken from Chiang et al. (1995).

It should be pointed out that in these simulations, the BCU method again consistently gives slightly conservative estimates of CCTs. These estimates are in compliance with the analytical results of the controlling UEP method, confirming that the critical energy value based on the controlling UEP should give slightly conservative stability assessments if exact energy functions are used. The simulation results also reveal that the CCTs estimated by the other methods can be either overestimates or underestimates. Overestimating CCT is undesirable because it could lead to classifying an unstable contingency as stable. Numerical energy functions were used in these simulations.

15.8 CONCLUDING REMARKS

There are several possible ways to numerically implement the conceptual BCU method for power system transient stability models. This chapter has presented such a numerical implementation of the conceptual network-reduction BCU method. This numerical network-reduction BCU method includes several numerical procedures, such as a reliable numerical procedure, to detect the exit point, a stability-boundary-following procedure for computing the MGP, a nonlinear solver to compute the controlling UEP, and a numerical energy function evaluation procedure.

Numerical studies indicate that the network-reduction BCU method consistently gives slightly conservative results in the estimated CCTs. This conservative nature is in compliance with the controlling UEP method despite the fact that numerical

Table 15.7 Simulation Results of Several Methods for Direct Stability Analysis of the 50-Generator, 145-Bus IEEE Test System

Faulted bus	Opened line	Time–domain method	BCU method	Error (%)	MOD method	Error (%)	Exit point method	Error (%)
7	7–6	108.2	102.0	–5.73	112.5	3.9	112.5	3.9
7*	7–6	107.5	106.2	–1.2	127.5	18.6	142.5	32.6
59	59–72	224.5	224.1	–0.18	242.5	8.0	197.5	–12.0
73	73–84	215.5	194.2	–9.88	**	**	**	**
112	112–69	248.6	235.4	–5.63	237.5	–4.4	237.5	–4.4
66	66–67	171.0	160.2	–6.31	**	**	**	**
115	115–116	292.5	288.3	–1.43	287.5	–1.7	287.5	–1.7
100	100–72	260.0	253.6	–2.46	252.9	–2.9	252.9	–2.9
101	101–73	248.0	238.2	–3.95	237.5	–3.5	237.5	–3.5
91	91–75	188.0	187.7	–0.15	187.5	–0.3	187.5	–0.3
6	6–1	171.0	155.0	–9.3	**	**	237.5	38.9
12	12–14	173.5	163.0	–6.1	**	**	**	**
6	6–10	177.0	165.0	–6.8	**	**	**	**
33	33–39	386.0	385.0	–0.3	**	**	347.5	–10.0
33	33–49	387.5	387.0	–0.1	432.5	11.6	347.5	–10.3
66	66–111	175.5	163.5	–6.8	**	**	**	**
106	106–74	185.5	172.0	–7.3	–172.5	–7.0	172.5	–7.0
69	69–32	205.3	186.2	–9.3	**	**	**	**
69	69–112	205.1	185.0	–9.8	**	**	**	**
105	105–73	213.5	206.5	–3.3	**	**	**	**
73	73–75	215.1	196.0	–8.9	**	**	**	**
67	67–65	233.7	227.4	–2.7	**	**	**	**
59	59–103	222.6	220.0	–1.7	**	**	242.5	8.6
12	12–14, 12–14	169.7	160.0	–5.7	**	**	**	**
105	105–73, 105–73	120.0	114.0	–5.0	127.5	6.3	127.5	6.3
66	66–8, 66–8	178.5	171.0	–4.2	**	**	**	**
6	6–1, 6–2, 6–7	39.4	39.2	–0.5	72.5	84.0	52.5	33.2
6	6–9, 6–10, 6–12, 6–12	81.5	77.0	–5.5	**	**	>500	**
33	33–37, 33–38, 33–39, 33–40, 33–49, 33–50	360.5	.355.0	–1.4	>600	**	>700	**
33	33–37, 33–38, 33–39, 33–40	378	373.4	–1.2	352.5	–6.7	>700	**
66	66–111, 66–111, 66–111	83.0	80.0	–3.6	**	**	112.5	35.5
73	73–26, 73–72, 73–82, 73–101	214.5	195.0	–9.1	**	**	**	**
73	73–69, 73–75, 73–91, 73–96, 73–109	190.5	190.1	–0.20	**	**	77.5	–59.3

The case marked * is more stressed than the other cases. The sign ** means that the corresponding method does not converge in that case.

Table 15.8 Simulation Results of Several Methods for Direct Stability Analysis of the 50-Generator, 145-Bus IEEE Test System

Faulted bus	Opened line	Time–domain method	BCU method	Error (%)	Hybrid method	Error (%)	DEEAC method	Error (%)
7	7–6	108.2	102.0	−5.73	107.5	−0.7	108	−0.18
7*	7–6	107.5	106.2	−1.2	107.5	0.0	118	9.7
59	59–72	224.5	224.1	−0.18	222.5	−0.9	227	1.3
73	73–84	215.5	194.2	−9.88	207.5	−3.7	209	−3.0
112	112–69	248.6	235.4	−5.63	247.5	−0.4	246	−1.0
66	66–67	171.0	160.2	−6.31	162.5	−5.0	165	−3.5
115	115–116	292.5	288.3	−1.43	292.5	0.0	291	−0.5
100	100–72	260.0	253.6	−2.46	257.5	−1.0	261	0.4
101	101–73	248.0	238.2	−3.95	247.5	0.6	244	−0.8
91	91–75	188.0	187.7	−0.15	187.5	−0.3	190	1.06
6	6–1	171.0	155.0	−9.3	167.5	−2.0	162	−5.2
12	12–14	173.5	163.0	−6.1	172.5	−0.6	168	−3.1
6	6–10	177.0	165.0	−6.8	172.5	−2.5	173	−2.2
33	33–39	386.0	385.0	−0.3	382.5	−0.9	383	−0.8
33	33–49	387.5	387.0	−0.1	382.5	−1.3	383	−0.8
66	66–111	175.5	163.5	−6.8	167.5	−4.6	168	−4.2
106	106–74	185.5	172.0	−7.3	182.5	−1.6	181	−2.4
69	69–32	205.3	186.2	−9.3	202.5	−1.4	203	−1.1
69	69–112	205.1	185.0	−9.8	207.5	1.2	199	−2.9
105	105–73	213.5	206.5	−3.3	207.5	−2.8	207	−3.0
73	73–75	215.1	196.0	−8.9	212.5	−1.2	209	−2.8
67	67–65	233.7	227.4	−2.7	232.5	−0.5	199	−14.8
59	59–103	222.6	220.0	−1.7	222.5	−0.0	227	2.0
12	12–14, 12–14	169.7	160.0	−5.7	167.5	−1.3	165	−2.8
105	105–73, 105–73	120.0	114.0	−5.0	117.5	−2.0	127	5.8
66	66–8, 66–8	178.5	171.0	−4.2	167.5	−6.2	173	−3.1
6	6–1, 6–2, 6–7	39.4	39.2	−0.5	22.5	−42.9	—	—
6	6–9, 6–10, 6–12, 6–12	81.5	77.0	−5.5	77.5	−4.9	69	−15.3
33	33–37, 33–38, 33–39, 33–40, 33–49, 33–50	360.5	.355.0	−1.4	352.5	−2.2	344	−4.5
33	33–37, 33–38, 33–39, 33–40	378	373.4	−1.2	372.5	−1.5	368	−2.6
66	66–111, 66–111, 66–111	83.0	80.0	−3.6	82.5	−0.6	82	−1.2
73	73–26, 73–72, 73–82, 73–101	214.5	195.0	−9.1	207.5	−3.3	209	−2.5
73	73–69, 73–75, 73–91, 73–96, 73–109	190.5	190.1	−0.20	187.5	−1.6	179	−6.0

The case marked * is more stressed than the other cases.

energy functions were used in these simulations. Even though the network-reduction transient stability model is not used in practice, development of the numerical network-reduction BCU method is useful and provides guidelines for the development of the numerical network-preserving BCU method, which will be detailed in the next two chapters.

Chapter 16

Network-Preserving BCU Method and Its Theoretical Foundation

16.1 INTRODUCTION

We present in this chapter a network-preserving BCU method for direct stability analysis of the following generic network-preserving transient stability model:

$$
\begin{aligned}
0 &= -\frac{\partial U}{\partial u}(u, w, x, y) + g_1(u, w, x, y) \\
0 &= -\frac{\partial U}{\partial w}(u, w, x, y) + g_2(u, w, x, y) \\
T\dot{x} &= -\frac{\partial U}{\partial x}(u, w, x, y) + g_3(u, w, x, y) \\
\dot{y} &= z \\
M\dot{z} &= -Dz - \frac{\partial U}{\partial y}(u, w, x, y) + g_4(u, w, x, y),
\end{aligned}
\tag{16.1}
$$

where $u \in R^k$ and $w \in R^l$ are instantaneous variables, and $x \in R^m$, $y \in R^n$, and $z \in R^n$ are state variables. T is a positive definite matrix, and M and D are diagonal positive definite matrices. Here, the differential equations describe generator and/or load dynamics, while the algebraic equations express the power flow equations at each bus. The function $U(u, w, x, y)$ is a scalar function. The physical limitations of variables are not expressed for the sake of convenience. Existing network-preserving transient stability models can be rewritten as a set of the above general differential and algebraic equations (DAEs) (Chu and Chiang, 2005).

Direct Methods for Stability Analysis of Electric Power Systems, by Hsiao-Dong Chiang

A structure-preserving BCU method (i.e., a BCU method for structure-preserving transient stability models) and its analytical basis will be presented in this chapter. The fundamental ideas behind the development of the BCU method can be explained as follows. Given a power system transient stability model called the original model, which admits an energy function, the BCU method first explores the special properties of the original model to define an artificial, reduced-state model so that certain static and dynamic properties of the original model are captured by the reduced-state model. We discuss a reduced-state model associated with the original model (Eq. 16.1) in the next section.

16.2 REDUCED-STATE MODEL

When developing a BCU method for a given power system stability model, the associated artificial, reduced-state model must be defined. Regarding the original model (Eq. 16.1), we choose the following differential–algebraic system as the artificial, reduced-state model:

$$\begin{aligned}
0 &= -\frac{\partial U}{\partial u}(u, w, x, y) + g_1(u, w, x, y) \\
0 &= -\frac{\partial U}{\partial w}(u, w, x, y) + g_2(u, w, x, y) \\
T\dot{x} &= -\frac{\partial U}{\partial x}(u, w, x, y) + g_3(u, w, x, y) \\
\dot{y} &= -\frac{\partial U}{\partial y}(u, w, x, y) + g_4(u, w, x, y).
\end{aligned} \tag{16.2}$$

It will be shown that the original model (Eq. 16.1) and the artificial, reduced-state model (Eq. 16.2) satisfy the following static and dynamic properties.

Static Properties

(S1) The locations of the equilibrium points of the reduced-state model (Eq. 16.2) correspond to the locations of the equilibrium points of the original model (Eq. 16.1). For example, $(\bar{u}, \bar{w}, \bar{x}, \bar{y})$ is an equilibrium point of the reduced-state model (Eq. 16.2) if and only if $(\bar{u}, \bar{w}, \bar{x}, \bar{y}, 0)$ is an equilibrium point of the original model (Eq. 16.1), where $0 \in R^m$ and m is an appropriate positive integer.

(S2) The types of equilibrium points of the reduced-state model (Eq. 16.2) are the same as those of the original model (Eq. 16.1). For example, (u_s, w_s, x_s, y_s) is a stable equilibrium point of the reduced-state model (Eq. 16.2) if and only if $(u_s, w_s, x_s, y_s, 0)$ is a stable equilibrium point of the original model (Eq. 16.1). $(\bar{u}, \bar{w}, \bar{x}, \bar{y})$ is a type-k equilibrium point of the reduced-state model (Eq. 16.2) if and only if $(\bar{u}, \bar{w}, \bar{x}, \bar{y}, 0)$ is a type-k equilibrium point of the original model (Eq. 16.1).

Dynamic Properties

(D1) There exists an energy function for the artificial, reduced-state model (Eq. 16.2).

(D2) An equilibrium point, say, $(\bar{u}, \bar{w}, \bar{x}, \bar{y})$, is on the stability boundary $\partial A(u_s, w_s, x_s, y_s)$ of the reduced-state model (Eq. 16.2) if and only if the equilibrium point $(\bar{u}, \bar{w}, \bar{x}, \bar{y}, 0)$ is on the stability boundary $\partial A(u_s, w_s, x_s, y_s, 0)$ of the original model (Eq. 16.1).

(D3) It is computationally feasible to detect efficiently the point at which the projected fault-on trajectory $(u(t), w(t), x(t), y(t))$ intersects the stability boundary $\partial A(u_s, w_s, x_s, y_s)$ of the postfault reduced-state model (Eq. 16.2) without resorting to an iterative time–domain procedure to compute the exit point of the postfault reduced-state model (Eq. 16.2).

The dynamic property (D3) plays an important role in the development of the BCU method. This property allows us to circumvent the difficulty of requiring an iterative time–domain procedure to compute the exit point on the original model. The BCU method finds the controlling unstable equilibrium point (UEP) of the artificial, reduced-state model (Eq. 16.2) by exploring the special structure of the stability boundary and the energy function of the reduced-state model (Eq. 16.2). The BCU method then relates the controlling UEP of the reduced-state model (Eq. 16.2) to the controlling UEP of the original model (Eq. 16.1).

Given a power system stability model, there exists a corresponding version of the BCU method. The BCU method does not directly compute the controlling UEP of the original model because computing the exit point of the original model, which is a key to computing the controlling UEP, requires an iterative time–domain procedure. Instead, the BCU method computes the controlling UEP of the original model (Eq. 16.1) via computing the controlling UEP of the artificial, reduced-state model (Eq. 16.2).

We propose the following seven steps to establish the static properties (S1) and (S2) and the dynamic property (D2) between the original model (Eq. 16.1) and the reduced-state model (Eq. 16.2):

Step 1. Determine the static as well as dynamic relationships between the reduced-state system (Eq. 16.2) and the following singularly perturbed system:

$$
\begin{aligned}
\varepsilon_1 \dot{u} &= -\frac{\partial}{\partial u} U(u, w, x, y) + g_1(u, w, x, y) \\
\varepsilon_2 \dot{w} &= -\frac{\partial}{\partial w} U(u, w, x, y) + g_2(u, w, x, y) \\
T\dot{x} &= -\frac{\partial}{\partial x} U(u, w, x, y) + g_3(u, w, x, y) \\
\dot{y} &= -\frac{\partial}{\partial y} U(u, w, x, y) + g_4(u, w, x, y).
\end{aligned}
\tag{16.3}
$$

Step 2. Determine the static as well as dynamic relationships between

$$
\begin{aligned}
\varepsilon_1 \dot{u} &= -\frac{\partial}{\partial u} U(u, w, x, y) + g_1(u, w, x, y) \\
\varepsilon_2 \dot{w} &= -\frac{\partial}{\partial w} U(u, w, x, y) + g_2(u, w, x, y) \\
T\dot{x} &= -\frac{\partial}{\partial x} U(u, w, x, y) + g_3(u, w, x, y) \\
\dot{y} &= -\frac{\partial}{\partial y} U(u, w, x, y) + g_4(u, w, x, y)
\end{aligned}
$$

and the following nonlinear dynamical system:

$$
\begin{aligned}
\varepsilon_1 \dot{u} &= -\frac{\partial}{\partial u} U(u, w, x, y) \\
\varepsilon_2 \dot{w} &= -\frac{\partial}{\partial w} U(u, w, x, y) \\
T\dot{x} &= -\frac{\partial}{\partial x} U(u, w, x, y) \\
\dot{y} &= -\frac{\partial}{\partial y} U(u, w, x, y).
\end{aligned}
\tag{16.4}
$$

Step 3. Determine the static as well as dynamic relationships between

$$
\begin{aligned}
\varepsilon_1 \dot{u} &= -\frac{\partial}{\partial u} U(u, w, x, y) \\
\varepsilon_2 \dot{w} &= -\frac{\partial}{\partial w} U(u, w, x, y) \\
T\dot{x} &= -\frac{\partial}{\partial x} U(u, w, x, y) \\
\dot{y} &= -\frac{\partial}{\partial y} U(u, w, x, y)
\end{aligned}
$$

and the following system:

$$\begin{aligned}
\varepsilon_1 \dot{u} &= -\frac{\partial}{\partial u} U(u, w, x, y) \\
\varepsilon_2 \dot{w} &= -\frac{\partial}{\partial w} U(u, w, x, y) \\
T\dot{x} &= -\frac{\partial}{\partial x} U(u, w, x, y) \\
\dot{y} &= -\frac{\partial}{\partial y} U(u, w, x, y) \\
M\dot{z} &= -Dz.
\end{aligned} \tag{16.5}$$

Step 4. Determine the static as well as dynamic relationship between

$$\begin{aligned}
\varepsilon_1 \dot{u} &= -\frac{\partial}{\partial u} U(u, w, x, y) \\
\varepsilon_2 \dot{w} &= -\frac{\partial}{\partial w} U(u, w, x, y) \\
T\dot{x} &= -\frac{\partial}{\partial x} U(u, w, x, y) \\
\dot{y} &= -\frac{\partial}{\partial y} U(u, w, x, y) \\
M\dot{z} &= -Dz
\end{aligned}$$

and the following one-parameter dynamical system $d(\lambda)$:

$$\begin{aligned}
\varepsilon_1 \dot{u} &= -\frac{\partial}{\partial u} U(u, w, x, y) \\
\varepsilon_2 \dot{w} &= -\frac{\partial}{\partial w} U(u, w, x, y) \\
T\dot{x} &= -\frac{\partial}{\partial x} U(u, w, x, y) \\
\dot{y} &= (1-\lambda)z - \lambda \frac{\partial}{\partial y} U(u, w, x, y) \\
M\dot{z} &= -Dz - (1-\lambda)\frac{\partial}{\partial y} U(u, w, x, y).
\end{aligned} \tag{16.6}$$

Step 5. Determine the static as well as dynamic relationships between

$$\varepsilon_1 \dot{u} = -\frac{\partial}{\partial u} U(u, w, x, y)$$

$$\varepsilon_2 \dot{w} = -\frac{\partial}{\partial w} U(u, w, x, y)$$

$$T\dot{x} = -\frac{\partial}{\partial x} U(u, w, x, y)$$

$$\dot{y} = (1-\lambda) z - \lambda \frac{\partial}{\partial y} U(u, w, x, y)$$

$$M\dot{z} = -Dz - (1-\lambda) \frac{\partial}{\partial y} U(u, w, x, y)$$

and the following nonlinear dynamical system:

$$\begin{aligned}
\varepsilon_1 \dot{u} &= -\frac{\partial}{\partial u} U(u, w, x, y) \\
\varepsilon_2 \dot{w} &= -\frac{\partial}{\partial w} U(u, w, x, y) \\
T\dot{x} &= -\frac{\partial}{\partial x} U(u, w, x, y) \\
\dot{y} &= z \\
M\dot{z} &= -Dz - \frac{\partial}{\partial y} U(u, w, x, y).
\end{aligned} \tag{16.7}$$

Step 6. Determine the static as well as dynamic relationships between

$$\varepsilon_1 \dot{u} = -\frac{\partial}{\partial u} U(u, w, x, y)$$

$$\varepsilon_2 \dot{w} = -\frac{\partial}{\partial w} U(u, w, x, y)$$

$$T\dot{x} = -\frac{\partial}{\partial x} U(u, w, x, y)$$

$$\dot{y} = z$$

$$M\dot{z} = -Dz - \frac{\partial}{\partial y} U(u, w, x, y)$$

and the following intermediate system:

$$
\begin{aligned}
\varepsilon_1 \dot{u} &= -\frac{\partial}{\partial u} U(u, w, x, y) + g_1(u, w, x, y) \\
\varepsilon_2 \dot{w} &= -\frac{\partial}{\partial w} U(u, w, x, y) + g_2(u, w, x, y) \\
T\dot{x} &= -\frac{\partial}{\partial x} U(u, w, x, y) + g_3(u, w, x, y) \\
\dot{y} &= z \\
M\dot{z} &= -Dz - \frac{\partial}{\partial y} U(u, w, x, y) + g_4(u, w, x, y).
\end{aligned}
\tag{16.8}
$$

Step 7. Determine the static as well as dynamic relationships between

$$
\begin{aligned}
\varepsilon_1 \dot{u} &= -\frac{\partial}{\partial u} U(u, w, x, y) + g_1(u, w, x, y) \\
\varepsilon_2 \dot{w} &= -\frac{\partial}{\partial w} U(u, w, x, y) + g_2(u, w, x, y) \\
T\dot{x} &= -\frac{\partial}{\partial x} U(u, w, x, y) + g_3(u, w, x, y) \\
\dot{y} &= z \\
M\dot{z} &= -Dz - \frac{\partial}{\partial y} U(u, w, x, y) + g_4(u, w, x, y)
\end{aligned}
$$

and the original system (Eq. 16.1).

We then apply the overall seven-step procedure to show that the original system (Eq. 16.1) and the reduced-state system (Eq. 16.2) satisfy static properties (S1) and (S2) as well as dynamic property (D2). These seven steps and their relationships are summarized in Figure 16.1.

16.3 STATIC AND DYNAMIC PROPERTIES

We present in this section some analytical results showing that, under certain conditions, the original system (Eq. 16.1) and the reduced-state system (Eq. 16.2) satisfy the static properties (S1) and (S2) as well as the dynamic property (D2). There exists an energy function for the reduced-state model (Eq. 16.2); hence, dynamic property (D1) is satisfied. A computational scheme will be developed and incorporated into the network-preserving BCU method to satisfy the dynamic relationship (D3).

In the following, we derive the properties stated in each step:

Steps 1 and 7. In these two steps, we use analytical results from the singular perturbation technique to show that, for sufficiently small values of ε_1, ε_2, the following properties hold:

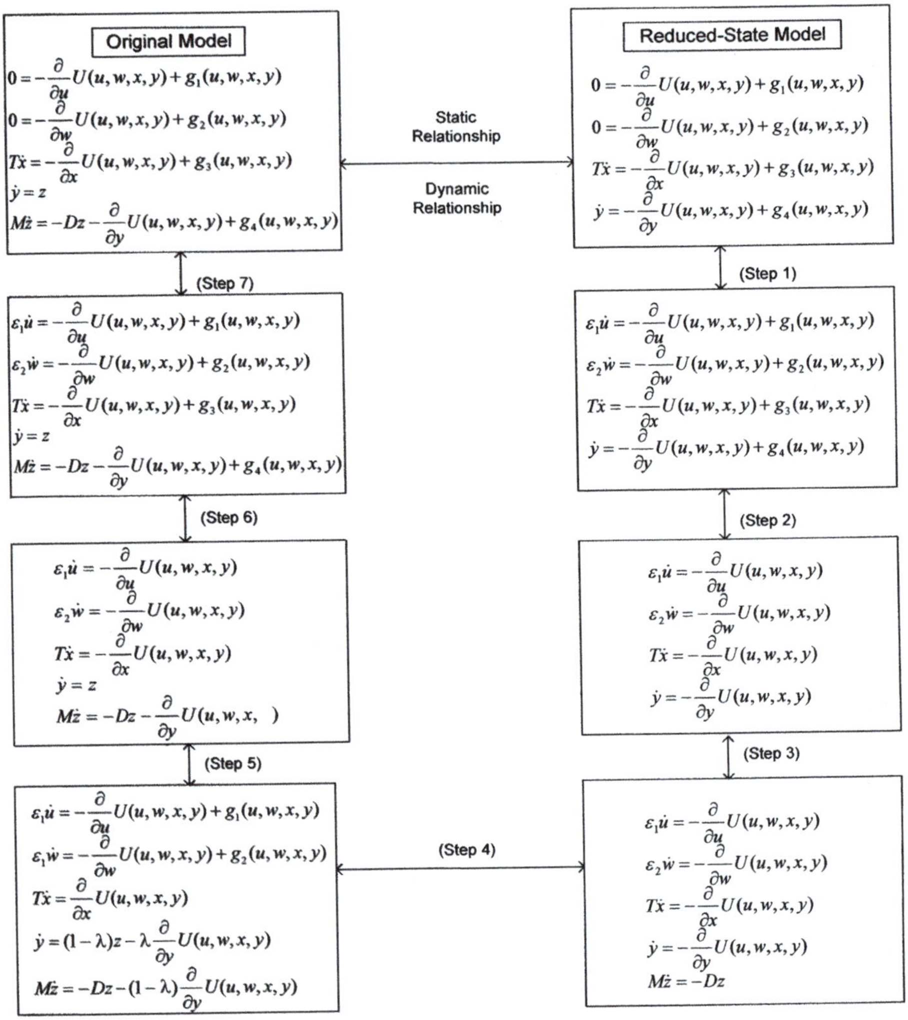

Figure 16.1 A seven-step procedure to show the static properties (S1) and (S2) as well as the dynamic properties (D1) and (D2) between the original model (Eq. 16.1) and the reduced-state model (Eq. 16.2).

[(Static property)] $(\bar{u}, \bar{w}, \bar{x}, \bar{y})$ is a type-k equilibrium point of the model (Eq. 16.2) if and only if $(\bar{u}, \bar{w}, \bar{x}, \bar{y})$ is a type-k equilibrium point of the model (16.3). In particular, (u_s, w_s, x_s, y_s) is a stable equilibrium point (SEP) of model (Eq. 16.2) if and only if (u_s, w_s, x_s, y_s) is a SEP of the model (Eq. 16.3).

[(Static property)] $(\bar{u}, \bar{w}, \bar{x}, \bar{y}, 0)$ is a type-k equilibrium point of model (Eq. 16.1) if and only if $(\bar{u}, \bar{w}, \bar{x}, \bar{y}, 0)$ is a type-k equilibrium point of model

(Eq. 16.8). In particular, $(u_s, w_s, x_s, y_s, 0)$ is a SEP of model (Eq. 16.1) if and only if $(u_s, w_s, x_s, y_s, 0)$ is a SEP of model (Eq. 16.8).

[(Dynamic property)] (u_i, w_i, x_i, y_i) is an equilibrium point on the stability boundary $\partial A(u_s, w_s, x_s, y_s)$ of model (Eq. 16.2) if and only if the equilibrium point (u_i, w_i, x_i, y_i) is on the stability boundary $\partial A(u_s, w_s, x_s, y_s)$ of model (Eq. 16.3).

[(Dynamic property)] $(u_i, w_i, x_i, y_i, 0)$ is an equilibrium point on the stability boundary $\partial A(u_s, w_s, x_s, y_s, 0)$ of model (Eq. 16.7) if and only if the equilibrium point $(u_i, w_i, x_i, y_i, 0)$ is on the stability boundary $\partial A(u_s, w_s, x_s, y_s, 0)$ of model (Eq. 16.8).

Step 3. In Step 3, the following properties are derived by taking into account the fact that the dynamics of variable z are decoupled from the other variables:

[(Static property)] $(\bar{u}, \bar{w}, \bar{x}, \bar{y})$ is a type-k equilibrium point of model (Eq. 16.4) if and only if $(\bar{u}, \bar{w}, \bar{x}, \bar{y}, 0)$ is a type-k equilibrium point of model (16.5), where $0 \in R^m$ and m is an appropriate positive integer.

[(Dynamic property)] $(\bar{u}, \bar{w}, \bar{x}, \bar{y})$ is an equilibrium point on the stability boundary $\partial A(u_s, w_s, x_s, y_s)$ of model (Eq. 16.4) if and only if $(\bar{u}, \bar{w}, \bar{x}, \bar{y}, 0)$ is an equilibrium point on the stability boundary $\partial A(u_s, w_s, x_s, y_s, 0)$ of model (Eq. 16.5).

Steps 4 and 5. In these two steps, we shall show that, under certain conditions, the following properties hold:

[(Static property)] $(\bar{u}, \bar{w}, \bar{x}, \bar{y}, 0)$ is a type-k equilibrium point of model (Eq. 16.7) if and only if $(\bar{u}, \bar{w}, \bar{x}, \bar{y}, 0)$ is a type-k equilibrium point of model (Eq. 16.5). In particular, $(u_s, w_s, x_s, y_s, 0)$ is a SEP of model (Eq. 16.7) if and only if $(u_s, w_s, x_s, y_s, 0)$ is a SEP of model (Eq. 16.5).

[(Dynamic property)] $(u_i, w_i, x_i, y_i, 0)$ is an equilibrium point on the stability boundary $\partial A(u_s, w_s, x_s, y_s, 0)$ of model (Eq. 16.7) if and only if the equilibrium point $(u_i, w_i, x_i, y_i, 0)$ is on the stability boundary $\partial A(u_s, w_s, x_s, y_s, 0)$ of model (Eq. 16.5).

Steps 2 and 6. In these two steps, we shall show that, if the transfer conductances are sufficiently small, then the following properties hold:

[(Static property)] $(\bar{u}, \bar{w}, \bar{x}, \bar{y})$ is a type-k equilibrium point of model (Eq. 16.3) if and only if $(\bar{u}, \bar{w}, \bar{x}, \bar{y})$ is a type-k equilibrium point of model (Eq. 16.4). In particular, (u_s, w_s, x_s, y_s) is a SEP of model (Eq. 16.3) if and only if (u_s, w_s, x_s, y_s) is a SEP of model (Eq. 16.4).

[(Static property)] $(\bar{u}, \bar{w}, \bar{x}, \bar{y}, 0)$ is a type-k equilibrium point of model (Eq. 16.7) if and only if $(\bar{u}, \bar{w}, \bar{x}, \bar{y}, 0)$ is a type-k equilibrium point of model (Eq. 16.8). In particular, $(u_s, w_s, x_s, y_s, 0)$ is a SEP of model (Eq. 16.7) if and only if $(u_s, w_s, x_s, y_s, 0)$ is a SEP of model (Eq. 16.8).

[(Dynamic property)] (u_i, w_i, x_i, y_i) is an equilibrium point on the stability boundary $\partial A(u_s, w_s, x_s, y_s)$ of model (Eq. 16.3) if and only if the equilibrium point (u_i, w_i, x_i, y_i) is on the stability boundary $\partial A(u_s, w_s, x_s, y_s)$ of model (Eq. 16.4).

[(Dynamic property)] $(u_i, w_i, x_i, y_i, 0)$ is an equilibrium point on the stability boundary $\partial A(u_s, w_s, x_s, y_s, 0)$ of model (Eq. 16.7) if and only if the equilibrium point $(u_i, w_i, x_i, y_i, 0)$ is on the stability boundary $\partial A(u_s, w_s, x_s, y_s, 0)$ of model (Eq. 16.8).

The overall relationship between the original model (Eq. 16.1) and the reduced-state model (Eq. 16.2) can be established by combining the analytical results to be derived in Steps 1 through 7.

16.4 ANALYTICAL RESULTS

We show the following static and dynamic properties stated in Step 3.

Theorem 16.1: Static and Dynamic Properties

(u_i, w_i, x_i, y_i) is a type-k equilibrium point of model (Eq. 16.4) if and only if $(u_i, w_i, x_i, y_i, 0)$ is a type-k equilibrium point of model (Eq. 16.5). Moreover, (u_i, w_i, x_i, y_i) is an equilibrium point on the stability boundary $\partial A(u_s, w_s, x_s, y_s)$ of model (Eq. 16.4) if and only if $(u_i, w_i, x_i, y_i, 0)$ is an equilibrium point on the stability boundary $\partial A(u_s, w_s, x_s, y_s, 0)$ of model (Eq. 16.5).

Proof: Variables z and (u, w, x, y) are decoupled in the vector field of Equation 16.5, and the vector field of Equation 16.5 involving (u, w, x, y) is exactly the same as the vector field (Eq. 16.4). Also, the state variable z in Equation 16.5 is completely stable because both matrix M and matrix D are positively definite. Hence, this theorem follows.

We show the following static and dynamic properties for Steps 4 and 5.

Theorem 16.2: Static Property

If zero is a regular value of

$$\frac{\partial^4 U(u_i, w_i, x_i, y_i)}{\partial u \partial w \partial x \partial y}$$

for all the equilibrium points of model (Eq. 16.4), then (u_i, w_i, x_i, y_i) is a type-k equilibrium point of model (Eq. 16.4) if and only if $(u_i, w_i, x_i, y_i, 0)$ is a type-k equilibrium point of model (Eq. 16.7).

Proof: Let $J_\lambda(u_e, w_e, x_e, y_e, 0)$ be the Jacobian matrix of the system evaluated at the equilibrium point $(u_e, w_e, x_e, y_e, z_e)$; then,

$$J_\lambda(u_e, w_e, x_e, y_e, 0) = \begin{bmatrix} \varepsilon_1^{-1} & 0 & 0 & 0 & 0 \\ 0 & \varepsilon_2^{-1} & 0 & 0 & 0 \\ 0 & 0 & T^{-1} & 0 & 0 \\ 0 & 0 & 0 & \lambda & (\lambda-1)M^{-1} \\ 0 & 0 & 0 & (\lambda-1)M^{-1} & M^{-1}DM^{-1} \end{bmatrix}.$$

Let

$$H(u_e, w_e, x_e, y_e) = \begin{bmatrix} -\nabla^2 U(u_e, w_e, x_e, y_e) & 0 \\ 0 & -M^{-1} \end{bmatrix}.$$

It is clear that $H(u_e, w_e, x_e, y_e)$ is a nonsingular Hermitian matrix, and

$$J_\lambda(u_e, w_e, x_e, y_e, 0)H(u_e, w_e, x_e, y_e) + H(u_e, w_e, x_e, y_e)^T J_\lambda(u_e, w_e, x_e, y_e, 0)^T$$
$$= \text{block diag}\left[2\varepsilon_1^{-1}, 2\varepsilon_2^{-1}, 2T^{-1}, 2\lambda I, 2M^{-1}DM^{-1}\right] \geq 0 \text{ for all } \varepsilon \in I.$$

BySylvester's inertia theorem,

$$In(J_\lambda(u_e, w_e, x_e, y_e, 0)) = In\langle H(u_e, w_e, x_e, y_e)\rangle.$$

This completes the proof.

Theorem 16.3: Dynamic Property

Let $(u_s, w_s, x_s, y_s, 0)$ be a SEP of model (Eq. 16.5). Suppose zero is a regular value of

$$\frac{\partial^4 U(u_i, w_i, x_i, y_i)}{\partial u \partial w \partial x \partial y}$$

for all the UEP (u_i, w_i, x_i, y_i), $i = 1, 2, \ldots, k$. If all the intersections of the stable and unstable manifolds of the equilibrium points on the stability boundary $\partial A(u_s, w_s, x_s, y_s, 0)$ of the one-parameterized model $d(\lambda)$ (Eq. 16.6) satisfy the transversality condition for $\lambda \in [0, 1]$, then

1. the equilibrium point $(u_i, w_i, x_i, y_i, 0)$ is on the stability boundary $\partial A(u_s, w_s, x_s, y_s, 0)$ of model (Eq. 16.5) if and only if the equilibrium point $(u_i, w_i, x_i, y_i, 0)$ is on the stability boundary $\partial A(u_s, w_s, x_s, y_s, 0)$ of model (Eq. 16.7);
2. the stability boundary $\partial A(u_s, w_s, x_s, y_s, 0)$ of model (Eq. 16.5) is the union of the stable manifold of all the equilibrium points $(u_i, w_i, x_i, y_i, 0)$, $i = 1, 2, \ldots$ on the stability boundary $\partial A(u_s, w_s, x_s, y_s, 0)$; that is, $\partial A(u_s, w_s, x_s, y_s, 0) = \cup W^s(u_i, w_i, x_i, y_i, 0)$; and
3. the stability boundary $\partial A(u_s, w_s, x_s, y_s, 0)$ of model (Eq. 16.7) is the union of the stable manifold of all the equilibrium points $(u_i, w_i, x_i, y_i, 0)$, $i = 1, 2, \ldots$ on the stability boundary $\partial A(u_s, w_s, x_s, y_s, 0)$; that is, $\partial A(u_s, w_s, x_s, y_s, 0) = \cup W^s(u_i, w_i, x_i, y_i, 0)$.

Steps 2 and 6. From a mathematical viewpoint, we study in these two steps the following properties:

- The types of equilibrium points remain the same under a small perturbation of the vector field.
- The equilibrium points on the stability boundary persist under a small perturbation of the vector field.

By applying Theorem 14.1 to model (Eq. 16.4) and to model (Eq. 16.7), we have the following results.

Theorem 16.4: Persistence Property

Let (u_s, w_s, x_s, y_s) be a SEP of model (Eq. 16.4), and its stability boundary $\partial A(u_s, w_s, x_s, y_s)$ contains a finite number of UEPs, (u_i, w_i, x_i, y_i), $i = 1, 2, \ldots n$. If the following two conditions are satisfied:

(i) all the intersections of the stable and unstable manifolds of the equilibrium points on the stability boundary satisfy the transversality condition, and

(ii) zero is a regular value of

$$\frac{\partial^4 U(u_i, w_i, x_i, y_i)}{\partial u \partial w \partial x \partial y}$$

for all the UEPs on the stability boundary,

then the following results hold:

1. There exist two positive numbers, $\delta > 0$ and $\varepsilon > 0$, such that if the transfer conductance of model (Eq. 16.3) satisfies $G_{jk} < \delta$ for all j and k, there exists a unique SEP $(\hat{u}_s, \hat{w}_s, \hat{x}_s, \hat{y}_s)$ of model (Eq. 16.3) with $\|(u_s, w_s, x_s, y_s) - (\hat{u}_s, \hat{w}_s, \hat{x}_s, \hat{y}_s)\| < \varepsilon$.
2. For each UEP (u_i, w_i, x_i, y_i), there exists a unique hyperbolic equilibrium point $(\hat{u}_i, \hat{w}_i, \hat{x}_i, \hat{y}_i)$ of model (Eq. 16.3) with $\|(u_i, w_i, x_i, y_i) - (\hat{u}_i, \hat{w}_i, \hat{x}_i, \hat{y}_i)\| < \varepsilon$. Moreover, (1) (u_i, w_i, x_i, y_i) and $(\hat{u}_i, \hat{w}_i, \hat{x}_i, \hat{y}_i)$ are the same type of UEPs; (2) $(\hat{u}_i, \hat{w}_i, \hat{x}_i, \hat{y}_i)$ lies on the stability boundary $\partial A(\hat{u}_s, \hat{w}_s, \hat{x}_s, \hat{y}_s)$ of model (Eq. 16.3); and (3) there are no other UEPs other than $(\hat{u}_i, \hat{w}_i, \hat{x}_i, \hat{y}_i)$, $i = 1, 2, \ldots, n$ lying on the stability boundary $\partial A(\hat{u}_s, \hat{w}_s, \hat{x}_s, \hat{y}_s)$ of model (Eq. 16.3).

Theorem 16.5: Persistence Property

Let $(u_s, w_s, x_s, y_s, 0)$ be a SEP of model (Eq. 16.7) and its stability boundary $\partial A(u_s, w_s, x_s, y_s, 0)$ contains a finite number of UEPs $(u_i, w_i, x_i, y_i, 0)$, $i = 1, 2, \ldots, n$. If the following two conditions are satisfied:

(i) all the intersections of the stable and unstable manifolds of the equilibrium points on the stability boundary satisfy the transversality condition, and

(ii) zero is a regular value of

$$\frac{\partial^4 U(u_i, w_i, x_i, y_i)}{\partial u \partial w \partial x \partial y}$$

for all the UEPs on the stability boundary,

then the following results hold:

1. There exist two positive numbers, $\delta > 0$ and $\varepsilon > 0$, such that if the transfer conductance of model (Eq. 16.8) satisfies $G_{jk} < \delta$ for all j and k, there exists a unique SEP $(\hat{u}_s, \hat{w}_s, \hat{x}_s, \hat{y}_s, 0)$ of model (Eq. 16.8) with $\|(u_s, w_s, x_s, y_s, 0) - (\hat{u}_s, \hat{w}_s, \hat{x}_s, \hat{y}_s, 0)\| < \varepsilon$.
2. For each UEP $(u_i, w_i, x_i, y_i, 0)$, there exists a unique hyperbolic equilibrium point $(\hat{u}_i, \hat{w}_i, \hat{x}_i, \hat{y}_i, 0)$ of model (Eq. 16.8) with $\|(u_i, w_i, x_i, y_i, 0) - (\hat{u}_i, \hat{w}_i, \hat{x}_i,$

$\hat{y}_i, 0)\| < \varepsilon$. Moreover, (1) $(u_i, w_i, x_i, y_i, 0)$ and $(\hat{u}_i, \hat{w}_i, \hat{x}_i, \hat{y}_i, 0)$ are the same type of UEPs; (2) $(\hat{u}_i, \hat{w}_i, \hat{x}_i, \hat{y}_i, 0)$ lies on the stability boundary $\partial A(\hat{u}_s, \hat{w}_s, \hat{x}_s, \hat{y}_s, 0)$ of model (Eq. 16.8); and (3) there are no other UEPs other than $(\hat{u}_i, \hat{w}_i, \hat{x}_i, \hat{y}_i, 0)$, $i = 1, 2, \ldots, n$ lying on the stability boundary $\partial A(\hat{u}_s, \hat{w}_s, \hat{x}_s, \hat{y}_s, 0)$ of model (Eq. 16.8).

Steps 1 and 7. In Step 1, we investigate the static properties (S1) and (S2) and the dynamic property (D2) between the reduced-state model (Eq. 16.2) and the corresponding singularly perturbed system (Eq. 16.3). It can be shown that there exists an $\varepsilon^* > 0$ such that for all $\varepsilon \in (0, \varepsilon^*)$, the reduced-state model (Eq. 16.2) and the corresponding singularly perturbed system (Eq. 16.3) satisfy the static properties (S1) and (S2) and contain the same UEPs on the intersection of the stability boundary and the stable component of the algebraic manifold. In Step 7, we investigate the static properties (S1) and (S2) and the dynamic property (D2) between the original system (Eq. 16.1) and the corresponding singularly perturbed system (Eq. 16.8). It can be shown that there exists an $\varepsilon^* > 0$ such that for all $\varepsilon \in (0, \varepsilon^*)$, the reduced-state model (Eq. 16.2) and the corresponding singularly perturbed system (Eq. 16.8) satisfy the static properties (S1) and (S2) and contain the same UEPs on the intersection of the stability boundary and the stable component of the algebraic manifold (Alberto and Chiang, 2009; Fekih-Ahmed, 1991; Zou et al., 2003).

The static properties (S1) and (S2) between the original system (Eq. 16.1) and the corresponding singularly perturbed system (Eq. 16.8) and that between the reduced-state model (Eq. 16.2) and the corresponding singularly perturbed system (Eq. 16.3) are established in the following two theorems.

Theorem 16.6: Static Property

If an equilibrium point, say, $(u_i, w_i, x_i, y_i, 0)$, of the original system (Eq. 16.1) lies on one stable component, Γ_s, of the constraint manifold, then there exists an $\hat{\varepsilon} > 0$ such that for all $\varepsilon \in (0, \hat{\varepsilon})$, it follows that

1. $(u_i, w_i, x_i, y_i, 0)$ is a hyperbolic equilibrium point of the original system (Eq. 16.1) if and only if $(u_i, w_i, x_i, y_i, 0)$ is a hyperbolic equilibrium point of the singularly perturbed system (Eq. 16.8); moreover,
2. $(u_i, w_i, x_i, y_i, 0)$ is a type-k equilibrium point of the original system (Eq. 16.1) if and only if $(u_i, w_i, x_i, y_i, 0)$ is a type-k equilibrium point of the singularly perturbed system (Eq. 16.8).

Theorem 16.7: Static Property

If an equilibrium point, say, (u_i, w_i, x_i, y_i), of the reduced-state model (Eq. 16.2) lies on one stable component, Γ_s, of the constraint manifold, then there exists an $\hat{\varepsilon} > 0$ such that for all $\varepsilon \in (0, \hat{\varepsilon})$, it follows that

1. (u_i, w_i, x_i, y_i) is a hyperbolic equilibrium point of the reduced-state model (Eq. 16.2) if and only if (u_i, w_i, x_i, y_i) is a hyperbolic equilibrium point of the singularly perturbed system (Eq. 16.3); moreover,
2. (u_i, w_i, x_i, y_i) is a type-k equilibrium point of the reduced-state model (Eq. 16.2) if and only if (u_i, w_i, x_i, y_i) is a type-k equilibrium point of the singularly perturbed system (Eq. 16.3).

The above results show that the type of each equilibrium point of the original system (Eq. 16.1) (respectively, reduced-state model [Eq. 16.2]) is the same as the type of the corresponding equilibrium point of the singularly perturbed system (Eq. 16.8) (respectively, the singularly perturbed system [Eq. 16.3), provided ε is sufficiently small. In the following two theorems, the dynamic relationship that the stability boundaries of these two systems contain the same set of equilibrium points on stable components will be shown.

Theorem 16.8: Dynamic Relationship

Let $(u_s, w_s, x_s, y_s, 0)$ and $(u_i, w_i, x_i, y_i, 0)$ be the SEP and UEP of the original system (Eq. 16.1) on a stable component, Γ_s, respectively. Suppose that, for each $\varepsilon > 0$, the corresponding singularly perturbed system (Eq. 16.8) has an energy function and its equilibrium points are isolated. Then there exists an $\hat{\varepsilon} > 0$ such that, for all $\varepsilon \in (0, \hat{\varepsilon})$, $(u_i, w_i, x_i, y_i, 0)$ lies on the stability boundary $\partial A_0(u_s, w_s, x_s, y_s, 0)$ of the original system (Eq. 16.1) if and only if $(u_i, w_i, x_i, y_i, 0)$ lies on the stability boundary $\partial A_\varepsilon(u_s, w_s, x_s, y_s, 0)$ of the singularly perturbed system (Eq. 16.8).

Theorem 16.9: Dynamic Relationship

Let (u_s, w_s, x_s, y_s) and (u_i, w_i, x_i, y_i) be the SEP and UEP of the reduced-state model (Eq. 16.2) on a stable component, Γ_s, respectively. Suppose that, for each $\varepsilon > 0$, the corresponding singularly perturbed model (Eq. 16.3) has an energy function and its equilibrium points are isolated. Then there exists an $\hat{\varepsilon} > 0$ such that, for all $\varepsilon \in (0, \hat{\varepsilon})$, (u_i, w_i, x_i, y_i) lies on the stability boundary $\partial A_0(u_s, w_s, x_s, y_s)$ of the reduced-state model (Eq. 16.2) if and only if (u_i, w_i, x_i, y_i) lies on the stability boundary $\partial A_\varepsilon(u_s, w_s, x_s, y_s)$ of the singularly perturbed model (Eq. 16.3).

16.5 OVERALL STATIC AND DYNAMIC RELATIONSHIPS

By combining the relationships derived in Steps 1 through 7, the following analytical results regarding the static as well as dynamic relationships between the original system (Eq. 16.1) and the reduced-state model (Eq. 16.2) are established. We assume that the SEP and UEP of the original system (Eq. 16.1) and those of the reduced-state model (Eq. 16.2) that are of interest in our study are lying on a stable component, Γ_s, respectively.

Theorem 16.10: Static Relationship

Let (u_s, w_s, x_s, y_s) be a SEP on a stable component, Γ_s, of the reduced-state model (Eq. 16.2). If the following conditions are satisfied:

1. zero is a regular value of

$$\frac{\partial^4 U(u_i, w_i, x_i, y_i)}{\partial u \partial w \partial x \partial y}$$

for all the UEPs (u_i, w_i, x_i, y_i), $i = 1, 2, \ldots, k$ on the stable component Γ_s, and

2. the transfer conductance of the reduced-state model (Eq. 16.2) is sufficiently small,

then, $(\hat{u}, \hat{w}, \hat{x}, \hat{y})$ is a type-k equilibrium point of the reduced-state model (Eq. 16.2) if and only if $(\hat{u}, \hat{w}, \hat{x}, \hat{y}, 0)$ is a type-k equilibrium point of the original system (Eq. 16.1).

Theorem 16.10 asserts that, under the stated conditions, the static properties (S1) and (S2) between the original system (Eq. 16.1) and the reduced-state model (Eq. 16.2) hold.

Regarding dynamic property (D1), it can be shown that there exists a numerical energy function for the reduced-state model (Eq. 16.2). More specifically, it can be shown that for any compact set S of the state-space of model (Eq. 16.2), there is a positive number, α, such that, if the transfer conductance of the model satisfies $|G| < \alpha$, then there is an energy function defined on this compact set S. Hence, the dynamic property (D1) is satisfied.

To examine the dynamic property (D2), the following main result follows from the analytical results derived in Step 1 through Step 7. Indeed, Theorem 16.11 asserts that, under the stated conditions, the dynamic property (D2) is satisfied. Furthermore, the stability boundaries of both the original model and the state-reduced model are completely characterized, and they contain the "same" equilibrium points.

Theorem 16.11: Dynamic Relationship

Let (u_s, w_s, x_s, y_s) be a SEP of the reduced-state model (Eq. 16.2). If the following conditions are satisfied:

(1) zero is a regular value of

$$\frac{\partial^4 U(u_i, w_i, x_i, y_i)}{\partial u \partial w \partial x \partial y}$$

for all the UEPs (u_i, w_i, x_i, y_i), $i = 1, 2, \ldots, k$ on the stability boundary $\partial A(u_s, w_s, x_s, y_s)$;

(2) the transfer conductance of the reduced-state model (Eq. 16.2) is sufficiently small; and

(3) all the intersections of the stable and unstable manifolds of the equilibrium points on the stability boundary $\partial A(u_s, w_s, x_s, y_s, 0)$ of the one-parameterized model $d(\lambda)$ (Eq. 16.6) satisfy the transversality condition for $\lambda \in [0, 1]$,

then the following results hold:

1. The equilibrium point (u_i, w_i, x_i, y_i) is on the stability boundary $\partial A(u_s, w_s, x_s, y_s)$ of the model (Eq. 16.2) if and only if the equilibrium point $(u_i, w_i, x_i, y_i, 0)$ is on the stability boundary $\partial A(u_s, w_s, x_s, y_s, 0)$ of the model (Eq. 16.1).
2. The stability boundary $\partial A(u_s, w_s, x_s, y_s)$ of the reduced-state model (Eq. 16.2) is the union of the stable manifold of all the equilibrium points (u_i, w_i, x_i, y_i), $i = 1, 2, \ldots$ on the stability boundary $\partial A(u_s, w_s, x_s, y_s)$; that is, $\partial A(u_s, w_s, x_s, y_s) = \cup W^s(u_i, w_i, x_i, y_i)$.
3. The stability boundary $\partial A(u_s, w_s, x_s, y_s, 0)$ of the original model (Eq. 16.1) is the union of the stable manifold of all the equilibrium points $(u_i, w_i, x_i, y_i, 0)$, $i = 1, 2, \ldots$ on the stability boundary $\partial A(u_s, w_s, x_s, y_s, 0)$; that is, $\partial A(u_s, w_s, x_s, y_s, 0) = \cup W^s(u_i, w_i, x_i, y_i, 0)$.

16.6 DYNAMIC PROPERTY (D3)

We present a numerical scheme for efficiently detecting the intersection point, which is a projected fault-on trajectory $(u_f(t), w_f(t), x_f(t), y_f(t))$ that intersects with the stability boundary $\partial A_0(u_s, w_s, x_s, y_s)$ of the postfault reduced-state model, without resort to a detailed time–domain numerical simulation. The existence of such a numerical scheme ensures that the reduced-state model (Eq. 16.2) satisfies the dynamic property (D3). Recall that a numerical function for the original system can be expressed as follows, which is the summation of a numerical potential energy and a kinetic energy:

$$\begin{aligned} W_{num}(u, w, x, y, z) &= W_{ana}(u, w, x, y, z) + U_{path}(u, w, x, y) \\ &= \frac{1}{2} z^T M z + U(u, w, x, y) + U_{path}(u, w, x, y) \\ &= \frac{1}{2} z^T M z + U_{num}(u, w, x, y). \end{aligned} \tag{16.9}$$

The numerical scheme proceeds as follows:

Step 1. Along the fault-on trajectory $(u_f(t), w_f(t), x_f(t), y_f(t), z_f(t))$, detect the point, say, (u^*, w^*, x^*, y^*), at which the projected fault-on trajectory $(u_f(t), w_f(t), x_f(t), y_f(t))$ reaches the first local maximum (along the fault-on trajectory) of the numerical potential energy $U_{num}(u, w, x, y)$ of Equation 16.9.

Step 2. The point (u^*, w^*, x^*, y^*) is a simulated exit point, which is an approximated point for the intersection point between the stability boundary of the (postfault) reduced-state model (Eq. 16.2) and the projected fault-on trajectory $(u_f(t), w_f(t), x_f(t), y_f(t))$.

Another numerical scheme for detecting the exit point of the reduced-state stability boundary is the following:

Step 1. Along the fault-on trajectory $(u_f(t), w_f(t), x_f(t), y_f(t), z_f(t))$, compute the dot product of the following two vectors: (1) the fault-on generator speed vector and (2) the postfault power mismatch vector at each integration step. When the sign of the dot product changes from positive to negative, the exit point is detected.

The above numerical schemes can efficiently detect the exit point and does not require detailed time–domain simulations of the postfault system. Hence, the dynamic property (D3) is satisfied.

16.7 CONCEPTUAL NETWORK-PRESERVING BCU METHOD

A conceptual network-preserving BCU method for computing the controlling UEP of the generic network-preserving model (Eq. 16.1), along with its analytical basis, is presented in this section.

A Conceptual Network-Preserving BCU Method

Step 1. Integrate the (sustained) fault-on trajectory $(u_f(t), w_f(t), x_f(t), y_f(t), z_f(t))$ of the network-preserving original model (Eq. 16.1) until the exit point (u^*, w^*, x^*, y^*) is detected, at which the projected trajectory $(u_f(t), w_f(t), x_f(t), y_f(t))$ intersects the stability boundary of the postfault reduced-state model (Eq. 16.2).

Step 2. Use the exit point (u^*, w^*, x^*, y^*) as the initial condition and integrate the postfault reduced-state model (Eq. 16.2) to a UEP, say, $(u_{co}, w_{co}, x_{co}, y_{co})$.

Step 3. The controlling UEP with respect to the fault-on trajectory $(u_f(t), w_f(t), x_f(t), y_f(t), z_f(t))$ of the network-preserving model (Eq. 16.1) is $(u_{co}, w_{co}, x_{co}, y_{co}, 0)$.

In essence, the BCU method is to compute the controlling UEP of the original model (Eq. 16.1) via computing the controlling UEP of the reduced-state model (Eq. 16.2), whose controlling UEP can be effectively computed. Steps 1 and 2 of the conceptual network-preserving BCU method find the controlling UEP of the reduced-state model (i.e., the controlling UEP relative to the projected fault-on trajectory). Step 3 relates the controlling UEP of the reduced-state model (Eq. 16.2) to the controlling UEP of the original model.

According to Result 2 of Theorem 16.11, the exit point (u^*, w^*, x^*, y^*), computed in Step 1, must lie on the stable manifold of a UEP of the reduced-state system

(Eq. 16.2) since the stability boundary $\partial A(u_s, w_s, x_s, y_s)$ of the reduced-state model (Eq. 16.2) is the union of the stable manifold of all the equilibrium points on the stability boundary. Hence, the trajectory of the reduced-state system starting from the exit point (u^*, w^*, x^*, y^*) computed in Step 2 must converge to a UEP, say, $(u_{co}, w_{co}, x_{co}, y_{co})$, of the reduced-state model (Eq. 16.2). Result 1 of Theorem 16.11 asserts that, under the conditions stated, the equilibrium point $(u_{co}, w_{co}, x_{co}, y_{co})$ is on the stability boundary $\partial A(u_s, w_s, x_s, y_s)$ of the reduced-state system (Eq. 16.2) if and only if the equilibrium point $(u_{co}, w_{co}, x_{co}, y_{co}, 0)$ is on the stability boundary $\partial A(u_s, w_s, x_s, y_s, 0)$ of the original model (Eq. 16.1). Step 3 of the conceptual network-preserving BCU method hence relates the controlling UEP $(u_{co}, w_{co}, x_{co}, y_{co})$ of the reduced-state system (Eq. 16.2) to the controlling UEP $(u_{co}, w_{co}, x_{co}, y_{co}, 0)$ of the original model (Eq. 16.1).

It should be pointed out that Step 2 of the BCU method is a dynamic search step to find the controlling UEP of the state-reduced system. This step explores the structure of the (relevant) stability boundary to find the controlling UEP. In the literature, several attempts to modify the BCU method have been proposed, but without much success. For example, it was proposed to replace Step 2 by the following step:

- Apply the Newton method using the point (u^*, w^*, x^*, y^*) as the initial condition to compute a UEP.

One key problem with this proposal is that the Newton method using the point (u^*, w^*, x^*, y^*) will encounter one of the following scenarios:

- It may converge to another UEP lying outside the stability region, since this proposal is a static search and does not take the dynamic characterization of the controlling UEP into account.
- It is likely to diverge because the exit point may not lie inside the convergence region of a UEP under the Newton method.
- Even if it converges, the converged UEP may not be the controlling UEP unless under some special conditions.

Another proposal is to replace Step 2 by the following Step:

- Apply an optimization method using the point (u^*, w^*, x^*, y^*) as the initial condition to compute a UEP.

One key problem with this proposal is that an optimization method with the point (u^*, w^*, x^*, y^*) as an initial condition may behave in one of the following ways:

- It is a static search and does not take the dynamic characterization of the controlling UEP into account.
- Even if it converges, this UEP may not be the controlling UEP since it may be located far away from the exit point.

Regarding Step 3 of the conceptual network-preserving BCU method, the issue of whether the UEP $(u^*_{co}, w^*_{co}, x^*_{co}, y^*_{co}, 0)$ is truly the controlling UEP of the fault-on trajectory. One scheme to address this issue is to derive the stable manifold of $(u_{co}^*, w_{co}^*, x_{co}^*, y_{co}^*, 0)$ as well as the stable manifolds of, in theory, all the UEPs and to identify the UEP whose stable manifold is first hit by the fault-on trajectory. This is a formidable task since a computational method of deriving the stable manifold of an equilibrium point is still unavailable. We will present a verification scheme for checking whether or not $(u_{co}^*, w_{co}^*, x_{co}^*, y_{co}^*, 0)$ lies on the stability boundary of the original model (Eq. 16.1) in a later chapter.

There are several possible ways to numerically implement the conceptual network-preserving BCU method for power system models. A numerical implementation of this method for computing the controlling UEP will be presented in the next chapter.

To illustrate the computational procedure involved in the network-preserving BCU method, we numerically simulate the following objects related to the BCU method for two different contingencies:

- prefault SEP and postfault SEP,
- the stability boundary of the postfault reduced-state model,
- UEPs on the stability boundary of the postfault reduced-state model,
- the projected fault-on trajectory, and
- the controlling UEP relative to the projected fault-on trajectory.

Again, these objects are chosen because the projected fault-on trajectory intersects the stability boundary of the postfault reduced-state model at the exit point, which lies on the stable manifold of the controlling UEP relative to the projected fault-on trajectory.

For Contingency 10, a fault occurs near Bus 8, and the line between Buses 7 and 8 is tripped. The exact stability boundary of the postfault reduced-state model is shown in Figure 16.2. The projected fault-on trajectory intersects the stable manifold of its controlling UEP, a type-one UEP, as shown in the figure. There are four type-one UEPs and four type-two UEP lying on the stability boundary of the reduced-state system. The stable manifold of the controlling UEP relative to the projected fault-on trajectory contains the exit point. This exit point is at some distance away from the controlling UEP, while it is very close to a type-two UEP lying on the stability boundary. It is, however, incorrect to use this type-two UEP as the controlling UEP. We note that in this numerical study, the controlling UEP relative to the projected fault-on trajectory corresponds to the controlling UEP relative to the original fault-on trajectory.

For Contingency 2, a fault occurs near Bus 7, and the line between Buses 7 and 8 is tripped. The exact stability boundary of the postfault reduced-state model is shown in Figure 16.3. There are four type-one UEPs and four type-two UEP lying on the stability boundary of the reduced-state system. The stable manifold of the type-one UEP, which is the controlling UEP, contains the exit point. This exit point is close to the controlling UEP, while it is far away from the other type-one UEP. The controlling UEP of the reduced-state system corresponds to the controlling UEP of the original system. The computed exit point is slightly off the stability boundary

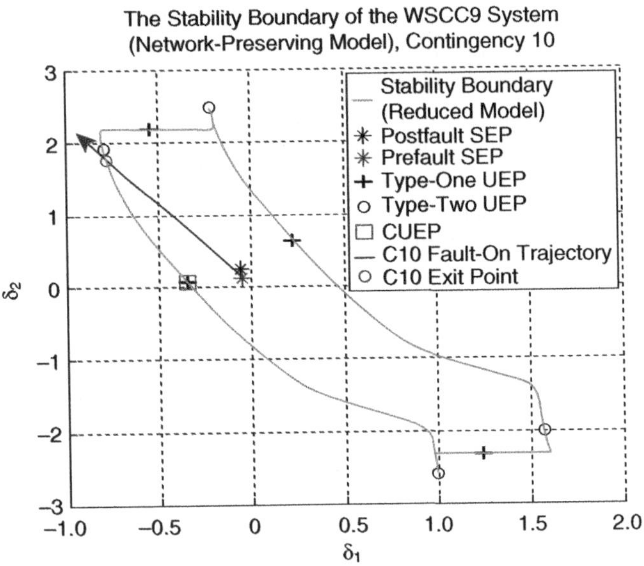

Figure 16.2 For Contingency 10, the prefault and postfault SEPs are shown. The exact stability boundary of the postfault reduced-state model is highlighted. The projected fault-on trajectory intersects the stable manifold of its controlling UEP, a type-one UEP, at the exit point, highlighted in the figure. The controlling UEP of the reduced-state model corresponds to the controlling UEP of the original model.

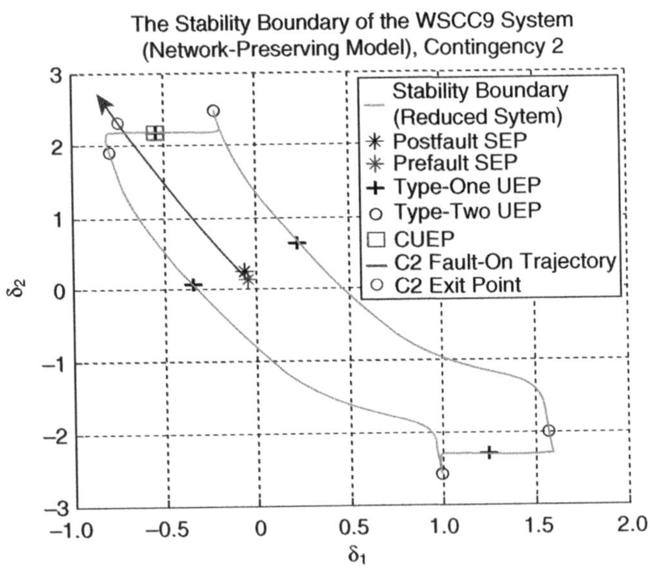

Figure 16.3 For Contingency 2, the prefault and postfault SEPs are shown. The exact stability boundary of the postfault reduced-state model is highlighted. There are four type-one UEPs and four type-two UEPs. The controlling UEP of the reduced-state system corresponds to the controlling UEP of the original system.

of the reduced-state system; however, the degree of accuracy in computing the exit point can be improved.

16.8 CONCLUDING REMARKS

The computational challenges of computing the controlling UEP question the correctness of any attempt to directly compute the controlling UEP of an original power system stability model. These challenges serve to explain why many methods proposed in the literature fail to compute the controlling UEP. These methods attempt to directly compute the controlling UEP of the power system stability model, which is difficult, if not impossible, to compute without using the time–domain approach.

The ability to compute the controlling UEP is vital to direct stability analysis. In this chapter, we have proven that it is fruitful to develop a tailored method for finding the controlling UEP by exploiting special properties, as well as some physical and mathematical insights, of the underlying power system transient stability model. Along this line, this chapter has focused on the development of such a method: the network-preserving BCU method. Given a generic network-preserving transient stability model, the network-preserving BCU method defines a reduced-state model, which is designed to capture the static and dynamic properties of the original model. The BCU method then computes the controlling UEP of the reduced-state model by exploring the special structure of the stability boundary and the energy function of the reduced-state model. It then relates the controlling UEP of the reduced-state model to the controlling UEP of the original model. Effective numerical methods for implementing this conceptual network-preserving BCU method will be described in the next chapter. We will conduct an analytical study and propose numerical methods for checking the static and dynamic relationship between the original model (Eq. 16.1) and the state-reduced model (Eq. 16.2) in Chapters 18 and 19.

Chapter 17

Numerical Network-Preserving BCU Method

17.1 INTRODUCTION

Given a power system transient stability model, there exists a corresponding version of the BCU method. There are, however, several ways to define a reduced-state model satisfying static properties (S1) and (S2) and dynamic properties (D1), (D2), and (D3). The reduced-state model proposed in the previous chapter is an effective one, though other reduced-state models exist. In addition, the dynamic property (D2) can be relaxed to (D2), which states that the (reduced-state) controlling unstable equilibrium point (CUEP) of the reduced-state system corresponds to the CUEP of the original system.

The BCU method does not compute the CUEP of the original model directly since the task of computing the exit point of the original model, necessary for the computation of the CUEP, is very difficult and usually requires the time–domain approach. The BCU method first explores the special structure of the underlying stability model so as to define an artificial, reduced-state model that captures all the equilibrium points on the stability boundary of the original power system stability model:

$$
\begin{aligned}
0 &= -\frac{\partial U}{\partial u}(u, w, x, y) + g_1(u, w, x, y) \\
0 &= -\frac{\partial U}{\partial w}(u, w, x, y) + g_2(u, w, x, y) \\
T\dot{x} &= -\frac{\partial U}{\partial x}(u, w, x, y) + g_3(u, w, x, y) \qquad (17.1) \\
\dot{y} &= z \\
M\dot{z} &= -Dz - \frac{\partial U}{\partial y}(u, w, x, y) + g_4(u, w, x, y).
\end{aligned}
$$

Direct Methods for Stability Analysis of Electric Power Systems, by Hsiao-Dong Chiang

Then CUEP of the original model is computed by computing the CUEP of the following reduced-state model (Eq. 17.2). This can be computed without resorting to the time–domain approach:

$$\begin{aligned}
0 &= -\frac{\partial U}{\partial u}(u, w, x, y) + g_1(u, w, x, y) \\
0 &= -\frac{\partial U}{\partial w}(u, w, x, y) + g_2(u, w, x, y) \\
T\dot{x} &= -\frac{\partial U}{\partial x}(u, w, x, y) + g_3(u, w, x, y) \\
\dot{y} &= -\frac{\partial U}{\partial y}(u, w, x, y) + g_4(u, w, x, y).
\end{aligned} \tag{17.2}$$

A numerical energy function for the original model is as follows. It is the summation of a numerical potential energy and kinetic energy:

$$\begin{aligned}
W_{num}(u, w, x, y, z) &= W_{ana}(u, w, x, y, z) + U_{path}(u, w, x, y) \\
&= \frac{1}{2} z^T M z + U(u, w, x, y) + U_{path}(u, w, x, y) \\
&= \frac{1}{2} z^T M z + U_{num}(u, w, x, y).
\end{aligned} \tag{17.3}$$

There are several possible ways to numerically implement the conceptual BCU method for network-preserving power system models. A numerical implementation of this method is presented below.

A Network-Preserving BCU Method

Step 1. Integrate the fault-on system of the original model (Eq. 17.1) to derive the fault-on trajectory ($u_f(t)$, $w_f(t)$, $x_f(t)$, $y_f(t)$, $z_f(t)$) for a few time steps. At each time step, solve for the initial condition of the postfault system, say, $(u(t_{cl}^+), \omega(t_{cl}^+), x(t_{cl}^+), y(t_{cl}^+), z(t_{cl}^+))$.

Step 2. Construct a numerical energy function for the postfault system using Equation 17.3 and compute the numerical potential energy at each time step of Step 1.

$$\begin{aligned}
W_{num}(u, w, x, y, z) &= W_{ana}(u, w, x, y, z) + U_{path}(u, w, x, y) \\
&= \frac{1}{2} z^T M z + U(u, w, x, y) + U_{path}(u, w, x, y) \\
&= \frac{1}{2} z^T M z + U_{num}(u, w, x, y).
\end{aligned}$$

BCU Steps

Step 3. Detect the exit point (u^*, w^*, x^*, y^*) at which the projected fault-on trajectory ($u_f(t)$, $w_f(t)$, $x_f(t)$, $y_f(t)$) reaches the first local maximum (along the fault-on trajectory) of the numerical potential energy $U_{num}(u, w, x, y)$.

Step 4. Use the exit point (u^*, w^*, x^*, y^*) as the initial condition and integrate the postfault reduced-state system (Eq. 17.2) until the trajectory reaches the first local minimum of the following norm of the postfault reduced-state system (Eq. 17.2), that is, the first local minimum of the following norm along the postfault reduced-state trajectory:

$$\left\|\frac{\partial U}{\partial u}(u, w, x, y)+g_1(u, w, x, y)\right\| + \left\|\frac{\partial U}{\partial w}(u, w, x, y)+g_2(u, w, x, y)\right\| + \left\|\frac{\partial U}{\partial x}(u, w, x, y)+g_3(u, w, x, y)\right\| + \left\|\frac{\partial U}{\partial y}(u, w, x, y)+g_4(u, w, x, y)\right\|.$$

Let the local minimum be denoted as $(u_0^*, w_0^*, x_0^*, y_0^*)$ and termed the minimum gradient point (MGP).

Step 5. Use the MGP $(u_0^*, w_0^*, x_0^*, y_0^*)$ as the initial guess and solve the following set of nonlinear algebraic equations:

$$\left\|\frac{\partial U}{\partial u}(u, w, x, y)+g_1(u, w, x, y)\right\| + \left\|\frac{\partial U}{\partial w}(u, w, x, y)+g_2(u, w, x, y)\right\| + \left\|\frac{\partial U}{\partial x}(u, w, x, y)+g_3(u, w, x, y)\right\| + \left\|\frac{\partial U}{\partial y}(u, w, x, y)+g_4(u, w, x, y)\right\| = 0.$$

Let the solution be $(u_{co}, w_{co}, x_{co}, y_{co})$.

Step 6. The CUEP with respect to the fault-on trajectory $(u(t), w(t), x(t), y(t))$ of the original system is $(u_{co}, w_{co}, x_{co}, y_{co}, 0)$.

Direct Stability Assessment Steps

Step 7. The critical energy, v_{cr}, is the value of the numerical energy function $W_{num}(u, w, x, y, z)$ at the CUEP; that is, $v_{cr} = W_{num}(u_{co}, w_{co}, x_{co}, y_{co}, 0)$.

Step 8. Calculate the value of the numerical energy function $W_{num}(u, w, x, y, z)$ at the time of fault clearance (t_{cl}) by using the initial condition of the postfault trajectory; that is,

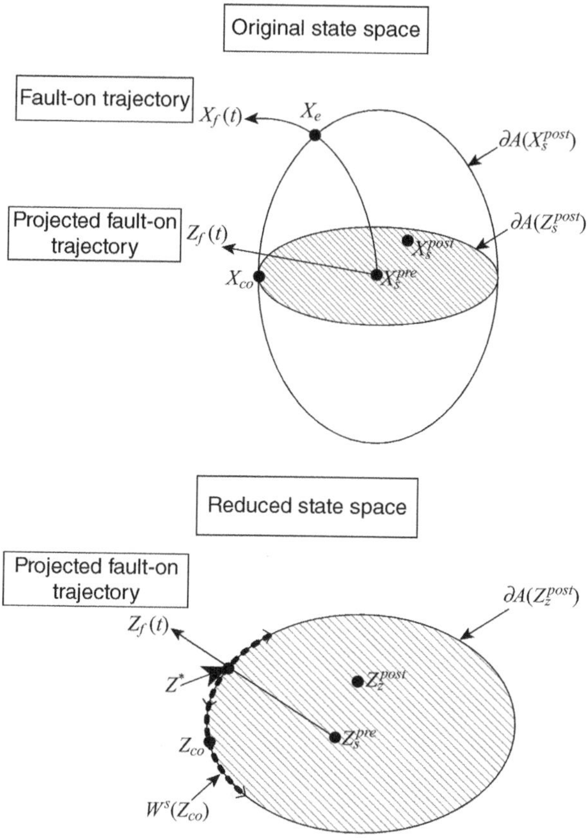

Figure 17.1 Steps 1 and 3 of the network-preserving BCU method.

$$v_f = W_{num}\left(u\left(t_{cl}^+\right), \omega\left(t_{cl}^+\right), x\left(t_{cl}^+\right), y\left(t_{cl}^+\right), z\left(t_{cl}^+\right)\right).$$

Step 9. If $v_f < v_{cr}$, then the initial condition of the postfault trajectory $(u(t_{cl}^+), \omega(t_{cl}^+), x(t_{cl}^+), y(t_{cl}^+), z(t_{cl}^+))$ is located inside the stability boundary of the postfault system; hence, the postfault trajectory is stable. Otherwise, it may be unstable.

Steps 3–5 compute the CUEP of the reduced-state system (Eq. 17.2). Note that in Step 4, the postfault reduced-state trajectory starting from the exit point (u^*, w^*, x^*, y^*) will move along the stable manifold of the CUEP of the reduced-state system (Eq. 17.2). The stable manifold of $z_{co} = (u_{co}, w_{co}, x_{co}, y_{co})$ contains the exit point (u^*, w^*, x^*, y^*) (see Figure 17.1). Step 6 relates the CUEP relative to the projected fault-on trajectory to the CUEP $(u_{co}, w_{co}, x_{co}, y_{co}, 0)$ relative to the original fault-on trajectory $X_f(t) = (Z_f(t), z_f(t)) = (u_f(t), w_f(t), x_f(t), y_f(t), z_f(t))$ (see Figure 17.2).

We discuss several computational issues associated with the above network-preserving BCU method and present several numerical schemes to address these issues. These numerical schemes will be incorporated into the network-preserving

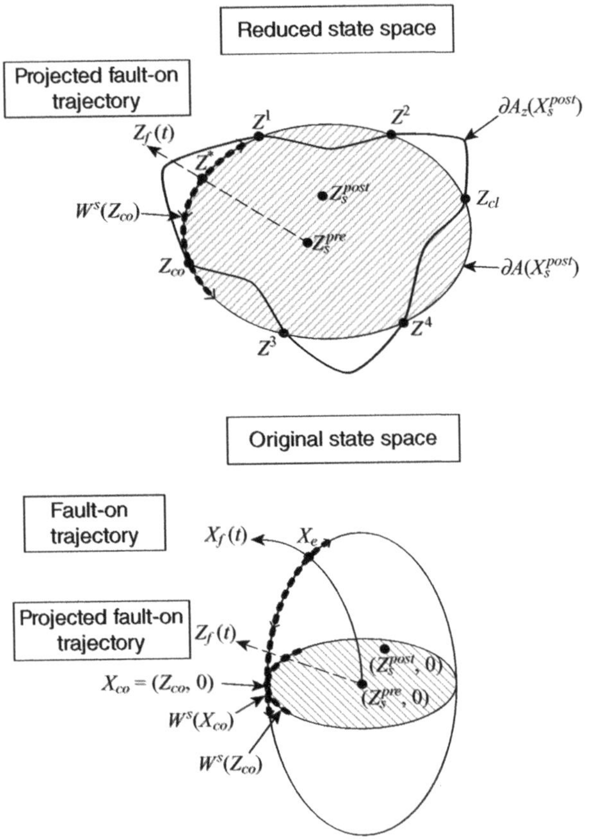

Figure 17.2 Steps 4 and 5 of the BCU method.

BCU method to form a numerical network-preserving BCU method. Finally, the numerical BCU method will be illustrated on two test systems.

17.2 COMPUTATIONAL CONSIDERATIONS

The network-preserving BCU method for direct stability analysis presented in the previous section contains nine steps. Computational issues associated with each step are addressed in this section.

A numerical implementation of Step 1 is straightforward; it requires a numerical integration of the fault-on system and the initialization of the postfault system. To derive a numerical energy function value along the fault-on trajectory in Step 2, the following procedure is required:

- Compute the postfault stable equilibrium point (SEP).
- The energy function value at each point of the simulated fault-on trajectory is obtained by performing a path-dependent approximation between the point and the postfault SEP.

In Steps 1 and 2, the fault-on trajectory needs to be simulated even after the fault is cleared if the exit point of the projected fault-on trajectory is not reached yet. In this case, the (sustained) fault-on trajectory is simulated until the exit point is reached. Steps 1 and 2 require the solution of two sets of differential and algebraic equations (DAEs), which can be solved by a partition–solution approach. At each time step, the set of differential equations is solved independently using a variable-step integration algorithm, and then the set of network algebraic equations is solved using the factored Y-matrix in a current injection algorithm.

A challenge in Step 3 is how to quickly detect the exit point without resorting to time–domain simulation. The direct scheme presented in the previous chapter will be extended and employed for implementing Step 3. This extension will be discussed in the next section. A major challenge in Step 4 is how to generate a "trajectory" so that it moves along the stability boundary of the reduced-state system and approaches the reduced-state CUEP. Roughly speaking, Step 4 is a dynamic search for computing the reduced-state CUEP. It involves the following:

- Starting from the exit point (u^*, w^*, x^*, y^*), move along the (relevant) stability boundary, which is part of the stable manifold of the reduced-state CUEP.
- Detect the MGP, which is close to the reduced-state CUEP.

A robust nonlinear algebraic solver needed for computing the CUEP of the reduced-state model in Step 5 will be presented. In this step, the computed MGP, computed in Step 4, is usually close to the reduced-state CUEP.

17.3 NUMERICAL SCHEME TO DETECT EXIT POINTS

To effectively detect the exit point of the reduced-state system, which is the desired dynamic property (D3), one can compute the dot product of the fault-on speed vector and the postfault power mismatch vector at each integration step of the fault-on trajectory. When the sign of the dot product changes from positive to negative, the exit point is detected. The derivative of the potential energy defined for the postfault system with respect to time and along the fault-on trajectory can be calculated as

$$\frac{dV_{PE}}{dt} = -\sum_{i=1}^{n}(P_{mi} - P_{ei})\frac{d\delta_i}{dt}.$$

The right-hand side of the above equation is termed the *dot product*. In some cases where the corresponding CUEP is very sensitive to the accuracy of the exit point, the BCU method may encounter a numerical convergence problem in computing the CUEP. To overcome this problem, the accuracy of computing the exit point can be improved via the linear interpolation method, as described in Chapter 15.

Some numerical ill-conditions may still occur. For instances, during the procedure of performing exit-point detection based on the dot product's sign change (from negative to positive), an ill-conditioned situation arises when the dot product is very small (i.e., close to zero in magnitude); its sign change is usually caused by a round-off error, not as an indication of the actual exit point. A safeguard step against this false detection of an exit point is to add a threshold checking before the sign checking. The checking of the dot product's sign change will be performed only if the magnitude of the dot product is larger than a given threshold value. This threshold should be large enough to mask the unfavorable effect of a round-off error. A value of 0.0001 has been found to be satisfactory.

From a computational viewpoint, the integration method can use the Adams method for nonstiff systems and the BDF method for stiff systems (Brenan et al., 1989). The algebraic equation is solved by the implicit impedance matrix method. In this method, the sparse LDU factors of the Y-matrix are obtained for one time. The bus voltages are solved in terms of current injection by the popular forward and backward substitution scheme. The solved bus voltages are then reinserted into the differential equations for evaluating derivatives. In addition to integrating the fault-on trajectory in which the fault-on network equations are required to solve, the postfault network equations must also be solved at each time step in order to complete the dot product computation for detecting the exit point.

To compute an adequate MGP so that the CUEP can be reliably computed, the stability-boundary-following procedure to guide the search process of an improved MGP is necessary. We next present a stability-boundary-following procedure based on the characterizations of the stability boundary of the reduced-state system described in Theorem 16.11.

A Stability-Boundary-Following Scheme

Step 1. Integrate the postfault reduced-state model (Eq. 17.2) for four or five time-step points of the postfault reduced-state trajectory and term the last point as the current point.

Step 2. Check whether the postfault reduced-state trajectory reaches a relative local minimum of the vector field of the reduced-state model along the trajectory. If yes, then the point with a local minimum of the vector field is the MGP, and then go to the next step; otherwise, go to Step 6.

Step 3. Apply a nonlinear algebraic solver, such as the Newton method, starting from the MGP to compute the reduced-state CUEP.

Step 4. Check the validity of the computed CUEP. Discard it and go to the next step if it is a postfault SEP or if the distance between the computed CUEP and the MGP is large; otherwise, return the computed CUEP as the reduced-state CUEP and stop the procedure.

Step 5. If the number of adjustments in the stability-boundary-following procedure is greater than a threshold, then stop the procedure and go back to the exit-point detection procedure to compute an improved exit point and go to Step 1; otherwise, use the MGP as the initial condition and go to Step 1.

Step 6. Draw a ray connecting the current point on the postfault reduced-state trajectory with the SEP of the postfault reduced-state system.

Step 7. Start from the current point on the postfault reduced-state trajectory and move along the ray to find the point on the ray that has the first local maximum value of the energy function for the postfault original system (Eq. 17.1). Replace the current point by the point found in this step and then go to Step 1.

A practical implementation of Step 7 can be executed by checking the zero crossing of the dot product between the power mismatch and the speed. The process of checking the zero crossing can be sped up by starting the search from the current point on the "trajectory" instead of starting from the postfault SEP. The sign of the dot product at the current point determines the starting direction of the local maximal search.

The theoretical basis for the above stability-boundary-following procedure is the structure of the stability boundary of the reduced-state system, which is composed of the stable manifold of the UEPs lying on the stability boundary. Steps 6 and 7 are based on the computational scheme developed to implement the detection of the exit point. Hence, these two steps serve to ensure the stability-boundary-following procedure moves along the stability boundary of the postfault reduced-state system. Since the CUEP is of type-one (the UEP whose corresponding Jacobian matrix contains only one eigenvalue with a positive real part), it has the nice feature of being a "relative SEP" within the stability boundary.

The stability-boundary-following procedure described above yields a search process close to the stability boundary. Replacing the point generated in Step 1 of the procedure (along the postfault reduced-state trajectory) with the point having a local maximal energy function along the ray generated in Step 7 amounts to confining the MGP search process close to the stability boundary.

A numerical implementation of the stability-boundary-following procedure may encounter oscillation around the CUEP when the norm of the sequence produced by the stability-boundary-following procedure reaches a small number. When the norm is smaller than a threshold value, we stop the procedure and use a robust nonlinear algebraic equation solver, such as the Newton method, using the current point as an initial guess to compute the reduced-state CUEP. The convergence process to the CUEP is usually smooth without numerical difficulties due to the closeness of the MGP to the CUEP.

17.4 COMPUTING THE MGP

From a mathematical viewpoint, the MGP in the BCU method is not a well-defined state. Instead, it is a computational by-product. The MGP is a good initial point for computing the CUEP. Both the coordinates and the energy function value of the MGP are not used in BCU methods. Instead, the information of the MGP is used in the BCU method for the following two purposes:

1. to stop the stability-boundary-following procedure and
2. to start the search for the reduced-state CUEP.

The stability-boundary-following procedure can be computationally intensive. We have observed that this procedure can compute tens of current points. For each current point, the stability-boundary-following procedure requires four to five time steps of the postfault reduced-state trajectory. For achieving computational speed, a variable time-step integration method can be implemented for simulating the postfault reduced-state trajectory. However, there is a trade-off between the accuracy of the simulated postfault reduced-state trajectory and the required computational efforts. In addition, the postfault reduced-state system can be stiff, the variable time-step integration method uses very small time steps for accuracy. This can significantly slow down the stability-boundary-following procedure.

It is hence desirable to develop an integrated scheme for computing the MGP and the reduced-state CUEP with the aim of balancing the trade-off. Since the accuracy of the postfault reduced-state trajectory is not very important, a fast numerical integration method can be applied to the stability-boundary-following procedure for computing the MGP. This computed MGP can be different from the MGP obtained by, say, a variable time-step integration method. The fact that different MGPs exist is not an issue for computing the CUEP since they only serve as the initial condition for a nonlinear algebraic solver for computing the CUEP, which is a unique state.

An ill-conditioned numerical problem may occur during the procedure of computing the MGP using the guideline of the first local minimum of the norm of the vector field. This ill-conditioned situation arises when the norm is very small (close to zero in magnitude), so that the local minimum is caused by the numerical round-off error. A safeguard step against this false detection of the MGP is to add a threshold checking of the variation in norm before checking the value of the norm of the vector field.

17.5 COMPUTATION OF EQUILIBRIUM POINTS

Two types of equilibrium points are required in the CUEP and BCU methods: one is the postfault SEP and the other is the CUEP. The techniques for solving these two types of equilibrium points are basically the same. The only difference is that for the postfault SEP computation, the prefault SEP is used as the initial point, whereas for the CUEP computation, the BCU method uses the MGP point as an initial guess.

In this section, we will present computational procedures for computing these two types of equilibrium points. Since the procedure of computing these two types of equilibrium points involves the same set of nonlinear algebraic equations characterizing the postfault power system, we present numerical methods for computing equilibrium points of large-scale transient stability models. For ease of exposition, the following dynamic equations for transient stability analysis are repeated below.

17.5.1 Motion Dynamic Equations

For generator i, the motion dynamic equations in the center of inertia (COI) coordinates are expressed as follows:

$$\frac{d\delta_i}{dt} = \omega_i \quad \text{and} \tag{17.4}$$

$$\frac{d\omega}{dt} = P_{mi} - P_{ei} - \frac{M_i}{M_T} P_{COI}$$

$$P_{COI} = \sum_i (P_{mi} - P_{ei}). \tag{17.5}$$

17.5.2 Generator Electric Dynamic Equations

The electric dynamic equations of generator i can be expressed as follows:

$$\frac{dX_{ei}}{dt} = f_e(X). \tag{17.6}$$

17.5.3 Excitation System and PSS

The excitation system and PSS of generator i can be expressed as follows:

$$\frac{dX_{exi}}{dt} = f_{ex}(X). \tag{17.7}$$

17.5.4 Network Equations

The network equations of general power systems are composed of real power network equations and reactive power network equations:

$$\begin{aligned} G_P(X) &= 0 \\ G_Q(X) &= 0. \end{aligned} \tag{17.8}$$

17.5.5 Equilibrium Points

The vector field of the postfault dynamical equation at a SEP or a UEP is zero. More specifically, a SEP or a UEP must satisfy the following equations:

$$\begin{aligned} P_{mi} - P_{ei} - \frac{M_i}{M_T} P_{COI} &= 0 \\ f_e(X) &= 0 \\ f_{ex}(X) &= 0 \\ G_P(X) &= 0 \\ G_Q(X) &= 0. \end{aligned} \tag{17.9}$$

If the excitation system is considered, then the effect of electrical dynamic and excitation systems is represented as follows (regardless of which detailed generator model is used). For the purpose of simplicity, we use the simplified one time constant, one gain exciter model as an example to describe these solution algorithms. Solution algorithms for computing UEPs and SEPs of the original exciter model can be similarly developed:

$$E_{fdi} = V_{qi} + I_{di} X_{di}$$

$$E_{fdi} = \begin{cases} E_{ef\max,i} & \text{if } K_{exi}(V_{refi} - V_i) > E_{ef\max,i} \\ E_{ef\max,i} & \text{if } K_{exi}(V_{refi} - V_i) < E_{ef\min,i} \\ K_{exi}(V_{refi} - V_i) & \text{otherwise,} \end{cases} \tag{17.10}$$

where V_{qi} is the q-axis component of the terminal voltage of generator i, I_{di} is the d-axis component of the current of generator i, and K_{exi} is the total gain of the excitation system and PSS of generator i. $E_{fdmax,i}$ and $E_{fdmin,i}$ are the upper and lower limits of the excitation system of generator i, respectively. We will term Equations 17.9 and 17.10 as the set of extended power flow equation.

Algorithm for Computing the SEP

Step 1. Compute E' behind the direct axis transient reactance x'_d from the base case power flow data (i.e., the prefault SEP).

Step 2. Expand the power system network to include the direct axis transient reactance x'_d of all generators. Treat each expanded bus as a PV bus with the corresponding E' as the specified voltage magnitude.

Step 3. Apply the Newton method or its variants to the extended power flow equations.

Step 4. Compute E_{fd} and adjust it for saturation according to Equation 17.10.

Step 5. Compute E' based on the updated E_{fd} from Step 4.

Step 6. Check the convergence criterion based on the computed value of E'. If it converges, then go to the next step; otherwise, go to Step 3.

Step 7. Output the extended power flow solution as the (postfault) SEP (Figure 17.3).

Algorithm for Computing the CUEP

Step 1. Use the voltage magnitude at the MGP as E'.

Step 2. Expand the power system network to include the direct axis transient reactance x'_d of all generators. Treat each expanded bus as a PV bus with the corresponding E' as the specified voltage magnitude.

Step 3. Apply the Newton method or its variants to the extended power flow equations.

Step 4. Compute E_{fd} and adjust it for saturation according to Equation 17.10.

Step 5. Compute E' based on the updated E_{fd} from Step 4.

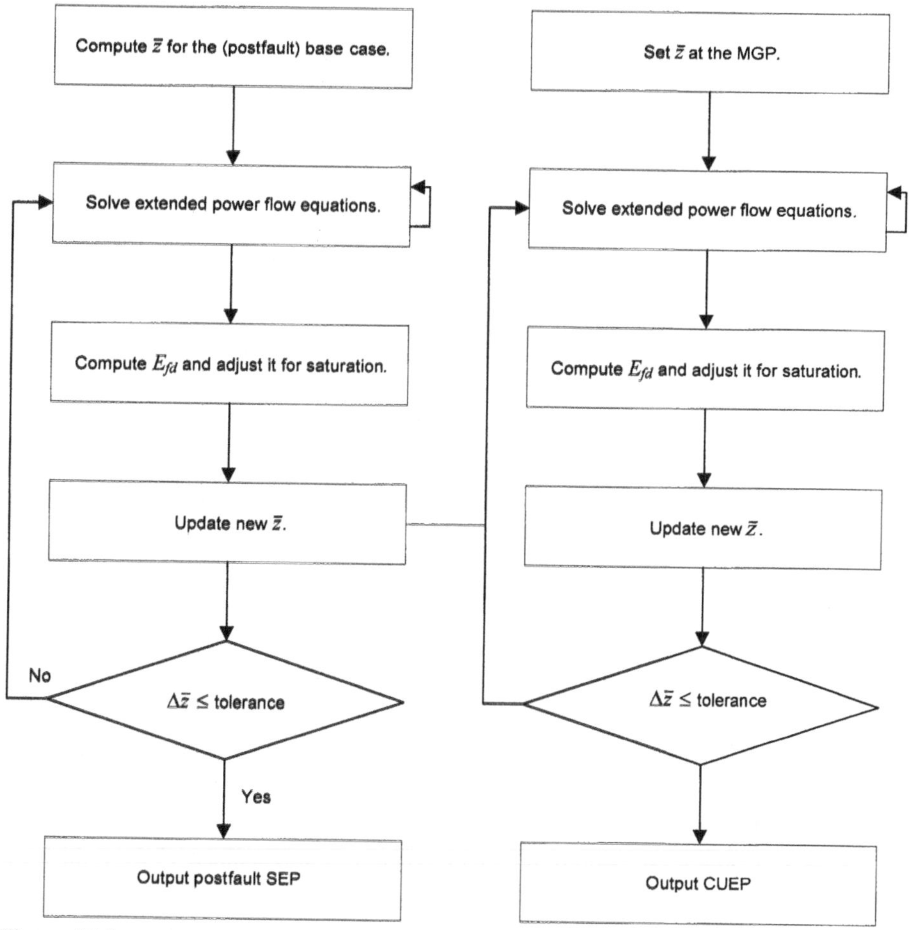

Figure 17.3 Algorithms for computing the SEP and the CUEP of the extended power flow equations.

Step 6. Check the convergence criterion based on the computed value of E'. If it converges, then go to the next step; otherwise, go to Step 3.

Step 7. Output the extended power flow solution as the CUEP.

Step 3 solves the extended power flow equations. This step can be numerically implemented via the following procedure. A generator terminal bus is chosen as the slack bus. The solution procedure for solving the CUEP is described as follows:

Step 1. From the initial MGP, compute the P_{COI}, E'_{qi}, E'_{di}, and E_{fdi} and allocate P_{COI} among the generators (note that P_{mi} is fixed, while P_{ei} is unknown):

$$P_{ei} = P_{mi} - \frac{M_i}{M_T} P_{COI} \quad i = 1, \ldots, n.$$

Step 2. Convert the two-axis generator model into the classical model and treat generator internal buses as PV buses, where the specified P_i is P_{ei} and the specified V_i is the equivalent internal bus voltage E'.

Step 3. Treat the internal buses as PV buses and solve the network equations by using the Newton method.

Step 4. Examine the slack bus's mismatch equation to check whether the extended power flow solution has converged. If yes, go to Step 6; otherwise (this means that the real power mismatch is still large), reallocate it among the n generators by the following equation:

$$P_{ei} = P_{mi} = -\frac{M_i}{M_T}(P_{COI} + P_{Slack}) \quad i = 1, \ldots, n$$

Step 5. Repeat Steps 3 and 4 until the slack bus's mismatch is sufficiently small.

Step 6. Check the equations associated with generator internal buses. If these equations also converge, stop the computation; otherwise, go back to Step 1.

After the postfault SEP and CUEP have been obtained, care should be taken to modify the rotor angles and voltage angles to satisfy the COI reference frame in order to compute the energy function value, in particular, the critical energy value. Note that the computation of the energy margin is performed in COI reference frame. The above two solution algorithms for computing accurate SEPs and UEPs can be slow. To overcome the slowness, improved solution algorithms with only one inner loop are developed as shown below (see Figure 17.4).

Improved Algorithm for Computing the Postfault SEP

Step 1. Compute E' behind the direct axis transient reactance x'_d from the base case power flow data (i.e., the prefault SEP).

Step 2. Expand the power system network to include the direct axis transient reactance x'_d of all generators. Treat each expanded bus as a PV bus with the corresponding E' as the specified voltage magnitude.

Step 3. Solve the extended power flow equations for one iteration.

Step 4. Check the convergence criterion using the current value of E'. If it converges, then go to Step 7; otherwise, go to the next step.

Step 5. Compute E_{fd} and adjust it for saturation according to Equation 17.10.

Step 6. Compute E' based on the updated E_{fd} from Step 5 and go to Step 3.

Step 7. Output the power flow solution as the (postfault) SEP.

Improved Algorithm for Computing the CUEP

Step 1. Use the voltage magnitude at the MGP as E'.

Step 2. Expand the power system network to include the direct axis transient reactance x'_d of all generators. Treat each expanded bus as a PV bus with the corresponding E'as the specified voltage magnitude.

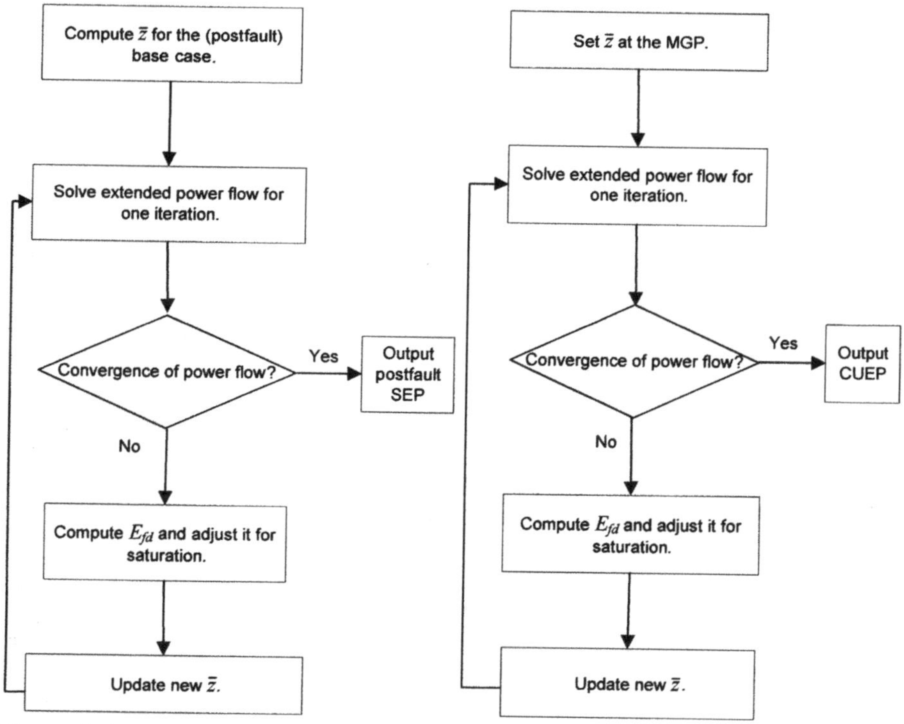

Figure 17.4 Improved algorithms for computing the SEP and the CUEP of the extended power flow equations.

Step 3. Solve the extended power flow equations for one iteration.

Step 4. Check the convergence criterion using the current value of E'. If it converges, then go to Step 7; otherwise, go to the next step.

Step 5. Compute E_{fd} and adjust it for saturation according to Equation 17.10.

Step 6. Compute E' based on the updated E_{fd} from Step 5 and go to Step 3.

Step 7. Output the current power flow solution as the CUEP.

The improved algorithms are fast. For most test cases, only one or two extra iterations are needed as compared to the original solution algorithms.

17.6 NUMERICAL EXAMPLES

In order to illustrate the network-preserving BCU method developed in the previous section in a simple context, we consider the following example, which nearly represents a three-machine system in an absolute angle coordinate. Here, the system model is described by the following DAEs:

$$
\begin{aligned}
\dot{\delta}_1 &= \omega_1 \\
\dot{\delta}_2 &= \omega_2 \\
\dot{\delta}_3 &= \omega_3 \\
m_1\dot{\omega}_1 &= -d_1\omega_1 + P_{m_1} - P_{G_1}(\delta_1, \theta_1, V_1) \\
m_2\dot{\omega}_2 &= -d_2\omega_2 + P_{m_2} - P_{G_2}(\delta_2, \theta_2, V_2) \\
m_3\dot{\omega}_3 &= -d_3\omega_3 + P_{m_3} - P_{G_3}(\delta_3, \theta_3, V_3) \\
0 &= \sum_{k=1}^{9} V_1V_k(G_{1k}\cos(\theta_1-\theta_k)+B_{1k}\sin(\theta_1-\theta_k)) - P_{G_1}(\delta_1, \theta_1, V_1) \\
0 &= \sum_{k=1}^{9} V_2V_k(G_{2k}\cos(\theta_2-\theta_k)+B_{2k}\sin(\theta_2-\theta_k)) - P_{G_2}(\delta_2, \theta_2, V_2) \\
0 &= \sum_{k=1}^{9} V_3V_k(G_{3k}\cos(\theta_3-\theta_k)+B_{3k}\sin(\theta_3-\theta_k)) - P_{G_3}(\delta_3, \theta_3, V_3) \\
0 &= \sum_{k=1}^{9} V_4V_k(G_{4k}\cos(\theta_4-\theta_k)+B_{4k}\sin(\theta_4-\theta_k)) \\
&\cdots \\
&\cdots \\
&\cdots \\
0 &= \sum_{k=1}^{9} V_9V_k(G_{9k}\cos(\theta_9-\theta_k)+B_{9k}\sin(\theta_9-\theta_k)) \\
0 &= \sum_{k=1}^{9} V_1V_k(G_{1k}\sin(\theta_1-\theta_k)+B_{1k}\cos(\theta_1-\theta_k)) - Q_{G_1}(\delta_1, \theta_1, V_1) \\
0 &= \sum_{k=1}^{9} V_2V_k(G_{2k}\sin(\theta_2-\theta_k)+B_{2k}\cos(\theta_2-\theta_k)) - Q_{G_2}(\delta_2, \theta_2, V_2) \\
0 &= \sum_{k=1}^{9} V_3V_k(G_{3k}\sin(\theta_3-\theta_k)+B_{3k}\cos(\theta_3-\theta_k)) - Q_{G_3}(\delta_3, \theta_3, V_3) \\
0 &= \sum_{k=1}^{9} V_4V_k(G_{4k}\sin(\theta_4-\theta_k)+B_{4k}\cos(\theta_4-\theta_k)) \\
&\cdots \\
&\cdots \\
&\cdots \\
0 &= \sum_{k=1}^{9} V_9V_k(G_{9k}\sin(\theta_9-\theta_k)+B_{9k}\cos(\theta_9-\theta_k)),
\end{aligned} \tag{17.11}
$$

where the uniform damping of 0.1 is assigned to each generator:

$$P_{G_i}(\delta, \theta, V) = \begin{cases} \dfrac{E'_{qi} V_i \sin(\delta_i - \theta_i)}{X'_{di}} & \text{if } i = 1, \dots, 3 \\ 0 & \text{if } i = 4, \dots, 9 \end{cases}$$

$$Q_{G_i}(\delta, \theta, V) = \begin{cases} -\dfrac{V_i^2}{X'_{di}} + \dfrac{E'_{qi} V_i \sin(\delta_i - \theta_i)}{X'_{di}} & \text{if } i = 1, \dots, 3 \\ 0 & \text{if } i = 4, \dots, 9. \end{cases}$$

Denote the system equations as

$$\begin{aligned} 0 &= h_i^1(V, \theta, \delta) && \text{—real power balance;} \\ 0 &= h_i^2(V, \theta, \delta) && \text{—reactive power balance;} \\ \dot{\delta}_i &= f_i(w) && \text{—angle dynamics; and} \\ \dot{w}_i &= g_i(V, \theta, \delta, w) && \text{—angular speed dynamics,} \end{aligned}$$

where

$$h_i^1 : \Re^{21} \to \Re, \quad f_i : \Re^3 \to \Re$$

$$h_i^2 : \Re^{21} \to \Re, \quad g_i : \Re^{24} \to \Re.$$

The corresponding reduced-state model for the network-preserving BCU method is

$$\begin{aligned} 0 &= h_i^1(V, \theta, \delta) \\ 0 &= h_i^2(V, \theta, \delta) \\ \dot{\delta}_1 &= P_{m_1} - P_{G_1}(\delta_1, \theta_1, V_1) \\ \dot{\delta}_2 &= P_{m_2} - P_{G_2}(\delta_2, \theta_2, V_2) \\ \dot{\delta}_3 &= P_{m_3} - P_{G_3}(\delta_3, \theta_3, V_3). \end{aligned} \tag{17.12}$$

17.6.1 COI Framework

The three-machine system represented in the COI angle coordinate is expressed as follows:

$$0 = \sum_{k=1}^{9} V_1 V_k \left(G_{1k} \cos(\tilde{\theta}_1 - \tilde{\theta}_k) + B_{1k} \sin(\tilde{\theta}_1 - \tilde{\theta}_k) \right) - P_{G_1}(\tilde{\delta}_1, \tilde{\theta}_1, V_1)$$

$$0 = \sum_{k=1}^{9} V_2 V_k \left(G_{2k} \cos(\tilde{\theta}_2 - \tilde{\theta}_k) + B_{2k} \sin(\tilde{\theta}_2 - \tilde{\theta}_k) \right) - P_{G_2}(\tilde{\delta}_2, \tilde{\theta}_2, V_2)$$

$$
\begin{aligned}
0 &= \sum_{k=1}^{9} V_3 V_k\left(G_{3k}\cos\left(\tilde{\theta}_3-\tilde{\theta}_k\right)+B_{3k}\sin\left(\tilde{\theta}_3-\tilde{\theta}_k\right)\right)-P_{G3}\left(\tilde{\delta}_3,\tilde{\theta}_3,V_3\right)\\
0 &= \sum_{k=1}^{9} V_4 V_k\left(G_{4k}\cos\left(\tilde{\theta}_4-\tilde{\theta}_k\right)+B_{4k}\sin\left(\tilde{\theta}_4-\tilde{\theta}_k\right)\right)\\
&\cdots\\
&\cdots\\
&\cdots\\
0 &= \sum_{k=1}^{9} V_9 V_k\left(G_{9k}\cos\left(\tilde{\theta}_9-\tilde{\theta}_k\right)+B_{9k}\sin\left(\tilde{\theta}_9-\tilde{\theta}_k\right)\right)\\
0 &= \sum_{k=1}^{9} V_1 V_k\left(G_{1k}\sin\left(\tilde{\theta}_1-\tilde{\theta}_k\right)+B_{1k}\cos\left(\tilde{\theta}_1-\tilde{\theta}_k\right)\right)-Q_{G1}\left(\tilde{\delta}_1,\tilde{\theta}_1,V_1\right)\\
0 &= \sum_{k=1}^{9} V_2 V_k\left(G_{2k}\sin\left(\tilde{\theta}_2-\tilde{\theta}_k\right)+B_{2k}\cos\left(\tilde{\theta}_2-\tilde{\theta}_k\right)\right)-Q_{G2}\left(\tilde{\delta}_2,\tilde{\theta}_2,V_2\right)\\
0 &= \sum_{k=1}^{9} V_3 V_k\left(G_{3k}\sin\left(\tilde{\theta}_3-\tilde{\theta}_k\right)+B_{3k}\cos\left(\tilde{\theta}_3-\tilde{\theta}_k\right)\right)-Q_{G3}\left(\tilde{\delta}_3,\tilde{\theta}_3,V_3\right)\\
0 &= \sum_{k=1}^{9} V_4 V_k\left(G_{4k}\sin\left(\tilde{\theta}_4-\tilde{\theta}_k\right)+B_{4k}\cos\left(\tilde{\theta}_4-\tilde{\theta}_k\right)\right)\\
&\cdots\\
&\cdots\\
&\cdots\\
0 &= \sum_{k=1}^{9} V_9 V_k\left(G_{9k}\sin\left(\tilde{\theta}_9-\tilde{\theta}_k\right)+B_{9k}\cos\left(\tilde{\theta}_9-\tilde{\theta}_k\right)\right)\\
\dot{\tilde{\delta}}_1 &= \tilde{W}_1\\
\dot{\tilde{\delta}}_2 &= \tilde{W}_2\\
\dot{\tilde{\delta}}_3 &= \tilde{W}_3\\
M_1\dot{\tilde{\omega}} &= P_{m1}-P_{e1}\left(\tilde{\delta}_1,\tilde{\theta}_1,V_1\right)-\frac{M_1}{M_T}P_{COI}-D_1\omega_1\\
M_2\dot{\tilde{\omega}} &= P_{m2}-P_{e2}\left(\tilde{\delta}_2,\tilde{\theta}_2,V_2\right)-\frac{M_2}{M_T}P_{COI}-D_2\omega_2\\
M_3\dot{\tilde{\omega}} &= P_{m3}-P_{e3}\left(\tilde{\delta}_3,\tilde{\theta}_3,V_3\right)-\frac{M_3}{M_T}P_{COI}-D_3\omega_3,
\end{aligned}
\tag{17.13}
$$

where $\delta_0 = \frac{1}{M_T}\sum_{i=1}^{3} M_i\delta_0$, $w_0 = \frac{1}{M_T}\sum_{i=1}^{3} M_i w_i$, $M_T = \frac{1}{M_T}\sum_{i=1}^{3} M_i$, $\tilde{\delta}_i = \delta_i - \delta_0$ for $i = 1, \ldots, 3$, $\tilde{w}_i = w_i - w_0$ for $i = 1, \ldots, 3$, $\tilde{\theta}_i = \theta_i - \theta_0$ for $i = 1, \ldots, 3$,

$$P_{ei}(\tilde{\delta}) = \sum_{j=1}^{3} E'_{qi}E'_{qj}\left(G_{ij}\cos\left(\tilde{\delta}_i - \tilde{\delta}_j\right) + B_{ij}\sin\left(\tilde{\delta}_i - \tilde{\delta}_j\right)\right) \quad \text{and}$$

$$P_{COI} = \sum_{i=1}^{3} P_{m_i} - \sum_{i=1}^{3}\sum_{j=1}^{3} E'_{qi}E'_{qj}\left(G_{ij}\cos\left(\tilde{\delta}_i - \tilde{\delta}_j\right) + B_{ij}\sin\left(\tilde{\delta}_i - \tilde{\delta}_j\right)\right).$$

The associated reduced-state model is the following:

$$\begin{aligned}
0 &= h^1_{COI_i}\left(V, \tilde{\theta}, \tilde{\delta}\right) \\
0 &= h^2_{COI_i}\left(V, \tilde{\theta}, \tilde{\delta}\right) \\
\dot{\tilde{\delta}}_1 &= P_{m_1} - P_{e1}\left(\tilde{\delta}_1, \tilde{\theta}_1, V_1\right) - \frac{M_1}{M_T}P_{COI} \\
\dot{\tilde{\delta}}_2 &= P_{m_2} - P_{e2}\left(\tilde{\delta}_2, \tilde{\theta}_2, V_2\right) - \frac{M_2}{M_T}P_{COI} \\
\dot{\tilde{\delta}}_3 &= P_{m_3} - P_{e3}\left(\tilde{\delta}_3, \tilde{\theta}_3, V_3\right) - \frac{M_3}{M_T}P_{COI}.
\end{aligned} \tag{17.14}$$

The network-preserving BCU method is applied to a list of 10 contingencies. The prefault SEP is summarized in Table 17.1 The types of faults are three-phase faults with fault locations at both generators and load buses.

Table 17.2 lists these 10 contingencies and the postfault SEP after each contingency. The second line of Table 17.2 states that a three-phase fault occurs at Bus 7, and the postfault system is the system with the transmission line between Buses 7 and 5 tripped due to the openings of circuit breakers at both ends of the line. The

Table 17.1 The Coordinate of the Prefault SEP

E'_{q1}	1.108278	V_1	1.1	θ_1	−0.09302
E'_{q2}	1.107144	V_2	1.097355	θ_2	−0.00761
E'_{q3}	1.060563	V_3	1.08662	θ_3	−0.0363
δ_1	−0.04822	V_4	1.094222	θ_4	−0.136
δ_2	0.125232	V_5	1.071755	θ_5	−0.17357
δ_3	0.11242	V_6	1.084448	θ_6	−0.16252
ω_1	0	V_7	1.1	θ_7	−0.07721
ω_2	0	V_8	1.089489	θ_8	−0.1139
ω_3	0	V_9	1.1	θ_9	−0.0825

Table 17.2 Contingency List and the Postfault SEP after Each Contingency

Contingency no.	Faulted bus	Description of fault clearing		Postfault SEP	Postfault SEP
		From bus	To bus	$[\delta_1, \delta_2, \delta_3]$, $[\theta_1, \theta_2, \ldots, \theta_9]$	$[V_1, V_2, \ldots, V_9]$
1	7	7	5	[−0.1204, 0.3394, 0.2239], [−0.1584, 0.2091, 0.0756, −0.1963, −0.2816, −0.1621, 0.1396, 0.0589, 0.0284]	[1.0781, 1.0845, 1.0669, 1.0511, 1.0066, 1.0402, 1.0802, 1.0657, 1.0740]
2	7	8	7	[−0.0655, 0.2430, 0.0024], [−0.1048, 0.1129, −0.1534, −0.1430, −0.1361, −0.2089, 0.0443, −0.2823, −0.2022]	[1.0930, 1.0940, 1.0538, 1.0801, 1.0547, 1.0615, 1.0947, 1.0235, 1.0568]
3	5	7	5	[−0.1204, 0.3394, 0.2239], [−0.1584, 0.2091, 0.0756, −0.1963, −0.2816, −0.1621, 0.1396, 0.0589, 0.0284]	[1.0781, 1.0845, 1.0669, 1.0511, 1.0066, 1.0402, 1.0802, 1.0657, 1.0740]
4	4	4	6	[−0.0319, 0.0949, 0.0492], [−0.0718, −0.0362, 0.1005, −0.1102, −0.1665, −0.2695, −0.1056, −0.1577, 0.1482]	[1.0986, 1.0898, 1.0654, 1.0910, 1.0672, 1.0205, 1.0884, 1.0710, 1.0720]
5	9	6	9	[−0.0967, 0.2180, 0.2958], [−0.1352, 0.0868, 0.1464, −0.1734, −0.1657, −0.2388, 0.0164, 0.0160, 0.0985]	[1.0836, 1.0805, 1.0614, 1.0617, 1.0360, 1.0294, 1.0742, 1.0600, 1.0667]
6	9	9	8	[−0.0462, 0.0728, 0.2082], [−0.0868, −0.0599, 0.0604, −0.1262, −0.1860, −0.1223, −0.1313, −0.1887, 0.0141]	[1.0942, 1.0786, 1.0802, 1.0825, 1.0563, 1.0710, 1.0716, 1.0457, 1.0915]
7	8	9	8	[−0.0462, 0.0728, 0.2082], [−0.0579, −0.0843, 0.0649, −0.0911, −0.3085, −0.1435, −0.1559, −0.1715, 0.1111]	[1.0942, 1.0786, 1.0802, 1.0825, 1.0563, 1.0710, 1.0716, 1.0457, 1.0915]
8	8	8	7	[−0.0655, 0.2430, 0.0024], [−0.1048, 0.1129, −0.1534, −0.1430, −0.1361, −0.2089, 0.0443, −0.2823, −0.2022]	[1.0930, 1.0940, 1.0538, 1.0801, 1.0547, 1.0615, 1.0947, 1.0235, 1.0568]
9	6	4	6	[−0.0319, 0.0949, 0.0492], [−0.0718, −0.0362, 0.1005, −0.1102, −0.1665, −0.2695, −0.1056, −0.1577, 0.1482]	[1.0986, 1.0898, 1.0654, 1.0910, 1.0672, 1.0205, 1.0884, 1.0710, 1.0720]
10	6	6	9	[−0.0967, 0.2180, 0.2958], [−0.1352, 0.0868, 0.1464, −0.1734, −0.1657, −0.2388, 0.0164, 0.0160, 0.0985]	[1.0836, 1.0805, 1.0614, 1.0617, 1.0360, 1.0294, 1.0742, 1.0600, 1.0667]

Table 17.3 The Computed Exit Point, Minimum Gradient Point, and Controlling Unstable Equilibrium Point (CUEP) of Contingency 1 by the Proposed Network-Preserving BCU Method

	$[\delta_1, \delta_2, \delta_3]$	$[V_1, V_2, \ldots, V_9]$	$[\theta_1, \theta_2, \ldots, \theta_9]$
Exit point	[−0.8387, 2.6561, 0.9392]	[0.9074, 0.7281, 0.5341, 0.7172, 0.6868, 0.4487, 0.5556, 0.3860, 0.3640]	[−0.8499, 2.4488, 0.9226, −0.8663, −0.9516, −0.6509, 2.2331, 1.8114, 0.9069]
MGP	[−0.7614, 1.9630, 1.8063]	[0.8648, 0.9065, 0.6722, 0.6342, 0.6073, 0.3201, 0.8102, 0.6931, 0.5583]	[−0.7589, 1.8323, 1.6007, −0.7547, −0.8401, −0.3364, 1.7393, 1.6179, 1.4750]
CUEP	[−0.7589, 1.9528, 1.8079]	[0.8652, 0.9073, 0.6730, 0.6349, 0.6080, 0.3222, 0.8113, 0.6945, 0.5598]	[−0.7557, 1.8228, 1.5978, −0.7504, −0.8357, −0.3298, 1.7305, 1.6101, 1.4697]

Table 17.4 The Computed Exit Point, Minimum Gradient Point, and CUEP of Contingency 2 by the Proposed Network-Preserving BCU Method

	$[\delta_1, \delta_2, \delta_3]$	$[V_1, V_2, \ldots, V_9]$	$[\theta_1, \theta_2, \ldots, \theta_9]$
Exit point	[−0.7694, 2.4080, 0.9225]	[0.8146, 0.6600, 0.7192, 0.5398, 0.2085, 0.5061, 0.4315, 0.6285, 0.6490]	[−0.7232, 2.3240, 0.5265, −0.6332, −0.5674, −0.3118, 2.2115, 0.2412, 0.3213]
MGP	[−0.5453, 2.1827, −0.3584]	[0.9042, 0.6390, 0.9236, 0.7109, 0.3714, 0.7643, 0.4191, 0.8574, 0.8853]	[−0.5463, 1.9943, −0.5175, −0.5477, −0.3625, −0.6033, 1.7331, −0.6590, −0.5789]
CUEP	[−0.5424, 2.1802, −0.3755]	[0.9045, 0.6391, 0.9241, 0.7114, 0.3721, 0.7647, 0.4195, 0.8577, 0.8856]	[−0.5447, 1.9911, −0.5301, −0.5481, −0.3621, −0.6079, 1.7293, −0.6699, −0.5898]

location of the SEP of the postfault system is $[\delta_1, \delta_2, \delta_3] = [-0.1204, 0.3394, 0.2239]$, $[\theta_1, \theta_2, \ldots, \theta_9] = [-0.1584, 0.2091, 0.0756, -0.1963, -0.2816, -0.1621, 0.1396, 0.0589, 0.0284]$, and $[V_1, V_2, \ldots, V_9] = [1.0781, 1.0845, 1.0669, 1.0511, 1.0066, 1.0402, 1.0802, 1.0657, 1.0740]$.

For each contingency, the network-preserving BCU method computes the three important states: exit point, MGP, and CUEP. These three states of each contingency are summarized in Tables 17.3–17.12, respectively for the list of 10 contingencies.

17.7 LARGE TEST SYSTEMS

The numerical network-preserving BCU method has also been tested on several power systems, modeled by a network-preserving classical generator with static

Table 17.5 The Computed Exit Point, Minimum Gradient Point, and CUEP of Contingency 3 by the Proposed Network-Preserving BCU Method

	$[\delta_1, \delta_2, \delta_3]$	$[V_1, V_2, \ldots, V_9]$	$[\theta_1, \theta_2, \ldots, \theta_9]$
Exit point	[−0.7785, 2.0851, 1.6806]	[0.8669, 0.8913, 0.6679, 0.6383, 0.6113, 0.3260, 0.7911, 0.6725, 0.5470]	[−0.7766, 1.9271, 1.5336, −0.7734, −0.8588, −0.3687, 1.8120, 1.6557, 1.4417]
MGP	[−0.7627, 1.9683, 1.8053]	[0.8646, 0.9061, 0.6718, 0.6338, 0.6070, 0.3190, 0.8096, 0.6924, 0.5575]	[−0.7605, 1.8372, 1.6021, −0.7569, −0.8423, −0.3397, 1.7438, 1.6218, 1.4776]
CUEP	[−0.7589, 1.9528, 1.8079]	[0.8652, 0.9073, 0.6730, 0.6349, 0.6080, 0.3222, 0.8113, 0.6945, 0.5598]	[−0.7557, 1.8228, 1.5978, −0.7504, −0.8357, −0.3298, 1.7305, 1.6101, 1.4697]

Table 17.6 The Computed Exit Point, Minimum Gradient Point, and CUEP of Contingency 4 by the Proposed Network-Preserving BCU Method

	$[\delta_1, \delta_2, \delta_3]$	$[V_1, V_2, \ldots, V_9]$	$[\theta_1, \theta_2, \ldots, \theta_9]$
Exit point	[−0.8298, 2.0463, 2.1662]	[0.8451, 0.7347, 0.8223, 0.5960, 0.2366, 0.7184, 0.5506, 0.6286, 0.7547]	[−0.8191, 1.9147, 1.9717, −0.8002, −0.5119, 1.7625, 1.7766, 1.7952, 1.8838]
MGP	[−0.8263, 2.0693, 2.0900]	[0.8456, 0.7336, 0.8265, 0.5970, 0.2396, 0.7221, 0.5516, 0.6314, 0.7586]	[−0.8142, 1.9211, 1.9138, −0.7929, −0.4953, 1.7132, 1.7658, 1.7654, 1.8345]
CUEP	[−0.8256, 2.0830, 2.0549]	[0.8458, 0.7324, 0.8280, 0.5973, 0.2403, 0.7232, 0.5512, 0.6318, 0.7597]	[−0.8133, 1.9274, 1.8882, −0.7917, −0.4936, 1.6920, 1.7643, 1.7541, 1.8133]

Table 17.7 The Computed Exit Point, Minimum Gradient Point, and CUEP of Contingency 5 by the Proposed Network-Preserving BCU Method

	$[\delta_1, \delta_2, \delta_3]$	$[V_1, V_2, \ldots, V_9]$	$[\theta_1, \theta_2, \ldots, \theta_9]$
Exit point	[−0.5084, 0.5168, 2.8945]	[0.9768, 0.7264, 0.4160, 0.8522, 0.6829, 0.8262, 0.5374, 0.3018, 0.2558]	[−0.5069, 0.3896, 2.5487, −0.5051, −0.3925, −0.5705, 0.2528, 0.5637, 2.0773]
MGP	[−0.7672, 1.6695, 2.4758]	[0.8588, 0.7249, 0.7662, 0.6225, 0.3010, 0.6036, 0.5276, 0.5711, 0.6912]	[−0.7574, 1.6111, 2.1939, −0.7409, −0.4017, −0.8063, 1.5472, 1.7582, 2.0554]
CUEP	[−0.7576, 1.8583, 1.9986]	[0.8530, 0.7528, 0.8390, 0.6116, 0.2914, 0.5929, 0.5806, 0.6515, 0.7739]	[−0.7406, 1.7099, 1.8361, −0.7115, −0.3019, −0.7769, 1.5622, 1.6264, 1.7644]

Table 17.8 The Computed Exit Point, Minimum Gradient Point, and CUEP of Contingency 6 by the Proposed Network-Preserving BCU Method

	$[\delta_1, \delta_2, \delta_3]$	$[V_1, V_2, \ldots, V_9]$	$[\theta_1, \theta_2, \ldots, \theta_9]$
Exit point	[−0.4687, 0.4965, 2.6252]	[0.8977, 0.9257, 0.4509, 0.6996, 0.7115, 0.3752, 0.8589, 0.8382, 0.2713]	[−0.4362, 0.2556, 2.4131, −0.3875, −0.2550, −0.3129, 0.0945, 0.0371, 2.1439]
MGP	[−0.2912, −0.1006, 2.5014]	[0.9326, 0.9937, 0.4548, 0.7665, 0.8090, 0.4446, 0.9422, 0.9195, 0.2990]	[−0.3082, −0.2294, 2.1711, −0.3314, −0.3764, −0.1976, −0.3082, −0.3656, 1.7902]
CUEP	[−0.2910, −0.1011, 2.5008]	[0.9326, 0.9938, 0.4550, 0.7666, 0.8091, 0.4447, 0.9423, 0.9196, 0.2993]	[−0.3080, −0.2298, 2.1701, −0.3312, −0.3764, −0.1971, −0.3085, −0.3659, 1.7890]

Table 17.9 The Computed Exit Point, Minimum Gradient Point, and CUEP of Contingency 7 by the Proposed Network-Preserving BCU Method

	$[\delta_1, \delta_2, \delta_3]$	$[V_1, V_2, \ldots, V_9]$	$[\theta_1, \theta_2, \ldots, \theta_9]$
Exit point	[−0.7753, 1.7588, 2.3490]	[0.7574, 0.7138, 0.5445, 0.4276, 0.2471, 0.1603, 0.5325, 0.5196, 0.3848]	[−0.7373, 1.5592, 2.2012, −0.6441, −0.0386, −0.3873, 1.3423, 1.2849, 2.0696]
MGP	[−0.2936, −0.0920, 2.5014]	[0.9325, 0.9934, 0.4549, 0.7663, 0.8086, 0.4444, 0.9419, 0.9192, 0.2991]	[−0.3095, −0.2225, 2.1713, −0.3314, −0.3742, −0.1976, −0.3024, −0.3598, 1.7906]
CUEP	[−0.2910, −0.1011, 2.5008]	[0.9326, 0.9938, 0.4550, 0.7666, 0.8091, 0.4447, 0.9423, 0.9196, 0.2993]	[−0.3080, −0.2298, 2.1701, −0.3312, −0.3764, −0.1971, −0.3085, −0.3659, 1.7890]

Table 17.10 The Computed Exit Point, Minimum Gradient Point, and CUEP of Contingency 8 by the Proposed Network-Preserving BCU Method

	$[\delta_1, \delta_2, \delta_3]$	$[V_1, V_2, \ldots, V_9]$	$[\theta_1, \theta_2, \ldots, \theta_9]$
Exit point	[−0.7730, 1.7539, 2.3416]	[0.7545, 0.7335, 0.5334, 0.4229, 0.2369, 0.1729, 0.5545, 0.3647, 0.3766]	[−0.7290, 1.5896, 2.1378, −0.6198, 0.0313, −0.2920, 1.4185, 1.8724, 1.9526]
MGP	[−0.3515, 0.0793, 2.5920]	[0.9313, 1.0082, 0.4277, 0.7639, 0.8063, 0.4436, 0.9625, 0.2558, 0.2641]	[−0.3406, −0.0349, 2.2696, −0.3257, −0.2914, −0.2205, −0.1033, 1.7658, 1.8459]
CUEP	[−0.3495, 0.0745, 2.5864]	[0.9318, 1.0088, 0.4284, 0.7647, 0.8074, 0.4451, 0.9633, 0.2576, 0.2660]	[−0.3387, −0.0384, 2.2596, −0.3239, −0.2914, −0.2163, −0.1060, 1.7530, 1.8332]

Table 17.11 The Computed Exit Point, Minimum Gradient Point, and CUEP of Contingency 9 by the Proposed Network-Preserving BCU Method

	$[\delta_1, \delta_2, \delta_3]$	$[V_1, V_2, \ldots, V_9]$	$[\theta_1, \theta_2, \ldots, \theta_9]$
Exit point	[−0.8296, 2.0153, 2.2308]	[0.8455, 0.7351, 0.8171, 0.5967, 0.2381, 0.7136, 0.5494, 0.6251, 0.7497]	[−0.8186, 1.8968, 2.0171, −0.7992, −0.5100, 1.7987, 1.7720, 1.8106, 1.9200]
MGP	[−0.8262, 2.0695, 2.0886]	[0.8456, 0.7335, 0.8266, 0.5970, 0.2397, 0.7222, 0.5517, 0.6314, 0.7587]	[−0.8141, 1.9210, 1.9127, −0.7927, −0.4948, 1.7122, 1.7654, 1.7647, 1.8335]
CUEP	[−0.8256, 2.0830, 2.0549]	[0.8458, 0.7324, 0.8280, 0.5973, 0.2403, 0.7232, 0.5512, 0.6318, 0.7597]	[−0.8133, 1.9274, 1.8882, −0.7917, −0.4936, 1.6920, 1.7643, 1.7541, 1.8133]

Table 17.12 The Computed Exit Point, Minimum Gradient Point, and CUEP of Contingency 10 by the Proposed Network-Preserving BCU Method

	$[\delta_1, \delta_2, \delta_3]$	$[V_1, V_2, \ldots, V_9]$	$[\theta_1, \theta_2, \ldots, \theta_9]$
Exit Point	[−0.7584, 1.8363, 2.0522]	[0.8532, 0.7524, 0.8341, 0.6119, 0.2914, 0.5933, 0.5782, 0.6468, 0.7687]	[−0.7419, 1.6910492, 1.8745, −0.7135, −0.3073, −0.7789, 1.5622, 1.6414, 1.7956]
MGP	[−0.7579, 1.8486, 2.0216]	[0.8531, 0.7527, 0.8370, 0.6117, 0.2915, 0.5931, 0.5797, 0.6497, 0.7718]	[−0.7411, 1.7051, 1.8525, −0.7122, −0.3038, −0.7776, 1.5620, 1.6326, 1.7777]
CUEP	[−0.7576, 1.8583, 1.9986]	[0.8530, 0.7528, 0.8390, 0.6116, 0.2914, 0.5929, 0.5806, 0.6515, 0.7739]	[−0.7406, 1.7099, 1.8361, −0.7115, −0.3019, −0.7769, 1.5622, 1.6264, 1.7644]

nonlinear load representations. The static nonlinear load model ZIP is a combination of constant power, constant impedance, and constant current. The simulation results presented in this section are on the IEEE 50-generator, 145-bus test system (IEEE Committee Report, 1992) and a 202-generator, 1293-bus system (Chiang, 1999).

Table 17.13 displays the estimated CCT of the IEEE 50-generator, 145-bus test system with different locations of three-phase faults using the time–domain method and the network-preserving BCU method. The load model used in this numerical study is a combination of 20% constant power, 20% constant current, and 60% constant impedance. The estimated CCTs by the time–domain method are used as a bench mark. Table 17.13 is explained as follows. The second line of Table 17.13 states that a three-phase fault occurs at Bus 7, and the postfault system is the system with the transmission line between Buses 6 and 7 tripped due to the openings of circuit breakers at both ends of the line. The CCT estimated by the network-preserving BCU method is 0.097 s, while the CCT estimated by the time–domain

Table 17.13 Simulation Results of the Network-Preserving BCU Method on the 50-Generator, 145-Bus IEEE Test System Modeled by the Network-Preserving Model with Nonlinear Static Load Models (20% Constant Power, 20% Constant Current, and 60% Constant Impedance)

Faulted bus	Opened line	Estimated CCTs by the BCU method (s)	Estimated CCTs by the time–domain method (s)	Relative error (%)
7	7–6	0.097	0.103	–5.8
59	59–72	0.208	0.222	–6.3
73	73–74	0.190	0.215	–11.8
112	112–69	0.235	0.248	–5.2
66	66–69	0.156	0.168	–7.1
115	115–116	0.288	0.292	–1.3
110	110–72	0.245	0.260	–5.7
101	101–73	0.232	0.248	–6.4
91	91–75	0.187	0.189	–1.1
6	6–1	0.153	0.170	–10.0
12	12–14	0.163	0.173	–5.8
6	6–10	0.162	0.177	–9.4
66	66–111	0.157	0.172	–9.7
106	106–74	0.170	0.186	–9.6
69	69–32	0.186	0.202	–7.9
69	69–112	0.110	0.118	–6.7
105	105–73	0.191	0.211	–9.4
73	73–75	0.194	0.210	–7.6
67	67–65	0.230	0.231	–0.4
59	59–103	0.221	0.223	–0.9
12	12–14, 12–14	0.156	0.167	–6.5
105	105–73, 105–73	0.110	0.118	–6.7
66	66–8, 66–8	0.167	0.174	–4.0
66	66–111, 66–111, 66–111	0.070	0.080	–12.5
73	73–26, 73–72, 73–82, 73–101	0.192	0.212	–9.4
73	73–69, 73–75, 73–96, 73–109	0.182	0.190	–4.2

simulation method is 0.103 s. The relative error is –5.8%; meaning the estimated CCT by the network-preserving BCU is 5.8% conservative, as compared to the exact CCT. It should be pointed out that in these simulation results, the network-preserving BCU method consistently gives slightly conservative results in estimating the CCTs. The range of conservativeness is between 1.1% and 11.8%. This is consistent with the analytical results of the CUEP method that the critical energy value based on

Table 17.14 Simulation Results of the Network-Preserving BCU Method on the 202-Generator, 1293-Bus IEEE Test System Modeled by the Network-Preserving Model with (1) Constant Impedance Load and (2) Nonlinear Static Load Models (20% Constant Power, 20% Constant Current, and 60% Constant Impedance)

Faulted bus	Opened line	Load model (1)			Load model (2)		
		CCT by time–domain	CCTs by the BCU method	Relative error (%)	CCT by time–domain	CCTs by the BCU method	Relative error (%)
77	77–124	0.325	0.320	−1.6	0.325	0.320	−1.6
74	74–76	0.343	0.313	−9.7	0.336	0.305	−9.2
75	75–577	0.210	0.212	−0.9	0.210	0.212	−0.9
136	136–103	0.262	0.260	−0.7	0.262	0.260	−0.7
248	248–74	0.165	0.160	−3.1	0.165	0.160	−3.1
360	360–345	0.273	0.262	−4.1	0.271	0.261	−3.6
559	559–548	0.215	0.197	−9.3	0.212	0.190	−11.3
634	634–569	0.212	0.197	−7.1	0.205	0.188	−9.2
661	661–669	0.123	0.103	−16.3	0.123	0.103	−16.3
702	702–1376	0.233	0.224	−3.8	0.230	0.222	−3.4
221	221–223	0.220	0.214	−3.2	0.216	0.209	−3.2
175	175–172	0.276	0.269	−3.5	0.272	0.262	−3.6
198	198–230	0.156	0.145	−7.1	0.155	0.145	−6.4
245	245–246	0.230	0.224	−2.6	0.228	0.220	−3.5
319	319–332	0.230	0.198	−13.9	0.229	0.198	−13.5
360	360–362	0.272	0.262	−3.6	0.272	0.261	−4.1

the CUEP gives accurate and yet slightly conservative stability assessments, despite the fact that numerical energy functions were used in these simulations. These simulation results also support the fact that the network-preserving BCU method computes the correct CUEPs of these contingencies.

Table 17.14 summarizes the simulation results of the 202-generator, 1293-bus system. The simulation results are the estimated CCTs with different locations of three-phase faults using the exact time–domain method and the BCU method with the following two load models: (1) constant impedance loads and (2) a nonlinear static load, which is a combination of 20% constant power, 20% constant current, and 60% constant impedance. The estimated CCTs by the time–domain method are used as a bench mark. Table 17.14 is explained as follows. The last line of Table 17.14 states that a three-phase fault occurs at Bus 360, and the postfault system is the system with the transmission line between Buses 360 and 362 tripped due to the openings of circuit breakers at both ends of the line. For the constant impedance load, the CCT estimated by the network-preserving BCU method is 0.272 s, while the CCT estimated by the time–domain simulation method is 0.262 s. The relative error is –3.6%, meaning the estimated CCT by the network-preserving BCU is 3.6%

conservative, as compared to the exact CCT. For the constant ZIP load, the CCT estimated by the network-preserving BCU method is 0.261 s, while the CCT estimated by the time–domain simulation method is 0.272 s. The relative error is –4.1%, meaning the estimated CCT by the network-preserving BCU is 4.1% conservative, as compared to the exact CCT. It should be pointed out that in these simulation results, the network-preserving BCU method consistently gives slightly conservative results in estimating the CCTs. The range of conservativeness is between 0.7% and 16.3% for both the constant impedance load and the constant ZIP load. This is consistent with the analytical results of the CUEP method that the critical energy value based on the CUEP gives accurate and yet slightly conservative stability assessments despite the fact that numerical energy functions were used in these simulations.

In summary, the network-preserving BCU method consistently gives slightly conservative results in estimated CCT, which is in compliance with the CUEP method despite the fact that numerical energy functions were used in these simulations. Moreover, the performance of the network-preserving BCU method seems to be very consistent when applied to different static load models.

17.8 CONCLUDING REMARKS

There are several possible ways to numerically implement the conceptual BCU method for power system transient stability models. This chapter has presented a numerical network-preserving BCU method. This numerical method includes several numerical procedures such as a reliable numerical procedure to detect the exit point, an efficient stability-boundary-following procedure, the MGP and CUEP computation procedure, and the energy value calculation procedure.

We have presented computational procedures for computing these two types of equilibrium points: the postfault SEP and the CUEP. The techniques for solving these two types of equilibrium points are basically the same: solving a large set of constrained nonlinear algebraic equations. The only difference is that for computing the postfault SEP, the prefault SEP is used as the initial point, whereas for computing the CUEP, the BCU method uses the MGP point as the initial condition. After the SEP and the CUEP have been obtained, the rotor angles and voltage angles need to be modified in order to satisfy the COI reference frame since computation of the energy margin is performed in a COI reference frame.

Numerical studies indicate that the network-preserving BCU method consistently gives slightly conservative results for the estimated CCT. This conservative nature is in compliance with the CUEP method despite the fact that numerical energy functions were used in these simulations. In addition, the level of conservativeness and the computational performance of the network-preserving BCU method seem to be very consistent when applied to different static load models.

Chapter 18

Numerical Studies of BCU Methods from Stability Boundary Perspectives

The BCU method computes the reduced-state controlling unstable equilibrium point (UEP) and explores the dynamic relationship between the controlling UEP of the original system and the reduced-state controlling UEP of the reduced-state system to achieve its key objective: computing the controlling UEP of the original system. Hence, the dynamic relationship and computation of the reduced-state controlling UEP of the reduced-state system play a key role in the success of the BCU method. In other words, the successful computation of the controlling UEP by the BCU method depends on the computation of the reduced-state controlling UEP and the correspondence between the reduced-state controlling UEP and the controlling UEP.

In this chapter, we present numerical studies on the BCU method computational procedure for computing the reduced-state controlling UEP of the reduced-state system and for computing the dynamic relationship between the controlling UEP of the original system and the reduced-state controlling UEP. The computational procedure of the BCU method will be numerically demonstrated using the stability boundary of the original system and that of the reduced-state system. These numerical studies also shed light on dynamic property (D2) and lead to an improved version of the dynamic property (D2) needed in the BCU method.

18.1 INTRODUCTION

In developing a BCU method for a given power system stability model, an associated reduced-state model must first be defined. We consider the general network-preserving transient stability model repeated below:

Direct Methods for Stability Analysis of Electric Power Systems, by Hsiao-Dong Chiang

$$
\begin{aligned}
0 &= -\frac{\partial U}{\partial u}(u, w, x, y) + g_1(u, w, x, y) \\
0 &= -\frac{\partial U}{\partial w}(u, w, x, y) + g_2(u, w, x, y) \\
T\dot{x} &= -\frac{\partial U}{\partial x}(u, w, x, y) + g_3(u, w, x, y) \\
\dot{y} &= z \\
M\dot{z} &= -Dz - \frac{\partial U}{\partial y}(u, w, x, y) + g_4(u, w, x, y),
\end{aligned}
\tag{18.1}
$$

where $U(u, w, x, y)$ is a scalar function. Regarding the original model (Eq. 18.1), we choose the following differential–algebraic system as the associated reduced-state model:

$$
\begin{aligned}
0 &= -\frac{\partial U}{\partial u}(u, w, x, y) + g_1(u, w, x, y) \\
0 &= -\frac{\partial U}{\partial w}(u, w, x, y) + g_2(u, w, x, y) \\
T\dot{x} &= -\frac{\partial U}{\partial x}(u, w, x, y) + g_3(u, w, x, y) \\
\dot{y} &= -\frac{\partial U}{\partial y}(u, w, x, y) + g_4(u, w, x, y).
\end{aligned}
\tag{18.2}
$$

It has been shown in the previous chapters that, under certain conditions, the original system (Eq. 18.1) and the reduced-state system (Eq. 18.2) satisfy the static properties (S1) and (S2) as well as the dynamic property (D2). Moreover, the reduced-state system must have dynamic properties (D1) and (D3).

We perform numerical studies to examine the above static and dynamic properties from the perspectives of stability regions and stability boundaries. These numerical studies are performed for the following purposes:

1. to reveal the dynamic relationship between the stability boundary of the original system and the stability boundary of the reduced-state system,
2. to shed light on the algorithmic aspects of the BCU method from the viewpoint of state spaces,
3. to examine how well the dynamic property (D3) is implemented in the BCU method, and
4. to illustrate how the dynamic property (D2) is explored in the BCU method to compute the controlling UEP of the original system via computation of the reduced-state controlling UEP of the reduced-state system.

18.2 STABILITY BOUNDARY OF NETWORK-REDUCTION MODELS

We perform numerical studies on the static and dynamic relationships of the network-reduction model of a nine-bus test system with several contingencies. We first consider Contingency 7, where the faulted bus is #9 and the line between Buses 6 and 9 is tripped to clear the fault. Hence, the fault-on and postfault systems are specified and the corresponding reduced-state fault-on and postfault systems are also specified.

The static relationships between the postfault (original) system and the postfault reduced-state system are summarized in Table 18.1. It can be seen that, within the state space of interest, the static relationships (S1) and (S2) are satisfied.

Next, the following key dynamic objects of the BCU method are computed.

For the Original Model

- The fault-on trajectory
- The point whose projection on the state space of the reduced-state model is the exit point
- The controlling UEP

For the Reduced-State Model

- The projected fault-on trajectory
- The exit point (lying in the state space of the reduced-state system)
- The reduced-state controlling UEP (relative to the projected fault-on trajectory)

The exit point, controlling UEP, and the reduced-state controlling UEP relative to the projected fault-on trajectory are summarized in Table 18.2. It is clear that the reduced-state controlling UEP computed by the BCU method corresponds to the

Table 18.1 The Static Relationship between the Postfault (Original) System and the Postfault Reduced-State System under Contingency 7

Reduced-state system		Original system	
	$(\delta_1, \delta_2, \delta_3)$		$(\delta_1, \delta_2, \delta_3, \omega_1, \omega_2, \omega_3)$
Prefault SEP	(−0.0482, 0.1252, 0.1124)	Prefault SEP	(−0.0482, 0.1252, 0.1124, 0, 0, 0)
Postfault SEP	(−0.09669, 0.218, 0.2959)	Postfault SEP	(−0.09669, 0.218, 0.2959, 0, 0, 0)
Type-one UEP #1	(−0.7576, 1.858, 1.9995)	Type-one UEP #1	(−0.7576, 1.858, 1.9995, 0, 0, 0)
Type-one UEP #2	(1.604, −2.064, −8.209)	Type-one UEP #2	(1.604, −2.064, −8.209, 0, 0, 0)

Table 18.2 The Relationship between Exit Points and the Controlling UEP of the Postfault (Original) System and Those of the Postfault Reduced-State System under Contingency 7

Reduced-state system		Original system	
	$(\delta_1, \delta_2, \delta_3)$		$(\delta_1, \delta_2, \delta_3, \omega_1, \omega_2, \omega_3)$
Exit point	(−0.5085, 0.5169, 2.8945)	Exit point in the original state space	(−0.5085, 0.5169, 2.8945, −2.3737, 1.1341, 16.2316)
Reduced-state controlling UEP	(−0.7576, 1.858, 1.9995)	Controlling UEP	(−0.7576, 1.858, 1.9995, 0, 0, 0)

controlling UEP relative to the fault-on trajectory. We note that the exit point is the intersection point between the stability boundary of the reduced-state system and the projected fault-on trajectory; hence, its dimension is the same as that of the reduced-state system. The corresponding point of the exit point on the original fault-on trajectory is also shown in this table.

The existence of an energy function for the reduced-state system ensures that the dynamic relationship (D1) is satisfied. We next examine the dynamic properties (D2) and (D3) using the numerical simulation of the stability boundary of the reduced-state system and the stability boundary of the original system using the time–domain simulation approach. In particular, the following stability boundaries and the associated dynamic entities will be simulated.

For the Original Model (See Figure 18.1)

- The intersecting stability region, which is the intersection between the stability region of the original system and its angle subspace
- The equilibrium points (of the original system) lying on the intersecting stability boundary
- The prefault stable equilibrium point (SEP), the postfault SEP, and the exit point lying on the angle subspace

For the Reduced-State Model (See Figure 18.2)

- The stability boundary of the reduced-state system
- The projected fault-on trajectory
- The controlling UEP relative to the projected fault-on trajectory
- The equilibrium points (of the reduced-state system) lying on the stability boundary
- The prefault and postfault SEPs of the reduced-state system

From these numerical simulations, we have the following observations. There is only one type-one UEP lying on the stability boundary of the original system. This UEP is also the controlling UEP of the fault-on trajectory, and it is noted that

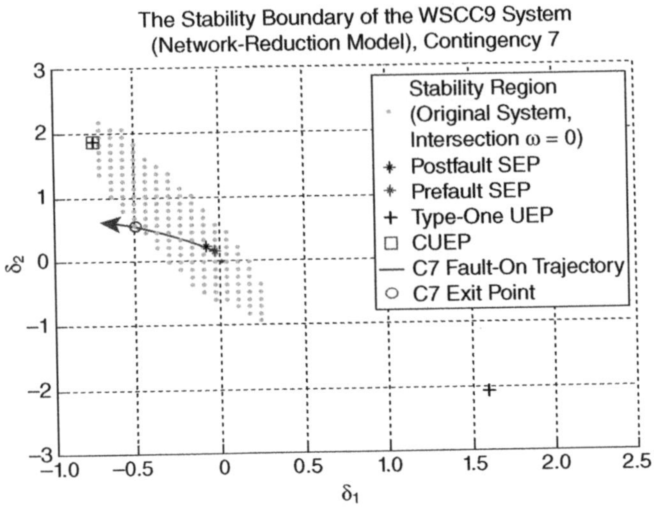

Figure 18.1 The relationship among the (projected) fault-on trajectory, exit point, controlling UEP, intersecting stability region of the original system, and other type-one UEPs lying outside the stability region (for Contingency 7).

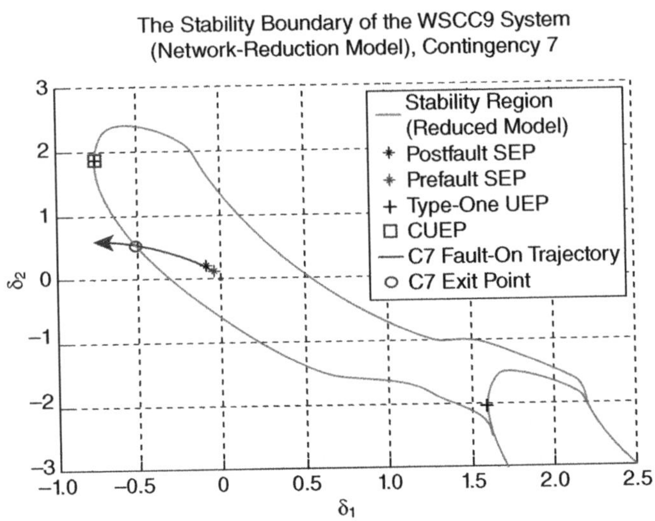

Figure 18.2 The projected fault-on trajectory, exit point, reduced-state controlling UEP (i.e., controlling UEP relative to the projected fault-on trajectory), stability boundary of the reduced-state system, and other type-one UEPs lying on the stability boundary (for Contingency 7).

the prefault SEP lies inside the stability region of the postfault SEP. Hence, the fundamental assumption of the controlling UEP method is satisfied. The projected fault-on trajectory intersects the stability boundary of the reduced-state system at the exit point (see Figure 18.2).

It can be seen that the computed exit point lies near the stability boundary of the reduced-state system (see Figure 18.2). Thus, the direct scheme for detecting the exit point and the construction of the reduced-state model satisfy the dynamic property (D3). Indeed, the computed exit point is located very close to the exact location of the exit point, which is the intersection between the projected fault-on trajectory and the stability boundary of the reduced-state system. There are two type-one equilibrium points lying on the stability boundary of the reduced-state system, while there is only one type-one equilibrium point on the stability boundary of the original system. Hence, the dynamic property (D2) is not satisfied. However, the controlling UEP relative to this contingency, the only type-one equilibrium point lying on the stability boundary of the original system (see Figure 18.1), indeed corresponds to the type-one equilibrium point lying on the stability boundary of the reduced-state system.

Even though the dynamic property (D2) is not satisfied for Contingency 7, the BCU method still computes the correct controlling UEP. The reason for is that the reduced-state controlling UEP satisfies the boundary property (i.e., the reduced-state controlling UEP lies on the stability boundary of the reduced-state system if and only if the corresponding controlling UEP lies on the stability boundary of the original system). The other UEPs do not need to satisfy the boundary property. We emphasize that the dynamic property (D2) is only a sufficient condition for the UEP computed by the BCU method lying on the stability boundary of the original system. It is not a necessary condition. Hence, the dynamic property (D2) can be replaced by the following without damaging the ability of the BCU method to compute the correct controlling UEP:

Dynamic Property (D2′): the reduced-state controlling UEP lies on the stability boundary of the reduced-state system if and only if the corresponding controlling UEP lies on the stability boundary of the original system.

To illustrate how the dynamic relationship is explored in the BCU method for computing the controlling UEP of the original system, we employ the simulated fault-on trajectory, the intersecting stability region of the original system along with the UEPs lying on its boundary, and the stability boundary of the reduced-state system along with the UEPs lying on its boundary in Figure 18.3. As is clearly shown in the figure, after successfully computing the exit point, the BCU method computes the reduced-state controlling UEP by "moving" along the stable manifold of the reduced-state controlling UEP which corresponds to the controlling UEP of the original system.

Regarding the relationship between the stability boundary of the original system and the stability boundary of the reduced-state system, we have the following observations:

- For Contingency 7, the intersection between the stability boundary of the original system and the angle space does not lie in a small neighborhood of the stability boundary of the reduced-state system.
- For Contingency 7, the intersection between the relevant stability boundary of the original system and the angle space lies in a small neighborhood of the relevant stability boundary of the reduced-state system.

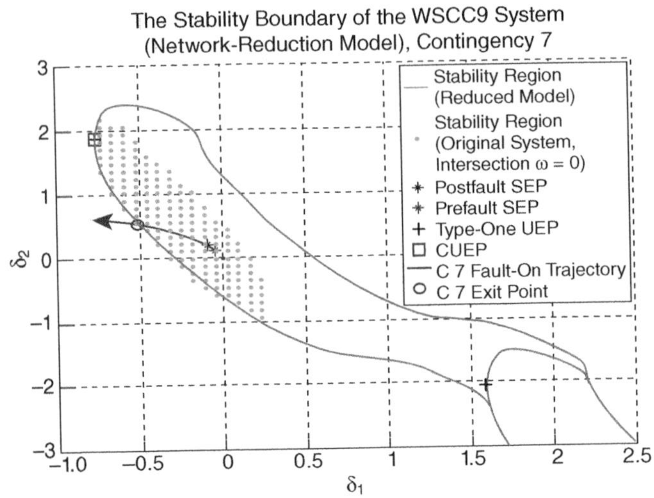

Figure 18.3 The projected fault-on trajectory, exit point, reduced-state controlling UEP, controlling UEP, stability region, stability boundary of the reduced-state system containing two type-one UEPs on the stability boundary, and the stability boundary of the original system on the angle subspace (for Contingency 7).

The relevant stability boundary depends on the fault-on trajectory. The relevant stability boundary of the reduced-state system is the stable manifold of the reduced-state controlling UEP relative to the projected fault-on trajectory. The relevant stability boundary of the original system is the stable manifold of the controlling UEP relative to the fault-on trajectory. One sufficient condition for computing the correct controlling UEP using the BCU method is the following. The relevant stability boundary of the reduced-state system and the relevant stability boundary of the original system and the angle space must be sufficiently close.

We next consider Contingency 12, which is specified by the events that the faulted bus is #6 and that the line between Buses 6 and 9 is tripped to clear the fault. Hence, both the fault-on and postfault systems and their corresponding reduced-state fault-on and postfault systems are also specified. Comparing Contingency 12 with Contingency 7, the corresponding two postfault systems are the same and therefore have the same postfault stability regions. In addition, since the corresponding fault-on systems are different, these two contingencies lead to two different exit points. For Contingency 12, we numerically illustrate the following relationships:

- for the original model, the prefault SEP, postfault SEP, (projected) fault-on trajectory, exit point, controlling UEP, and intersecting stability region of the original system on the angle subspace (see Figure 18.4); and
- for the reduced-state model, the stability boundary of the reduced-state system, projected fault-on trajectory, exit point, reduced-state controlling UEP, and the other type-one UEP lying on the boundary (see Figure 18.5). A collection of these simulated objects is shown in Figures 18.4–18.6.

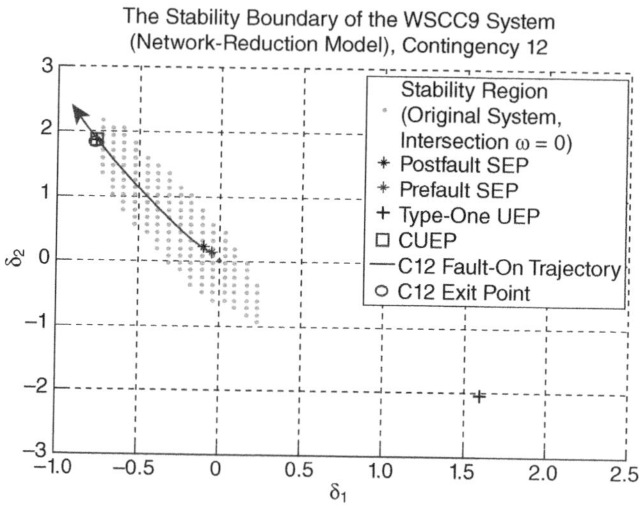

Figure 18.4 The relationship among the projected fault-on trajectory, exit point, controlling UEP, another off-stability-boundary type-one UEP of the original system, and the intersecting stability region of the original system on the angle subspace (for Contingency 12). The prefault and postfault SEPs are also listed.

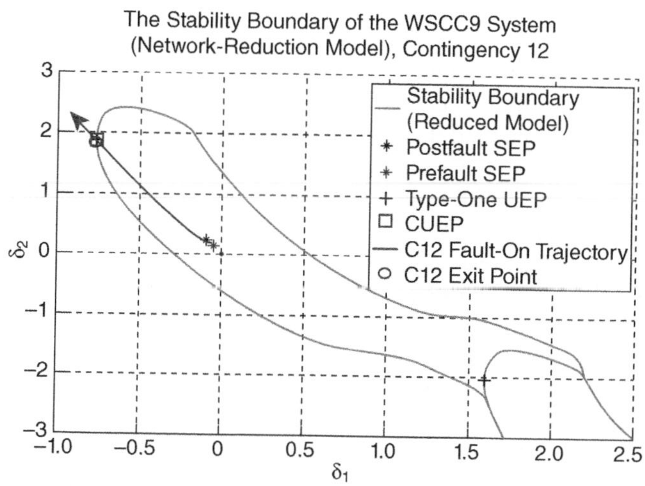

Figure 18.5 The projected fault-on trajectory, exit point, reduced-state controlling UEP, stability boundary of the reduced-state system, and other type-one UEPs on the stability boundary (for Contingency 12). Since the exit point is very close to the reduced-state controlling UEP, these two points are unclearly separated.

We observe from Figure 18.5 that the fault-on trajectory projected on the angle space intersects the stability boundary of the reduced-state system at the exit point. On the other hand, the computed exit point by the BCU method lies very close to the exit point. Hence, the dynamic property (D3) is satisfied. The computed exit

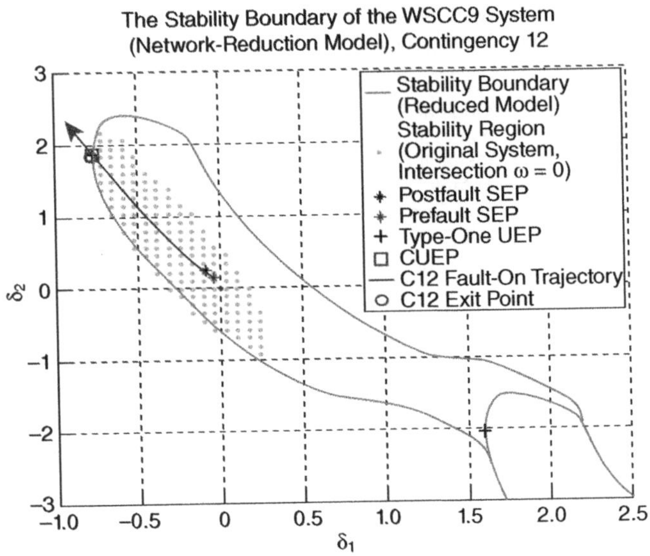

Figure 18.6 The (projected) fault-on trajectory, exit point, controlling UEP, stability boundary of the reduced-state system, other type-one UEPs on the stability boundary, and the stability region of the original system on the angle subspace (for Contingency 12).

point lies near the intersecting stability boundary of the original system on the angle subspace, and it is very close to the reduced-state controlling UEP. Starting from the computed exit point, the BCU method then computes the reduced-state controlling UEP by moving along the stable manifold in which the computed exit point is located. There is only one type-one equilibrium point lying on the stability boundary of the original system, which is the controlling UEP of the original system. Again, the reduced-state controlling UEP corresponds to the controlling UEP of the original system. It is noted that the prefault SEP lies inside the stability region of the postfault SEP. Hence, the fundamental assumption of the controlling UEP method is satisfied.

While the dynamic property (D2) is not satisfied for this case, the dynamic property (D2′) is satisfied. The BCU method explores this dynamic property to compute the correct controlling UEP. Indeed, the dynamic property (D2′), instead of the dynamic property (D2), ensures the correctness of the BCU method when computing the controlling UEP.

18.3 NETWORK-PRESERVING MODEL

We now perform a numerical study on the network-preserving model of the nine-bus test system with contingencies. The dimension of the state space for the nine-bus network-preserving model is 21. The focus of this numerical study is on the static and dynamic properties between the reduced-state system and the original system.

Table 18.3 The Static Relationships between the Network-Preserving Model of the Postfault (Original) System and That of the Postfault Reduced-State System under Contingency 2

Reduced-state system		Original system	
$\begin{pmatrix} \delta_1, \delta_2, \delta_3, V_1, V_2, V_3, V_4, V_5, V_6, V_7 \\ V_8, V_9, \theta_1, \theta_2, \theta_3, \theta_4, \theta_5, \theta_6, \theta_7, \theta_8, \theta_9 \end{pmatrix}$		$\begin{pmatrix} \delta_1, \delta_2, \delta_3, \omega_1, \omega_2, \omega_3, V_1, V_2, V_3, V_4, V_5, V_6 \\ V_7, V_8, V_9, \theta_1, \theta_2, \theta_3, \theta_4, \theta_5, \theta_6, \theta_7, \theta_8, \theta_9 \end{pmatrix}$	
Prefault SEP	(−0.0482, 0.1252, 0.1124, 1.1, 1.0974, 1.0866, 1.0942, 1.0718, 1.0844, 1.1, 1.0895, 1.1, −0.093, −0.0076, −0.0363, −0.136, −0.1736, −0.1625, −0.0772, −0.1139, −0.0825)	Prefault SEP	(−0.0482, 0.1252, 0.1124, 0, 0, 0, 1.1, 1.0974, 1.0866, 1.0942, 1.0718, 1.0844, 1.1, 1.0895, 1.1, −0.093, −0.0076, −0.0363, −0.136, −0.1736, −0.1625, −0.0772, −0.1139, −0.0825)
Postfault SEP	(−0.0655, 0.243, −0.0024, 1.093, 1.094, 1.0538, 1.0801, 1.0547, 1.0615, 1.097, 1.0235, 1.0568, −0.1048, 0.1129, −0.1534, −0.143, −0.1361, −0.2089, 0.0443, −0.2823, −0.2022)	Postfault SEP	(−0.0655, 0.243, −0.0024, 0, 0, 0, 1.093, 1.094, 1.0538, 1.0801, 1.0547, 1.0615, 1.0947, 1.0235, 1.0568, −0.1048, 0.1129, −0.1534, −0.143, −0.1361, 0.2089, 0.0443, −0.2823, −0.2022)

We first consider Contingency 2, which is specified such that the faulted bus is #7 and that the line between Buses 7 and 8 is tripped to clear the fault. Hence, the fault-on and postfault systems are specified and the corresponding reduced-state fault-on and postfault systems (whose state space has 21 dimensions) are also specified.

The static relationships between the SEPs of the prefault and postfault (original) systems and the SEP of the prefault and postfault reduced-state systems are summarized in Table 18.3. It is clear that, within the state space of interest, the static relationship (S1) is satisfied. It is also shown that the static relationship (S2) is satisfied.

Next, the following key dynamic objects of the BCU method are computed:

- the fault-on trajectory and the projected fault-on trajectory,
- the exit point and the point lying on the fault-on trajectory whose projection is the exit point, and
- the reduced-state controlling UEP and the controlling UEP.

The exit point, reduced-state controlling UEP, and controlling UEP are summarized in Table 18.4. It is clear that the computed reduced-state controlling UEP computed by the BCU method corresponds to the controlling UEP. Hence, the BCU method computes the correct controlling UEP.

Table 18.4 The Exit Point, the Point on the Fault-On Trajectory Whose Projected Point Is the Exit Point, the Controlling UEP, and the Reduced-State Controlling UEP

Reduced-state system		Original system	
$\begin{pmatrix} \delta_1, \delta_2, \delta_3, V_1, V_2, V_3, V_4, V_5, V_6, V_7 \\ V_8, V_9, \theta_1, \theta_2, \theta_3, \theta_4, \theta_5, \theta_6, \theta_7, \theta_8, \theta_9 \end{pmatrix}$		$\begin{pmatrix} \delta_1, \delta_2, \delta_3, \omega_1, \omega_2, \omega_3, V_1, V_2, V_3, V_4, V_5, V_6 \\ V_7, V_8, V_9, \theta_1, \theta_2, \theta_3, \theta_4, \theta_5, \theta_6, \theta_7, \theta_8, \theta_9 \end{pmatrix}$	
Exit point	(−0.7418, 2.3101, 0.914, 0.8201, 0.6643, 0.7346, 0.5527, 0.2331, 0.5281, 0.4432, 0.6475, 0.6685, −0.6807, 2.1883, 0.5202, −0.5644, −0.3569, 0.2664, 2.0294, 0.2421, 0.3222)	The point whose projection is the exit point	(−0.7418, 2.3101, 0.914, 3.0407, 10.7634, 0.9956, 0.8201, 0.6643, 0.7346, 0.5527, 0.2331, 0.5281, 0.4432, 0.6475, 0.6685, −0.6807, 2.1883, 0.5202, −0.5644, −0.3569, −0.2664, 2.0294, 0.2421, 0.3222)
Reduced-state CUEP	(−0.5424, 2.1802, −0.3755, 0.9045, 0.6391, 0.9241, 0.7114, 0.3721, 0.7647, 0.4195, 0.8577, 0.8856, −0.5447, 1.9911, −0.5301, −0.5481, −0.3621, −0.6079, 1.7293, −0.6699, −0.5898)	CUEP	(−0.5424, 2.1802, −0.3755, 0.0, 0.0, 0.0, 0.9045, 0.6391, 0.9241, 0.7114, 0.3721, 0.7647, 0.4195, 0.8577, 0.8856, −0.5447, 1.9911, −0.5301, −0.5481, −0.3621, −0.6079, 1.7293, −0.6699, −0.5898)

We next illustrate how the dynamic relationship is explored in the BCU method to compute the controlling UEP. We will numerically study the stability boundary of the reduced-state system and the stability boundary of the original system. In particular, the following stability boundaries are simulated:

- the intersecting stability boundary, which is the intersection between the stability boundary of the original system and its angle subspace, and the equilibrium points (of the original system) lying on the intersecting stability boundary (see Figure 18.7), and
- the stability boundary of the reduced-state system and the equilibrium points (of the reduced-state system) lying on the boundary (see Figure 18.8).

The fault-on trajectory projected on the angle space intersects the stability boundary of the reduced-state system at the exit point (see Figure 18.8). The computed exit point lies very close to the stability boundary of the reduced-state system. Hence, the procedure for exit point detection in the BCU method satisfies the dynamic property (D3).

There are two type-one equilibrium points lying on the stability boundary of the original system, one of which is the controlling UEP of the fault-on trajectory. In addition, there is one type-two UEP lying on the stability boundary. Of the two type-one UEPS lying on the stability boundary, the fault-on trajectory is closer to the controlling UEP. It is noted that the prefault SEP lies inside the stability region

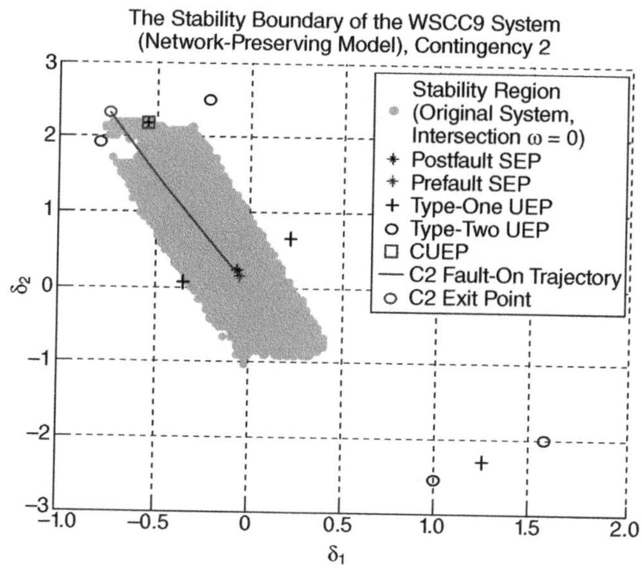

Figure 18.7 The intersecting stability region, (projected) fault-on trajectory, exit point, controlling UEP, and other UEPs on and off the stability boundary (for Contingency 2).

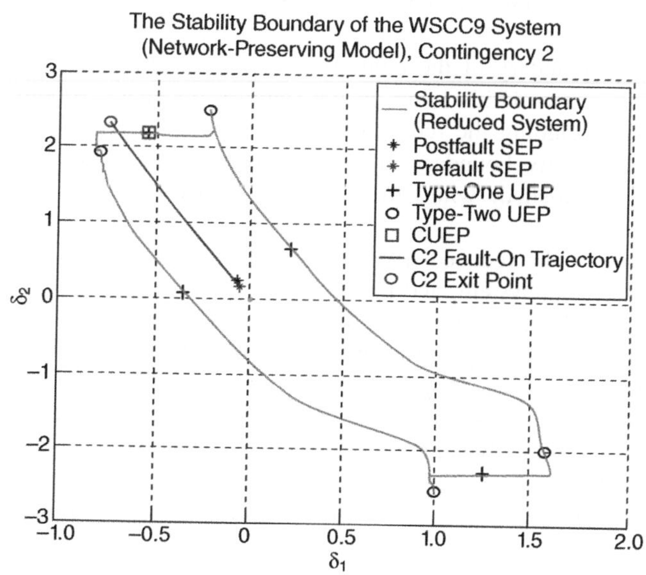

Figure 18.8 The projected fault-on trajectory, exit point, reduced-state controlling UEP, stability boundary of the reduced-state system, and other type-one UEPs and type-two UEPs on the stability boundary (for Contingency 12).

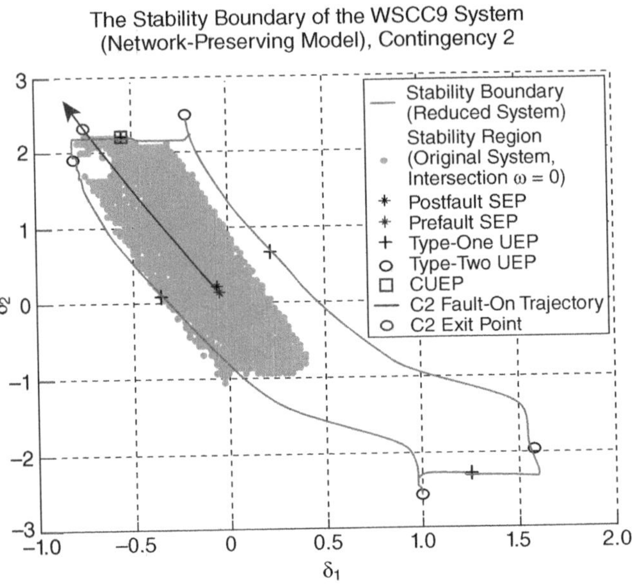

Figure 18.9 The (projected) fault-on trajectory, exit point, controlling UEP, stability boundary of the reduced-state system, other type-one and type-two UEPs on the stability boundary, and the stability region of the original system on the angle subspace (for Contingency 2).

of the postfault SEP. Hence, the fundamental assumption of the controlling UEP method is satisfied.

Regarding the stability boundary of the reduced-state system, there are four type-one UEPs and four type-two UEPs lying on the stability boundary. Among them, two type-one UEPs, including the reduced-state controlling UEP, and one type-two UEP lie on the stability boundary of the original system. Hence, the dynamic property (D2′) is satisfied, while the dynamic property (D2) is not satisfied.

The stability boundary of the reduced-state system, the stability region, the stability boundary of the original system, and the projected fault-on trajectory are presented in Figure 18.9. For this case, the controlling UEP indeed corresponds to the reduced-state controlling UEP computed by the BCU method, and this correspondence is used to compute the controlling UEP. Again, even though the dynamic property (D2) is not satisfied for Contingency 2, the BCU method still computes the correct controlling UEP because the dynamic property (D2′) is satisfied.

Regarding the relationship between the stability boundary of the original system and the stability boundary of the reduced-state system, we have the following observations:

- For Contingency 2, the intersection between the stability boundary of the original system and the angle space does not lie in a small neighborhood of the stability boundary of the reduced-state system.

- For Contingency 2, the intersection between the relevant stability boundary of the original system and the angle space lies in a small neighborhood of the relevant stability boundary of the reduced-state system.

18.4 ONE DYNAMIC PROPERTY OF THE CONTROLLING UEP

It should be stressed that the dynamic property of controlling UEP must be explored so that the controlling UEP can be correctly computed; otherwise, a UEP other than the controlling UEP will be computed. As we observe in the numerical studies performed in the previous sections, the exit point may be very close to one equilibrium point and far away from the controlling UEP. In this case, computing the controlling UEP using a nonlinear algebraic solver starting from the exit point will fail.

In the literature, several attempts have been made to compute the controlling UEP using the Newton method starting from the exit point. These attempts did not consider the dynamic properties of controlling UEP. Consequently, these attempts compute incorrect controlling UEPs when the exit points are far away from controlling UEPs. We next illustrate that the exit point can be far away from the controlling UEP on both network-reduction and network-preserving models.

We consider Contingency 10, which is specified by the events such that the faulted bus is #8 and that the line between Buses 7 and 8 is tripped to clear the fault. Hence, the fault-on and postfault systems are specified, and the corresponding reduced-state fault-on and postfault systems are also specified. The static relationships (S1) and (S2) between the postfault (original) system and the postfault reduced-state system are satisfied.

For this contingency, the fault-on trajectory projected on the angle space intersects the stability boundary of the reduced-state system at the exit point. The exit point lies near a type-two UEP, which is not the controlling UEP. In fact, the exit point lies far away from the controlling UEP, which is a type one UEP. Indeed, the coordinate of the exit point (on the angle-space) is (–0.7751, 1.7587, 2.3478), and it is close to the type-two UEP (–0.7919, 1.907, 2.1647), which lies on the stability boundaries of both the reduced-state system and the original system. The coordinate of the controlling UEP is (–0.3495, 0.0745, 2.5865, 0, 0, 0), which is far away from the exit point.

It is obvious that starting from the exit point, the Newton method or its variants will not converge to the controlling UEP. This failure is due to the fact that the dynamic property of controlling UEP is not incorporated into the underlying solution method. By contrast, the BCU method explores this property to find the correct controlling UEP.

For the stability boundary of the reduced-state system, we have the following observations (see Figure 18.10):

- There are four type-one equilibrium points lying on the boundary.
- There are four type-two equilibrium points lying on the boundary.

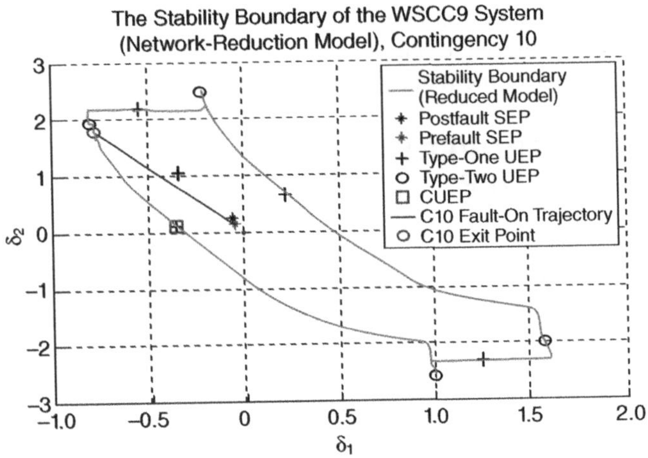

Figure 18.10 The (projected) fault-on trajectory, exit point, controlling UEP, prefault SEP, postfault SEP, stability boundary of the reduced-state system, and other type-one UEPs on the stability boundary (for Contingency 10).

- Each type-two equilibrium point lies on the boundary of the stable manifolds of two type-one equilibrium points.
- The stability boundary of the reduced-state system equals the union of the stable manifolds of these four type-one equilibrium points and the stable manifolds of these four type-two equilibrium points (i.e., the type-two equilibrium points themselves).

For the stability boundary of the original system, we have the following observations (see Figure 18.11):

- There are only two type-one equilibrium points and one type-two equilibrium point lying on the stability boundary of the original system.
- One of the two type-one equilibrium points is the controlling UEP of the fault-on trajectory.
- The two type-one equilibrium points lying on the stability boundary of the original system correspond to two type-one equilibrium points lying on the stability boundary of the reduced-state system.
- One type-two equilibrium point lying on the stability boundary of the original system corresponds to one type-two equilibrium point lying on the stability boundary of the reduced-state system.

Hence, the exit point lies far away from the reduced-state controlling UEP and close to a type-two UEP. The BCU method detects the exit point and moves along the stable manifold of the reduced-state controlling UEP to find the reduced-state controlling UEP. This movement along the stable manifold is implemented using

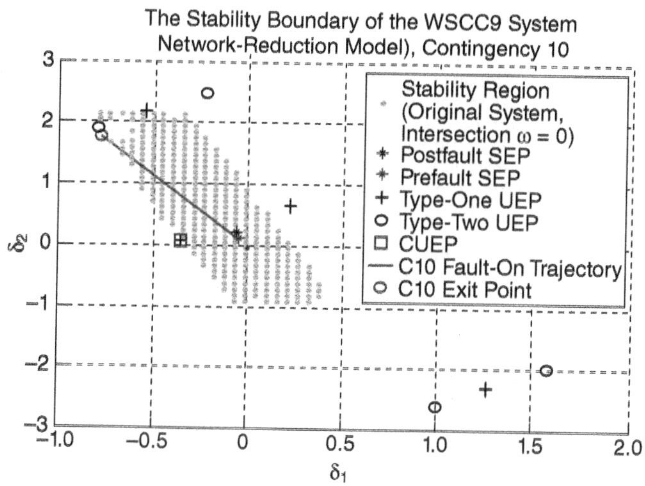

Figure 18.11 The relationship among the projected fault-on trajectory, exit point, controlling UEP, other type-one UEPs, and the stability region of the original system on the angle subspace (for Contingency 10). The exit point is close to a type-two equilibrium point and is far away from the controlling UEP.

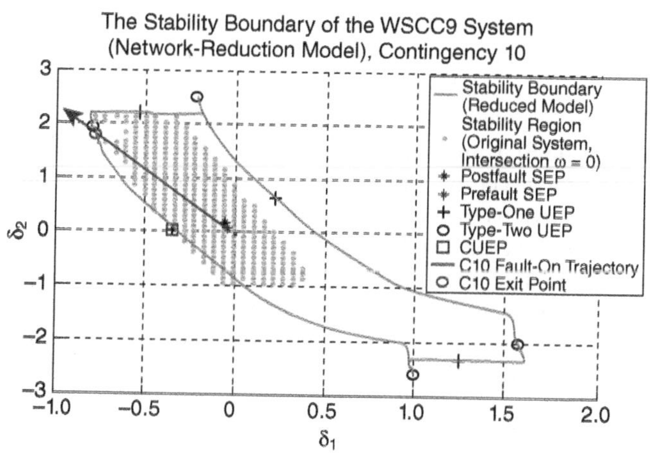

Figure 18.12 The projected fault-on trajectory, exit point, controlling UEP, stability boundary of the reduced-state system, other type-one UEPs on the stability boundary, and the stability region of the original system on the angle subspace (for Contingency 10).

the stability-boundary-following procedure. The BCU method then relates it to the controlling UEP of the original system (see Figure 18.12).

We next consider the network-preserving model with Contingency 10, which is specified by the events that the faulted bus is #8 and that the line between Buses

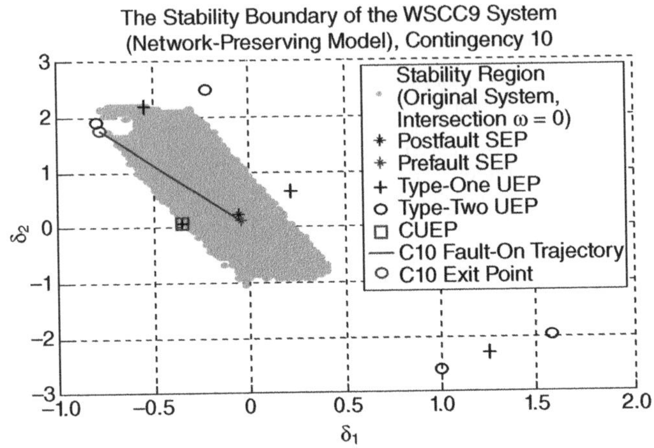

Figure 18.13 The projected fault-on trajectory reaches the exit point, which is close to a type-two equilibrium point. The highlighted area represents the stability region of the original system on the angle subspace for Contingency 10.

7 and 8 is tripped to clear the fault. The prefault and postfault SEPs are computed and the static relationship (S1) is confirmed.

Next, the following key objects of the BCU method are computed: the fault-on trajectory, exit point, and the reduced-state controlling UEP. For this contingency, the projected fault-on trajectory intersects the stability boundary of the reduced-state system at the exit point. The computed exit point is very close to the exit point. Hence, exit point detection of the BCU method satisfies the dynamic property (D3). The computed exit point lies near a type-two UEP but far away from the controlling UEP, which is a type one. It is obvious that starting from the exit point, the Newton method or its variants will not converge to the controlling UEP. The BCU method explores both the stable manifold of the reduced-state controlling UEP and the dynamic property (D2) to compute the correct controlling UEP. The process of computing the exit point is illustrated in Figure 18.13, while the process of computing the controlling UEP is illustrated in Figure 18.14.

18.5 CONCLUDING REMARKS

We have conducted several numerical studies to examine the dynamic relationship between the stability boundary of the original system and the stability boundary of the reduced-state system. This relationship illustrates the algorithmic aspects of the BCU method from the perspective of state spaces.

The fault-on trajectory projected on the angle space intersects the stability boundary of the reduced-state system at the exit point. On the other hand, for each contingency under study, the computed exit point by the BCU method lies very close

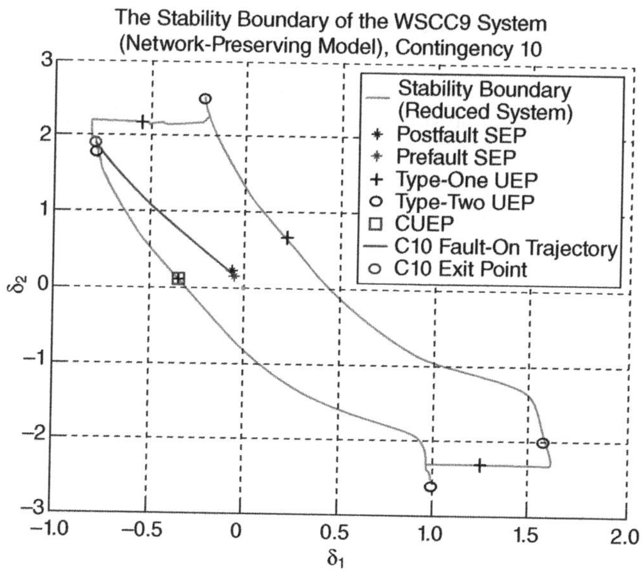

Figure 18.14 The stability-boundary-following procedure starts from the exit point and reaches a minimum gradient point (MGP), from which the reduced-state controlling UEP is found using the Newton method. The exit point is located, unlike the type-two equilibrium point, far away from the controlling UEP. The highlighted boundary represents the stability boundary of the reduced-state system (for Contingency 10).

to the exit point. Hence, the dynamic property (D3) is well implemented in the BCU method.

Even though the dynamic property (D2) may not be satisfied for each contingency, the BCU method still computes the correct controlling UEP. The reason for this is that the reduced-state controlling UEP satisfies the boundary property (i.e., the reduced-state controlling UEP lies on the stability boundary of the reduced-state system if and only if the corresponding controlling UEP lies on the stability boundary of the original system). We emphasize that the dynamic property (D2) is only a sufficient condition for the UEP computed by the BCU method lying on the stability boundary of the original system. It is not a necessary condition. Hence, dynamic property (D2) can be replaced by the following condition without damaging the ability of the BCU method to compute the correct controlling UEP:

Dynamic Property (D2′): the reduced-state controlling UEP lies on the stability boundary of the reduced-state system if and only if the corresponding controlling UEP lies on the stability boundary of the original system.

We have illustrated how the dynamic relationship is explored in the BCU method for computing the controlling UEP of the original system. The BCU method computes the reduced-state controlling UEP relative to the projected fault-on trajectory by moving along the stable manifold of the reduced-state controlling UEP to compute the reduced-state controlling UEP. Then, by exploring the dynamic property (D2′), the BCU method relates the reduced-state controlling UEP to the controlling UEP.

We have also derived several conditions for the BCU method to compute the correct controlling UEP. For example, one sufficient condition for the BCU method to compute the correct controlling UEP is the following. If the relevant stability boundary of the reduced-state system and the relevant stability boundary of the original system and the angle space are sufficiently close, the BCU computes the correct controlling UEP.

Chapter 19

Study of the Transversality Conditions of the BCU Method

19.1 INTRODUCTION

The transversality condition plays a significant role in the development of the dynamical system theory. The condition is one of the three ingredients in the so-called Morse–Smale systems. The violation of the transversality condition often leads to a global bifurcation, making the task of verifying the transversality condition important. In the past, significant efforts have been directed to the development of tools for verifying the transversality condition. Despite these efforts, little progress has been made on this matter. It is now well recognized that the task of verifying the transversality condition is very challenging. There are no useful tools available for verifying the transversality condition except for computationally extensive numerical methods.

The one-parameter transversality conditions play an important role in the theoretical foundation of the conceptual BCU method. Violation of the one-parameter transversality conditions can cause a violation of dynamic property (D2) required in the BCU method; consequently, it may cause incorrectness in the BCU method when computing the controlling unstable equilibrium point (UEP). However, due to the complexity of practical power system models, the one-parameter transversality conditions may not be always satisfied. Several counterexamples showing that the BCU method may fail to give correct stability assessments, predicting stability for a three-bus power system that in fact suffers from second-swing instability, appeared in Llamas et al. (1995). These counterexamples raise questions regarding the practical application of the BCU method to general power system models. Ejebe and Tong argued that lightly damped systems, such as the three-bus power system in Llamas et al. (1995), are rare and typical faults tend to move the power system in directions where the BCU method works (Ejebe and Tong, 1995).

Direct Methods for Stability Analysis of Electric Power Systems, by Hsiao-Dong Chiang

It should be stressed that the violation of the dynamic property (D2) does not necessarily lead the BCU method to compute incorrect CUEPs. As shown in the previous chapter, a weaker condition of (D2), that is, the dynamic property (D2′), is sufficient for the BCU method to compute the correct CUEP.

In Paganini and Lesieutre (1997, 1999), the following three points regarding the BCU method are raised and investigated:

1. A requirement for the BCU method to be able to find the controlling UEP is that the following boundary property is satisfied: the full model (i.e., the original model) and the gradient model (i.e., the reduced-state model) should have the same UEPs on their stability boundaries (in other words, the dynamic property [D2] is satisfied).
2. The preceding boundary property is not a generic property of power system models.
3. The one-parameter deformation is not an appropriate tool to provide theoretical justifications for the boundary property: it merely translates the problem into an unverifiable transversality condition.

In this chapter, these issues will be addressed. In addition, analytical results on the transversality condition will be developed and applied to the one-parameter transversality condition. The one-parameter deformation is an appropriate analytical tool, though not a computational tool, to provide theoretical justifications for the boundary property. A computational tool to verify the boundary property and the dynamic property (D2′) will be presented in the next chapter.

19.2 A PARAMETRIC STUDY

In this section, the boundary property on a single salient pole generator connected to an infinite bus through a lossless transmission line is examined:

$$\begin{cases} \dot{\delta} = \dfrac{d\delta}{dt} = \omega \\ \dot{\omega} = -D\omega + P_m - P_{e1}\sin(\delta) + P_{e2}\sin(2\delta). \end{cases} \tag{19.1}$$

This model is a differential equation with five parameters. Since these parameters are physically motivated, it is natural to study the genericity of the boundary property in the parameter space, which is five-dimensional.

The reduced-state model associated with the original model (Eq. 19.1) is

$$\dot{\delta} = P_m - P_{e1}\sin(\delta) + P_{e2}\sin(2\delta). \tag{19.2}$$

Note that $\bar{\delta}$ is an equilibrium point of the reduced-state model (Eq. 19.2) if and only if $(\bar{\delta}, 0)$ is an equilibrium point of the original model (Eq. 19.1). This property is independent of the values of the damping factor D. It can be further shown that, for any positive damping factor D, δ_s is a stable equilibrium point (SEP) of the reduced-state model (Eq. 19.2) if and only if $(\delta_s, 0)$ is a SEP of the original model (Eq. 19.1).

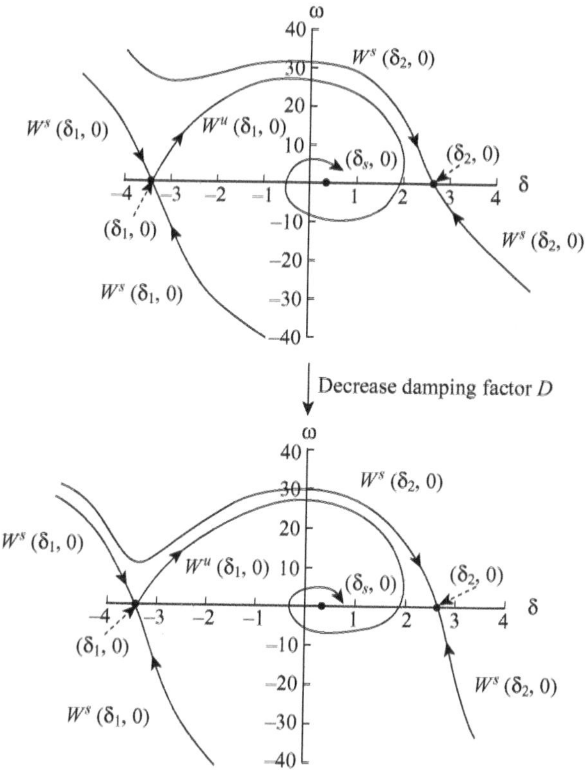

Figure 19.1 For large values of D, both UEPs (δ_1, 0) and (δ_2, 0) lie on the stability boundary $\partial A(\delta_s, 0)$ of the original model (Eq. 19.1). The stability boundary $\partial A(\delta_s, 0)$ is equal to the union of the stable manifolds of (δ_1, 0) and (δ_2, 0). Both the transversality condition and the boundary property are satisfied for large D.

We study the static and dynamic relationship between the reduced-state model (Eq. 19.2) and the original model (Eq. 19.1) for various values of the damping factor D while the other parameters are fixed at the values $P_{e1} = 3.02$, $P_{e2} = 0.416$, $P_m = 0.91$, and $M - 0.0138$.

For a positive D, there is a SEP, say, (δ_s, 0), surrounded by two UEPs, (δ_1, 0) and (δ_2, 0). The dynamic relationship between the SEP (δ_s, 0) and the two UEPs (δ_1, 0) and (δ_2, 0) of the original model (Eq. 19.1) depends on damping factor D. On the other hand, the dynamic relationship between the SEP δ_s and the two UEPs δ_1 and δ_2 of the reduced-state model (Eq. 19.2) does not depend on damping factor D. Both δ_1 and δ_2 always lie on the stability boundary $\partial A(\delta_s)$ of the reduced-state model (Eq. 19.2).

For large values of D, (δ_1, 0) and (δ_2, 0) lie on the stability boundary $\partial A(\delta_s)$ of the original model (Eq. 19.1). The stability boundary $\partial A(\delta_s, 0)$ is equal to the union of the stable manifolds of (δ_1, 0) and (δ_2, 0), and the boundary property is satisfied for large D. For large D, the intersections between the stable and unstable manifolds of (δ_s, 0) and (δ_1, 0), (δ_2, 0) satisfy the transversality condition (see Figure 19.1). Both the original model (Eq. 19.1) and the reduced-state model (Eq. 19.2) are structurally stable for large D.

It should be pointed out that for large values of D, the structures of stability boundary $\partial A(\delta_s, 0)$ remain similar under small variations of D. In particular, the stability boundary $\partial A(\delta_s, 0)$ is equal to the union of the stable manifolds of $(\delta_1, 0)$ and $(\delta_2, 0)$. The structures of stability boundary $\partial A(\delta_s, 0)$ remain "stable" with respect to small variations of D in the sense that the stability boundary $\partial A(\delta_s, 0)$ remains as the union of the stable manifolds of $(\delta_1, 0)$ and $(\delta_2, 0)$ until damping factor D is decreased to the critical value $D_{cr} = 0.04687$.

The structures of the stability boundary $\partial A(\delta_s, 0)$ encounter a dramatic change when damping factor D passes through its critical value. For $D = D_{cr}$, both $(\delta_1, 0)$ and $(\delta_2, 0)$ still lie on the stability boundary $\partial A(\delta_s, 0)$, and the stability boundary $\partial A(\delta_s, 0)$ remains the union of the stable manifolds of $(\delta_1, 0)$ and $(\delta_2, 0)$. However, the intersection between the unstable manifold of $(\delta_1, 0)$ and the stable manifold of $(\delta_2, 0)$ does not satisfy the transversality condition. The boundary property still holds for $D = D_{cr}$.

There is a saddle–saddle connection trajectory between $(\delta_1, 0)$ and $(\delta_2, 0)$ when $D = D_{cr}$. The structure of the saddle–saddle connection trajectory is destroyed when damping factor D changes its value from D_{cr} (see Figure 19.2). The original model

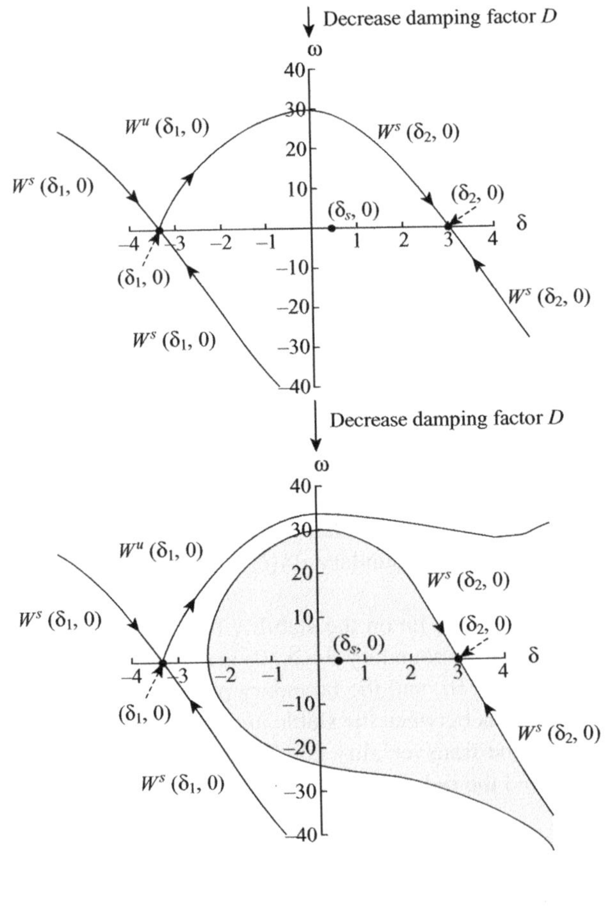

Figure 19.2 The structures of the stability boundary $\partial A(\delta_s, 0)$ encounter a dramatic change when damping factor D passes through its critical value. For $D = D_{cr}$, both $(\delta_1, 0)$ and $(\delta_2, 0)$ still lie on the stability boundary $\partial A(\delta_s, 0)$, and the stability boundary $\partial A(\delta_s, 0)$ remains as the union of the stable manifolds of $(\delta_1, 0)$ and $(\delta_2, 0)$. The boundary property still holds, while the transversality condition is violated for $D = D_{cr}$.

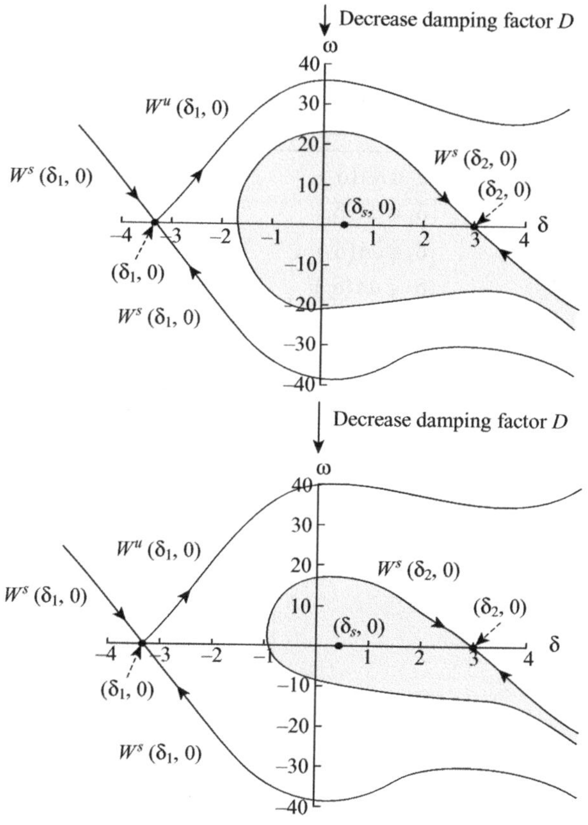

Figure 19.3 For $D < D_{cr}$, the UEP $(\delta_2, 0)$ still lies on the stability boundary $\partial A(\delta s, 0)$, while $(\delta_1, 0)$ lies outside the stability boundary $\partial A(\delta_s, 0)$. The stability boundary $\partial A(\delta_s, 0)$ is composed only of the stable manifold of $(\delta_2, 0)$. In this case, δ_1 lies on the stability boundary $\partial A(\delta_s)$, while $(\delta_1, 0)$ does not lie on the stability boundary $\partial A(\delta_s, 0)$. The boundary property is violated, while the transversality condition is satisfied.

(Eq. 19.1) undergoes a global bifurcation when damping factor D passes through its critical value, D_{cr}. More specifically, the structure of the stability boundary $\partial A(\delta_s, 0)$ undergoes a dramatic change when D passes through D_{cr}. For $D < D_{cr}$, the UEP $(\delta_2, 0)$ still lies on the stability boundary $\partial A(\delta_s, 0)$, but the UEP $(\delta_1, 0)$ does not (see Figure 19.3); moreover, the stability boundary $\partial A(\delta_s, 0)$ is composed only of the stable manifold of $(\delta_2, 0)$. In this case, δ_1 lies on the stability boundary $\partial A(\delta_s)$, while $(\delta_1, 0)$ does not lie on the stability boundary $\partial A(\delta_s, 0)$. From the above analysis, it can be concluded that the boundary property between the original model (Eq. 19.1) and the reduced-state model (Eq. 19.2) holds only for large damping factors, in particular, only if the damping factor satisfies $D \geq D_{cr}$. We summarize the above analysis in Table 19.1.

We next allow the machine inertia M to vary, in addition to D, and plot the parameter values on the $M - D$ plane at which the transversal intersection between the stable and unstable manifolds of the UEPs $(\delta_1, 0)$ and $(\delta_2, 0)$ does not hold. The parameter values at which a global bifurcation occurs due to the violation of the transversal intersection are shown in Figure 19.4. It can be seen from the figure that the boundary property holds in the areas of $\frac{D}{M}$ above the curve. The boundary

Table 19.1 A Relationship among the Damping Factor, Boundary Property, and Transversality Condition

	The original model	The reduced-state model	Boundary property	Transversality condition
Large damping	$(\delta_1, 0), (\delta_2, 0) \in \partial A(\delta_s, 0) = W^s(\delta_1, 0) \cup W^s(\delta_2, 0)$	$\begin{cases}\delta_1 \in \partial A(\delta_s) \\ \delta_2 \in \partial A(\delta_s)\end{cases}$	Yes	Yes
Critical damping	$(\delta_1, 0), (\delta_2, 0) \in \partial A(\delta_s, 0) = W^s(\delta_1, 0) \cup W^s(\delta_2, 0)$	$\begin{cases}\delta_1 \in \partial A(\delta_s) \\ \delta_2 \in \partial A(\delta_s)\end{cases}$	Yes	No
Low damping	$\begin{cases}(\delta_2, 0) \in \partial A(\delta_s, 0) = W^s(\delta_2, 0) \\ (\delta_1, 0) \notin \partial A(\delta_s, 0)\end{cases}$	$\begin{cases}\delta_1 \in \partial A(\delta_s) \\ \delta_2 \in \partial A(\delta_s)\end{cases}$	No	Yes

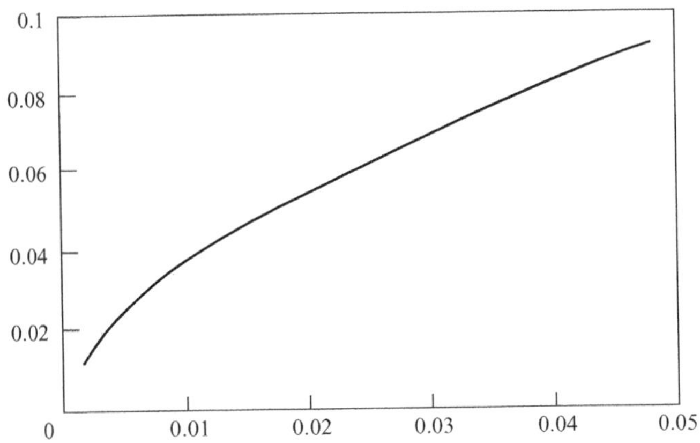

Figure 19.4 Parameter values (M, D) at which the stable and unstable manifolds of the UEPs $(\delta_1, 0)$ and $(\delta_2, 0)$ do not intersect transversally. The horizontal axis represents the machine inertia and the vertical axis represents the damping factor.

property holds for the high damping systems whose $\frac{D}{M}$ values are above the curve, while the boundary property does not hold for the low damping systems whose $\frac{D}{M}$ values are below the curve.

The boundary property holds for high damping systems, while it may not hold for low damping systems. The issue of how to determine the critical damping value above which the boundary property holds remains open. The critical damping value depends on a variety of factors including network topology, loading condition, and system models used, among others. It should be further pointed out that, even if the boundary property is satisfied, this still does not guarantee that the BCU method will always find the correct controlling UEP by studying the projected fault-on trajectory. This is because the exit point of the projected fault-on trajectory and the exit point of the fault-on trajectory may not be related by the projection.

Despite the facts that the boundary property is not a generic property of general power system models and that the BCU method fails to give correct stability assessments for a lightly damped three-bus power system, which suffers from second-swing instability, extensive numerical simulations still favor that

- the BCU method always works for first-swing stability/instability of damped power systems;
- the boundary property depends on the level of system damping;
- the BCU method *may* give incorrect stability assessments for power systems, which suffer from multiswing instability; and
- even if the boundary property is not satisfied, depending on the fault clearing time, the BCU method may still give correct stability assessments for power systems suffering from multiswing stability/instability.

It will become clear that the development of the BCU–exit point method and group-based BCU methods will remedy these problems and give reliable stability assessments. The BCU–exit point method will be presented in Chapter 20, while the group-based BCU method will be presented in Chapter 24.

19.3 ANALYTICAL INVESTIGATION OF THE BOUNDARY PROPERTY

In this section, we will develop some analytical results for verifying the transversality condition for simple, low-dimension power system models. We will employ tools such as the self-indexing energy function, the symmetrical properties of the models, and the structural stability to analyze the invariant properties of the UEPs on the stability boundary and the boundary property for low-dimension, simple power system stability models.

We consider the following one-machine infinite bus (OMIB) system:

$$\begin{aligned}\dot{\delta} &= \omega \\ M\dot{\omega} &= -D\omega + P_m - B\sin\delta,\end{aligned} \tag{19.3}$$

where $P_m \geq 0$, $M > 0$, $D > 0$, and $B > 0$. The corresponding artificial, reduced-state system is defined as

$$\dot{\delta} = -P_m - B\sin\delta, \tag{19.4}$$

and the corresponding one-parameter system $\Sigma(\lambda)$ is defined as

$$\begin{aligned}\dot{\delta} &= \lambda(P_m - B\sin\delta) + (1-\lambda)\omega \\ M\dot{\omega} &= -D\omega + (1-\lambda)(P_m - B\sin\delta).\end{aligned} \tag{19.5}$$

We start the analysis of the OMIB system with zero mechanical power injection (i.e., $P_m = 0$). First, we examine the location of the equilibrium points. Let E be the set of all the equilibrium points. It is easy to see that all of the equilibrium points of the one-parameter system $\Sigma(\lambda)$ are hyperbolic and are located at $(n\pi, 0)$; n is an integer. Moreover, we have the following:

(1) $(\delta_s, 0) = (2k\pi, 0)$ are SEPs.

(2) $(\delta_u, 0) = ((2k + 1)\pi, 0)$ are type-one UEPs;.

(3) There are no other higher-type UEPs.

The following function is an energy function of the OMIB system:

$$W(\delta, \omega) = \frac{1}{2} M\omega^2 - B\cos\delta. \tag{19.6}$$

We have the following observations:

1. For SEP $(\delta_s, 0) = (2k\pi, 0)$, $W(\delta_s, 0) = -B$.
2. For UEP $(\delta_u, 0) = ((2k + 1)\pi, 0)$, $W(\delta_u, 0) = B$.
3. $W(\delta_u, 0) > W(\delta_s, 0)$ for all $(\delta_s, 0)$, $(\delta_u, 0) \in E$.
4. $W(\delta, \omega)$ is self-indexing, which is symmetrical to its state variables.

We next show the invariant property of all the UEPs on the stability boundary of the one-parameter system $\Sigma(\lambda)$.

Proposition 19.1

If the OMIB system (Eq. 19.3) has zero mechanical power injection P_m, then

(i) there are only two UEPs, $(\pi, 0)$ and $(-\pi, 0)$, lying on the stability boundary $\partial A(0, 0)$; and

(ii) the UEPs on the stability boundary $(2k\pi, 0)$, $k = 0, \pm 1$ of the one-parameter system $\Sigma(\lambda)$ are invariant for all $J \in [0, 1]$.

Next, we show that the transversality condition and the boundary property hold for the OMIB system (Eq. 19.3) with zero mechanical power injection.

Proposition 19.2: Transversality Condition and Boundary Property

If the OMIB system (Eq. 19.3) has zero mechanical power injection, then

(i) the stable and unstable manifolds of the UEPs of the corresponding one-parameter system $\Sigma(\lambda)$ intersect transversally for all $\lambda \in [0, 1]$; and

(ii) the UEPs $(\pi, 0)$ and $(-\pi, 0)$ lie on the stability boundary $\partial A(0, 0)$ of the original system (Eq. 19.3). The UEPs (π) and $(-\pi)$ lie on the stability boundary $\partial A(0)$ of the reduced-state system (Eq. 19.4).

The analytical results in Proposition 19.2 can be extended to the OMIB system with a small mechanical power injection and small transfer conductances. The property of structural stability will be explored for this extension. The structural stability of a nonlinear dynamical system ensures that its "nearby" system (i.e., its perturbed system) is topologically equivalent to the system itself. The task of checking whether or not a given nonlinear system is structurally stable is challenging. One sufficient condition for structural stability is that the given nonlinear system is Morse–Smale.

Proposition 19.3: Structural Stability

If the OMIB system (Eq. 19.3) has zero mechanical power injection, then the corresponding one-parameter system $\Sigma(\lambda)$ is structurally stable.

The property of structural stability for the OMIB system (Eq. 19.3) with zero mechanical power injection can be extended to the OMIB system with a small mechanical power injection and small transfer conductances. For this, we consider the following OMIB system with a transfer conductance:

$$\begin{cases} \dot{\delta} = \omega \\ M\dot{\omega} = -D\omega + P_m - B\sin\delta - G\cos\delta, \end{cases} \tag{19.7}$$

where $G \geq 0$ represents the transfer conductance. The corresponding artificial, reduced-state system is defined as

$$\dot{\delta} = P_m - B\sin\delta - G\cos\delta, \tag{19.8}$$

and the corresponding one-parameter system $\Sigma(\lambda)$ is defined as

$$\begin{aligned} \dot{\delta} &= \lambda(P_m - B\sin\delta - G\cos\delta) + (1-\lambda)\omega \\ M\dot{\omega} &= -D\omega + (1-\lambda)(P_m - B\sin\delta - G\cos\delta). \end{aligned} \tag{19.9}$$

Proposition 19.4: Boundary Property

If the mechanical power injection and the transfer conductance G of the OMIB system (Eq. 19.7) are sufficiently small with $(\delta_s, 0)$ being a SEP, then

(i) the stable and unstable manifolds of the UEPs of the corresponding one-parameter system (Eq. 19.9) intersect transversally for each $\lambda \in [0, 1]$, and

(ii) $(\delta_i, 0)$ is a UEP on the stability boundary $\partial A(\delta_s, 0)$ of the original system (Eq. 19.7) if and only if (δ_i) is a UEP on the stability boundary $\partial A(\delta_s)$ of the reduced-state system (Eq. 19.8).

Hence, Proposition 19.4 ensures that the OMIB system with a small mechanical power injection and small transfer conductance satisfies the boundary property.

19.4 THE TWO-MACHINE INFINITE BUS (TMIB) SYSTEM

We consider the following TMIB:

$$\begin{aligned} \dot{\delta}_1 &= \omega_1 \\ \dot{\delta}_2 &= \omega_2 \\ M_1\dot{\omega}_1 &= -D_1\omega_1 - B_{12}\sin(\delta_1-\delta_2) - B_{13}\sin(\delta_1-\delta_3) + P_{m1} \\ M_2\dot{\omega}_2 &= -D_2\omega_2 - B_{21}\sin(\delta_2-\delta_1) - B_{23}\sin(\delta_2-\delta_3) + P_{m2}, \end{aligned} \tag{19.10}$$

where $M_i > 0$, $D_i > 0$, $B_{ij} \geq 0$, and $P_{mi} \geq 0$. The corresponding artificial, reduced-state system is defined as

$$\begin{aligned} \dot{\delta}_1 &= -B_{12}\sin(\delta_1-\delta_2) - B_{13}\sin(\delta_1-\delta_3) + P_{m1} \\ \dot{\delta}_2 &= -B_{12}\sin(\delta_2-\delta_1) - B_{23}\sin(\delta_2-\delta_3) + P_{m2}, \end{aligned} \tag{19.11}$$

and the corresponding one-parameter system $\Sigma(\lambda)$ is defined as

$$\begin{aligned} \dot{\delta}_1 &= -\lambda(B_{12}\sin(\delta_1-\delta_2) + B_{13}\sin(\delta_1-\delta_3) - P_{m1}) + (1-\lambda)\omega_1 \\ \dot{\delta}_2 &= -\lambda(B_{12}\sin(\delta_2-\delta_1) + B_{23}\sin(\delta_2-\delta_3) - P_{m2}) + (1-\lambda)\omega_2 \\ M_1\dot{\omega}_1 &= -D_1\omega_1 - (1-\lambda)(B_{12}\sin(\delta_1-\delta_2) + B_{13}\sin(\delta_1-\delta_3) - P_{m1}) \\ M_2\dot{\omega}_2 &= -D_2\omega_2 - (1-\lambda)(B_{12}\sin(\delta_2-\delta_1) + B_{23}\sin(\delta_2-\delta_3) - P_{m2}). \end{aligned} \tag{19.12}$$

We first examine the location of the equilibrium points of the TMIB (Eq. 19.10) with zero mechanical power injections. In addition to the elementary equilibrium points $E_e = (k_1\pi, k_2\pi, 0, 0)$, the system may contain extra equilibrium points.

We define

$$L(B_{12}, B_{13}, B_{23}) = B_{12}^4B_{13}^4 + B_{12}^4B_{23}^4 + B_{13}^4B_{23}^4 - 2B_{12}^4B_{13}^4B_{23}^2 - 2B_{12}^2B_{13}^4B_{23}^2 - 2B_{12}^2B_{13}^2B_{23}^4.$$

It can be shown that if $L(B_{12}, B_{13}, B_{23}) < 0$, then the one-parameter system (Eq. 19.12) contains the following extra equilibrium points:

$$E_x = \{e_x(k_1, k_2) = (\delta_1 \pm 2k_1\pi, \delta_2 \pm 2k_2\pi, 0, 0) : k_i \in Z\},$$

where

$$\delta_1 = arc\tan\frac{\sqrt{-L(B_{12}, B_{13}, B_{23})}}{B_{13}^2B_{23}^2 - B_{12}^2B_{23}^2 + B_{12}^2 + B_{13}^2} \text{ and}$$

$$\delta_2 = arc\cos\frac{-B_{13}^2B_{23}^2 - B_{12}^2B_{23}^2 + B_{12}^2B_{13}^2}{2B_{12}B_{13}B_{23}^2}.$$

According to their locations and their energy function values, all the equilibrium points of the TMIB (Eq. 19.10) are put into five classes, as shown in Table 19.2, where the parameter B_{12} is normalized to 1 without loss of generality.

Table 19.2 Classification of Equilibrium Points and Their Corresponding Energy Values

Class	Location of equilibrium points	Energy value
I	$(2k_1\pi, 2k_2\pi)$	$-1 - B_{13} - B_{23}$
II	$((2k_1 + 1)\pi, 2k_2\pi)$	$1 + B_{13} - B_{23}$
III	$(2k_1\pi, (2k_2 + 1)\pi)$	$1 - B_{13} + B_{23}$
IV	$((2k_1 + 1)\pi, (2k_2 + 1)\pi)$	$-1 + B_{13} + B_{23}$
V	$e_x(k_1, k_2)$	$-\dfrac{3B_{12}^2B_{23}^2 - B_{13}^2 - B_{23}^2}{2B_{13}B_{23}}$

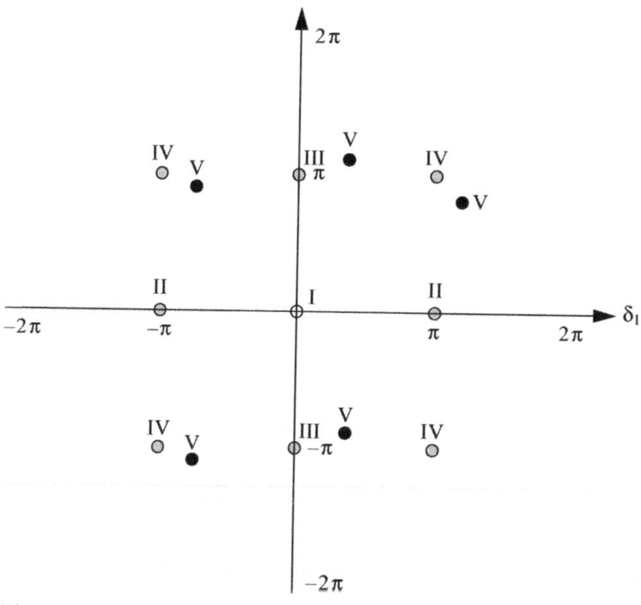

Figure 19.5 The locations of the five classes of equilibrium points in the angle space $\Re^2$.

The locations of the five classes of equilibrium points are displayed in Figure 19.5. Note that $(2k_1\pi, 2k_2\pi)$ are SEPs, while $e_x(k_1, k_2)$ are type-two equilibrium points independent of the parameter value B_{ij}. The type of the other three classes of equilibrium points cannot be predetermined without specific information on the parameter value B_{ij}. Different parameter values of B_{13} and B_{23} result in different types for the other three classes of equilibrium points. Note that if $L(B_{12}, B_{13}, B_{23}) \neq 0$, then all of the equilibrium points are hyperbolic. On the other hand, $L(B_{12}, B_{13}, B_{23}) = 0$ indicates that the linearized system has at least one zero eigenvalue. We next treat B_{13} and B_{23} as parameters. Then, the system will encounter a saddle-node bifurcation when $L(B_{12}, B_{13}, B_{23}) = 0$. Since the parameters B_{13} and B_{23} are commutable in $L(B_{12}, B_{13}, B_{23})$, we only need to consider the case $B_{23} > B_{13}$. Depending on the values of B_{23} and B_{13}, the equilibrium points of classes II–IV may be of type one or type two.

We will use the notation II (2) to denote type-two equilibrium point of class II and the notation III (1) to denote type-one equilibrium point of class III. Note that the equilibrium points of class I are always stable, $I(0)$, if $L(B_{12}, B_{13}, B_{23}) \neq 0$, while the equilibrium points of class V are always of type two, $V(2)$. It can be shown that the following results hold:

- If $B_{23} > \dfrac{B_{13}}{1-B_{13}}$ and $B_{23} > 1$, then (1) the equilibrium points of class II are of type one; (2) the equilibrium points of class III are of type two; and (3) the equilibrium points of class IV are of type one. Moreover,

$$W(III(2)) \geq W(IV(1)) \geq W(II(1)) \geq W(I(0)).$$

- If $B_{23} > \dfrac{B_{13}}{1-B_{13}}$ and $B_{23} < 1$, then (1) the equilibrium points of class II are of type one; (2) the equilibrium points of class III are of type two; and (3) the equilibrium points of class IV are of type one. Moreover,

$$W(III(2)) \geq W(II(1)) \geq W(IV(1)) \geq W(I(0)).$$

- If $B_{13} < 1$, $B_{23} > 1$ and $L(1, B_{13}, B_{23}) < 0$, then (1) the equilibrium points of class II are of type one; (2) the equilibrium points of class III are of type one; (3) the equilibrium points of class IV are of type one; and (4) the equilibrium points of class V are of type two. Moreover,

$$W(V(2)) \geq W(III(1)) \geq W(IV(1)) \geq W(II(1)) \geq W(I(0)).$$

- If $B_{23} > B_{13}$, $B_{13} < 1$, $B_{23} < 1$, and $L(1, B_{13}, B_{23}) < 0$, then (1) the equilibrium points of class II are of type one; (2) the equilibrium points of class III are of type one; (3) the equilibrium points of class IV are of type one; and (4) the equilibrium points of class V are of type two. Moreover,

$$W(V(2)) \geq W(III(1)) \geq W(II(1)) \geq W(IV(1)) \geq W(I(0)).$$

- If $B_{13} > 1$, $B_{23} > 1$, $B_{23} > B_{13}$, and $L(1, B_{13}, B_{23}) < 0$, then (1) the equilibrium points of class II are of type one; (2) the equilibrium points of class III are of type one; (3) the equilibrium points of class IV are of type one; and (4) the equilibrium points of class V are of type two. Moreover,

$$W(V(2)) \geq W(IV(1)) \geq W(III(1)) \geq W(II(1)) \geq W(I(0)).$$

- If $B_{23} > \dfrac{B_{13}}{B_{13}-1}$ and $B_{23} > B_{13}$, then (1) the equilibrium points of class II are of type one; (2) the equilibrium points of class III are of type one; and (3) the equilibrium points of class IV are of type two. Moreover,

$$W(IV(2)) \geq W(III(1)) \geq W(II(1)) \geq W(I(0)).$$

These results are summarized in Table 19.3 and the corresponding classification in parameter space B_{13}–B_{23} is shown in Figure 19.6. One important observation from

Table 19.3 Classifications of Parameter Regions and the Corresponding Ordering of Energy Values of Different Classes of Equilibrium Points

Region numbers	Characterization	Energy value of equilibrium point (class [type])
1, 7	$B_{23} > \frac{B_{13}}{1-B_{13}}$ and $B_{23} > 1$.	$W(III(2)) \geq W(IV(1)) \geq W(II(1)) \geq W(I(0))$
2, 8	$B_{23} > \frac{B_{13}}{1-B_{13}}$ and $B_{23} < 1$.	$W(III(2)) \geq W(II(1)) \geq W(IV(1)) \geq W(I(0))$
3, 9	$B_{13} < 1$, $B_{23} > 1$, and $L(1, B_{13}, B_{23}) < 0$.	$W(V(2)) \geq W(III(1)) \geq W(IV(1)) \geq W(II(1)) \geq W(I(0))$
4, 10	$B_{23} > B_{13}$, $B_{13} < 1$, $B_{23} < 1$, and $L(1, B_{13}, B_{23}) < 0$.	$W(V(2)) \geq W(III(1)) \geq W(II(1)) \geq W(IV(1)) \geq W(I(0))$
5, 11	$B_{13} > 1$, $B_{23} > 1$, $B_{23} > B_{13}$, and $L(1, B_{13}, B_{23}) < 0$.	$W(V(2)) \geq W(IV(1)) \geq W(III(1)) \geq W(II(1)) \geq W(I(0))$
6, 12	$B_{23} > \frac{B_{13}}{B_{13}-1}$ and $B_{23} > B_{13}$.	$W(IV(2)) \geq W(III(1)) \geq W(II(1)) \geq W(I(0))$

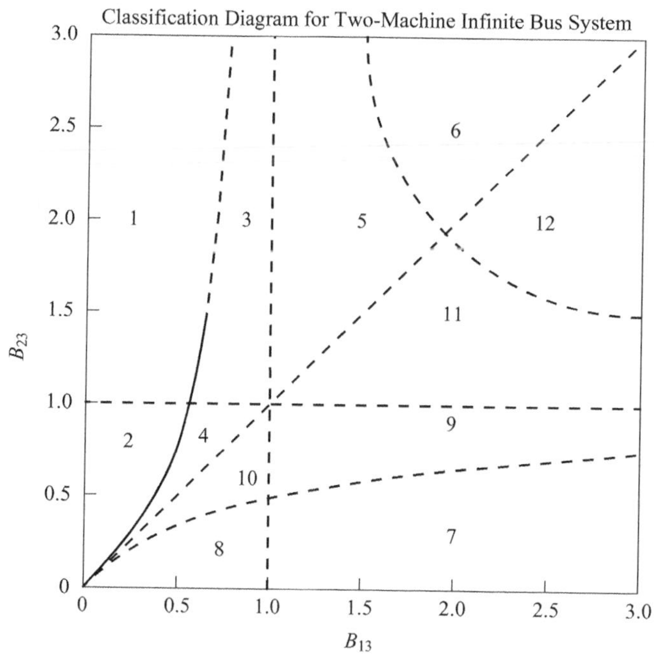

Figure 19.6 The parameter space B_{13}–B_{23} is divided into 12 regions. Each region is characterized by an algebraic relationship in the parameter space.

Table 19.3 is that the types of the equilibrium points provide an interesting linear ordering of equilibrium points in terms of their energy values:

$$W(\text{type}-2) \geq W(\text{type}-1) \geq W(\text{SEP}).$$

Note that this observation agrees with the analytical results that the closest UEP, which possesses the lowest energy value on the stability boundary, must be of type one.

Classification Diagram

We next perform a dynamic analysis of the TMIB system (Eq. 19.10). Our first goal is to establish the invariant property of the UEPs on the stability boundary $\partial A(0, 0, 0, 0)$ for the one-parameter dynamical system (Eq. 19.12).

The property of the linear order of the energy values of the UEPs will be explored in our analysis, which will start from the closest UEP and then will proceed to other type-one UEPs, and finally will proceed to the type-two UEPs on the stability boundary. Based on the different regions classified in Table 19.3, our analysis will consist of the following stages:

Stage 1: Perform a dynamic analysis in Regions 1 and 3.

Stage 2: Perform a dynamic analysis in Regions 5 and 6.

Stage 3: Perform a dynamic analysis in Regions 2 and 4.

Our analysis in each stage is composed of the following four steps:

Step 1. Locate the closest UEP of the SEP (0, 0, 0, 0).

Step 2. Locate the second closest UEP (0, 0, 0, 0), which is the UEP with the second lowest energy value on the stability boundary $\partial A(0, 0, 0, 0)$.

Step 3. Establish the invariant property of other type-one UEPs on the stability boundary $\partial A(0, 0, 0, 0)$.

Step 4. Establish the invariant property of type-two UEPs on the stability boundary $\partial A(0, 0, 0, 0)$.

We now summarize the main results in the following proposition.

Proposition 19.5: Invariant Property and Boundary Property

If the mechanical power injection of the TMIB system (Eq. 19.10) is zero, then for any values of $M_i > 0$, $D_i > 0$, $i = 1, 2$ and $B_{ij} \geq 0$, $i, j = 1, 2, 3$, the following results hold:

(i) Both the location and the number of the UEPs on the stability boundary $\partial A(0, 0, 0, 0)$ of the one-parameter systems (Eq. 19.12) are invariant as λ varies from $\lambda = 0$ to $\lambda = 1$.

(ii) $(\delta_1^u, \delta_2^u, 0, 0)$ is a UEP lying on the stability boundary $\partial A(0, 0, 0, 0)$ of system (Eq. 19.10) if and only if (δ_1^u, δ_2^u) is a UEP lying on the stability boundary $\partial A(0, 0)$ of the system (Eq. 19.11).

The analytical results derived in Proposition 19.5 can be extended to the TMIB system with a small mechanical power injection and small transfer conductances. The property of structural stability will be explored for this extension.

Proposition 19.6: Structural Stability

If the TMIB system (Eq. 19.10) has zero mechanical power injection and satisfies the following conditions:

(C1) $L(B_{12}, B_{13}, B_{23}) \neq 0$, and

(C2) the intersections between the stable and unstable manifolds satisfy the transversality condition,

then the TMIB system (Eq. 19.10) with zero mechanical power injection is structurally stable. In addition, if the reduced-state TMIB system (Eq. 19.11) satisfies (C2), then it is also structurally stable.

Condition (C1) is equivalent to the condition that all the equilibrium points of the TMIB system (Eq. 19.10) are hyperbolic. The task of checking the transversality condition (C2) for the TMIB system (Eq. 19.10) can be very involved since the system is four-dimensional. However, the transversality condition (C2) for the reduced-state TMIB system (Eq. 19.11) is easy to check since it is two-dimensional.

The property of structural stability for the TMIB system (Eq. 19.10) with zero mechanical power injection can be extended to the TMIB system with a small mechanical power injection and small transfer conductances. For this, we consider the following TMIB system with transfer conductances:

$$
\begin{aligned}
\dot{\delta}_1 &= \omega_1 \\
\dot{\delta}_2 &= \omega_2 \\
M_1\dot{\omega}_1 &= -D_1\omega_1 - B_{12}\sin(\delta_1-\delta_2) - G_{12}\cos(\delta_1-\delta_2) \\
&\quad - B_{13}\sin(\delta_1-\delta_3) - G_{13}\cos(\delta_1-\delta_3) + P_{m1} \\
M_2\dot{\omega}_2 &= -D_2\omega_2 - B_{21}\sin(\delta_2-\delta_1) - G_{21}\cos(\delta_2-\delta_1) \\
&\quad - B_{23}\sin(\delta_2-\delta_3) - G_{23}\cos(\delta_2-\delta_3) + P_{m2},
\end{aligned}
\tag{19.13}
$$

where $G \geq 0$ represents the transfer conductances. The corresponding artificial, reduced-state system is defined as

$$
\begin{aligned}
\dot{\delta}_1 &= -B_{12}\sin(\delta_1-\delta_2) - G_{12}\cos(\delta_1-\delta_2) \\
&\quad - B_{13}\sin(\delta_1-\delta_3) - G_{13}\cos(\delta_1-\delta_3) + P_{m1} \\
\dot{\delta}_2 &= -B_{21}\sin(\delta_2-\delta_1) - G_{21}\cos(\delta_2-\delta_1) \\
&\quad - B_{23}\sin(\delta_2-\delta_3) - G_{23}\cos(\delta_2-\delta_3) + P_{m2},
\end{aligned}
\tag{19.14}
$$

and the corresponding one-parameter system is defined as

$$\begin{aligned}
\dot{\delta}_1 &= -\lambda\left(B_{12}\sin(\delta_1-\delta_2)+B_{13}\sin(\delta_1-\delta_3)+G_{12}\cos(\delta_1-\delta_2)\right.\\
&\quad \left.+G_{13}\cos(\delta_1-\delta_3)-P_{m1}\right)+(1-\lambda)\omega_1\\
\dot{\delta}_2 &= -\lambda\left(B_{21}\sin(\delta_2-\delta_1)+B_{23}\sin(\delta_2-\delta_3)+G_{22}\cos(\delta_2-\delta_1)\right.\\
&\quad \left.+G_{23}\cos(\delta_2-\delta_3)-P_{m2}\right)+(1-\lambda)\omega_2\\
M_1\dot{\omega}_1 &= -D_1\omega_1-(1-\lambda)\left(B_{12}\sin(\delta_1-\delta_2)+B_{13}\sin(\delta_1-\delta_3)\right.\\
&\quad \left.G_{12}\cos(\delta_1-\delta_2)+G_{13}\cos(\delta_1-\delta_3)-P_{m1}\right)\\
M_2\dot{\omega}_2 &= -D_2\omega_2-(1-\lambda)\left(B_{21}\sin(\delta_2-\delta_1)+B_{23}\sin(\delta_2-\delta_3)\right.\\
&\quad \left.G_{21}\cos(\delta_2-\delta_1)+G_{23}\cos(\delta_2-\delta_3)-P_{m2}\right).
\end{aligned} \tag{19.15}$$

Proposition 19.7: Invariant Property and Boundary Property

If the TMIB system (Eq. 19.10) with zero mechanical power injection satisfies conditions (C1) and (C2) stated in Proposition 19.6 and the reduced-state TMIB system (Eq. 19.11) satisfies condition (C2), then, with any values of $M_i > 0$, $D_i > 0$, $i = 1, 2$ and $B_{ij} \geq 0$, $i, j = 1, 2, 3$, for the TMIB system (Eq. 19.13) with a sufficiently small mechanical power injection and transfer conductances, the following results hold:

(i) Both the location and the number of the UEPs on the stability boundary $\partial A(\delta_s^1, \delta_s^2, 0, 0)$, of the one-parameter systems (Eq. 19.15) are invariant as λ varies from $\lambda = 0$ to $\lambda = 1$.

(ii) $(\delta_1^u, \delta_2^u, 0, 0)$ is the UEP on the stability boundary $\partial A(\delta_1^s, \delta_2^s, 0, 0)$ of the system (Eq. 19.13) if and only if (δ_1^u, δ_2^u) is the UEP on the stability boundary $\partial A(\delta_1^s, \delta_2^s)$ of the system (Eq. 19.14).

19.5 NUMERICAL STUDIES

The boundary property is satisfied if the original model and the reduced-state model have the same UEPs on the stability boundary. It has been shown in the previous section that the boundary property is satisfied if the damping factors are sufficiently large. In this section, we perform a numerical study on the nine-bus test system network-preserving model to illustrate the relationship between the damping factor and the boundary property. The dimension of the state space of the nine-bus network-preserving model is 21. We first consider Contingency 4, which is specified by the events that the faulted bus is #5 and that the line between Buses 5 and 4 is tripped to clear the fault. Compared with the numerical studies presented in the previous chapters, the loading condition considered is heavier; that is, the loading condition is about 150% of the normal loading condition.

We consider the following three different ratios of damping over machine inertia:

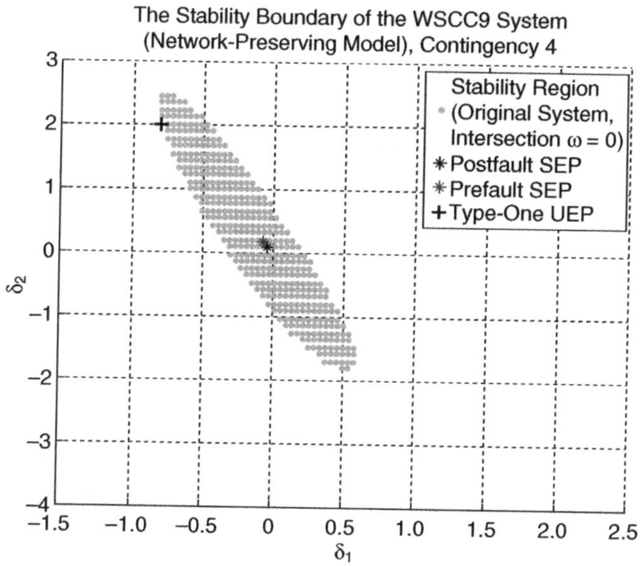

Figure 19.7 The intersecting stability region when the damping/machine ratio is 1.0.

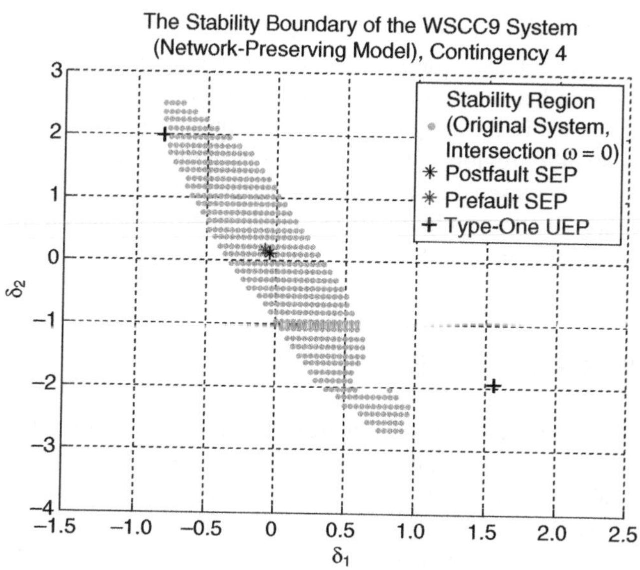

Figure 19.8 The intersecting stability region when the damping/machine ratio is 2.0.

$$\text{(i) } \frac{D_i}{M_i} = 1.0, \text{ (ii) } \frac{D_i}{M_i} = 2.0, \text{ and (iii) } \frac{D_i}{M_i} = 5.0.$$

The intersecting stability regions of the original systems with these three different damping rations are shown in Figures 19.7–19.9, respectively. It is confirmed in the

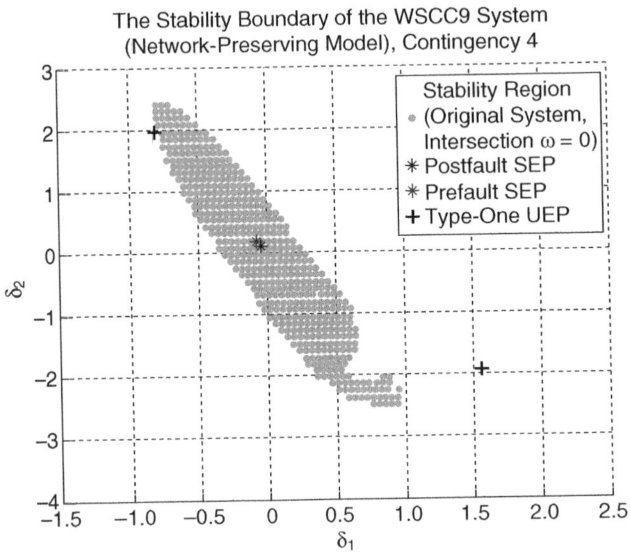

Figure 19.9 The intersecting stability region when the damping/machine ratio is 5.0.

numerical studies that as the damping factors increase, the size of the stability region of the original system increases. This is observed from the intersecting stability region, which is the intersection between the stability region of the original system and its angle subspace. As shown in these figures, as the damping factors increase, the size of the intersecting stability region of the original system increases. In addition, as the damping factors increase, the satisfiability of the boundary property improves as shown in Figure 19.10 and Figure 19.11. These improvements on the satisfiability of the boundary property confirm the theoretical developments presented in this chapter.

19.6 CONCLUDING REMARKS

We have applied the tools of the self-indexing energy function, the symmetrical properties, and the structural stability to analyze the boundary property of the UEPs on the stability boundary of low-dimension, simple power system stability models. In particular, for both the OMIB and the TMIB systems, we have shown that, under certain mild conditions, the original system and the reduced-state system contain the same UEPs on their stability boundaries. In particular, the boundary property is satisfied in these two simple power system models.

From a theoretical viewpoint, insufficient system damping leads to the occurrence of global bifurcation in the parameterized dynamical systems underlying the BCU method. On the other hand, it has been found that the BCU method performs very well if the boundary property is satisfied; in addition, the boundary property (D2) is satisfied if the system damping terms are sufficiently large. It is important

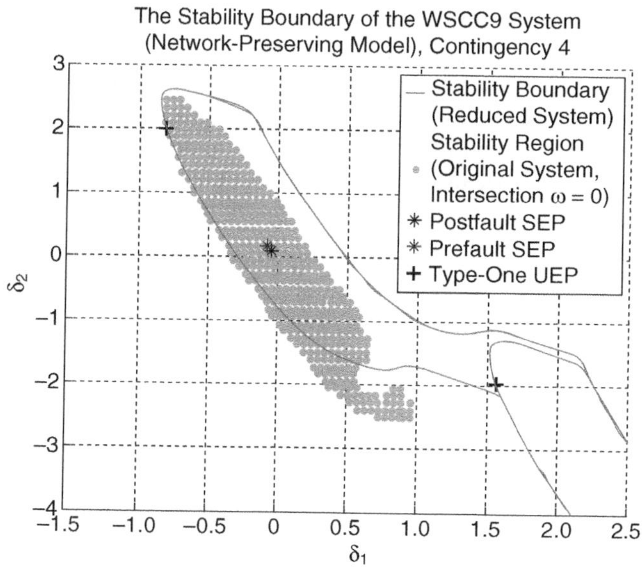

Figure 19.10 The stability boundary of the reduced-state system, other type-one UEPs on the stability boundary, and the stability region of the original system on the angle subspace (for Contingency 4) when the damping/machine ratio is 1.0.

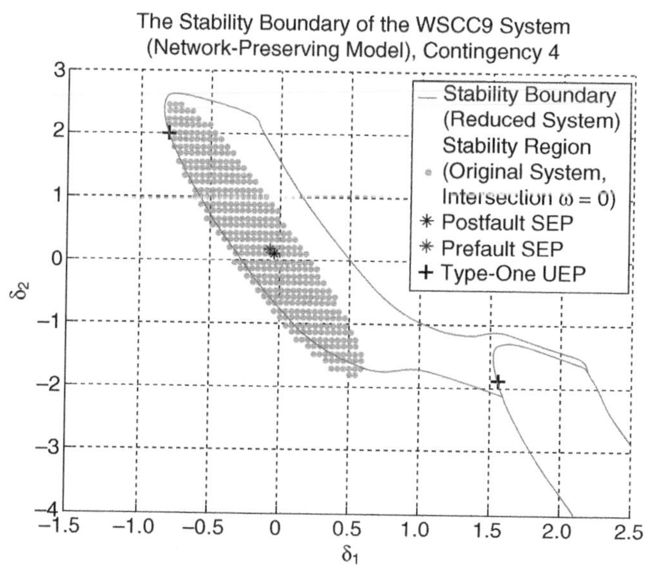

Figure 19.11 The stability boundary of the reduced-state system, other type-one UEPs on the stability boundary, and the stability region of the original system on the angle subspace (for Contingency 4) when the damping/machine ratio is 2.0.

to note that the BCU method still performs well if the boundary property (D2) is not satisfied, as long as the dynamic property (D2′) is satisfied.

As the size of power transient stability models grows, the number of UEPs on a stability boundary increases and the associated classification diagram becomes more complicated. More qualitative and quantitative analysis tools are needed for further developments. We believe that the tools used in this chapter can provide some theoretical contributions to the qualitative analysis of large-scale stability models.

Chapter 20

The BCU–Exit Point Method

20.1 INTRODUCTION

We present in this chapter an extended BCU method, termed the BCU–exit point method, to compute accurate critical energy values for direct stability analysis when the BCU method fails. The BCU–exit point method is built on the BCU method. Recall that a controlling unstable equilibrium point (UEP) computed by the BCU method is said to satisfy the boundary property if the controlling UEP lies on the stability boundary of the original system. If the controlling UEP computed by the BCU method satisfies the boundary property, then its energy value can be used as the critical energy. On the other hand, if the boundary property is not satisfied, then the computed controlling UEP is not correct and the energy value at the controlling UEP cannot be used as the critical energy.

In this chapter, we first present a verification scheme to check whether the boundary property of a computed controlling UEP is satisfied instead of checking the one-parameter transversality condition. This verification scheme will then be integrated into the BCU–exit point method to compute accurate critical energy values when the boundary property is not satisfied. Hence, the role of the BCU–exit point method is to determine an accurate critical energy value for direct stability analysis when the BCU method fails. The BCU–exit point method is "degenerated" into the BCU method when the boundary property of a computed controlling UEP by the BCU method is satisfied. If the boundary property is not satisfied, then the computation result obtained by the BCU method cannot be used since it may give an incorrect critical energy value. Under this situation, the BCU–exit point method determines an accurate critical energy.

20.2 BOUNDARY PROPERTY

Given a power system transient stability model, a study contingency, and a fault sequence, the BCU method computes a controlling UEP with respect to the contingency. The energy value at the computed controlling UEP is then used as the critical

Direct Methods for Stability Analysis of Electric Power Systems, by Hsiao-Dong Chiang

energy for the contingency. Under the one-parameter transversality condition, it has been shown that the computed controlling UEP is a UEP lying on the stability boundary of the original transient stability model. Since the one-parameter transversality condition is difficult to verify, we propose to directly verify the boundary property, which is equivalent to the following dynamic property (D2'):

> Dynamic Property (D2'): The reduced-state controlling UEP lies on the stability boundary of the reduced-state system if and only if the corresponding controlling UEP lies on the stability boundary of the original system.

We next define several terms related to the computational procedure of the BCU method.

Definition: BCU–Exit Point
For a computed UEP by the BCU method with respect to a contingency, the ray connecting the postfault stable equilibrium point (SEP) and the computed UEP will cross the stability boundary of the original system. The crossing point is called the *BCU–exit point.*

Note that if the computed UEP lies on the stability boundary of the original system, then the BCU–exit point is the computed UEP; otherwise, the BCU–exit point lies between the postfault SEP and the computed UEP. The BCU–exit point lies on the stability boundary of the original system.

Definition: BCU–Exit Point Distance
For a computed UEP with respect to a contingency, the *BCU–exit point distance* is the Euclidean distance between its BCU–exit point and the postfault SEP.

Definition: UEP Distance
For a computed UEP with respect to a contingency, the *UEP distance* is the Euclidean distance between this UEP and the postfault SEP.

Note that if the computed UEP lies on the stability boundary of the original system, then the BCU–exit point distance equals the UEP distance.

Definition: Boundary Distance
For the computed UEP with respect to a contingency, its *boundary distance* is the scalar, which is the BCU–exit point distance divided by the UEP distance; that is,

$$\text{Boundary distance} = \frac{\text{BCU} - \text{Exit point distance}}{\text{UEP distance}}.$$

Note that since a UEP will never lie inside the stability region, the boundary distance is not greater than 1.0. The BCU–exit point can be computed by using a time–domain simulation program. According to the definition, the distance of the computed UEP is the scalar, which is the Euclidean distance between the BCU–exit point and the postfault SEP divided by the Euclidean distance between the computed UEP and its postfault SEP (see Figure 20.1). If the boundary distance of a UEP is less than 1.0, then the computed UEP lies outside the stability boundary; otherwise, it lies on the stability boundary (see Figure 20.2).

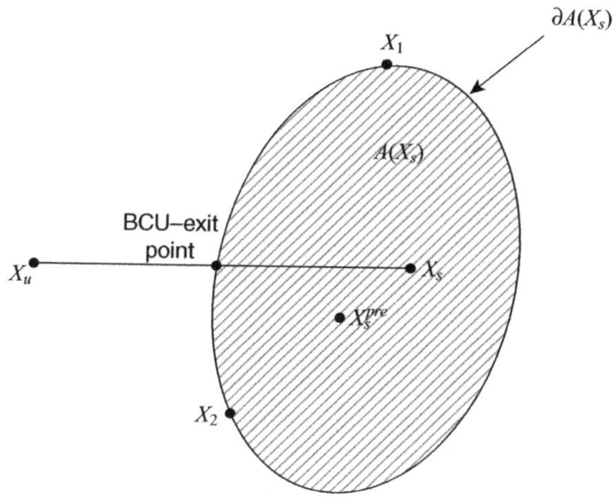

Figure 20.1 Given a computed UEP, we draw a ray between the computed UEP and its postfault SEP. The intersection point between the ray and the stability boundary is the BCU–exit point. If the boundary distance of a UEP is less than 1.0, then the UEP lies outside the stability boundary.

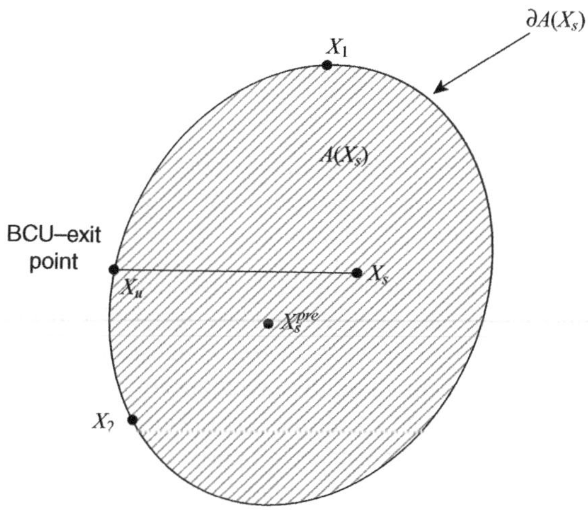

Figure 20.2 If the boundary distance of a UEP equals 1.0, then the UEP lies on the stability boundary.

Definition: Boundary Property
A computed UEP (with respect to a contingency) is said to satisfy the *boundary property* if the computed UEP lies on the stability boundary of the original postfault system; equivalently, the boundary distance of the computed UEP is 1.0.

Note that the boundary distance of a computed UEP is not greater than 1.0. A computed UEP satisfying the boundary property has a boundary distance of 1.0. The smaller the boundary distance of a UEP is, the farther the UEP is away from the stability boundary. With the introduction of a boundary property, one can check the correctness of the computed controlling UEP by the BCU method through checking its boundary property. By computing the boundary distance of the computed controlling UEP, one can verify whether or not the computed controlling UEP lies on the stability boundary of the original system.

20.2.1 Verification Scheme for the Boundary Property

A numerical verification scheme to check the boundary property of a computed UEP will be presented in this section. By computing the boundary distance of the computed UEP, one can verify whether or not the boundary property is satisfied, that is, whether the computed UEP lies on the stability boundary of the original system. If the boundary distance of the computed UEP is 1.0, then the UEP lies on the stability boundary of the original postfault system; otherwise, it does not.

We next present a scheme to check whether a UEP, say, an X^{UEP} of a general nonlinear dynamical system, lies on the stability boundary of a SEP, say, X_s^{post}.

Step 1. Selection step
Select a point (test vector). In practical implementation, we compute a test vector for each selected UEP, say X^{UEP}, using the following equation:

$$X^{test} = X_s^{post} + 0.99\left(X^{UEP} - X_s^{post}\right),$$

where X_s^{post} is the SEP.

Step 2. Checking step
Check the boundary property of X^{UEP} by simulating the system trajectory starting from X^{test}. If the ensuing system trajectory converges to X_s^{post}, then X^{UEP} lies on the stability boundary of X_s^{post}; otherwise, it does not.

If the test vector X^{test} lies inside the stability region of X_s^{post} (checked in Step 2), then the computed UEP, X^{UEP}, must lie on the stability boundary $\partial A\left(X_s^{post}\right)$. If the test vector X^{test} lies outside the stability region of X_s^{post}, then X^{UEP} must lie outside the stability boundary $\partial A\left(X_s^{post}\right)$. Under this situation, the ray connecting X_s^{post} and X^{UEP} must intersect the stability boundary $\partial A\left(X_s^{post}\right)$ at one point. This intersection point is the BCU–exit point. If X^{UEP} satisfies the boundary property, then X^{UEP} itself is the BCU–exit point; otherwise, the BCU–exit point is uniquely determined by the vectors X_s^{post}, X^{UEP}, and the stability boundary $\partial A\left(X_s^{post}\right)$. One significant aspect of the BCU–exit point is that its energy value can be used as the critical energy.

By checking the boundary property, one can ascertain whether or not the UEP computed by the BCU method lies on the stability boundary of the original (postfault) system without checking the one-parameter transversality condition. The theoretical basis of the above procedure rests on the following property of the stability boundary: if X^{UEP} lies on the stability boundary, then the test vector X^{test} must lie inside the stability region (see Figure 20.3); otherwise, X^{UEP} must lie outside the stability boundary (see Figure 20.4).

The computational effort required in the verification scheme roughly equals one time–domain trajectory simulation for each UEP computed by the BCU method. To improve the computational efficiency of this procedure, we will identify the existence of groups of coherent contingencies in a contingency list and their group

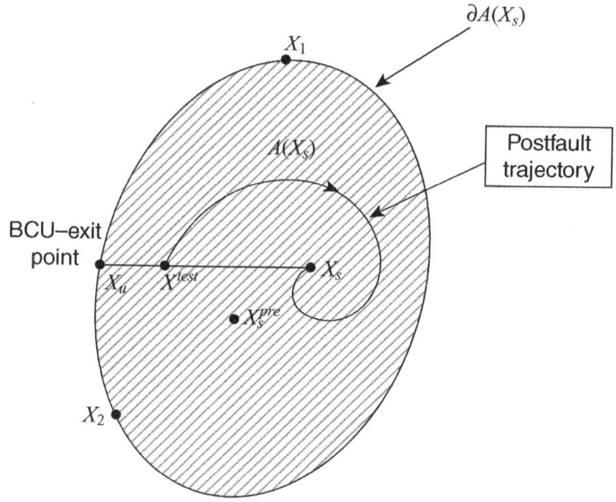

Figure 20.3 The postfault trajectory starting from X^{test} is simulated and assessed. If the postfault trajectory converges to X_s^{post}, then the UEP, X^{UEP}, lies on the stability boundary of the original system and therefore satisfies the boundary property.

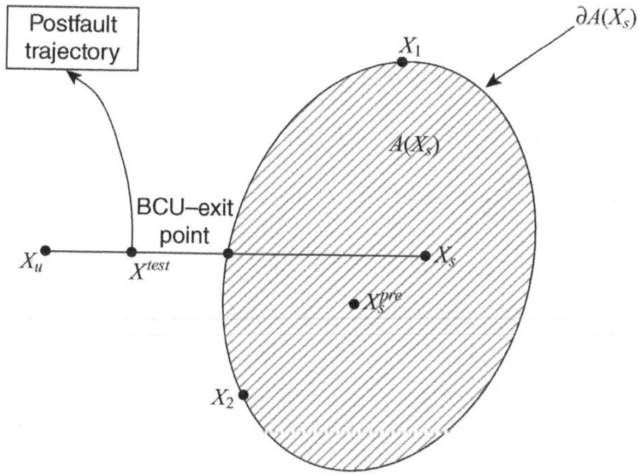

Figure 20.4 The postfault trajectory starting from X^{test} is simulated and assessed. If the postfault trajectory does not converge to X_s^{post}, then the UEP, X^{UEP}, lies outside the stability boundary of the original system, violating the boundary property.

properties. We will also explore these group properties to improve the verification procedure for each group of coherent contingencies. Specifically, we will develop a group-based verification scheme to quickly and accurately verify the boundary property for each computed UEP relative to each contingency in the group of coherent contingencies. The details will be presented later on.

To evaluate the above verification scheme, the BCU method is applied to analyze a contingency list of a 29-generator test system and a contingency list of a 134-generator test system. It is found that the BCU method works very well for most of the contingencies, but it fails to find the correct controlling UEP in some cases. We next present some cases in which the BCU method fails to find the correct controlling UEP and point out the reasons for this failure using the boundary property.

Table 20.1 Contingencies Wherein the BCU Method Finds the Correct Controlling UEP

Fault description	Fault clearing time (s)	Energy margin	BCU assessment	Time–domain assessment	Mode of instability	Boundary distance	Boundary property
Fault bus 709	0.07	1.948	Stable	Stable	Multiswing	1.0	Yes
Open line 104–709	0.10	−0.225	Unstable	Unstable		1.0	Yes
Fault bus 38	0.07	0.468	Stable	Stable	Multiswing	1.0	Yes
Open line 38–55	0.10	−0.551	Unstable	Unstable		1.0	Yes

Table 20.2 Contingencies Wherein the BCU Method Fails to Find the Correct Controlling UEP

Fault description	Fault clearing time (s)	Energy margin	BCU assessment	Time–domain assessment	Mode of instability	Boundary distance	Boundary property
Fault bus 536	0.07	1.1127	Stable	Stable	Multiswing	0.234	No
Open line 536–537	0.10	0.983	Stable	Unstable		0.234	No
Fault bus 707	0.07	1.17	Stable	Stable	Multiswing	0.234	No
Open line 707–708	0.10	0.9433	Stable	Unstable		0.234	No
Fault bus 521	0.05	0.3587	Stable	Unstable	Multiswing	0.232	No
Open line 521–522	0.07	0.2387	Stable	Unstable		0.232	No

Table 20.1 displays some cases for which the BCU method gives correct stability assessments. All these cases exhibit multiswing phenomena, and the boundary property is satisfied in every case. However, the BCU method gives incorrect stability assessments for all the cases listed in Table 20.2. These cases all exhibit multiswing phenomena, but the BCU method fails to find the correct controlling UEP in these cases because the boundary property is not satisfied. It is interesting to note that the boundary distances for all the cases in Table 20.2 are similar. In fact, it will be shown later that the boundary distance is a group property. All the contingencies in Table 20.1 belong to a group of coherent contingencies, while all of the contingencies of Table 20.2 belong to another group of coherent contingencies.

20.2.2 Boundary Property and System Damping

The boundary property is an important property that ensures the reduced-state controlling UEP computed by the BCU method lies on the stability boundary of the original system. If the computed reduced-state controlling UEP satisfies the boundary property, then the corresponding controlling UEP lies on the stability boundary of the original system. The energy value at the computed controlling UEP by the BCU method can then be used as a critical energy for the study contingency. Hence, the boundary property ensures the accuracy of the BCU method in computing the controlling UEP. While the boundary property depends on several factors of power networks, high damping terms tend to ensure that the boundary property is satisfied. This is in agreement with the analytical results derived in Chapter 19.

There is a close relationship between the satisfiability of the boundary property and the system damping. As the system damping increases, the ratio of contingencies that satisfy the boundary property also tends to increase. The task of developing analytical results to confirm this relationship may be challenging. The analytical results developed in Chapter 19 shed some light on this relationship. Instead, we conduct numerical studies to reveal the relationship.

For the 134-generator test system with zero damping, some contingencies for which the BCU method gives incorrect stability assessments are listed in Table 20.3. Again, all these contingencies exhibit multiswing phenomena and the boundary property is not satisfied. We also list some contingencies for which the BCU method gives correct stability assessments in Table 20.4. Again, all these contingencies satisfy the boundary property.

As system damping increases, the performance of the BCU method improves in the sense that the ratio of contingencies satisfying the boundary property increases. For example, the performance of the BCU method on those cases listed in Table 20.3 is improved as the system damping factor increases. The performance of the BCU method on the same 134-generator system, but with a small, nonzero damping factor, is summarized in Table 20.5. It can be seen from this table that the boundary distance of each contingency moves closer to 1.0 as the damping factor increases. From the viewpoint of state space, the computed controlling UEP moves closer to the stability boundary of the original system. Moreover, the boundary property is now satisfied with the contingency (with a fault bus at 107), which now becomes multiswing stable.

As we continue to increase the damping effect, the performance of the BCU method is further improved. Table 20.6 lists the performance of the BCU method on the same contingencies as those in Table 20.5, except that the system damping factors are now doubled. From the viewpoint of stability assessment accuracy, direct stability assessment by the BCU method is now exactly the same as the step-by-step time–domain simulation method. However, from the viewpoint of boundary property, the BCU method computes the controlling UEPs of only three out of the four contingencies.

These numerical studies indicate that as the system damping factor increases, the boundary distance of the controlling UEP computed by the BCU method also

Table 20.3 Cases Wherein the BCU Method Fails (without Damping)

Fault description	Fault clearing time (s)	BCU margin	BCU assessment	Time–domain assessment	Mode of instability	Boundary distance/ UEP group	Boundary property
Fault bus 3036	0.05	0.0211	Stable	Unstable	Multiswing	0.200/27	No
Open line 3036–3037	0.07	−0.0695	Unstable	Unstable		0.200/27	No
Fault bus 4021	0.05	1.9706	Stable	Unstable	Multiswing	0.310/39	No
Open line 4021–4022	0.07	1.9420	Stable	Unstable		0.310/39	No
Fault bus 4029	0.07	0.5595	Stable	Unstable	Multiswing	0.289/35	No
Open line 4026–4029	0.27	−0.3527	Unstable	Unstable		0.289/35	No
Fault bus 107	0.07	0.4302	Stable	Unstable	Multiswing	0.234/52	No
Open line 107–4124	0.15	−0.1049	Unstable	Unstable		0.234/52	No

Table 20.4 Cases Wherein the BCU Method Succeeds (without Damping)

Fault description	Fault clearing time (s)	BCU margin	BCU assessment	Time–domain assessment	Mode of instability	Boundary distance/ UEP group	Boundary property
Fault bus 204	0.07	0.706	Stable	Stable	Single swing	1.00/50	Yes
Open line 204–4005	0.15	−0.392	Unstable	Unstable		1.00/50	Yes
Fault bus 5169	0.07	0.345	Stable	Stable	Multiswing	1.00/21	Yes
Open line 4123–5169	0.15	−0.738	Unstable	Unstable		1.00/21	Yes

Table 20.5 Improvements in the Boundary Property and in the Boundary Distances as System Damping Increases

Fault description	Fault clearing time (s)	BCU margin	BCU assessment	Time–domain assessment	Mode of instability	Boundary distance/ UEP group	Boundary property
Fault bus 3036	0.07	0.177	Stable	Stable	Multiswing	0.701/29	No
Open line 3036–3037	0.17	–0.303	Unstable	Unsable		0.701/29	No
Fault bus 4021	0.07	2.727	Stable	Stable	Multiswing	0.824/34	No
Open line 4021–4022	0.77	–0.157	Unstable	Unstable		0.824/34	No
Fault bus 4029	0.07	0.563	Stable	Stable	Multiswing	0.867/36	No
Open line 4026–4029	0.37	–0.048	Unstable	Unstable		0.867/36	No
Fault bus 107	0.07	1.081	Stable	Sable	Multiswing	1.00/48	Yes
Open line 107–4124	0.30	–0.245	Stable	Stable		1.00/48	Yes

Damping Increases

increases and moves towards the value 1.0. Consequently, the controlling UEP computed by the BCU method tends to satisfy the boundary property as the damping factor increases.

20.3 COMPUTATION OF THE BCU–EXIT POINT

The task of computing the BCU–exit point requires an iterative time–domain simulation between the computed UEP and the postfault SEP. To speed up this computation, we present a golden bisection-based method for fast computation of the BCU–exit point.

The golden bisection method is a one-dimensional search method used to find the optimal solution to a real-valued unimodal function. A unimodal function, $F(x)$, has the property that there is a unique x^* on a given interval, $[a, b]$, such that $F(x^*)$ is the only minimum of $F(x)$ on the interval, and $F(x)$ is strictly decreasing for $x \leq x^*$ and strictly increasing for $x \geq x^*$. The significance of this property is that it enables

Table 20.6 Further Improvements on Contingencies in which the BCU Method Fails as the System Damping Increases

Fault description	Fault clearing time (s)	BCU margin	BCU assessment	Time–domain assessment	Mode of instability	Boundary distance/ UEP group	Boundary property
Fault bus 3036	0.07	0.934	Stable	Stable	Multiswing	1.00/35	Yes
Open line 3036–3037	0.17	0.525	Stable	Stable		1.00/35	Yes
Fault bus 4021	0.07	2.813	Stable	Stable	Multiswing	1.00/7	Yes
Open line 4021–4022	0.77	1.750	Stable	Stable		1.00/7	Yes
Fault bus 4029	0.07	0.588	Stable	Stable	Multiswing	0.948/9	No
Open line 4026-4029	0.37	0.279	Stable	Stable		0.948/9	No
Fault bus 107	0.07	1.182	Stable	Sable	Multiswing	1.00/1	Yes
Open line 107–4124	0.30	0.306	Stable	Stable		1.00/1	Yes

The BCU method computes the correct controlling UEP for three contingencies out of the four multiswing contingencies.

us to refine an interval containing a solution by computing sample values of the solution within the interval and by discarding portions of the interval according to the function values obtained.

Suppose x_1 and x_2 are two points in the interval $[a, b]$ with $x_1 \leq x_2$. Comparing the function values $F(x_1)$ and $F(x_2)$ using the unimodality property will enable us to discard a subinterval, either (x_2, b) or (a, x_1), and to know that the minimum of the function lies within the remaining subinterval. In particular, if $F(x_1) < F(x_2)$, then the minimum cannot lie in the interval (x_2, b); and if $F(x_1) > F(x_2)$, then the minimum cannot lie in the interval (a, x_1). Thus, we are left with a shorter interval, either $[a, x_2]$ or $[x_1, b]$, within which we have already computed one function value, either $F(x_1)$ or $F(x_2)$, respectively. Hence, we will need to compute only one new function evaluation to repeat this process.

To make consistent progress in reducing the length of the interval containing the minimum, we would like for each new pair of points to have the same relationship with respect to the new interval that the previous pair had with respect to the previous interval. Such an arrangement will enable us to reduce the length of the interval by a fixed fraction at each iteration. To accomplish this objective, we choose the relative positions of the two points as λ and $1-\lambda$. Then, we have

$$x_1 = \lambda a + (1-\lambda) b$$

$$x_2 = (1-\lambda) a + \lambda b.$$

The length relationship between the new and previous intervals can be mathematically described as

$$\frac{x_1 - a}{x_2 - a} = \frac{x_2 - a}{b - a}.$$

Hence, we have

$$\frac{(1-\lambda)(b-a)}{\lambda(b-a)} = \frac{\lambda(b-a)}{b-a},$$

from which we get the value of λ as $\lambda = (\sqrt{5}-1)/2 \approx 0.618$ and $1-\lambda = 0.382$. With this choice, no matter which subinterval is retained, its length will be λ relative to the previous interval, and the interior point retained will be at the position of either λ or $1-\lambda$ relative to the new interval. Thus, we need only to compute one new function value at the complementary point to continue the iteration. This choice of temporary points is called the golden bisection search. The complete algorithm is as follows:

Step 0. Initially input function $F(x)$, an interval $[a, b]$ on which $F(x)$ is unimodal, and an error tolerance, ε.

Step 1. If $|b-a| \leq \varepsilon$, the solution is found, stop the algorithm.

Step 2. Calculate the two interior points and the function values at the two points as

$$x_1 = \lambda a + (1-\lambda) b \quad F_1 = F(x_1) \quad \text{and}$$

$$x_2 = (1-\lambda) a + \lambda b \quad F_2 = f(x_2).$$

Step 3. If $F_1 > F_2$, then let

$$a = x_1$$

$$x_1 - x_2 \qquad F_1 = F_2$$

$$x_2 = (1-\lambda) a + vb \quad F_2 = f(x_2)$$

and go back to Step 1; otherwise go to next step.

Step 4. If $F_1 < F_2$, then let

$$b = x_2$$

$$x_2 = x_1 \quad F_2 = F_1$$

$$x_1 = \lambda a + (1-\lambda)b \quad F_1 = F(x_1)$$

and go back to Step 1.

The golden bisection method has an excellent reliability reputation with fast convergence and has been widely used in many commercial software packages for performing a one-dimensional optimal search. We next discuss applying the golden bisection method to compute the critical fault clearing time at which the energy margin is zero.

The golden bisection-based method to determine the critical clearing time is presented as follows:

Step 1. Establish an initial fault clearing time interval, $[t_1, t_2]$, where fault clearing time t_1 corresponds to a stable case and fault clearing time t_2 corresponds to an unstable case.

Step 2. Use the golden bisection interpolation method to calculate two fault clearing time instants in the interval $[t_1, t_2]$: $t_0^{(1)} = 0.618t_1 + 0.382t_2$ and $t_0^{(2)} = 0.618t_2 + 0.382t_1$.

Step 3. Set index counter $k = 1$.

Step 4. Perform the usual time–domain simulation according to the fault clearing time $t_0^{(k)}$. If the postfault system is stable, set $t_1 = t_0^{(k)}$. If the postfault system is unstable, set $t_2 = t_0^{(k)}$.

Step 5. Check for convergence: if $\|t_1 - t_2\| \leq \varepsilon$, go to Step 8. Otherwise, go to Step 6.

Step 6. If $k = 1$, then go to Step 7; if $k = 2$, then go to Step 2.

Step 7. Set the index counter $k = 2$ and go to Step 4.

Step 8. The critical clearing time is set as t_1, and the critical energy is calculated as the energy injected into the system at this fault clearing time.

When applying the golden bisection method to compute the energy margin, the first task is to estimate the lower and upper bounds of the initial fault clearing time interval because the golden bisection method will be applied to this interval. The shorter the interval is, the less computational effort will be involved. The above method can also be applied to compute the BCU–exit point with the computed UEP and the postfault SEP at both end points.

20.4 BCU–EXIT POINT AND CRITICAL ENERGY

For the BCU method, all of the computed controlling UEPs violating the boundary property share one common property—they do not lie on the stability boundary of the original stability model. Instead, they lie outside the stability region of the corresponding postfault SEP. The energy value at the computed UEP cannot be used as the critical energy. When the boundary property is not satisfied, the corresponding BCU–exit point is different from the UEP computed by the BCU method. In this situation, it is necessary to develop remedy methods to determine an accurate critical energy value when the boundary property is violated. To this end, the following issues arise:

- How can one make use of the BCU–exit point to compute the critical energy value?
- What is the relationship, if any, between the computed UEP and the corresponding BCU–exit point?
- What is the relationship between the energy value at the computed UEP and the energy value at the corresponding BCU–exit point?
- Can one compute the correct controlling UEP for each contingency starting from the corresponding BCU–exit point?
- How can one develop a fast method to compute the BCU–exit point?

If the computed UEP corresponding to a contingency does not satisfy the boundary property, then the energy value at the computed UEP cannot be used as a critical energy for the contingency. To compute accurate critical energy values for those contingencies violating the boundary property, we propose to use the energy at the BCU–exit point as the critical energy if the following condition is satisfied:

(C1) The energy at the BCU–exit point must not be greater than the exact critical energy.

This condition is needed in order to preserve the conservative spirit of the controlling UEP method. We next show that the above condition (C1) is always satisfied using numerical simulation on a study power system. In particular, we evaluate the accuracy and reliability of using the energy at the BCU–exit point as the critical energy for direct stability assessment. For the purpose of comparison, we compute the exact critical energy using the time–domain simulation to compute the critical clearing time at which the energy gives the exact critical energy, and compute the critical energy based on the energy at the BCU–exit point. We compare the exact critical energy with the critical energy for evaluating the satisfaction of condition (C1).

The numerical simulation results are summarized in Table 20.7, wherein we list the case number, boundary distance, boundary property, critical energy using the time–domain simulation method, and the critical energy based on the energy at the BCU–exit point. It can be seen from this table that the energy at the BCU–exit point is always less than the exact critical energy and that the difference is small. For example, the boundary distance of Contingency 1174 is 0.964, indicating that the computed UEP does not satisfy the boundary property. The exact critical energy based on the time–domain simulation to find the exit point (at the critical clearing time) for this contingency is 0.351, while the critical energy based on the BCU–exit point method is 0.35. It is clearly shown in this table that the difference between the exact critical energy and the critical energy based on the BCU–exit point is quite small and that condition (C1) is satisfied. Hence, the energy value at the BCU–exit point can be used as the critical energy if the computed controlling UEP by the BCU method violates the boundary property.

Table 20.7 Evaluation of the BCU–Exit Point in Determining Critical Energy Values

Case number	Boundary distance	Boundary property	Critical energy based on	
			Exact time–domain	BCU–exit point
1174	0.964	No	0.351	0.35
1173	0.964	No	0.351	0.35
642	0.814	No	0.358	0.357
31	0.814	No	0.358	0.356
730	0.711	No	0.277	0.271
729	0.711	No	0.277	0.271
595	0.948	No	0.609	0.592
1353	0.948	No	0.61	0.591
715	0.88	No	0.764	0.755
1220	0.88	No	0.763	0.754
1357	0.948	No	0.802	0.8
174	0.948	No	0.886	0.852
357	0.969	No	0.835	0.82
356	0.969	No	0.838	0.821
419	0.804	No	0.279	0.269

20.5 BCU–EXIT POINT METHOD

We present the following BCU–exit point method for computing the BCU–exit point.

BCU–Exit Point Method

Given: a power system transient stability model, a study contingency, and an energy function for the postfault power system model

Function: Compute the BCU–exit point.

Step 1. Use the BCU method to compute the controlling UEP relative to the study contingency.

Step 2. Apply the verification procedure to the computed controlling UEP in Step 1. If the boundary property is satisfied, then the critical energy value of the study contingency is the energy value at the computed controlling UEP; then, go to Step 5; otherwise, go to the next step.

Step 3. Apply an effective time–domain-based method, such as the golden bisection-based method, to compute the corresponding BCU–exit point.

Step 4. The critical energy value of the study contingency is the energy value at the BCU–exit point.

Step 5. Based on the critical energy value, perform a direct stability assessment and calculate the energy margin for the study contingency.

The BCU–exit point method is a corrective method for the BCU method when the boundary property of the computed controlling UEP is violated. To illustrate the accuracy of the BCU–exit point method, it will be numerically shown that the BCU–exit point method satisfies condition (C1) on a practical power system model with a large number of contingencies. To provide benchmark results, golden bisection-based time–domain simulation is used to determine the exact exit point (i.e., the intersection point between the stability boundary of the original system and the fault-on trajectory) whose energy value is the exact critical energy.

Numerical simulation results are summarized in Tables 20.8 and 20.9, which contain five columns; the first column lists the number of contingencies, while the second lists the boundary distance of each computed UEP associated with each contingency in column one. The third column states whether or not the boundary property is satisfied. The forth column is the exact energy value, while the fifth column is the energy value at the BCU–exit point. The contingencies in each table belong to the same group of coherent contingencies.

It is clear from the boundary distance values in Tables 20.8 and 20.9 that all the computed UEPs do not satisfy the boundary property. Hence, the energy value at the computed UEP cannot be used as a critical energy. Instead, the energy value at the BCU–exit point can be used as a critical energy value. We make a comparison between the critical energy values based on the BCU–exit point method and the energy values based on the exact time–domain method and focus on the computed energy margins by these two methods. The critical energy value obtained using the exact time–domain method is listed in the fourth column of the table, while the critical energy value obtained using the BCU–exit point method is listed in the fifth column.

One important observation from these numerical studies is that the energy value at the BCU–exit point is always less than the energy value at the critical clearing time point (i.e., the exact critical energy). Hence, condition (C1) is satisfied. This indicates that the BCU–exit point method, following the spirit of the controlling UEP method, yields accurate yet conservative stability assessments.

20.6 CONCLUDING REMARKS

Given a study contingency, the BCU method computes a controlling UEP relative to the study contingency. The question of whether the computed controlling UEP lies on the stability boundary of the (original) postfault system naturally arises. To address this issue, we have presented in this chapter a numerical verification scheme to check the boundary property of the computed controlling UEP. This verification scheme overcomes the difficult task of checking the one-parameter transversality condition.

When the boundary property is not satisfied, the energy at the computed controlling UEP cannot be used as the critical energy for the contingency. In this situation, the proposed BCU–exit point method computes an accurate critical energy for the study contingency. The BCU–exit point method can also be viewed as an extended

Table 20.8 Evaluation of the BCU–Exit Point Method to Determine Critical Energy Values on a Group of Coherent Contingencies Whose Boundary Distance Is Around 0.948

Case number	Boundary distance	Boundary property	Critical energy based on	
			Time–domain	BCU–exit point
1293	0.948	No	0.611	0.568
1295	0.948	No	0.611	0.568
1291	0.948	No	0.611	0.568
512	0.948	No	0.561	0.563
1355	0.948	No	0.611	0.594
595	0.948	No	0.609	0.592
1353	0.948	No	0.61	0.591
1210	0.948	No	0.608	0.59
1351	0.948	No	0.61	0.587
1329	0.948	No	0.611	0.588
1429	0.948	No	0.557	0.555
601	0.948	No	0.607	0.584
405	0.948	No	0.608	0.584
394	0.948	No	0.604	0.579
396	0.948	No	0.604	0.579
398	0.948	No	0.605	0.58
245	0.948	No	0.601	0.576
247	0.948	No	0.599	0.572
1248	0.948	No	0.526	0.524
1247	0.948	No	0.528	0.525
1244	0.948	No	0.526	0.524
1243	0.948	No	0.528	0.525
1246	0.948	No	0.526	0.524
1245	0.948	No	0.528	0.525
742	0.948	No	0.523	0.522
741	0.948	No	0.524	0.523
740	0.948	No	0.523	0.522
739	0.948	No	0.524	0.523
744	0.948	No	0.523	0.522
743	0.948	No	0.524	0.522

BCU method including an effective verification procedure to check the boundary property. It gives accurate stability assessments and energy margin calculations for the study contingency regardless of whether the boundary condition is satisfied.

The computational effort required by the BCU–exit point method can be considerably greater than that of the BCU method. Roughly speaking, the BCU–exit point method contains the following major computational steps:

Table 20.9 Evaluation of the BCU–Exit Point Method to Determine Critical Energy Values on One Group of Coherent Contingencies

Case number	Boundary distance	Boundary property	Critical energy based on	
			Time–domain	BCU–exit point
1216	0.88	No	0.763	0.742
1214	0.88	No	0.763	0.742
1218	0.88	No	0.763	0.754
712	0.88	No	0.758	0.751
710	0.88	No	0.758	0.751
711	0.88	No	0.763	0.754
709	0.88	No	0.763	0.754
715	0.88	No	0.764	0.755
1220	0.88	No	0.763	0.754
716	0.88	No	0.759	0.752
1237	0.88	No	0.758	0.751
714	0.88	No	0.758	0.751
713	0.88	No	0.758	0.751

Table 20.10 A Comparison between the Computing Functions and Computational Efforts of the BCU Method and the BCU–Exit Point Method

Methods/computation	BCU method	BCU–exit point method
Computing the controlling UEP	Yes	Yes
Verification	No	Yes (one time–domain simulation)
Computing the BCU–exit point	No	Yes if boundary property is violated (four or five time–domain simulations)

- Compute the controlling UEP, which equals the computational effort required by the BCU method.
- Verify the boundary condition, which requires one time–domain simulation.
- If necessary, compute the BCU–exit point, which requires three to four time–domain simulations (using the BCU-guided time–domain procedure).

A comparison of the computational efforts required by the BCU method and those of the BCU–exit point method reveals that the required computational effort in the BCU–exit point method is about one to four or five time–domain simulations more than that required by the BCU method (see Table 20.10).

It should be noted that the computational efforts spent in the BCU–exit point method will prove fruitful when we explore the group properties of contingencies. To improve the verification procedure, we will develop a group-based verification procedure that entails only one or two time–domain simulations for each group of coherent contingencies instead of one time–domain simulation for every contingency in the group. By exploring the group properties of contingencies, we will develop a group-based BCU–exit point method to significantly reduce the computational requirement. Details of the group-based BCU–exit point method will be presented in Chapter 21.

Chapter 21

Group Properties of Contingencies in Power Systems

21.1 INTRODUCTION

Group properties in general power networks will be explored in this chapter. The concept of a group of coherent contingencies will be formulated and demonstrated (Chiang et al., 2007, 2009; Tada and Chiang, 2008). The associated group properties for each group of coherent contingencies will be presented. These group properties include static group properties and dynamic group properties. These group properties will be explored and incorporated into the development of group-based BCU methods and will be presented in the next few chapters.

We define a group property with respect to a group containing several members. A group property is a property that every member of the group satisfies. To identify a group property, it is necessary to form a group of members. We will show that the boundary property is a group property provided that the group of contingencies is properly and accurately formed. Using group properties, we will show that it is not necessary to compute the boundary distance for every computed unstable equilibrium point (UEP) in a group of coherent contingencies. Instead, it is sufficient to compute the boundary distance of one UEP in a group of coherent contingencies in order to check the boundary property for all contingencies in that group. This exploration leads to a significant reduction in the computation and development of effective preventive control actions against a set of insecure contingencies and enhancement control actions for a set of critical contingencies.

The concept of coherent contingencies is not related to the concept of coherent generators (Podmore, 1978; Wu and Murlhi, 1983). Coherency in generators is an observed phenomenon in power systems where certain generators tend to swing together after a disturbance. A "disturbance" can include generator tripping, line

Direct Methods for Stability Analysis of Electric Power Systems, by Hsiao-Dong Chiang

switching, load shedding, and sudden variations in generated or absorbed power. These generators are referred to as a group of coherent generators. A coherent group of generators can be aggregated to form a single, equivalent generator model that has the same effect on system dynamic modes as the original generator set. The process of aggregation removes the high-frequency, intergenerator modes from the aggregated model, and the controls of the generators in the group must be compatible. To achieve this, coherent groups are split, when necessary, into subgroups that have compatible controls. Then, each of these subgroups is replaced by an equivalent detailed generator model.

Mathematically speaking, a group $I \subset \{1, 2, \ldots, N\}$ of generators is said to be coherent in $x^0 = x(t_0)$ with respect to a perturbation (disturbance) occurring at time t_0 if the rotor angles of any pair of generators, $\{p, q\} \in I$, satisfies the following conditions:

$$\delta_p(t) - \delta_q(t) = \delta_p(t_0) - \delta_q(t_0), t \geq t_0 \text{ or}$$
$$\delta_p(t) - \delta_q(t) = \varepsilon + \delta_p(t_0) - \delta_q(t_0), t \geq t_0,$$

where ε is a small number. Once the set of coherent generators is identified, there remains the problem of deriving a reduced-order approximation model of the external system, known as the "dynamic equivalent." To this end, an aggregated generator model is computed and inserted into the system while the generators in the corresponding groups are removed. The network is modified to maintain the same power flow conditions as the external system seen from the coherent group. The purpose of approximating the effects of external system dynamics via the dynamic network reduction is to reduce the computational burden, data requirements, and storage associated with time–domain stability simulations. The process of dynamic network reduction consists of the following three steps: (1) identifying groups of coherent generators, (2) aggregating coherent generators, and (3) performing network reduction.

While the application of coherent generators is from a modeling viewpoint (i.e., to derive a dynamic equivalent), the application of the group of coherent contingencies is from a computational and control viewpoint. Under the proposed concept of a group of coherent contingencies, the underlying transient stability model will not be simplified. Instead, the number of contingencies that needs to be examined will be greatly reduced. Moreover, the group properties will facilitate the design of the required control actions such as preventive control (against insecure contingencies) and enhancement control (against critical contingencies).

The group of coherent contingencies and associated group properties are studied in this chapter with the purpose of (1) checking whether the controlling UEP computed by the BCU method lies on the stability boundary of the original system in a computationally effective manner and (2) computing accurate and yet slightly conservative critical energy values in a computationally effective manner provided the boundary property is not maintained in the group. These group properties will be explored and incorporated into the development of a group-based BCU method, which is presented in the next two chapters. This chapter focuses on the concept of coherent contingencies and the discovery of their group properties and applications. These group properties include both static and dynamic group properties.

21.2 GROUPS OF COHERENT CONTINGENCIES

Some definitions related to the group of coherent contingencies will be presented in this section. Given a contingency, the corresponding postfault system is well-defined. The postfault stable equilibrium point (SEP), whose stability region contains the prefault SEP and the controlling UEP relative to the fault-on trajectory of the contingency, is unique. Given a contingency, it makes sense to relate a contingency to the corresponding postfault SEP and the controlling UEP computed by the BCU method. In other words, given a contingency, the corresponding postfault SEP is unique and the corresponding controlling UEP computed by the BCU method is also unique. Hence, the concept of a group of coherent contingencies is defined by the corresponding postfault SEP and/or the controlling UEP.

Definition: Contingency List
For a given power system dynamic model, the entire group of contingencies (say, L) to be studied by a study program in one execution is said to form a *contingency list.*

Definition: SEP Separation
Given a contingency, the *SEP separation* of the contingency is the infinite norm between its prefault SEP, say, δ_s^{pre}, and its postfault SEP, say, δ_s^{post}; that is,

$$\text{SEP separation} = \left\| \delta_s^{pre} - \delta_s^{post} \right\|_\infty .$$

Definition: Coherent Contingencies
Two contingencies are said to be *coherent* if the generator rotor angles at their corresponding SEPs and computed controlling UEPs are close to each other. Mathematically speaking, we say contingency i and contingency j are *coherent contingencies* if the following conditions are satisfied:

$$\left\| \delta_i^{cuep} - \delta_j^{cuep} \right\|_\infty < \varepsilon^{cuep}$$

$$\left\| \delta_i^{sep} - \delta_j^{sep} \right\|_\infty < \varepsilon^{sep},$$

where (1) δ_i^{sep} and δ_j^{sep} are the generator rotor angle vectors of the postfault SEPs for contingency i and contingency j, respectively, and ε^{sep} is a specified angle tolerance; and (2) δ_i^{cuep} and δ_j^{cuep} are the generator rotor angle vectors of the computed controlling UEPs for contingency i and contingency j, respectively, and ε^{cuep} is a specified angle tolerance.

Note that the norm used in this definition is the infinite norm. Hence, this definition is valid for both large and small power systems. The above definitions were motivated by the following analytical basis. The SEP separation can be viewed as a measure of the static severity of a contingency. The "separation" between the computed UEP and the postfault SEP of a contingency can be viewed as a measure of the dynamic severity of a contingency. As a result, we define these two measures and apply them to develop schemes for grouping the list of contingencies into groups of coherent contingencies.

21.3 IDENTIFICATION OF A GROUP OF COHERENT CONTINGENCIES

Given a contingency list, we can group all contingencies in the contingency list into groups of coherent contingencies based on the difference between the coordinates of the computed controlling UEP and their SEP separations. The grouping scheme proceeds as follows:

Step 1. Compute the SEP separation for each contingency in the contingency list.

Step 2. Apply the BCU method to compute the controlling UEP for each contingency in the contingency list.

Step 3. (Grouping step) Classify all the contingencies into groups of coherent contingencies based on the closeness of the coordinates of each computed controlling UEP associated with each contingency.

Step 4. (Regrouping step) Divide each group of contingencies (obtained in Step 3) into several subgroups of contingencies based on the SEP separation associated with each contingency until no more subgroups can be formed.

We next show the existence of groups of coherent contingencies and present their static and dynamic group properties in the next two sections. We apply the proposed concept of a group of coherent contingencies to lists of credible contingencies of two test systems to validate the concept. These two test systems are an IEEE test system and a 29-generator test system. The IEEE test system has 67 groups of coherent contingencies; some of them contain large numbers. For example, 305 coherent contingencies are contained in Group 1, while other groups contain a smaller number of contingencies. We list in Table 21.1 the top 10 groups of coherent contingencies.

The 29-generator test system has 19 groups of coherent contingencies; some of them contain large numbers. For example, 180 coherent contingencies are contained in Group 1, while other groups contain a smaller number of contingencies. In Table 21.2, we list 10 groups of coherent contingencies.

Several group properties have been discovered in each group of coherent contingencies. These group properties are classified into static group properties and dynamic group properties. We describe these two classes of group properties in the next two sections.

Table 21.1 Groups of Coherent Contingencies for the IEEE 50-Generator Test System

Group number	Number of coherent contingencies
1	305
2	83
3	54
4	17
5	16
6	15
7	11
8	11
9	10
10	10

Table 21.2 The Number of Contingencies in the Top 10 Groups of Coherent Contingencies in the 29-Generator Test System

Group number	Number of coherent contingencies
1	180
2	37
3	19
4	11
5	8
6	7
7	7
8	7
9	7
10	6

21.4 STATIC GROUP PROPERTIES

We will show that within each group of coherent contingencies, the UEP with the greatest SEP separation and the UEP with the smallest SEP separation define the boundary property for all UEPs in the group. In addition, each group of coherent contingencies shares some interesting properties as shown below.

Group Property 1: Locational Relationship

The fault locations at which each group of coherent contingencies occurs usually form a geographic cluster. In order to illustrate this static group property, we first list the coordinates of the computed controlling UEPs of a group of coherent contingencies from the IEEE test system in table format and then display their geographic relationship on a schematic online diagram. Since the number of coherent contingencies can be large, we only list the coordinates of the computed controlling UEP for some contingencies from the group in the table, while the locations of all the coherent contingencies are highlighted on the schematic online diagram in the figures. For instance, Table 21.3 lists the coordinates of the computed controlling UEPs of 12 contingencies from Group 7 of coherent contingencies, while the locations of all the coherent contingencies in the group, which has 17 contingencies, are highlighted on the schematic online diagram in Figure 21.1. Obviously, the coordinates of the computed controlling UEPs of the coherent contingencies in the group are indeed close to each other.

We notice that within the group, the coherent contingencies are usually geographically clustered. However, these locations can also be disconnected in a geographic sense, while they are "connected" in an electrical sense. For instance, the locations of all the coherent contingencies in Group 16, which has 10 contingencies, are highlighted on the schematic online diagram in Figure 21.2. The locations of these 10 contingencies are divided into two subregions. We note that the locations of these coherent contingencies are clustered into two subregions, which can be close together in electrical distance, while far apart in geographic distance. Table 21.4 lists the coordinates of the computed controlling UEP of 10 contingencies from Group 16 of the coherent contingencies. Again, the coordinates of the computed controlling UEPs of the coherent contingencies in the group are close to each other.

The locational relationship among the coherent contingencies in the group also occurs in the 29-generator test system. The locations of all the coherent contingencies can be clustered in one subregion or in two disconnected subregions. For instance, the locations of coherent contingencies of one group are clustered in one subregion as shown in Figure 21.3, while the locations of coherent contingencies of one group are clustered in two subregions as shown in Figure 21.4.

Group Property 2: Boundary Property

Recall that a computed UEP with respect to a contingency is said to satisfy the *boundary property* if the computed UEP lies on the stability boundary of the original postfault system. Hence, a computed UEP satisfying the boundary property has a boundary distance of 1.0, while it violates the boundary property if its boundary distance is less than 1.0. The smaller the boundary distance of a UEP, the farther the UEP is from the stability boundary of the original system.

We next illustrate that the boundary property is a group property through numerical studies. Several groups of coherent contingencies along their boundary distances are listed in Tables 21.5–21.8. The boundary property is a group property as seen, for example, from the fact that every contingency of Group 1 does not possess the boundary property (i.e., its boundary distance is less than 1.0), while every contingency of Groups 5, 11, and 32 all possess the boundary property.

Table 21.3 The Coordinates of the Computed CUEPs of 12 Coherent Contingencies in Group 7

Case/ generator	Case 450	Case 460	Case 154	Case 434	Case 481	Case 468	Case 507	Case 483	Case 501	Case 150	Case 505	Case 493
1	87.6	87.5	87.6	87.5	87.6	87.4	87.4	87.7	87.1	87.5	87.5	87.6
2	130.8	130.8	130.8	130.8	130.8	130.7	130.7	130.9	130.5	130.8	130.8	130.8
3	117	116.9	117	116.9	116.9	116.8	116.8	117.1	116.5	116.9	116.9	117
4	117.3	117.3	117.3	117.3	117.3	117.2	117.2	117.4	116.9	117.2	117.3	117.3
5	131.7	131.7	131.7	131.8	131.7	131.9	131.7	131.7	131.7	131.8	131.7	131.7
6	146.0	146.0	146.0	146.0	146.0	145.9	145.8	146.1	145.5	145.8	146.0	146.0
7	108.6	108.6	108.6	108.6	108.6	108.5	108.4	108.7	108.1	108.5	108.6	108.6
8	134.6	134.6	134.6	134.7	134.7	134.9	134.6	134.7	134.5	134.9	134.6	134.6
9	129.0	128.9	129.0	128.9	128.9	128.9	128.8	129.0	128.5	128.9	128.9	129.0
10	97.5	97.5	97.5	97.5	97.5	97.4	97.3	97.6	97	97.4	97.5	97.5
11	91.8	91.8	91.8	91.8	91.8	91.7	91.7	91.9	91.4	91.8	91.8	91.8
12	267.2	267.2	267.1	267.3	267.2	267.6	267	267.2	266.8	267.1	267.1	267.1
13	134.4	134.4	134.4	134.4	134.4	134.3	134.3	134.5	134.1	134.4	134.4	134.4
14	150.9	150.9	150.9	150.9	150.9	150.8	150.8	151.0	150.5	150.8	150.9	150.9
15	136.5	136.5	136.5	136.5	136.5	136.4	136.4	136.6	136.1	136.5	136.5	136.5
16	136.3	136.3	136.3	136.3	136.3	136.2	136.2	136.4	135.9	136.2	136.3	136.3
17	132.2	132.2	132.2	132.2	132.2	132.0	132.0	132.3	131.7	132.3	132.2	132.5
18	56.0	56.0	56.0	56.0	56.0	56.0	56.0	56.1	56.0	56.1	56.0	56.1
19	132.7	132.7	132.7	132.7	132.7	132.6	132.5	132.8	132.2	132.6	132.7	132.7
20	152.2	152.2	152.2	152.2	152.2	152.1	152	152.3	151.7	152.1	152.2	152.2
21	140.9	140.9	140.9	141	141.1	141.4	140.7	141.1	140.4	141.7	140.8	140.8
22	140.7	140.7	140.7	140.7	140.5	140.8	140.6	140.8	140.3	140.8	140.7	140.7
23	130.9	130.9	130.9	130.9	130.9	131	130.9	130.8	131.3	130.9	130.9	130.9
24	119.2	119.2	119.2	119.2	119.2	119.3	119.1	119.1	119.1	119.3	119.2	119.2

Table 21.3 Continued

Case/ generator	Case 450	Case 460	Case 154	Case 434	Case 481	Case 468	Case 507	Case 483	Case 501	Case 150	Case 505	Case 493
25	128.8	128.8	128.8	128.7	128.8	128.7	128.6	128.9	128.3	128.7	128.8	128.8
26	145.5	145.5	145.6	145.5	145.5	145.4	145.4	145.7	145.1	145.4	145.5	145.6
27	131.6	131.6	131.6	131.6	131.6	131.5	131.5	131.7	131.2	131.7	132	131.7
28	−3.2	−3.2	−3.2	−3.2	−3.2	−3.2	−3.2	−3.2	−3.1	−3.2	−3.2	−3.2
29	5.9	5.9	5.9	5.9	5.9	5.9	6.0	5.9	6.0	6.0	5.9	5.9
30	16.6	16.6	16.6	16.6	16.6	16.6	16.6	16.6	16.7	16.7	16.6	16.6
31	9.1	9.1	9.1	9.1	9.1	9.1	9.1	9.1	9.2	9.2	9.1	9.1
32	−40.2	−40.2	−40.2	−40.2	−40.2	−40.2	−40.2	−40.1	−40.3	−40.1	−40.2	−40.2
33	120.5	120.5	120.5	120.5	120.5	120.5	120.6	120.4	120.8	120.4	120.5	120.5
34	121.3	121.3	121.3	121.3	121.3	121.2	121.3	121.3	121.2	121.3	121.3	121.3
35	122.3	122.3	122.3	122.2	122.3	122.2	122.2	122.3	122	122.2	122.3	122.3
36	−9.1	−9.1	−9.1	−9.1	−9.1	−9.1	−9.2	−9.1	−9.3	−9.1	−9.1	−9.1
37	−33.5	−33.5	−33.5	−33.5	−33.5	−33.5	−33.5	−33.5	−33.5	−33.5	−33.5	−33.5
38	−9.4	−9.4	−9.4	−9.4	−9.4	−9.4	−9.4	−9.4	−9.4	−9.4	−9.4	−9.4
39	27.3	27.3	27.3	27.3	27.3	27.3	27.3	27.3	27.4	27.3	27.3	27.3
40	−10.6	−10.6	−10.6	−10.6	−10.6	−10.6	−10.6	−10.6	−10.5	−10.6	−10.6	−10.6
41	28.6	28.6	28.6	28.6	28.6	28.6	28.6	28.6	28.7	28.6	28.6	28.6
42	2.9	2.9	2.9	2.9	2.9	2.9	2.9	2.9	3.0	2.9	2.9	2.9
43	−100.0	−100.0	−99.9	−99.9	−100.0	−100.0	−99.9	−100.0	−99.8	−100.0	−99.9	−100.0
44	−37.4	−37.4	−37.4	−37.4	−37.4	−37.4	−37.4	−37.4	−37.4	−37.4	−37.4	−37.4
45	−4	−4	−4	−4	−4	−4	−4	−4	−4	−4	−4	−4
46	2.0	2.0	2.0	2.0	2.0	2.0	2.0	2.0	2.0	2.0	2.0	2.0
47	10.0	10.0	10.0	10.0	10.0	10.0	10.0	10.0	10.0	10.0	10.0	10.0
48	2.5	2.5	2.5	2.5	2.5	2.5	2.5	2.5	2.5	2.5	2.5	2.5
49	11.1	11.1	11.1	11.1	11.1	11.1	11.2	11.1	11.2	11.1	11.1	11.1
50	−16.6	−16.5	−16.6	−16.6	−16.5	−16.5	−16.5	−16.6	−16.5	−16.6	−16.5	−16.6

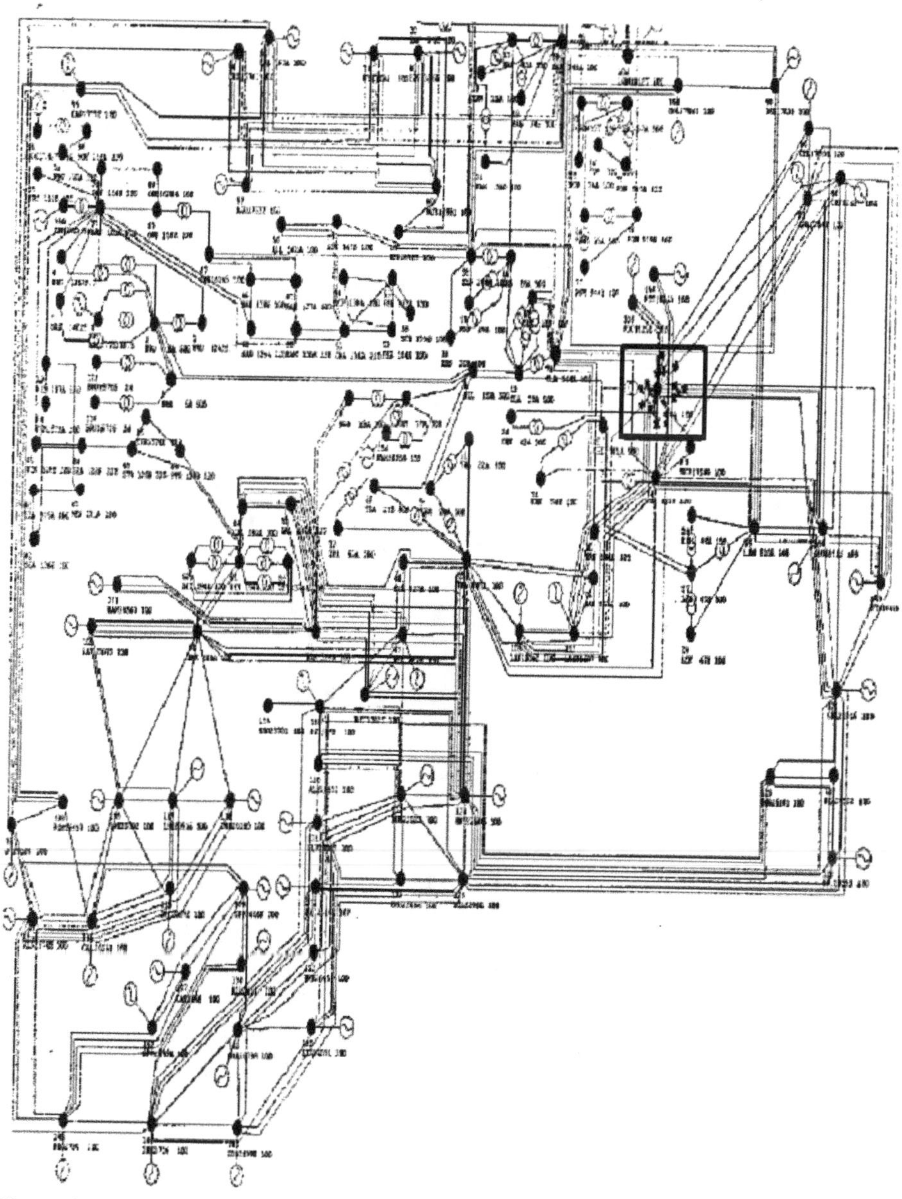

Figure 21.1 The highlighted portion of a transmission network of the IEEE test system represents the geographic relationship among the coherent contingencies from Group 7 (17 cases). The sign X indicates the location of a contingency on a transmission line.

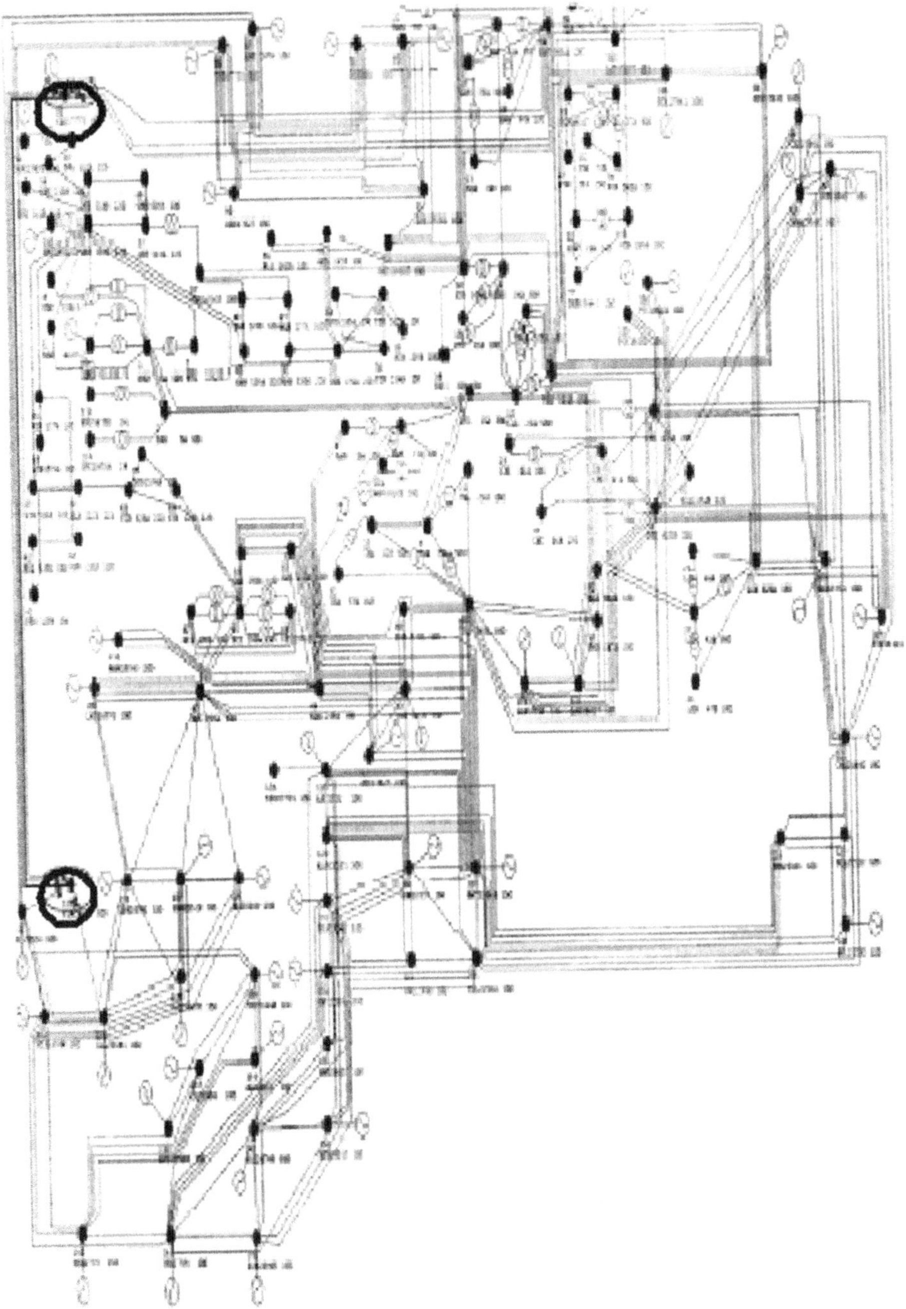

Figure 21.2 The highlighted portion of a transmission network of the IEEE test system represents the geographic relationship among the coherent contingencies from Group 16 (10 cases). The sign X indicates the location of a contingency on a transmission line.

Table 21.4 The Coordinates of the Computed CUEPs of 10 Coherent Contingencies in Group 16

Case/ generator	Case 822	Case 806	Case 558	Case 598	Case 316	Case 596	Case 314	Case 600	Case 599	Case 810
1	33.5	33.4	31.4	31.9	30.1	22.2	14.1	35.6	35.6	29.2
2	19.6	19.6	19.4	19.5	19.4	19	18.6	20	20	19.2
3	27.1	27.1	25.1	26.4	25.6	21.7	18.5	28.5	28.5	25.1
4	29.1	29.1	27.1	28.3	27.5	23.3	19.9	30.5	30.5	26.9
5	12.1	12.2	11.9	12.0	11.9	11.4	11.0	12.6	12.7	11.7
6	34.4	34.4	33.8	34.2	33.9	32.9	32.0	35.0	35.1	33.6
7	44.8	44.7	42.6	43.4	41.9	33.6	27.8	46.7	46.6	41.2
8	15.5	15.6	15.2	15.4	15.3	14.7	14.3	16.0	16.0	15.0
9	18.3	18.4	18.0	18.2	18.1	17.5	17.0	18.8	18.9	17.8
10	44.2	44.0	42.0	42.5	40.9	30.9	25.6	46.3	46.2	39.9
11	169.1	164.2	165.7	166.6	165.9	152.8	155	155	155	152.8
12	17.2	17.3	16.9	17.1	16.9	16.4	15.9	17.7	17.7	16.7
13	22.8	22.9	22.6	22.7	22.6	22.2	21.8	23.2	23.3	22.3
14	37.6	37.6	37.1	37.4	37.2	36.4	35.7	38.1	38.2	36.9
15	25.6	25.7	25.3	25.5	25.3	24.8	24.2	26.1	26.2	25.1
16	25.2	25.3	24.7	25.0	24.8	23.8	23.0	25.8	25.9	24.5
17	18.7	18.8	18.4	18.6	18.4	17.9	17.3	19.2	19.2	18.2
18	19.8	19.8	19.7	19.7	19.7	19.5	19.3	20	20.0	19.5
19	22.3	22.4	21.8	22.1	21.9	20.8	20.0	22.9	23.0	21.6
20	37.9	38.0	37.6	37.8	37.7	37.1	36.6	38.4	38.4	37.4
21	25.8	25.9	25.5	25.7	25.5	25	24.4	26.3	26.4	25.3
22	26.2	26.2	25.8	26.0	25.9	25.3	24.8	26.6	26.7	25.6
23	7.7	7.8	7.5	7.6	7.5	7.1	6.7	8.1	8.2	7.2
24	0.2	0.3	0	0.1	0	−0.5	−0.9	0.7	0.7	−0.3
25	18.6	18.7	18.3	18.5	18.4	17.8	17.3	19.1	19.2	18.1
26	32.2	32.2	31.9	32.1	31.9	31.4	31.0	32.6	32.7	31.7
27	18.2	18.3	17.9	18.1	17.9	17.4	16.8	18.7	18.7	17.7
28	−2.4	−2.4	−2.4	−2.4	−2.4	−2.3	−2.3	−2.4	−2.4	−2.4
29	−3.5	−3.5	−3.6	−3.5	−3.6	−3.6	−3.6	−3.4	−3.4	−3.6
30	−2.6	−2.5	−2.6	−2.6	−2.6	−2.7	−2.8	−2.4	−2.4	−2.7
31	−4.2	−4.2	−4.3	−4.3	−4.3	−4.3	−4.4	−4.1	−4.1	−4.4
32	−48.4	−48.3	−48.4	−48.4	−48.4	−48.4	−48.5	−48.1	−48	−48.6
33	−6.9	−6.8	−7.0	−6.9	−7.0	−7.2	−7.5	−6.5	−6.4	−7.2
34	13.0	13.0	12.8	12.9	12.8	12.6	12.3	13.3	13.4	12.6
35	11.9	12	11.7	11.9	11.8	11.4	11.1	12.3	12.4	11.5
36	−28.9	−28.8	−29.0	−28.9	−29.0	−29.0	−29.1	−28.6	−28.5	−29.2
37	−38.9	−38.8	−38.9	−38.9	−38.9	−38.9	−38.9	−38.6	−38.6	−39.1
38	−12.9	−12.8	−12.9	−12.9	−12.9	−12.8	−12.9	−12.7	−12.6	−13.0
39	5.5	5.5	5.5	5.5	5.5	5.4	5.3	5.7	5.8	5.3
40	2.1	2.2	2.2	2.2	2.2	2.3	2.4	2.2	2.2	2.1

Table 21.4 Continued

Case/ generator	Case 822	Case 806	Case 558	Case 598	Case 316	Case 596	Case 314	Case 600	Case 599	Case 810
41	43.3	42.7	43.0	42.9	43	43	43	41.6	41.6	44.4
42	17.3	17.3	17.4	17.4	17.4	17.5	17.5	16.9	16.9	17.7
43	−61.2	−61.2	−61.1	−61.2	−61.1	−60.9	−60.7	−61.0	−61.0	−61.3
44	−13.1	−13.1	−13.1	−13.1	−13.1	−12.9	−12.8	−13.1	−13.1	−13.1
45	2.8	2.9	2.9	2.9	2.9	2.9	3	2.8	2.8	2.9
46	2.4	2.4	2.4	2.4	2.4	2.4	2.4	2.5	2.5	2.3
47	2.6	2.7	2.6	2.6	2.6	2.6	2.6	2.8	2.8	2.5
48	3.9	4.0	3.9	3.9	3.9	4.0	4.0	4.1	4.1	3.8
49	27.2	27.2	27.2	27.2	27.3	27.3	27.4	27.2	27.1	27.3
50	2.2	2.1	2.2	2.2	2.2	2.3	2.4	2.1	2.1	2.3

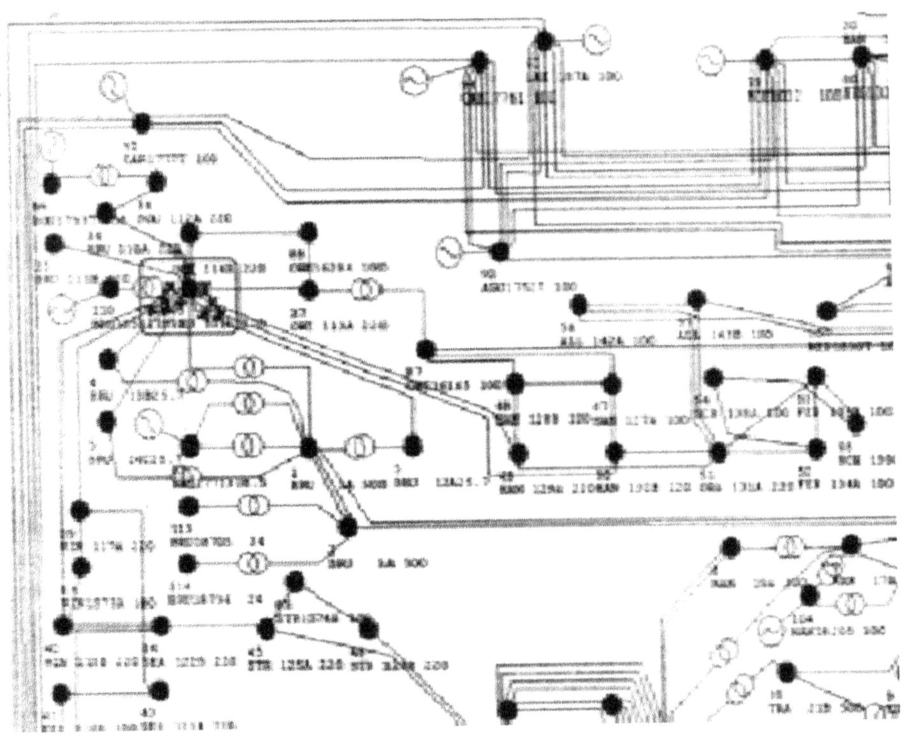

Figure 21.3 The highlighted portion of a transmission network of the 29-generator test system represents the geographic relationship among a group of coherent contingencies. The sign X indicates the location of a contingency on a transmission line. It is observed that the locations of coherent contingencies form a geographic cluster.

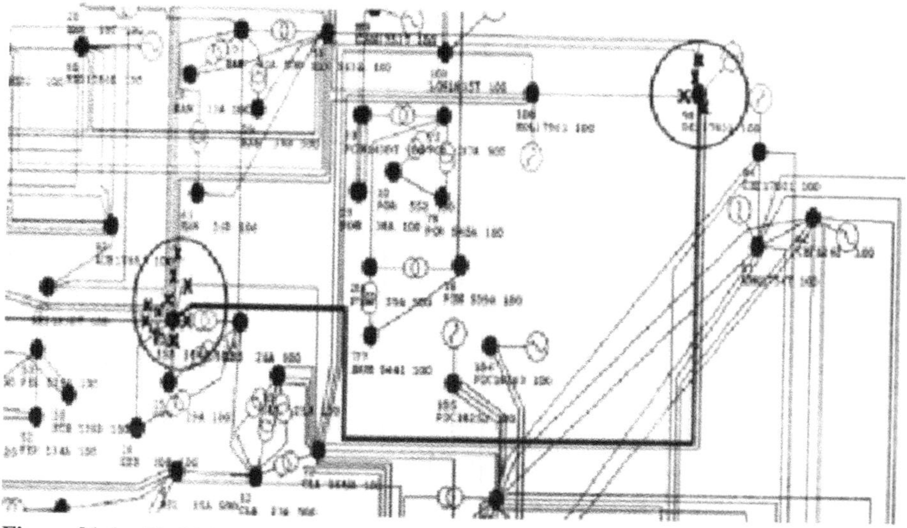

Figure 21.4 The highlighted portion of a transmission network of the 29-generator test system represents the geographic relationship among a group of coherent contingencies. The sign X indicates the location of a contingency on a transmission line. It is observed that the locations of coherent contingencies may be disconnected.

The boundary distances of contingencies in the group are close to each other. For instance, the boundary distances of all the contingencies in Group 1 are close to 0.612, while the boundary distances of all contingencies in Groups 5, 11, and 32 are equal to 1.0. The formation of a group of coherent contingencies enables the boundary property to be a group property.

Group Property 3: Boundary Distance

We next describe another static group property: the boundary distance. The boundary distance of a contingency is close to that of another contingency in the group of coherent contingencies. This group property is illustrated through numerical studies. Several groups of coherent contingencies along with their boundary distances are listed in Table 21.9–21.11. It should be noted that the boundary distance of every contingency in the group of coherent contingencies is close to each other. Hence, the boundary distance is a group property in the sense that under any two coherent contingencies, their corresponding boundary distances are similar in magnitude. For instance the boundary distance of coherent Group 1 is around 0.612, while that of Group 12 is around 0.378.

21.5 DYNAMIC GROUP PROPERTIES

Groups of coherent contingencies share several dynamic properties. Three dynamic group properties are described in this section. These dynamic group properties are

Table 21.5 The Boundary Distance of Some Coherent Contingencies in Group 1

Case number	537	530	529	294	293	10	9	541	578	533	130	6
Boundary distance	0.612	0.612	0.612	0.612	0.612	0.612	0.612	0.612	0.612	0.612	0.612	0.612

Table 21.6 The Boundary Property of All Coherent Contingencies in Group 5

Case number	513	286	285	61	510	509	406	405	230	229
Boundary property	Exact	Exact	Exact	Exact	Exact	Exact	Exact	Exact	Exact	Exact

Table 21.7 The Boundary Property of All Coherent Contingencies in Group 11

Case number	597	86	85	605	82	90	89	434	433
Boundary property	Exact	Exact	Exact	Exact	Exact	Exact	Exact	Exact	Exact

Table 21.8 The Boundary Property of All Coherent Contingencies in Group 32

Case number	382	381	706	705	702
Boundary property	Exact	Exact	Exact	Exact	Exact

related to the system stability property, system critical energy values, and their impacts on the modes of system instability. They are listed below:

Group Property 4: The stability/instability (with a preset fault clearing time) is a group property.

Group Property 5: The exact critical energy value is a group property in the sense that the exact critical energy value of each contingency in the group of coherent contingencies is close to each other.

Group Property 6: The multiswing stability/instability phenomenon is a group property of coherent contingencies.

Regarding Group Property 5, the critical energy is the energy value at the exit point (of the original model) relative to a fault-on trajectory, that is, the point on the fault-on trajectory with the fault cleared at the critical clearing time (CCT). This point is also the intersection point between the fault-on trajectory and the stability boundary of the postfault SEP of the original system. Dynamic Group Property 6 asserts that the multiswing stability/instability is a group property. Given a specified fault clearing time, the mode of single-swing or multiswing stability/instability is a group property.

We next use a 134-generator test system to illustrate some dynamic group properties. Table 21.12 lists the stability property, exact critical energy (i.e., the energy at the CCT by the time–domain simulation), the boundary distance, and the SEP separation of a group of coherent contingencies. The following dynamic properties are observed from this table:

- The stability property of each contingency in the group is the same.
- The critical energy value of each contingency in the group is close to each other.
- The boundary property of each contingency in the group is the same.

Hence, dynamic Group Properties 4 and 5 were observed in this 134-generator test system. We next illustrate dynamic Group Property 6 using the 29-generator test system. Table 21.13 lists the multiswing stability/instability of one group of coherent contingencies. Only two contingencies are listed in this table. The entire group of coherent contingencies with a specified fault clearing time (e.g., 0.07 s) shares the same multiswing mode of instability. We note that this group of coherent contingencies violates the boundary property.

Table 21.9 The Boundary Distance of Some Coherent Contingencies in Group 2

Case	714	690	689	682	554	553	306	305	202	201	194	198
Boundary distance	0.310	0.310	0.310	0.310	0.310	0.310	0.310	0.310	0.310	0.310	0.309	0.309

Table 21.10 The Boundary Distance of All Coherent Contingencies in Group 5

Case number	513	286	285	61	510	509	406	405	230	229
Boundary distance	1.0	1.0	1.0	1.0	1.0	1.0	1.0	1.0	1.0	1.0

Table 21.11 The Boundary Distance of Some Coherent Contingencies in Group 12

Case number	545	302	301	138	137	222	221	166	165	134	150	149
Boundary distance	0.378	0.378	0.378	0.378	0.378	0.378	0.378	0.378	0.378	0.378	0.378	0.378

Table 21.12 The Stability Property, Exact Critical Energy, and Boundary Distance are Group Properties, as Shown in This Table along with the SEP Separation of a Group of Coherent Contingencies

Case number	Faulted bus	Open line from–to	Stability of the postfault system	Critical energy	Boundary distance	SEP separation
1	3007	3007–107	Stable	0.509	Exact	11.061
2	3016	3015–3016	Stable	0.508	Exact	11.072
3	3015	3015–3016	Stable	0.508	Exact	11.072
4	3524	3007–3524	Stable	0.508	Exact	11.083
5	3007	3007–3524	Stable	0.508	Exact	11.083
6	3024	3007–3024	Stable	0.508	Exact	11.084
7	3007	3007–3024	Stable	0.508	Exact	11.084
8	3007	3007–207	Stable	0.508	Exact	11.109
9	3004	3004–404	Stable	0.504	Exact	11.248
10	3004	3004–304	Stable	0.504	Exact	11.25

21.6 CONCLUDING REMARKS

This chapter has introduced the concept of coherent contingencies and the group of coherent contingencies. Several static and dynamic group properties have been developed. It has been shown that a list of contingencies can be classified into groups of coherent contingencies. These groups share the same static group properties as well as the same dynamic group properties.

Given a contingency list, we compute the controlling UEP and the postfault SEP for each contingency in the contingency list. We then group all the contingencies in the contingency list into groups of coherent contingencies based on the difference between the coordinates of the computed UEPs and their SEP separations. In addition to the proposed grouping scheme, other grouping schemes can be developed as long as the static and dynamic group properties are shared by the group of coherent contingencies.

Group properties offer several practical applications. It is expected that new group properties will be discovered in the future and will be incorporated into computational schemes for practical applications. The concept of coherent contingencies will prove useful not only in the development of the group-based BCU method for accurate critical energy computation and for the exact calculation of the controlling UEP and critical energy, but it will also prove useful in the development of several application functions such as contingency analysis and enhancement and preventive control for power system static and dynamic stabilities.

Table 21.13 The Mode of Multiswing Instability Is a Group Property

Fault description	Fault clearing time (s)	Energy margin	BCU assessment	Time–domain assessment	Mode of instability	Boundary distance	Boundary property
Fault bus 536	0.07	1.1127	Stable	Stable	Multiswing	0.234	No
Open line 536–537	0.10	0.983	Stable	Unstable		0.234	No
Fault bus 707	0.07	1.17	Stable	Stable	Multiswing	0.234	No
Open line 707–708	0.10	0.9433	Stable	Unstable		0.234	No

These are cases in which the BCU method fails to find the correct controlling UEP.

Chapter 22

Group-Based BCU–Exit Method

22.1 INTRODUCTION

The correctness of the controlling unstable equilibrium point (controlling UEP, or CUEP) computed by the BCU method can be verified by checking the boundary property rather than by checking the one-parameter transversality condition. By computing the boundary distance of the computed unstable equilibrium point (UEP), one can verify whether the boundary property is satisfied. If the boundary distance of the computed UEP is 1.0, then the boundary property is satisfied; otherwise, it is not. It has been shown that the BCU method computes the correct CUEP if the boundary property is satisfied. When the boundary property is not satisfied, the BCU–exit point should be computed and its energy value will give the accurate critical energy. Since the tasks of verifying the boundary property and computing the BCU–exit point are rather time-consuming, the exploration of some group properties will prove very useful in reducing the required computational effort.

The boundary property is a group property. It is therefore unnecessary to compute the boundary distance of every computed UEP (relative to a contingency) in a group of coherent contingencies. Instead, it is sufficient to compute the boundary distance of a selected UEP in a group of coherent contingencies and to check the boundary property for all contingencies in that group. If a group of coherent contingencies does not satisfy the boundary property, then the UEPs computed by the BCU method for that group do not lie on the stability boundary of the corresponding postfault stable equilibrium point (SEP). The BCU method does not compute the correct CUEP for the group.

In this chapter, we will explore the boundary property to develop a group-based BCU–exit point method, which is designed to enhance the BCU method when it does not compute the correct CUEP due to a boundary property violation. In particular, we will

Direct Methods for Stability Analysis of Electric Power Systems, by Hsiao-Dong Chiang

- develop a group-based verification scheme for checking the boundary property;
- explore linear and nonlinear relationships among the exact critical energy, SEP separation, and the energy at the BCU–exit point;
- develop a fast scheme to determine accurate critical energy values for the entire group of coherent contingencies; and
- develop a group-based BCU–exit point method for determining accurate critical energy values when a group of coherent contingencies violate the boundary property.

Compared with verifying the BCU method based on a one-to-one contingency basis, the group-based verification scheme is far more effective. Regarding the BCU–exit point method, the group-based BCU–exit point method is also significantly faster.

22.2 GROUP-BASED VERIFICATION SCHEME

To improve the procedure of checking the boundary property on a per contingency basis, we will explore the group property of the boundary property to develop a group-based verification scheme for the entire group of coherent contingencies. The group-based verification scheme presented below is effective because it only requires one time–domain simulation for the entire group of coherent contingencies.

A Group-Based Procedure to Verify Computed CUEPs

Step 1. Selection step
For each group of coherent contingencies, we propose the following criterion to select one or two contingencies for verification. If the SEP separation of each contingency in the coherent contingencies is small, say, less than 3°, then the computed UEP with the largest SEP separation in the group is selected; otherwise, the computed UEP with the largest SEP separation and the one with the smallest SEP separation in the group are selected.

Step 2. Checking step
Check the boundary property of the computed UEPs that correspond to the selected contingency in the selection step. We compute a test vector for each selected UEP, say, X^{UEP}, using the following equation:

$$X^{test} = X_s^{post} + 0.99\left(X^{UEP} - X_s^{post}\right),$$

where X_s^{post} is the postfault SEP of the contingency that corresponds to X^{UEP}. The postfault trajectory starting from X^{test} is simulated and assessed. If the postfault trajectory converges to X_s^{post}, then X^{UEP} satisfies the boundary property; otherwise, it does not.

Step 3. Assessment step
Based on the results of Step 2, the following assessment results are obtained:

1. If the selected UEPs satisfy the boundary property, then the computed UEPs of the coherent contingencies in the entire group lie on the stability boundary of the original postfault system.
2. If none of the selected UEPs satisfy the boundary property, then the UEPs of the entire group lie outside the stability boundary of the original postfault system.
3. If one of the selected UEPs satisfies the boundary property while the others do not, then some of the UEPs in the group lie on the stability boundary of the original system, while the others in the group lie outside the stability boundary. (Note that this situation will not occur if a correct grouping scheme is employed.)

The above group-based verification procedure is composed of three steps: (1) a selection step, (2) a checking step, and (3) an assessment step. The most computational effort is required in the checking step, Step 2, in which one time–domain simulation result is performed. By checking the boundary property, one can verify whether or not the UEP computed by the BCU method lies on the stability boundary of the original postfault system without the need to check the one-parameter transversality condition. Moreover, the above verification scheme eliminates the need for a complete check of the boundary property for every contingency in the group of coherent contingencies.

As shown before, the energy value at the BCU–exit point can be used as a critical energy when the boundary property is violated. However, the task of computing the BCU–exit point is time-consuming. Hence, a faster scheme is needed to determine an accurate critical energy value for each contingency in the group of coherent contingencies. To this end, the following issues for each group of coherent contingencies violating the boundary property are relevant:

- What is the relationship between the energy value at the computed UEP and that at the corresponding BCU–exit point?
- Can we use the energy value at the BCU–exit point of some particular contingency from the group of coherent contingencies as the critical energy value for the entire group of coherent contingencies? If yes, which contingency should we choose to compute its BCU–exit point? If no, is there a scheme to adjust the critical energy value for each contingency in the group?
- Can we compute the correct CUEP for each contingency in the group of coherent contingencies starting from the corresponding BCU–exit point?
- What is the relationship between the energy value at the corresponding BCU–exit point and the SEP separation for the entire group of coherent contingencies?

22.3 LINEAR AND NONLINEAR RELATIONSHIPS

We consider a group of coherent contingencies that violate the boundary property. Our goal is to determine an accurate critical energy for each contingency in the

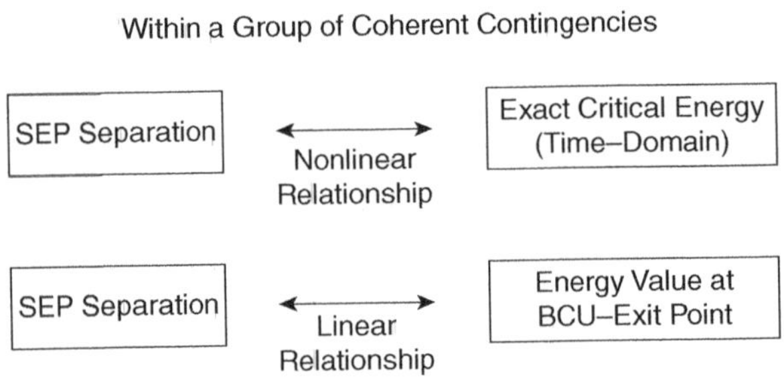

Figure 22.1 Within a coherent group of contingencies, there is a nonlinear relationship between its critical energy and its SEP separation, while there is a nearly linear relationship between the energy at the BCU–exit point and its SEP separation.

coherent group. Let L_l be the contingency whose computed UEP, say, X_l^{UEP}, has the largest SEP separation in the group. Let L_s be the contingency whose computed UEP, say, X_s^{UEP}, has the smallest SEP separation in the group. Let the BCU–exit point of X_l^{UEP} and X_s^{UEP} be X_l^{bcu} and X_s^{bcu}, respectively. The energy value at the BCU–exit point X_l^{bcu} can be used as the critical energy value for the contingency L_l. On the other hand, the energy value at the computed UEP, X_l^{UEP}, cannot be used as the critical energy value for the contingency L_l. It is also known that the energy value at the BCU–exit point X_s^{bcu} can be used as the critical energy value for the contingency L_s. On the other hand, the energy value at the computed UEP, X_s^{UEP}, cannot be used as the critical energy value for the contingency L_s.

We next move one step further by investigating the following relationships that may hold for each coherent group of contingencies (see Figure 22.1):

- For each contingency in a coherent group, what is the relationship between the exact critical energy and its SEP separation?
- For each contingency in a coherent group, what is the relationship between the energy at the BCU–exit point and its SEP separation?

Extensive numerical studies, which serve to verify the above linear and nonlinear relationships, will be presented in the rest of the section. In particular, we will illustrate the linear and nonlinear relationships for two large groups of coherent contingencies on a practical power system model.

We exhibit in Tables 22.1 and 22.2 the case number, boundary distance, SEP separation, and critical energy function based on the time–domain simulation and the BCU–exit point on the coherent contingencies of Groups 9 and 34. We plot in Figures 22.2 and 22.3 the following relationships for Groups 9 and 34, respectively:

- the relationship between the exact critical energy (which is obtained using the time–domain simulation method) and the SEP separation for all the contingencies in the group and
- the relationship between the critical energy based on the energy at the BCU–exit point and the SEP separation for all the contingencies in the group.

Table 22.1 The Case Number, Boundary Distance, SEP Separation, and Critical Energy Function Based on the Time–Domain Simulation and the BCU–Exit Point of Each Contingency in Group 9

Case number	Boundary distance	SEP separation	Critical energy based on	
			Time–domain	BCU–exit point
1303	0.948	10.815	0.884	0.799
1359	0.948	10.832	0.803	0.799
178	0.948	10.838	0.891	0.855
177	0.948	10.838	0.898	0.86
186	0.948	10.838	0.896	0.858
185	0.948	10.838	0.89	0.855
1305	0.948	10.84	0.88	0.849
1357	0.948	10.84	0.802	0.8
174	0.948	10.842	0.886	0.852
166	0.948	10.842	0.801	0.8
165	0.948	10.842	0.884	0.851
1361	0.948	10.844	0.803	0.801
180	0.948	10.849	0.883	0.85
179	0.948	10.849	0.883	0.85
1194	0.948	10.858	0.801	0.8
1193	0.948	10.858	0.898	0.86
1197	0.948	10.859	0.877	0.847
1318	0.948	10.867	0.838	0.823
1195	0.948	10.868	0.878	0.847
1107	0.948	10.868	0.895	0.858
1446	0.948	10.869	0.861	0.837
176	0.948	10.869	0.892	0.856
175	0.948	10.869	0.891	0.855
1472	0.948	10.869	0.882	0.85
1471	0.948	10.869	0.88	0.849
1191	0.948	10.869	0.88	0.849
1135	0.948	10.87	0.887	0.853
184	0.948	10.871	0.891	0.855
183	0.948	10.871	0.88	0.849
161	0.948	10.872	0.874	0.845
1129	0.948	10.873	0.89	0.855
1131	0.948	10.873	0.891	0.855
1315	0.948	10.874	0.802	0.8
182	0.948	10.875	0.891	0.855
181	0.948	10.875	0.89	0.855
1473	0.948	10.876	0.88	0.849
1125	0.948	10.877	0.896	0.858
649	0.948	10.878	0.875	0.845

Table 22.1 Continued

Case number	Boundary distance	SEP separation	Critical energy based on	
			Time–domain	BCU–exit point
1313	0.948	10.88	0.8	0.799
698	0.948	10.882	0.879	0.848
697	0.948	10.882	0.883	0.85
1127	0.948	10.883	0.895	0.858
686	0.948	10.883	0.891	0.855
684	0.948	10.883	0.891	0.855
696	0.948	10.884	0.877	0.847
695	0.948	10.884	0.879	0.848
648	0.948	10.886	0.885	0.852
692	0.948	10.887	0.877	0.847
691	0.948	10.887	0.879	0.848
1186	0.948	10.887	0.879	0.848
1188	0.948	10.887	0.879	0.848
694	0.948	10.893	0.877	0.847
693	0.948	10.893	0.878	0.847
678	0.948	10.893	0.892	0.856
1190	0.948	10.894	0.879	0.848
217	0.948	10.894	0.876	0.846
1178	0.948	10.895	0.882	0.85
702	0.948	10.895	0.878	0.847
701	0.948	10.895	0.882	0.85
1180	0.948	10.896	0.882	0.85
700	0.948	10.896	0.878	0.847
699	0.948	10.896	0.882	0.85
200	0.948	10.898	0.876	0.846
199	0.948	10.898	0.877	0.847
219	0.948	10.899	0.876	0.846
206	0.948	10.899	0.876	0.846
205	0.948	10.899	0.877	0.847
204	0.948	10.9	0.876	0.846
203	0.948	10.9	0.877	0.847
1176	0.948	10.9	0.882	0.85
202	0.948	10.902	0.876	0.846
201	0.948	10.902	0.877	0.847
682	0.948	10.904	0.892	0.856
680	0.948	10.904	0.892	0.856
704	0.949	10.912	0.874	0.845
703	0.949	10.912	0.876	0.846
706	0.949	10.912	0.874	0.845
705	0.949	10.912	0.876	0.846

Table 22.1 Continued

Case number	Boundary distance	SEP separation	Critical energy based on	
			Time–domain	BCU–exit point
1182	0.949	10.914	0.876	0.846
1184	0.949	10.915	0.876	0.846
1476	0.949	10.94	0.857	0.834
1475	0.949	10.94	0.861	0.837
252	0.949	10.943	0.863	0.838
237	0.949	10.959	0.862	0.837
64	0.949	11.022	0.87	0.842
63	0.949	11.022	0.871	0.843
66	0.949	11.025	0.87	0.842
65	0.949	11.025	0.871	0.843
38	0.95	11.091	0.872	0.843
40	0.95	11.093	0.872	0.843
352	0.967	13.478	0.826	0.814
351	0.967	13.478	0.825	0.814
350	0.967	13.493	0.825	0.814
349	0.967	13.493	0.825	0.814
348	0.967	13.513	0.825	0.814
347	0.967	13.513	0.825	0.814
364	0.968	13.571	0.832	0.818
363	0.968	13.571	0.832	0.818
362	0.968	13.579	0.832	0.818
361	0.968	13.579	0.832	0.818
360	0.968	13.592	0.831	0.818
359	0.968	13.592	0.831	0.818
235	0.969	13.775	0.814	0.807
239	0.969	13.822	0.859	0.834
358	0.969	14.337	0.839	0.822
357	0.969	14.337	0.835	0.82
356	0.969	14.341	0.838	0.821
355	0.969	14.341	0.834	0.819
354	0.969	14.364	0.838	0.801
353	0.969	14.364	0.833	0.796

These numerical simulations clearly indicate that the following:

- There is a nearly *linear* relationship between the energy at the BCU–exit point of each contingency in the group and its SEP separation.
- There is a very nonlinear relationship between the energy at the BCU–exit point of each contingency in the group and its SEP separation.

Table 22.2 The Case Number, Boundary Distance, SEP Separation, and Critical Energy Function Based on the Time–Domain Simulation and the BCU–Exit Point of Each Contingency in Group 34

Case number	Boundary distance	SEP separation	Critical energy based on	
			Time–domain	BCU–exit point
1236	0.803	10.863	0.282	0.253
1235	0.803	10.863	0.284	0.255
430	0.803	10.864	0.282	0.27
429	0.803	10.864	0.282	0.27
1347	0.803	10.865	0.283	0.271
432	0.804	10.869	0.282	0.27
431	0.804	10.869	0.282	0.27
1221	0.804	10.871	0.283	0.271
404	0.804	10.871	0.283	0.271
403	0.804	10.871	0.283	0.271
1371	0.804	10.872	0.283	0.271
460	0.804	10.872	0.279	0.269
459	0.804	10.872	0.28	0.269
434	0.804	10.873	0.283	0.271
433	0.804	10.873	0.282	0.27
428	0.804	10.874	0.28	0.269
427	0.804	10.874	0.279	0.269
424	0.804	10.874	0.28	0.269
423	0.804	10.874	0.279	0.269
437	0.804	10.875	0.278	0.268
1209	0.804	10.875	0.282	0.27
1329	0.804	10.876	0.285	0.272
422	0.804	10.878	0.28	0.269
421	0.804	10.878	0.279	0.269
440	0.804	10.881	0.278	0.268
426	0.804	10.882	0.28	0.269
425	0.804	10.882	0.279	0.269
420	0.804	10.887	0.277	0.267
419	0.804	10.887	0.279	0.269
406	0.804	10.888	0.281	0.27
405	0.804	10.888	0.283	0.271
1204	0.805	10.926	0.275	0.266
1203	0.805	10.926	0.279	0.268
1206	0.805	10.926	0.275	0.266
1205	0.805	10.926	0.279	0.268
1208	0.805	10.927	0.275	0.266
1207	0.805	10.927	0.279	0.268
738	0.807	10.985	0.274	0.265

Table 22.2 Continued

Case number	Boundary distance	SEP separation	Critical energy based on	
			Time–domain	BCU–exit point
737	0.807	10.985	0.274	0.265
736	0.807	10.988	0.274	0.265
735	0.807	10.988	0.274	0.265
734	0.807	10.988	0.274	0.265
733	0.807	10.988	0.274	0.265
108	0.823	11.556	0.269	0.259
416	0.844	17.363	0.177	0.177
415	0.844	17.363	0.177	0.177

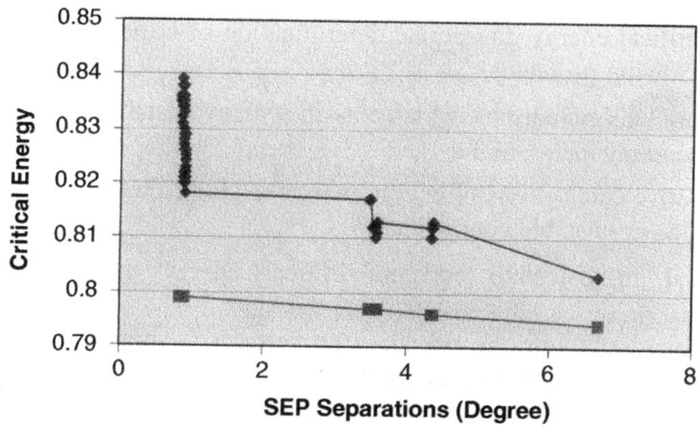

Figure 22.2 The nearly linear relationship between the energy at the BCU–exit point and the SEP separation for all the coherent contingencies in Group 9 and the nonlinear relationship between the exact critical energy value and the SEP separation for all coherent contingencies in Group 9.

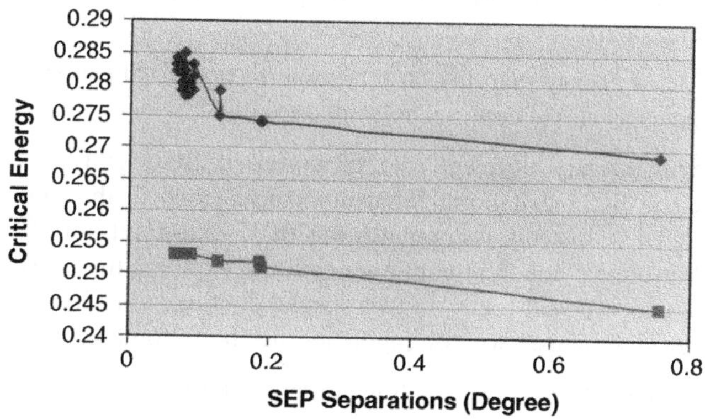

Figure 22.3 The nearly linear relationship between energy at the BCU–exit point and the SEP separation for all coherent contingencies in Group 34 and the nonlinear relationship between the exact critical energy value and the SEP separation for all of the coherent contingencies in Group 34.

The linear relationship between the critical energy based on the energy at the BCU–exit point and the SEP separation will prove useful for developing fast remedy schemes to determine accurate critical energy values when the boundary property is not satisfied. However, it is not recommended that the energy at the BCU–exit point be adopted as the critical energy for practical applications since the task of computing the BCU–exit point is still time-consuming. To resolve this issue, we next present a fast, group-based BCU–exit point method utilizing the linear relationship to determine accurate critical energy values for each contingency in the group of coherent contingencies violating the boundary property.

22.4 GROUP-BASED BCU–EXIT POINT METHOD

Direct methods determine whether or not a postfault power system will remain stable by comparing the system energy at the state immediately after the fault clearing point is reached with a critical energy. Inaccurate determination of critical energy values can lead to the following problems:

(P1) overcalculation of energy margins, which classifies many unstable contingencies as stable contingencies, and

(P2) unduly conservative calculations of energy margins, which classify many stable contingencies as unstable contingencies.

It is therefore essential to calculate accurate critical energy values for contingencies.

For those contingencies satisfying the boundary property, it is appropriate to use the energy value at the UEP computed by the BCU method as the critical energy. Our goal in this section is to present a fast method to determine accurate critical energy values for those contingencies whose corresponding UEPs calculated by the BCU method do not satisfy the boundary property. The issue is determining critical energy values for those contingencies violating the boundary property.

To address this issue, we will develop a fast, group-based BCU–exit point method with the following property:

(S1) no overcalculation of energy margins, in other words, no classification of unstable contingencies as stable contingencies.

In order to be computationally efficient, the group-based BCU–exit point method will not compute the BCU–exit point for every contingency in the same group of coherent contingencies. Instead, it computes the BCU–exit point only for two contingencies in each group of coherent contingencies and explores the nearly linear relationship. To this end, we will focus on two special contingencies in each group of coherent contingencies. These two contingencies are the following:

- Let L_l be the contingency whose corresponding computed UEP, say, X_l^{UEP}, has the largest SEP separation in the group.
- Let L_s be the contingency whose corresponding computed UEP, say, X_s^{UEP}, has the smallest SEP separation in the group.

We then make use of the energy at the BCU–exit points of these two contingencies and the nearly linear relationship to develop the fast method. Let the BCU–exit point of X_l^{UEP} and X_s^{UEP} be X_l^{ray} and X_s^{ray}, respectively. We restate that there exists a linear relationship on a plane with coordinates of the critical energy and the SEP separation of each contingency among the group of coherent contingencies.

A group-based BCU–exit point method is presented below for determining accurate critical energy values for each contingency in the group of coherent contingencies that do not satisfy the boundary property.

A Group-Based BCU–Exit Point Method

Given: a group of coherent contingencies that violate the boundary property

Function: a fast computation of the critical energy value for each contingency in the group

Step 1. Selection step
Select the computed UEP with the largest SEP separation in the group, say, X_l^{UEP}. Let the corresponding contingency be denoted as L_l. Also, select the computed UEP with the smallest SEP separation in the group, say, X_s^{UEP}. Let the corresponding contingency be denoted as L_s.

Step 2. Computing the BCU–exit point
Compute the BCU–exit point for X_l^{UEP} and X_s^{UEP}, respectively. Let them be denoted as X_l^{ray} and X_s^{ray}, respectively.

Step 3. Determining the critical energy
Use the potential energy at X_l^{ray} as the critical energy for the contingency L_l. Let its value be denoted as V_l^{ray}. Likewise, use the potential energy at X_s^{ray} as the critical energy for the contingency L_s. Let its value be denoted as V_s^{ray}.

Step 4. Determining the critical energy for other contingencies in the group
Let the SEP separation of a contingency, say, L_i, in the group of coherent contingencies be SEP_i. Then, the critical energy for the contingency L_i is

$$V_i^{cr} = a \times \text{SEP}_i + b,$$

where $a = \dfrac{V_l^{ray} - V_s^{ray}}{\text{SEP}_l - \text{SEP}_s}$ and $b = \dfrac{V_l^{ray} \times \text{SEP}_l - V_s^{ray} \times \text{SEP}_s}{\text{SEP}_l - \text{SEP}_s}$.

Remarks:

1. The nearly linear relationship between the critical energy and the SEP separation among the group of coherent contingencies is explored in Step 4.
2. The linear operations in Step 4, combined with the fact that the energy at the BCU–exit point is less than the exact critical energy, make the critical energy determined by the above group-based BCU–exit point method slightly conservative. This conservative nature follows the spirit of the CUEP method.
3. The most time-consuming step of the above method is Step 2, which involves computing two BCU–exit points, while the other steps only involve simple algebraic calculations.

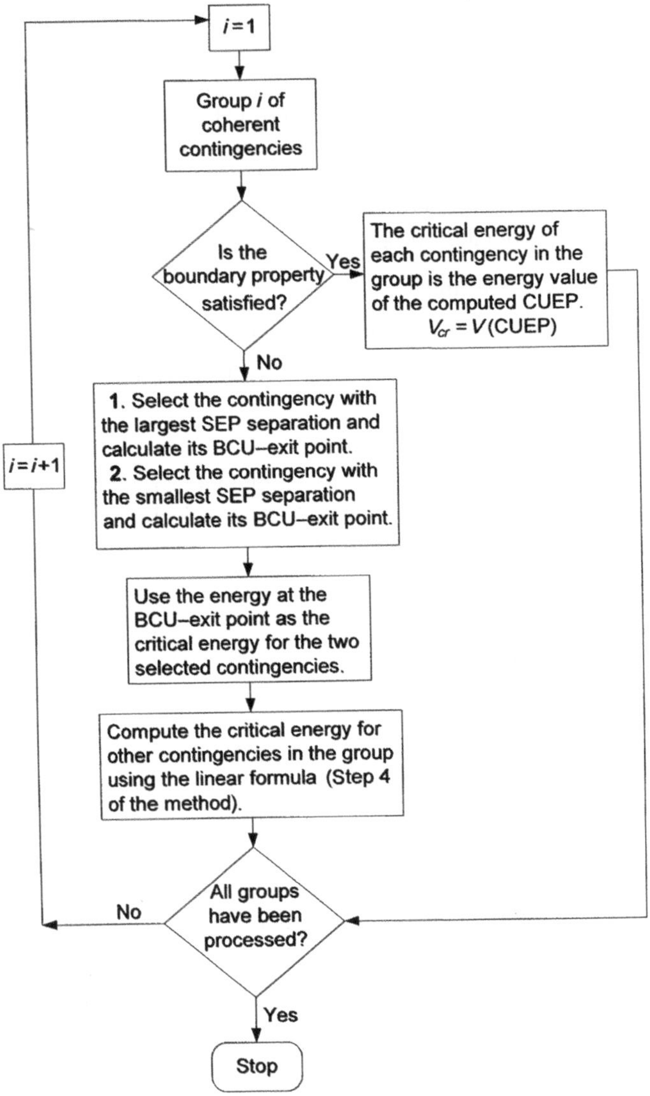

Figure 22.4 The flowchart of the group-based BCU–exit point method in determining critical energy values for each group of coherent contingencies.

4. A flowchart of the group-based BCU–exit point method is depicted in Figure 22.4.

22.5 NUMERICAL STUDIES

We evaluate the group-based BCU–exit point methods presented on a 134-machine test system. For the purpose of evaluation, we compare three different methods for computing the critical energy:

- exact critical energy at the exit point (of the original system), by applying the time–domain simulation to compute the critical clearing time (CCT);
- critical energy based on the energy value at the BCU–exit point; and
- critical energy based on the group-based BCU–exit point method.

The group-based BCU–exit point method has been applied to several groups of coherent contingencies. We present numerical results on two groups of coherent contingencies: Groups 9 and 34. The numerical results on Group 9 of the coherent contingencies are summarized in Table 22.3 including the case number, boundary distance, critical energy by the time–domain simulation method, critical energy based on the energy at the BCU–exit point, and critical energy by the group-based BCU–exit point method.

The nonlinear relationship between the exact critical energy at the exit point (of the original system) and the SEP separation and the linear relationship between the critical energy based on the energy at the BCU–exit point and the SEP separation are exhibited in Figure 22.5. The numerical results on Group 34 of the coherent contingencies are summarized in Table 22.4, and the linear and the nonlinear relationships are plotted in Figure 22.6. These numerical studies have confirmed the claim that there is a linear relationship between the critical energy based on the energy at the BCU–exit point and the SEP separation.

It is interesting to note the following from these simulations:

- The critical energy based on the group-based BCU–exit point method is always less than the critical energy at the BCU–exit point.
- The critical energy at the BCU–exit point is always less than the exact critical energy. This indicates that the critical energy based on the group-based BCU–exit point method is conservative in assessing stability and in estimating the CCT.
- The computational burden of the group-based BCU–exit point method is significantly less than that of the BCU–exit point method for every contingency in the group.
- A regrouping of Group 9 is possible based on the SEP separation. Under this regrouping, all the cases with the SEP separation greater than 11.2 will be classified into a new group, making the 20 contingencies at the bottom of Table 22.3 another group of coherent contingencies. With the formation of the new group, the new critical energy values of the remaining contingencies of Group 9 computed by the group-based BCU–exit point method will be closer to the exact critical energy and will be less conservative.

22.6 CONCLUDING REMARKS

The group property of the boundary property has been explored to develop a group-based verification scheme for the entire group of coherent contingencies. The group-based verification scheme is effective because it involves only one time–domain simulation for the entire group of coherent contingencies. The scheme also eliminates the need for a complete check of the boundary property for every contingency in every group of coherent contingencies.

There is a nearly linear relationship between the energy at the BCU–exit point of each contingency in the group and its SEP separation. In addition, there is a very nonlinear relationship between the energy at the BCU–exit point of each contingency in the group and its SEP separation. By exploring this linear relationship, we have developed a group-based BCU–exit point method for determining accurate critical energy values for each contingency in the group of coherent contingencies that do not satisfy the boundary property.

Extensive numerical studies examining the group-based BCU–exit point method have shown that the group-based BCU–exit point method has the following properties: reliability, accuracy, conservativeness, and modest computational burden. These properties make the group-based BCU–exit point method an ideal candidate for incorporation into the group-based BCU method for computing accurate critical energy values for all types of contingencies including those that both do and do not satisfy the boundary property.

Table 22.3 A Comparison of Critical Energy Values Computed by the Exact Time–Domain Simulation, the BCU–Exit Point Method, and the Group-Based BCU–Exit Point Method on Group 9 of the Coherent Contingencies

Case number	Boundary distance	SEP separation	Critical energy based on		
			Time–domain	BCU–exit point	Fast remedy scheme
1303	0.948	10.815	0.884	0.799	0.799
1359	0.948	10.832	0.803	0.799	0.799
178	0.948	10.838	0.891	0.855	0.799
177	0.948	10.838	0.898	0.86	0.799
186	0.948	10.838	0.896	0.858	0.799
185	0.948	10.838	0.89	0.855	0.799
1305	0.948	10.84	0.88	0.849	0.799
1357	0.948	10.84	0.802	0.8	0.799
174	0.948	10.842	0.886	0.852	0.799
166	0.948	10.842	0.801	0.8	0.799
165	0.948	10.842	0.884	0.851	0.799
1361	0.948	10.844	0.803	0.801	0.799
180	0.948	10.849	0.883	0.85	0.799
179	0.948	10.849	0.883	0.85	0.799
1194	0.948	10.858	0.801	0.8	0.799
1193	0.948	10.858	0.898	0.86	0.799
1197	0.948	10.859	0.877	0.847	0.799
1318	0.948	10.867	0.838	0.823	0.799
1195	0.948	10.868	0.878	0.847	0.799

Table 22.3 Continued

Case number	Boundary distance	SEP separation	Critical energy based on		
			Time–domain	BCU–exit point	Fast remedy scheme
1107	0.948	10.868	0.895	0.858	0.799
1446	0.948	10.869	0.861	0.837	0.799
176	0.948	10.869	0.892	0.856	0.799
175	0.948	10.869	0.891	0.855	0.799
1472	0.948	10.869	0.882	0.85	0.799
1471	0.948	10.869	0.88	0.849	0.799
1191	0.948	10.869	0.88	0.849	0.799
1135	0.948	10.87	0.887	0.853	0.799
184	0.948	10.871	0.891	0.855	0.799
183	0.948	10.871	0.88	0.849	0.799
161	0.948	10.872	0.874	0.845	0.799
1129	0.948	10.873	0.89	0.855	0.799
1131	0.948	10.873	0.891	0.855	0.799
1315	0.948	10.874	0.802	0.8	0.799
182	0.948	10.875	0.891	0.855	0.799
181	0.948	10.875	0.89	0.855	0.799
1473	0.948	10.876	0.88	0.849	0.799
1125	0.948	10.877	0.896	0.858	0.799
649	0.948	10.878	0.875	0.845	0.799
1313	0.948	10.88	0.8	0.799	0.799
698	0.948	10.882	0.879	0.848	0.799
697	0.948	10.882	0.883	0.85	0.799
1127	0.948	10.883	0.895	0.858	0.799
686	0.948	10.883	0.891	0.855	0.799
684	0.948	10.883	0.891	0.855	0.799
696	0.948	10.884	0.877	0.847	0.799
695	0.948	10.884	0.879	0.848	0.799
648	0.948	10.886	0.885	0.852	0.799
692	0.948	10.887	0.877	0.847	0.799
691	0.948	10.887	0.879	0.848	0.799
1186	0.948	10.887	0.879	0.848	0.799
1188	0.948	10.887	0.879	0.848	0.799
694	0.948	10.893	0.877	0.847	0.799
693	0.948	10.893	0.878	0.847	0.799
678	0.948	10.893	0.892	0.856	0.799
1190	0.948	10.894	0.879	0.848	0.799
217	0.948	10.894	0.876	0.846	0.799
1178	0.948	10.895	0.882	0.85	0.799
702	0.948	10.895	0.878	0.847	0.799

Table 22.3 Continued

Case number	Boundary distance	SEP separation	Critical energy based on		
			Time–domain	BCU–exit point	Fast remedy scheme
701	0.948	10.895	0.882	0.85	0.799
1180	0.948	10.896	0.882	0.85	0.799
700	0.948	10.896	0.878	0.847	0.799
699	0.948	10.896	0.882	0.85	0.799
200	0.948	10.898	0.876	0.846	0.799
199	0.948	10.898	0.877	0.847	0.799
219	0.948	10.899	0.876	0.846	0.799
206	0.948	10.899	0.876	0.846	0.799
205	0.948	10.899	0.877	0.847	0.799
204	0.948	10.9	0.876	0.846	0.799
203	0.948	10.9	0.877	0.847	0.799
1176	0.948	10.9	0.882	0.85	0.799
202	0.948	10.902	0.876	0.846	0.799
201	0.948	10.902	0.877	0.847	0.799
682	0.948	10.904	0.892	0.856	0.799
680	0.948	10.904	0.892	0.856	0.799
704	0.949	10.912	0.874	0.845	0.799
703	0.949	10.912	0.876	0.846	0.799
706	0.949	10.912	0.874	0.845	0.799
705	0.949	10.912	0.876	0.846	0.799
1182	0.949	10.914	0.876	0.846	0.799
1184	0.949	10.915	0.876	0.846	0.799
1476	0.949	10.94	0.857	0.834	0.799
1475	0.949	10.94	0.861	0.837	0.799
252	0.949	10.943	0.863	0.838	0.799
237	0.949	10.959	0.862	0.837	0.799
64	0.949	11.022	0.87	0.842	0.798
63	0.949	11.022	0.871	0.843	0.798
66	0.949	11.025	0.87	0.842	0.798
65	0.949	11.025	0.871	0.843	0.798
38	0.95	11.091	0.872	0.843	0.798
40	0.95	11.093	0.872	0.843	0.798
352	0.967	13.478	0.826	0.814	0.797
351	0.967	13.478	0.825	0.814	0.797
350	0.967	13.493	0.825	0.814	0.797
349	0.967	13.493	0.825	0.814	0.797
348	0.967	13.513	0.825	0.814	0.797
347	0.967	13.513	0.825	0.814	0.797
364	0.968	13.571	0.832	0.818	0.797

Table 22.3 Continued

Case number	Boundary distance	SEP separation	Critical energy based on		
			Time–domain	BCU–exit point	Fast remedy scheme
363	0.968	13.571	0.832	0.818	0.797
362	0.968	13.579	0.832	0.818	0.797
361	0.968	13.579	0.832	0.818	0.797
360	0.968	13.592	0.831	0.818	0.797
359	0.968	13.592	0.831	0.818	0.797
235	0.969	13.775	0.814	0.807	0.796
239	0.969	13.822	0.859	0.834	0.796
358	0.969	14.337	0.839	0.822	0.796
357	0.969	14.337	0.835	0.82	0.796
356	0.969	14.341	0.838	0.821	0.796
355	0.969	14.341	0.834	0.819	0.796
354	0.969	14.364	0.838	0.801	0.796
353	0.969	14.364	0.833	0.796	0.796

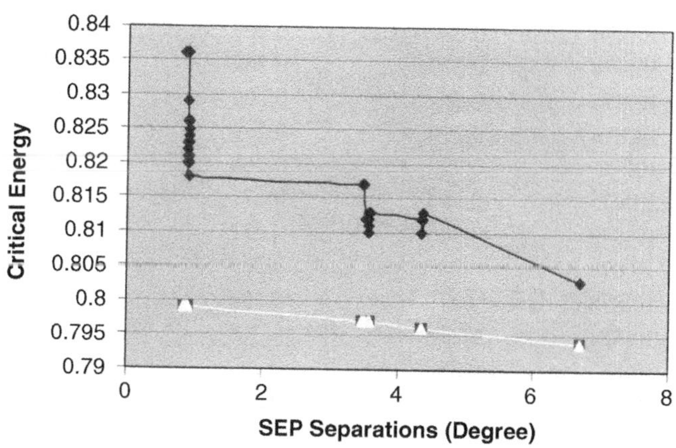

Figure 22.5 The nonlinear relationship between the exact critical energy value and the SEP separation and the linear relationship between the energy of the BCU–exit point and the SEP separation. The linear relationship between the critical energy determined by the group-based BCU–exit point and the SEP separation (Group 9) is also shown.

Table 22.4 A Comparison of Critical Energy Values Computed by the Exact Time–Domain Simulation, the BCU–Exit Point Method, and the Group-Based BCU–Exit Point Method on Group 34 of the Coherent Contingencies

Case number	Boundary distance	SEP separation	Critical energy based on		
			Time–domain	BCU–exit point	Fast remedy scheme
1236	0.803	10.863	0.282	0.253	0.253
1235	0.803	10.863	0.284	0.255	0.253
430	0.803	10.864	0.282	0.27	0.253
429	0.803	10.864	0.282	0.27	0.253
1347	0.803	10.865	0.283	0.271	0.253
432	0.804	10.869	0.282	0.27	0.253
431	0.804	10.869	0.282	0.27	0.253
1221	0.804	10.871	0.283	0.271	0.253
404	0.804	10.871	0.283	0.271	0.253
403	0.804	10.871	0.283	0.271	0.253
1371	0.804	10.872	0.283	0.271	0.253
460	0.804	10.872	0.279	0.269	0.253
459	0.804	10.872	0.28	0.269	0.253
434	0.804	10.873	0.283	0.271	0.253
433	0.804	10.873	0.282	0.27	0.253
428	0.804	10.874	0.28	0.269	0.253
427	0.804	10.874	0.279	0.269	0.253
424	0.804	10.874	0.28	0.269	0.253
423	0.804	10.874	0.279	0.269	0.253
437	0.804	10.875	0.278	0.268	0.253
1209	0.804	10.875	0.282	0.27	0.253
1329	0.804	10.876	0.285	0.272	0.253
422	0.804	10.878	0.28	0.269	0.253
421	0.804	10.878	0.279	0.269	0.253
440	0.804	10.881	0.278	0.268	0.253
426	0.804	10.882	0.28	0.269	0.253
425	0.804	10.882	0.279	0.269	0.253
420	0.804	10.887	0.277	0.267	0.253
419	0.804	10.887	0.279	0.269	0.253
406	0.804	10.888	0.281	0.27	0.253
405	0.804	10.888	0.283	0.271	0.253
1204	0.805	10.926	0.275	0.266	0.252
1203	0.805	10.926	0.279	0.268	0.252
1206	0.805	10.926	0.275	0.266	0.252
1205	0.805	10.926	0.279	0.268	0.252
1208	0.805	10.927	0.275	0.266	0.252
1207	0.805	10.927	0.279	0.268	0.252

Table 22.4 Continued

Case number	Boundary distance	SEP separation	Critical energy based on		
			Time–domain	BCU–exit point	Fast remedy scheme
738	0.807	10.985	0.274	0.265	0.252
737	0.807	10.985	0.274	0.265	0.252
736	0.807	10.988	0.274	0.265	0.251
735	0.807	10.988	0.274	0.265	0.251
734	0.807	10.988	0.274	0.265	0.251
733	0.807	10.988	0.274	0.265	0.251

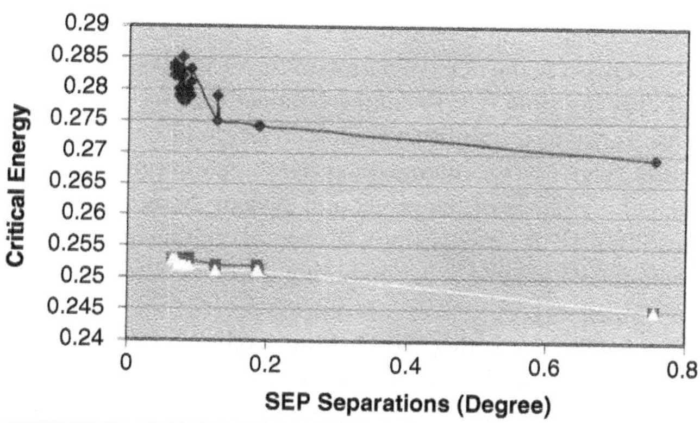

Figure 22.6 The nonlinear relationship between the critical energy found using the time–domain simulation method and the SEP separation and the linear relationship between the energy of the BCU–exit point and the SEP separation. The linear relationship between the critical energy determined by the group-based BCU–exit point and the SEP separation (Group 34) is also shown.

Chapter 23

Group-Based BCU–CUEP Methods

23.1 INTRODUCTION

The controlling unstable equilibrium point (controlling UEP, or CUEP) plays a key role in several applications, such as (1) the determination of the critical energy, from which an estimated critical clearing time (CCT) can be obtained, (2) the derivation of preventive control against transient instability, (3) the derivation of enhancement control for transient stability, and (4) the mode of system separations (i.e., the unstable mode) and the related diagnosis of the protection system. Nonetheless, the task of computing the controlling UEP for every power system contingency is very challenging. Several computational challenges in computing the controlling UEP are described in Chapter 12.

The BCU method computes the controlling UEP of the original model by computing the controlling UEP of an artificial reduced-state model. It has been shown that, under the boundary property or the one-parameter transversality condition, the UEP computed by the BCU method lies on the stability boundary of the original model. To check the boundary property, the boundary distance needs to be calculated, which amounts to a few time–domain simulations.

Since the boundary property is a group property, it is sufficient to compute the boundary distance of a selected contingency from a group of coherent contingencies. If the boundary distance of a UEP computed by the BCU method with respect to a contingency is 1.0, then the UEP is on the stability boundary of the original post-fault system. Otherwise, the UEP violates the boundary property and is not the controlling UEP relative to the contingency. Hence, a method is needed to compute the (exact) controlling UEP relative to a contingency when the boundary property is violated.

In this chapter, a group-based BCU–CUEP method will be presented to compute the exact controlling UEP relative to each contingency in a group of coherent contingencies violating the boundary property. We first present a time–domain-based method to compute the exact controlling UEP. We then explore a group property of

Direct Methods for Stability Analysis of Electric Power Systems, by Hsiao-Dong Chiang

the controlling UEP of coherent contingencies and incorporate the group property into the time–domain-based method to develop a group-based BCU–CUEP method. The goal of developing the group-based BCU–CUEP method is to recompute the UEPs of groups of coherent contingencies when the boundary property of the UEPs computed by the BCU method does not hold. Each newly computed UEP will be the exact controlling UEP relative to each contingency in the group. Hence, the group-based BCU–CUEP method recomputes the controlling UEP of each contingency so that the boundary property is satisfied. The group property explored in the group-based BCU–CUEP method is the coordinates of controlling UEPs (i.e., the controlling UEPs of any two contingencies in a coherent group of contingencies are close to each other).

Before we proceed with this chapter, it is appropriate to review some of the terminologies that will be used. The *exit point of the original model* is the intersection point between the fault-on trajectory and the stability boundary of the postfault stable equilibrium point (SEP). A computed UEP (with respect to a contingency) is said to satisfy the *boundary property* if the computed UEP lies on the stability boundary of the original postfault system. To compute the *boundary distance of a UEP*, we form a ray connecting the postfault SEP and the UEP and identify the intersection point between the ray and the stability boundary of the postfault SEP. The distance between this intersection point and the postfault SEP divided by the distance between the computed UEP and the postfault SEP is the boundary distance.

23.2 EXACT METHOD FOR COMPUTING THE CONTROLLING UEP

We review a time–domain-based method that can compute the exact controlling UEP relative to any contingency. The following method is an iterative time–domain-based procedure for computing the exit point of a general original model, which is the intersection point between the fault-on trajectory and the stability boundary of the original postfault model. This method is intended as a bench mark for computing the exact controlling UEPs.

Method for Computing the Exit Point of the Original Model

Step 1. Integrate the postfault system starting from a point on the fault-on trajectory, say, $x_f(t)$. If the resulting postfault trajectory does not converge to the postfault SEP, then the point, say, $x_f(k_1T)$, lies outside the stability region; then, go to Step 2. Otherwise, repeat the process of integrating the postfault system starting from the next point (in the forward time sense) along the fault-on trajectory until the first point along the fault-on trajectory whose corresponding postfault trajectory does not converge to the postfault SEP. Let the first point be denoted as $x_f(kT)$ and its previous point on the fault-on trajectory be denoted as $x_f((k-1)T)$, from which the corresponding postfault trajectory does converge to the postfault SEP. Go to Step 3.

Step 2. Integrate the postfault system starting from $x_f((k_1 - 1)T)$. If the resulting postfault trajectory does not converge to the postfault SEP, then set $k_1 = k_1 - 1$ and repeat Step 2; otherwise, set $k_2 = k_1$ and go to Step 3.

Step 3. The exit point, denoted as x_{ex}, lies between the point $x_f(k_1T)$ and the point $x_f(k_2T)$. Apply the golden bisection method to these two points to obtain an accurate exit point.

Remarks:

1. The above method is a state-space version of the "standard" time–domain method for finding the CCT. The method presented above is applicable to general power system stability models.
2. In the numerical BCU method, a procedure to fast compute the exit point of the artificial reduced-state system by employing the linear interpolation method has been implemented. This procedure explores a special property of the stability boundary of the reduced-state system, which is a quasi-gradient system. Unfortunately, the special property does not hold for general nonlinear systems.
3. Because of the lack of a good performance index for characterizing the stability boundary of the original model, such as the dot product for the reduced-state system, it is difficult to develop a procedure similar to the one used in the BCU method to accurately detect the exit point of the original model. One can instead implement an efficient one-dimensional search method such as the golden bisection method.

We next present a method to find the exact controlling UEP relative to a contingency via the exit point of the original model. This method is applicable to general power system transient stability models, and it does not require the existence of an energy function for the postfault power system. This method only requires the existence of a controlling UEP relative to a contingency under study. We recall the following time–domain-based method for computing the exact controlling UEP.

Method for Computing the Exact Controlling UEP

Step 1. Compute the exit point of the original model relative to a fault-on trajectory using the time–domain-based method.

Step 2. Use the exit point as an initial condition and integrate the postfault system for a few time steps to get an "end" point. If the norm of the end point is smaller than a threshold value, then go to Step 4; otherwise, go to Step 3.

Step 3. Draw a ray connecting the end point and the postfault SEP. Compute the exit point of the original model along the ray, that is, the intersection point between the ray and the stability boundary of the original model, using the time–domain-based method. Replace the current exit point by the exit point of the ray and go to Step 2.

Step 4. Use the end point as an initial guess and solve for an equilibrium point of the postfault system. Let the solution be x_{co}. The exact controlling UEP, with respect to the fault-on trajectory, is x_{co}.

Remarks:

1. The theoretical basis for the above method is the following: the exact controlling UEP is the UEP whose stable manifold intersects with the fault-on trajectory at the so-called exit point of the original model, that is, the point on the fault-on trajectory with the fault cleared at CCT. This point is also the intersection point between the fault-on trajectory and the stability boundary of the postfault SEP.
2. From a computational viewpoint, it is very difficult to compute the exact exit point because the simulated trajectory is composed of a sequence of points along the entire trajectory. Hence, Step 1 of the above method usually gives a point close to the exit point. If the point computed by Step 1 is not sufficiently close to the exact exit point, then a divergence problem might occur at Step 4. If this situation arises, it is recommended that the step size T involved in Step 1 be reduced.
3. Steps 1 and 3 are the most time-consuming steps because the task of computing the exact exit point of the original model is computationally intensive.

23.3 GROUP-BASED BCU–CUEP METHOD

The group-based BCU–CUEP method is designed to compute the exact controlling UEP of each contingency of a group of coherent contingencies that violate the boundary property. The method is composed of two steps:

Step 1. Compute the exact controlling UEP of a contingency selected from the group of coherent contingencies using a time–domain-based method.

Step 2. Use the exact controlling UEP computed at Step 1 as an initial guess to compute the exact controlling UEP of the remaining contingencies of the group of coherent contingencies by applying a nonlinear solver to the corresponding extended postfault power flow equations.

Step 2 explores the group property that the coordinates of the controlling UEPs of coherent contingencies are close to each other to compute the controlling UEP of other contingencies in the group. Step 1 is a straightforward time–domain-based method to compute the controlling UEP.

We next present a numerical group-based BCU–CUEP method to compute controlling UEPs of a group of coherent contingencies that violate the boundary property (Chiang et al., 2007, 2009; Tada and Chiang, 2008).

Group-Based BCU–CUEP Method

Input: a group of coherent contingencies that violate the boundary property

Output: the (exact) controlling UEP relative to each contingency in the group of coherent contingencies.

Step 1. Select a contingency from the group of coherent contingencies, say, the one with the largest SEP separation.

Step 2. Compute the exact controlling UEP for the contingency selected at Step 1 using the exact time–domain-based method.

Step 3. Use the computed controlling UEP in Step 2 as an initial guess to compute the controlling UEP of each remaining contingency in the group as follows:

Step 3.1. Update the system postfault Y-matrix to obtain the extended postfault power flow equations for each remaining contingency in the group.

Step 3.2. Use the computed controlling UEP in Step 2 as an initial guess and apply an effective nonlinear algebraic solver to the extended postfault power flow equations.

Step 3.3. Let the solution be x_{co}. The exact controlling UEP, with respect to the remaining contingency, is x_{co}.

Step 3 of the above method explores the group property to compute the exact controlling UEPs for the remaining contingencies in the group. The partitioned Newton method or its variants can be used as the nonlinear algebraic solver in Step 3.2. Since the controlling UEPs of any two contingencies in a coherent group of contingencies are close to each other, the above method usually works very well and the convergence problem usually does not arise in Step 3. The computational procedure of the group-based BCU–CUEP method is illustrated in Figure 23.1.

We next present an architecture for computing the exact controlling UEPs of a credible contingency list. This architecture is expressed in the flowchart shown in Figure 23.2, which represents an integration of the BCU method, grouping schemes, a group-based verification scheme, and a group-based BCU–CUEP method. This integrated group-based BCU–CUEP method is able to compute the exact controlling UEP of every contingency in the contingency list.

23.4 NUMERICAL STUDIES

The group-based BCU–CUEP method was applied to a 134-machine power system with a list of credible contingencies. After applying the BCU method, the lists of credible contingencies were grouped into several groups of coherent contingencies. Two groups of coherent contingencies were used for illustrational purposes; the first group was composed of four contingencies, while the second group consisted of 12 contingencies. The fault clearing time was set to 0.07 s.

The numerical results for the small group of coherent contingencies are summarized in Table 23.1, which lists the four contingencies, their corresponding energy margins, and their boundary distances. The boundary distance is less than one,

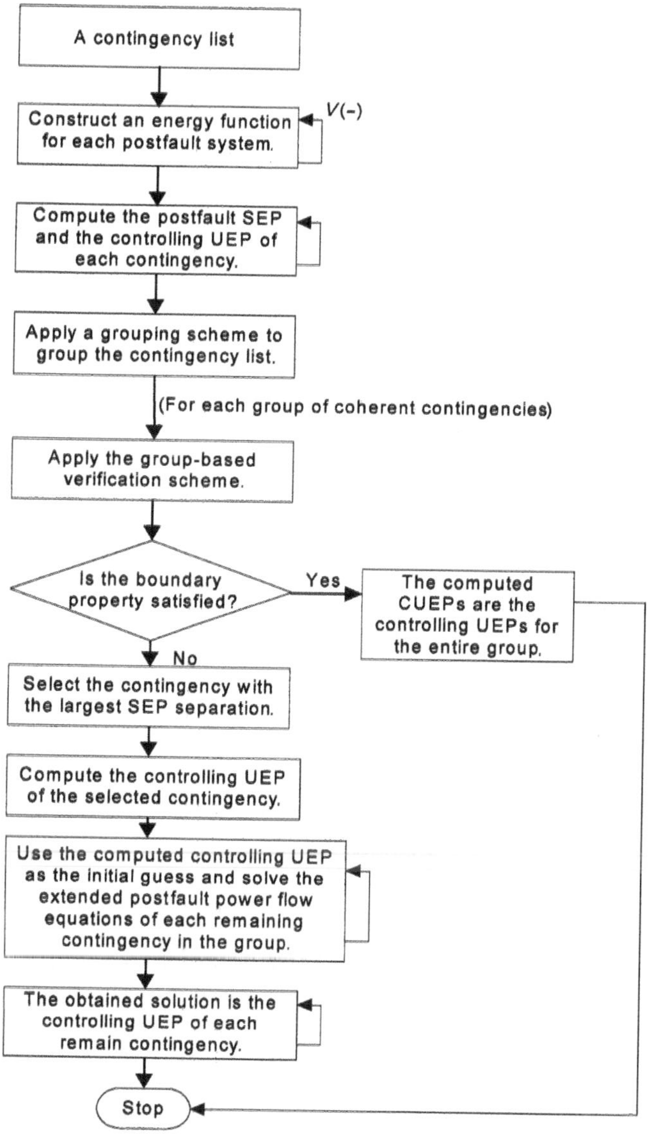

Figure 23.1 The flowchart of a group-based BCU–CUEP method for computing the exact controlling UEP of each contingency in the group that violates the boundary property.

indicating that the UEPs computed by the BCU method are not the controlling UEPs. In fact, each BCU–exit point of the computed UEP lies in the middle of the ray connecting the computed UEP and SEP. The energy margins shown in Table 23.1 were calculated based on the UEPs computed by the BCU method.

We have the following observations:

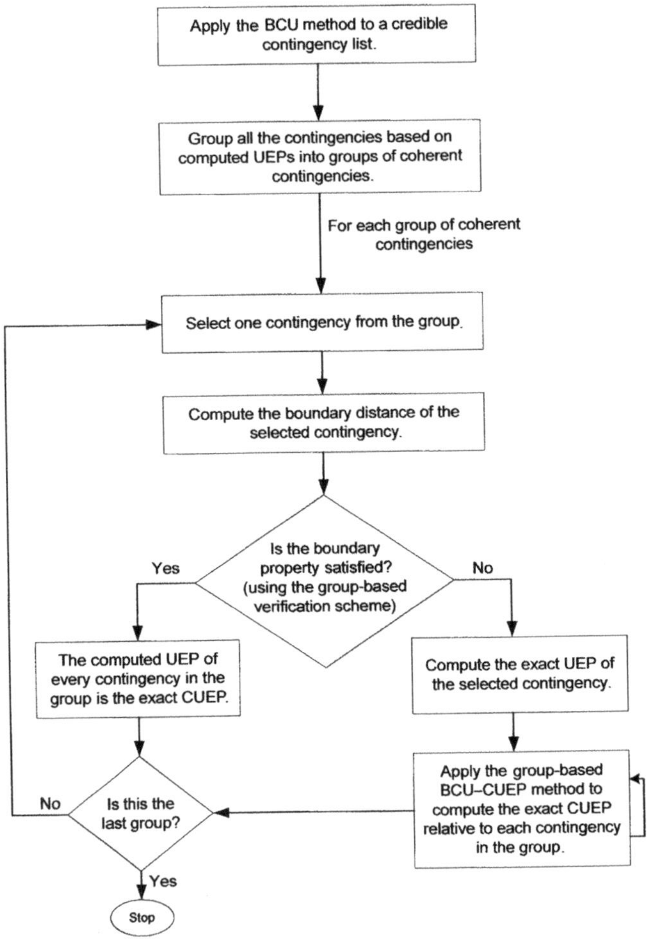

Figure 23.2 An architecture for computing the exact controlling UEPs for a credible contingency list. This flowchart is an integration of the BCU method, grouping schemes, a group-based verification scheme, and a group-based BCU–CUEP method.

Table 23.1 The Contingency, the Energy Margin Computed by the BCU Method, and the Corresponding Boundary Distance of a Group of Coherent Contingencies

Case number	Faulted bus	Open line from–to	Energy at the computed UEP	Boundary distance
0001	3010	3005–3010	1.9	0.785
0002	3007	3007–207	1.36	0.784
0003	3010	3010–3012	1.21	0.79
0004	3012	3010–3012	1.37	0.783

These computed UEPs do not lie on the stability boundary of the original model.

- The boundary distance of each contingency is smaller than 1.0, indicating that the controlling UEPs computed by the BCU method do not lie on the stability boundary of the original power system model.
- The energy at the computed controlling UEP cannot be used as the critical energy and hence the computed energy margin is not accurate.
- The energy margins of each contingency based on the computed controlling UEP are not close to each other.

The group-based BCU–CUEP method was applied to this group of coherent contingencies. The exact controlling UEPs relative to all the contingencies in the group of coherent contingencies are computed. The boundary distance and the corresponding energy margin of each contingency based on the (exact) controlling UEPs are listed in Table 23.2. We have the following observations and comments:

- The boundary distance of each contingency is now 1.0, indicating that the controlling UEP computed by the group-based BCU–CUEP method lies on the stability boundary of the original power system stability model.
- The energy margins of each contingency computed based on the exact controlling UEP are close to each other, indicating that the energy margin at the exact controlling UEP is a group property.
- The coordinates of the computed controlling UEPs are now close to each other, confirming that the closeness between controlling UEPs is a group property.
- The group property of the coordinates of the controlling UEPs deserves further explorations.

For the medium-size group of coherent contingencies with 12 contingencies, the energy margins, based on the computed UEPs by the BCU method, and their boundary distances are summarized in Table 23.3. The boundary distance of each computed UEP indicates that the computed UEPs are not exact controlling UEPs. In fact, the corresponding BCU–exit points were all positioned in the middle of the

Table 23.2 The Contingency, Energy Value at the UEP Computed by the Group-Based BCU–CUEP Method, and the Corresponding Boundary Distance of a Group of Coherent Contingencies

Case number	Faulted bus	Open line from–to	Energy value at the computed controlling UEP	Boundary distance
0001	3010	3005–3010	0.845	1.0
0002	3007	3007–207	0.834	1.0
0003	3010	3010–3012	0.832	1.0
0004	3012	3010–3012	0.845	1.0

These computed UEPs all lie on the stability boundary of the original model.

Table 23.3 The Contingency, the Energy Margin at the UEP Computed by the BCU Method, and the Corresponding Boundary Distance of Each Contingency of a Group of Coherent Contingencies

Case number	Faulted bus	Open line from–to	Energy value at the computed UEP	Boundary distance
01	3036	3036–136	0.75	0.539
02	136	3036–136	0.82	0.536
03	3036	3036–236	0.75	0.539
04	236	3036–236	0.82	0.537
05	3036	3036–336	0.75	0.539
06	336	3036–336	0.82	0.536
07	136	136–4110	0.82	0.538
08	4110	4110–136	0.83	0.537
09	236	236–4110	0.82	0.537
10	4110	4110–236	0.83	0.537
11	336	4110–336	0.82	0.537
12	4110	4110–336	0.83	0.537

ray connecting the computed UEP and the postfault SEP. The computed energy margins of these coherent contingencies are not close to each other.

We apply the group-based BCU–CUEP method to this medium-size group of coherent contingencies. The exact controlling UEP of a selected contingency is first computed. The exact controlling UEPs of the other contingencies in the group were then computed using the group-based BCU–CUEP method. The boundary distance and the corresponding energy margins of each contingency based on the newly computed controlling UEPs are listed in Table 23.4. We have the following observations:

- The boundary distance of each contingency is now 1.0, indicating that the newly computed UEP lies on the stability boundary of the original power system stability model.
- The energy values of the exact controlling UEPs within this group of coherent contingencies are close to each other.
- The coordinates of the computed controlling UEPs are now close to each other, confirming that the closeness between the controlling UEPs in the group of coherent contingencies is indeed a group property.

23.5 CONCLUDING REMARKS

In this chapter, a group-based BCU–CUEP method for computing the exact controlling UEP relative to each contingency in a group of coherent contingencies violating the boundary property has been presented. By exploring the group property

Table 23.4 The Contingency, the Energy at the UEP Computed by the Group-Based BCU–CUEP Method, and the Corresponding Boundary Distance of a Group of Coherent Contingencies

Case number	Faulted bus	Open line from–to	Energy at the computed exact controlling UEP	Boundary distance
01	3036	3036–136	0.28	1.0
02	136	3036–136	0.28	1.0
03	3036	3036–236	0.28	1.0
04	236	3036–236	0.28	1.0
05	3036	3036–336	0.28	1.0
06	336	3036–336	0.28	1.0
07	136	136–4110	0.28	1.0
08	4110	4110–136	0.28	1.0
09	236	236–4110	0.28	1.0
10	4110	4110–236	0.28	1.0
11	336	4110–336	0.28	1.0
12	4110	4110–336	0.28	1.0

These computed UEPs all lie on the stability boundary of the original model.

of closeness between the controlling UEPs of each contingency in the group of coherent contingencies, the group-based BCU–CUEP method effectively computes the exact controlling UEP for each contingency in the group. In addition, an integrated group-based BCU–CUEP method, which is an integration of the BCU method, grouping schemes, a group-based verification scheme, and a group-based BCU–CUEP method, has been presented. One significant feature of this integrated group-based BCU–CUEP method is that it computes the exact controlling UEP of every contingency in the contingency list of large-scale power systems.

The closeness between controlling UEPs is a group property. This group property of closeness will prove useful not only in the development of the group-based BCU method but also in several applications, such as enhancement control for critically stable contingencies and preventive control against insecure contingencies. In addition, the coordinates of the controlling UEP associated with a contingency in each group of coherent contingencies can provide useful information in other applications, such as designing controls to stabilize a power system or to enhance its stability with respect to a set of contingencies, and determining the primary and secondary spinning reserve with respect to a set of contingencies. Indeed, this group property of closeness between controlling UEPs deserves further exploration.

Chapter 24

Group-Based BCU Method

24.1 INTRODUCTION

An integrated group-based BCU method will be presented in this chapter. This integrated method is composed of the BCU method, the group-based verification scheme, the group-based BCU–exit point method, and the group-based BCU–CUEP method described in the previous chapters. Compared to the BCU method, the integrated group-based BCU method has the following features:

- The one-parameter transversality condition is not required.
- The boundary property is not required.
- The group-based BCU method computes the correct controlling unstable equilibrium point (controlling UEP, or CUEP) for each contingency.
- The group-based BCU method ensures the reliability and accuracy of the calculated critical energy.
- The group-based BCU method reduces the conservativeness of the BCU method.
- The group-based BCU method captures and explores the inherent characteristics of coherent contingencies. It reduces a large number of the contingencies whose corresponding UEP boundary property needs to be checked to a small number.
- Given a contingency list, an outstanding feature of the group-based BCU method is that it offers an effective means of controller design development for transient stability enhancement.
- The group-based BCU method offers useful information regarding the impact of system damping on the region of power system transient stability and on the energy margin.

Two versions of the group-based BCU method will be presented in this chapter. One version is designed to compute the controlling UEP of every contingency in the contingency list, and the other version is designed to compute the accurate critical energy value for every contingency in the contingency list.

24.2 GROUP-BASED BCU METHOD FOR ACCURATE CRITICAL ENERGY

This integrated group-based BCU method is designed to compute the accurate critical energy for every contingency in the contingency list. It is composed of the following methods described in the previous chapters:

- energy function method,
- BCU method,
- grouping scheme,
- group-based verification scheme, and
- group-based BCU–exit point method.

Given a study base case power system with a list of credible contingencies, this version of the group-based BCU method for accurate critical energy calculation and direct stability assessment proceeds as follows:

Step 1. Compute the postfault stable equilibrium point (SEP) of each contingency and construct an energy function for the corresponding postfault system.

Step 2. Compute the controlling UEP of each contingency using the BCU method.

Step 3. Repeat Steps 1 and 2 until all the contingencies have been processed.

Step 4. Group all the contingencies into groups of coherent contingencies.

For each group of coherent contingencies, do the following:

Step 5. Verify the boundary property using the group-based verification scheme. If the boundary property is satisfied, then the energy value at the computed UEP is an accurate critical energy for each contingency in the group; proceed to Step 7. Otherwise, go to the next step.

Step 6. Compute an accurate critical energy value for each contingency using the group-based BCU–exit point method until all the contingencies have been processed.

Step 7. Perform energy margin calculation and direct stability assessment. Proceed to the next group.

We notice that Step 2 is the BCU method step, while Step 4 is the grouping step. Step 5 is the group-based verification scheme step, while Step 6 is the group-based BCU–exit point method step. Step 7 is the overall step of direct stability assessment and energy margin calculation. From a functional viewpoint, the group-based BCU method for accurate critical calculation is depicted in Figure 24.1.

A numerical implementation of the group-based BCU–exit point method is described next.

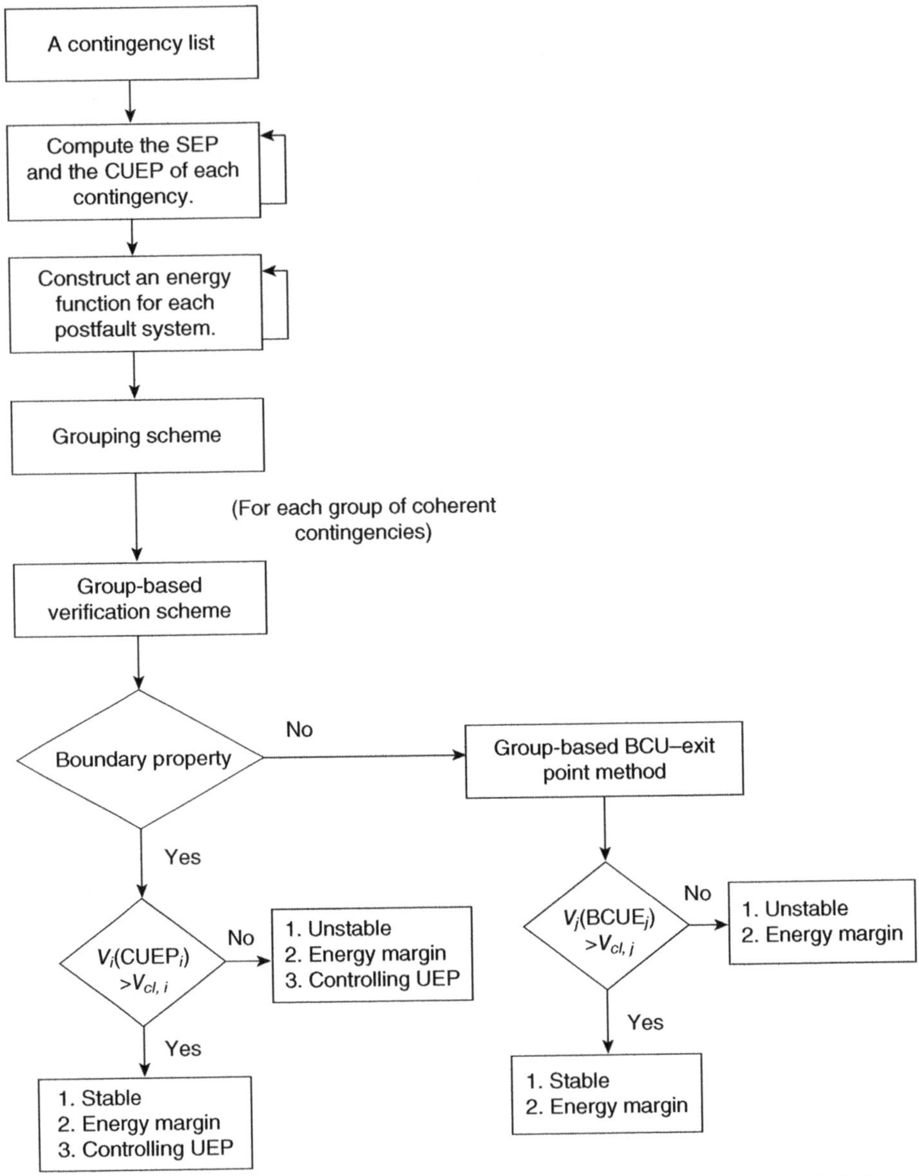

Figure 24.1 A flowchart of the group-based BCU method for accurate energy margin calculation and direct stability assessments. It is composed of the BCU method, the energy function methods, the group-based verification scheme for boundary property checking, the controlling UEP method, the group-based BCU–exit point method, and the step of direct stability analysis.

Group-Based BCU Method for Accurate Critical Energy

For each contingency of the contingency list, perform Steps 1 through 3:

Step 1. Apply the partitioned Newton method or its variants with the prefault SEP as the initial guess to compute the postfault SEP of each contingency. If the postfault SEP cannot be found, send the contingency to a time–domain simulation program for stability analysis and proceed to the next contingency.

Step 2. Construct a numerical energy function for the postfault system.

Step 3. Apply the BCU method to compute the controlling UEP of each contingency.

Step 4. Group all of the contingencies into groups of coherent contingencies based on the computed controlling UEPs and postfault SEPs.

For each group of coherent contingencies, do the following:

Step 5. Check the boundary property of each group.

Step 5.1. Selecting step

Select the computed UEP with the largest SEP separation, say, X^{UEP}.

Step 5.2. Checking step

Form the following test vector:

$$X^{test} = X_s^{post} + 0.99\left(X^{UEP} - X_s^{post}\right),$$

where X_s^{post} is the postfault SEP of the contingency corresponding to X^{UEP}. The postfault trajectory starting from X^{test} is simulated and assessed. If the postfault trajectory converges to X_s^{post}, then X^{UEP} satisfies the boundary property; otherwise, it does not.

Step 5.3. Boundary property

If the boundary property is satisfied, then the critical energy for each contingency is the energy value at the computed UEP; then, go to Step 7. Otherwise, go to Step 6.

Step 6. Determine the critical energy using the group-based BCU–exit point method.

Step 6.1. Selection step

Select the computed UEP with the largest SEP separation and the one with the smallest SEP separation, say, X_l^{UEP} and X_s^{UEP}. Let the corresponding contingencies be denoted as L_l and L_s, respectively.

Step 6.2. Computing the BCU–exit point

Compute the BCU–exit point for X_l^{UEP} and X_s^{UEP}, respectively, using the golden bisection-based time–domain method. Let them be denoted as X_l^{ray} and X_s^{ray}, respectively.

Step 6.3. Determining the critical energy

Use the energy at X_l^{ray} as the critical energy for contingency L_l. Let its value be denoted as V_s^{ray}. Use the energy at X_s^{ray} as the critical energy for contingency L_s. Let its value be denoted as V_s^{ray}.

Step 6.4. Determining the critical energy for other contingencies
Let the SEP separation of a contingency, say, L_i, in the group of coherent contingencies be SEP_i. Then, the critical energy for contingency L_i is

$$V_i^{cr} = a \times \text{SEP}_i + b,$$

where $a = \dfrac{V_l^{ray} - V_s^{ray}}{\text{SEP}_l - \text{SEP}_s}$ and $b = \dfrac{V_l^{ray} \times \text{SEP}_l - V_s^{ray} \times \text{SEP}_s}{\text{SEP}_l - \text{SEP}_s}$.

Step 7. Perform direct stability assessment based on the computed critical energy value and calculate the energy margin for each contingency in the group. Proceed to the next group.

Step 4 is the group-based verification scheme for the boundary property, while Step 5 is the group-based BCU–exit point method for accurate critical energy calculation. A flowchart for the group-based BCU method for accurate critical energy calculation is depicted in Figure 24.2.

The above version of the integrated group-based BCU method is useful for computing the accurate critical energy for the direct stability assessment of large-scale power systems, but it does not compute the controlling UEP of every contingency. When it demands the controlling UEP of each contingency in the contingency list, another version of the integrated group-based BCU method, presented in the next section, meets the demand.

24.3 GROUP-BASED BCU METHOD FOR CUEPs

This version of the group-based BCU method is composed of the following methods described in the previous chapters:

- energy function method,
- BCU method,
- grouping scheme,
- group-based verification scheme, and
- group-based BCU–CUEP method.

Given a study base case power system with a list of credible contingencies, this version of group-based BCU method for exact controlling UEP calculation is summarized as follows.

Group-Based BCU Method for Computing the Controlling UEPs

Step 1. Compute the postfault SEP of each contingency and construct an energy function for the corresponding postfault system.

Step 2. Compute the controlling UEP of each contingency using the BCU method.

Step 3. Repeat Steps 1 and 2 until all the contingencies have been processed.

Step 4. Group all the contingencies into groups of coherent contingencies.

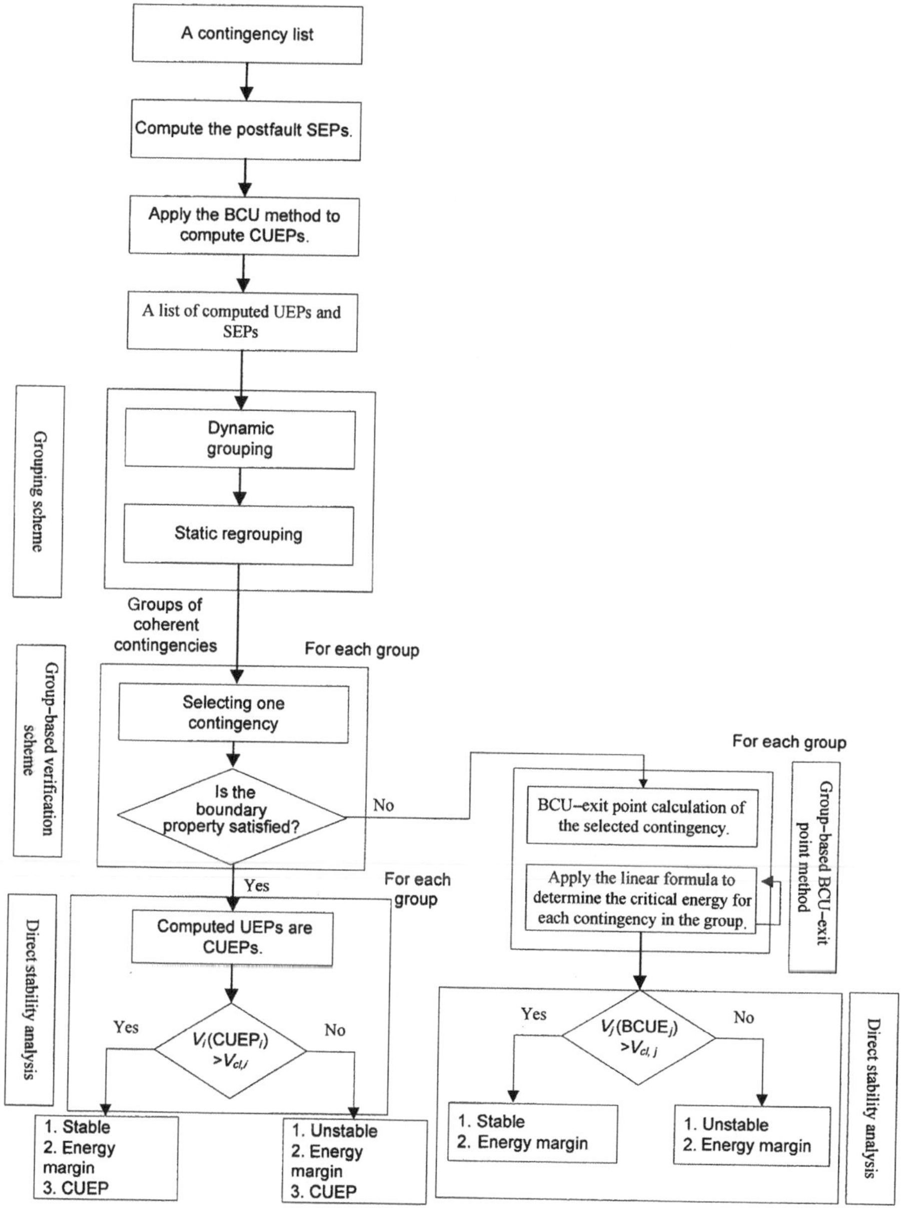

Figure 24.2 The overall computational procedure of the group-based BCU method for accurate critical energy calculation. It is composed of the following steps: computing the postfault SEP and the controlling UEP, grouping, group-based verification scheme for boundary property checking, group-based BCU–exit point method, and direct stability analysis.

For each group of coherent contingencies, do the following:

Step 5. Verify the boundary property using the group-based verification scheme. If the boundary property is satisfied, then the energy value at the computed UEP is an accurate critical energy for each contingency in the group; proceed to Step 7. Otherwise, go to the next step.

Step 6. Compute the exact controlling UEP of each contingency using the group-based BCU–CUEP method until all the contingencies have been processed. The energy value at the computed UEP is an accurate critical energy for each contingency in the group; then, proceed to Step 7.

Step 7. Perform energy margin calculation and direct stability assessment. Proceed to the next group of coherent contingencies.

We notice that Step 2 is the BCU method step, while Step 4 is the grouping step. Step 5 is the group-based verification scheme step, while Step 6 is the group-based BCU–CUEP method step. Step 7 is the overall step of direct stability assessment and energy margin calculation. From a functional viewpoint, the group-based BCU method for accurate critical calculation is depicted by Figure 24.3.

A numerical implementation of the group-based BCU–CUEP method is described in the following:

For each contingency of the contingency list, perform Steps 1 through 3:

Step 1. Apply the partitioned Newton method or its variants with the prefault SEP as the initial guess to compute the postfault SEP of each contingency. If the postfault SEP cannot be found, send the contingency to a time–domain simulation program for stability analysis and proceed to the next contingency.

Step 2. Construct a numerical energy function for the postfault system.

Step 3. Apply the BCU method to compute the controlling UEP of each contingency.

Step 4. Group all of the contingencies into groups of coherent contingencies based on the computed controlling UEPs and postfault SEPs.

For each group of coherent contingencies, do the following:

Step 5. Check the boundary property of each group.

Step 5.1. Selecting step
Select the computed UEP with the largest SEP separation, say, X^{UEP}.

Step 5.2. Checking step
Form the following test vector:

$$X^{text} = X_s^{post} + 0.99\left(X^{UEP} - X_s^{post}\right),$$

where X_s^{post} is the postfault SEP of the contingency corresponding to X^{UEP}. The postfault trajectory starting from X^{test} is simulated and assessed. If the postfault trajectory converges to X_s^{post}, then X^{UEP} satisfies the boundary property; otherwise, it does not.

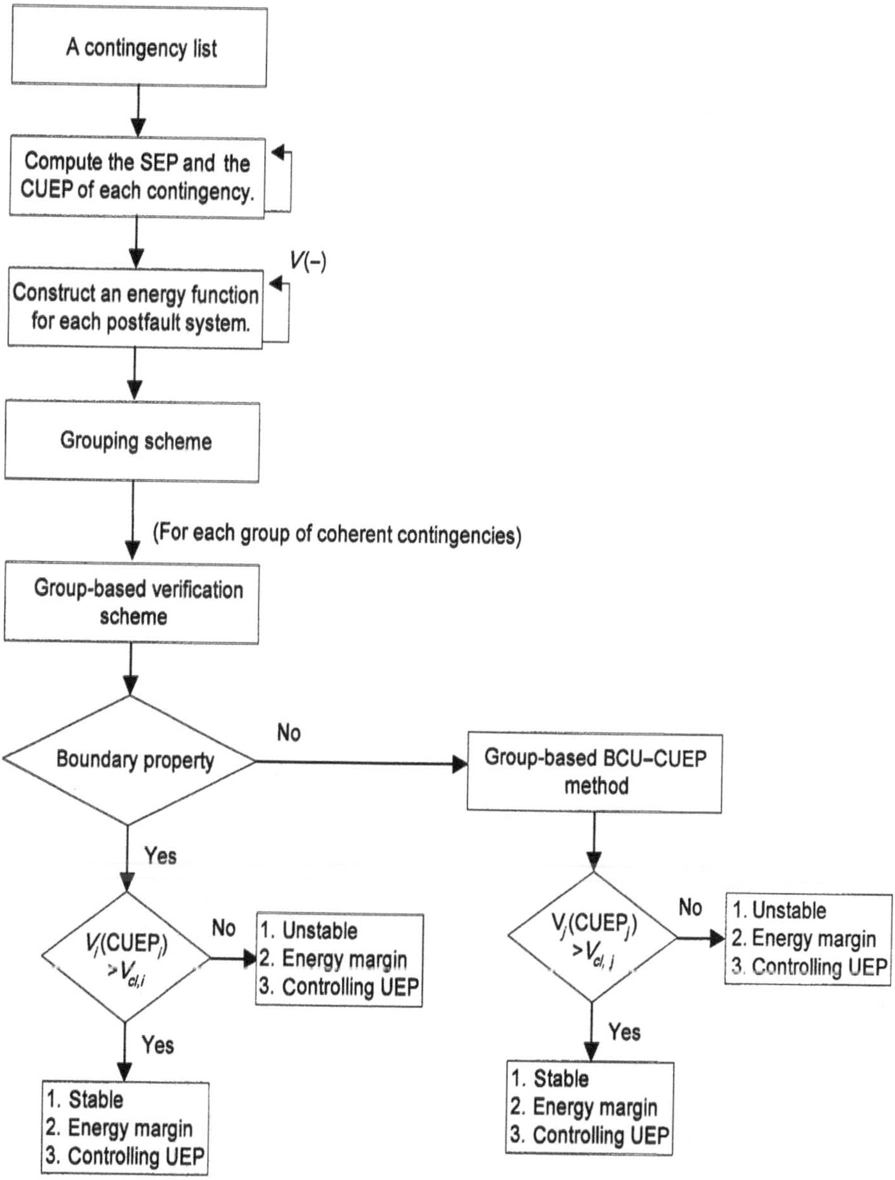

Figure 24.3 An overall architecture of the group-based BCU method for exact controlling UEP calculation. It is composed of the BCU method, the energy function method, the group-based verification scheme for boundary property checking, the controlling UEP method, the group-based BCU–CUEP method, and direct stability analysis.

Step 5.3. Boundary property

If the boundary property is satisfied, then the computed UEP by the BCU method of each contingency lies on the stability boundary of the original system and is considered the exact controlling UEP; then, go to Step 7. Otherwise, go to Step 6.

Step 6. Compute the exact controlling UEP of each contingency in the group using the group-based BCU–CUEP method.

Step 6.1. Selection step

Select a contingency from the group of coherent contingencies, say, the one with the largest SEP separation.

Step 6.2. Computing the exact CUEP

Compute the exact controlling UEP for the selected contingency using the exact time–domain controlling UEP method.

Step 6.3. Compute the controlling UEP of each remaining contingency in the group as follows:

Step 6.3.1. Update the system postfault *Y*-matrix to obtain the postfault power flow equations for each remaining contingency in the group.

Step 6.3.2. Use the computed controlling UEP in Step 6.2 as an initial guess and apply an effective nonlinear algebraic solver to the extended postfault power flow equations to compute the controlling UEP of the remaining contingency.

Step 6.3.3. Let the solution be x_{co}. The exact controlling UEP with respect to the remaining contingency is x_{co}.

Stage 7. Calculate the energy margin based on the critical energy at the exact controlling UEP and perform direct stability analysis.

The above version of the group-based BCU method computes the exact controlling UEP of each contingency. Step 5 is the group-based verification scheme for checking the boundary property, while Step 6 is the group-based BCU–CUEP method for exact CUEP calculation. Step 7 is the direct stability analysis and energy margin calculation. A flowchart of the numerical group-based BCU method for exact CUEP calculation is depicted in Figure 24.4.

24.4 NUMERICAL STUDIES

The group-based BCU method was applied to a 134-generator test power system with a list of credible contingencies. The purpose of these numerical studies is to show that while the BCU method may compute an incorrect controlling UEP for certain contingencies, the group-based BCU method computes the correct controlling UEP for each contingency. Furthermore, it will be shown that the group-based BCU method is less conservative in stability assessment than the BCU method. Detailed simulation results of multiple groups of coherent contingencies are presented.

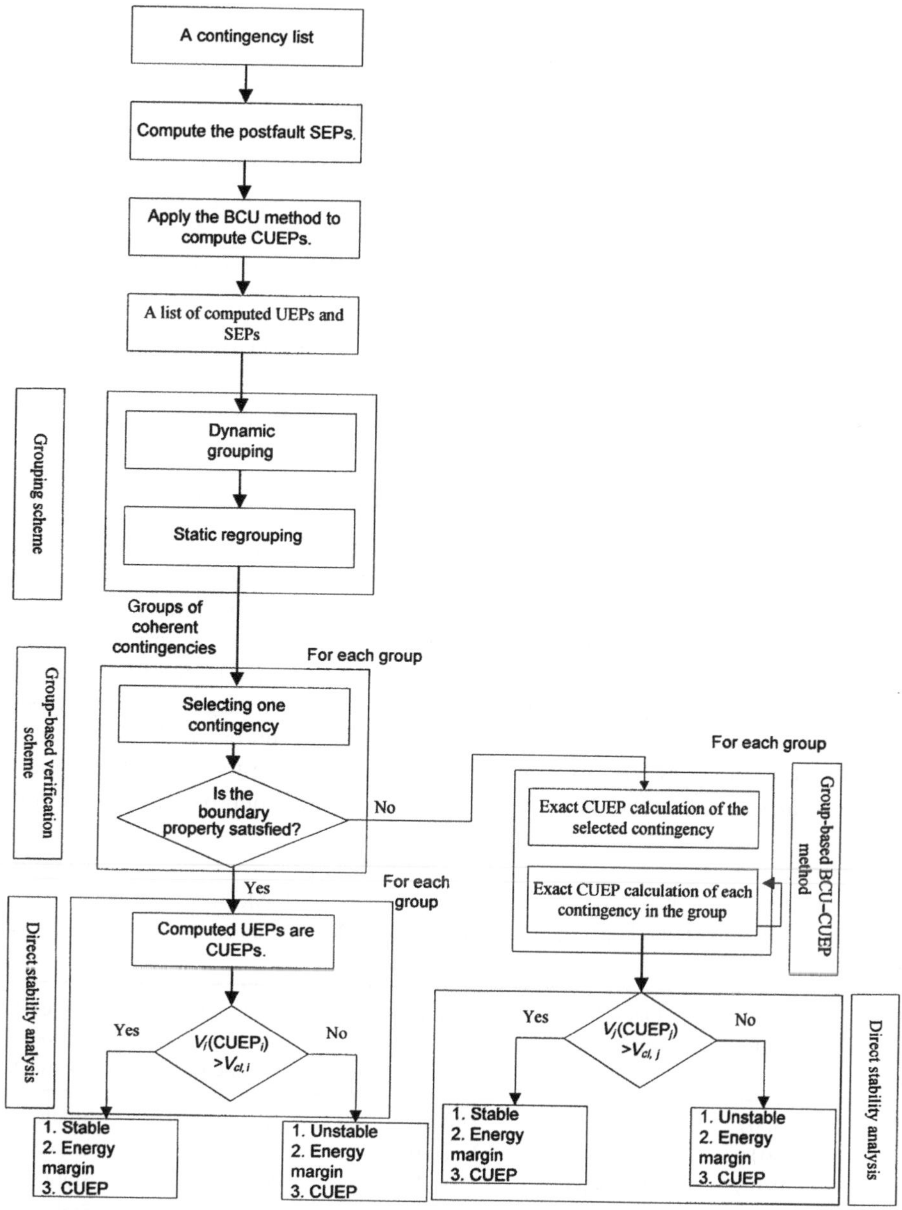

Figure 24.4 The overall computational procedure of the group-based BCU method for CUEP calculation. It is composed of the following steps: computing the postfault SEP and the controlling UEP, grouping, group-based verification scheme for boundary property checking, group-based BCU–CUEP, and direct stability analysis.

We display the stability assessment results and critical energy values obtained by different methods in the following tables. The following pieces of information are contained in these tables:

- a description of the contingency (The first three columns present the case number, faulted bus, and the location of the branch opened to clear the fault.),
- the dynamic security assessment (DSA) results (the assessment results by the BCU method, the assessment results by the detailed time–domain approach, and the assessment result by the group-based BCU method),
- the energy value at the fault clearing point,
- the energy value at the exit point computed by the PEBS (potential energy boundary surface) method,
- the energy value at the controlling UEP computed by the BCU method,
- the critical energy value based on the group-based BCU method, and
- the boundary property and the boundary distance of the computed UEP.

For the first group of 10 coherent contingencies, we summarize the dynamic stability assessments in Table 24.1. These contingencies are all stable, according to time–domain simulation. However, the BCU method gives a rather conservative assessment; all of them are unstable contingencies. It is interesting to note that the controlling UEPs of these 10 unstable contingencies computed by the BCU method all violate the boundary property. The group-based BCU method gives correct stability assessments of all contingencies. It is clear that the group-based BCU method is able to correct the DSA of 10 contingencies from unstable to stable. The critical energy values based on the BCU method and that on the group-based BCU method

Table 24.1 Dynamic Security Assessment Results by the BCU Method, the Time–Domain Simulation Method, and the Group-Based BCU Method on a Group of Coherent Contingencies

Case	Faulted bus	Open line from–to	BCU method assessment	Time–domain method assessment	Group–BCU method assessment
3	3007	3007–107	Unstable	Stable	Stable
4	3016	3015–3016	Unstable	Stable	Stable
5	3015	3015–3016	Unstable	Stable	Stable
6	3524	3007–3524	Unstable	Stable	Stable
7	3007	3007–3524	Unstable	Stable	Stable
8	3024	3007–3024	Unstable	Stable	Stable
9	3007	3007–3024	Unstable	Stable	Stable
10	3007	3007–207	Unstable	Stable	Stable
11	3004	3004–404	Unstable	Stable	Stable
12	3004	3004–304	Unstable	Stable	Stable

Table 24.2 Energy Values at the Fault Clearing Time, at the Computed UEP by the BCU Method, by the Group-Based BCU Method, and the Boundary Distance for Each Contingency

Case	Faulted bus	Open line from–to	Energy at fault clearing	Critical energy by the BCU method	Critical energy by G-BCU method	Boundary distance
3	3007	3007–107	0.407	0.362	0.509	Not exact
4	3016	3015–3016	0.306	0.36	0.508	Not exact
5	3015	3015–3016	0.388	0.375	0.508	Not exact
6	3524	3007–3524	0.405	0.373	0.508	Not exact
7	3007	3007–3524	0.428	0.332	0.508	Not exact
8	3024	3007–3024	0.405	0.373	0.508	Not exact
9	3007	3007–3024	0.428	0.332	0.508	Not exact
10	3007	3007–207	0.413	0.355	0.508	Not exact
11	3004	3004–404	0.494	0.355	0.504	Not exact
12	3004	3004–304	0.494	0.355	0.504	Not exact

along with the energy function values at the fault clearing state (cleared at 0.07 s) are summarized in Table 24.2.

It is clear from this table that the critical energy values determined by the BCU method were too conservative for these contingencies due to the violation of boundary property while the group-based BCU method significantly reduces the conservativeness. The critical energy values based on the BCU method range between 0.332 (for Case 7) and 0.375 (for Case 5), while the critical energy values based on the integrated group-based BCU method range between 0.504 and 0.509.

We study another group of coherent contingencies, a total of 156 coherent contingencies. The BCU method, the integrated group-based BCU method, and the time–domain simulation method are applied to these contingencies. The boundary distance of each contingency in this group is computed. This group of coherent contingencies does not satisfy the boundary property since the boundary distances all lie in the vicinity of 0.894. This implies that the controlling UEP computed by BCU is not correct, and hence the critical energy value provided by the BCU method is not accurate. Even though the BCU method gives correct stability assessment results for this group of coherent contingencies, it is necessary to correct the critical energy value using the group-based BCU method. The critical energy values of these 156 contingencies are corrected by the integrated group-based BCU method (see Table 24.3).

Table 24.4 summarizes the energy value at the fault clearing time, the critical energy value by the BCU method, and the critical energy value by the group-based BCU method. Since this group of coherent contingencies is large, only 36 contingencies were listed in these two tables. It is noted that for this group of coherent contingencies, the critical energy value by the group-based BCU method is smaller than that by the BCU method.

Table 24.3 Dynamic Security Assessment Results by the BCU Method, the Time–Domain Simulation Method, and the Group-Based BCU Method along with the Boundary Distance for a Large Group of 156 Coherent Contingencies

Case number	Faulted bus	Open line from–to	BCU assessment	Time–domain assessment	G-BCU assessment	Boundary distance
1	3020	3016–3020	Stable	Stable	Stable	0.893
2	316	3016–316	Stable	Stable	Stable	0.894
3	3016	3016–316	Stable	Stable	Stable	0.894
4	4139	116–4139	Stable	Stable	Stable	0.894
5	116	116–4139	Stable	Stable	Stable	0.894
6	4139	316–4139	Stable	Stable	Stable	0.894
7	316	316–4139	Stable	Stable	Stable	0.894
8	4525	4021–4525	Stable	Stable	Stable	0.894
9	4026	4026–4027	Stable	Stable	Stable	0.894
10	4033	4031–4033	Stable	Stable	Stable	0.894
11	4525	4024–4525	Stable	Stable	Stable	0.894
12	4026	4026–4029	Stable	Stable	Stable	0.894
13	4032	421–4032	Stable	Stable	Stable	0.894
14	116	3016–116	Stable	Stable	Stable	0.894
15	3016	3016–116	Stable	Stable	Stable	0.894
16	3021	3021–421	Stable	Stable	Stable	0.894
17	4139	216–4139	Stable	Stable	Stable	0.894
18	216	216–4139	Stable	Stable	Stable	0.894
19	216	3016–216	Stable	Stable	Stable	0.894
20	3016	3016–216	Stable	Stable	Stable	0.894
...	...	...	...	...	...	...
...	...	...	...	...	...	...
...	...	...	...	...	...	...
141	4050	4547–4050	Stable	Stable	Stable	0.895
142	4547	4547–4050	Stable	Stable	Stable	0.895
143	4898	4047–4898	Stable	Stable	Stable	0.896
144	4047	4047–4898	Stable	Stable	Stable	0.896
145	3522	3524–3522	Stable	Stable	Stable	0.896
146	3524	3524–3522	Stable	Stable	Stable	0.896
147	3022	3024–3022	Stable	Stable	Stable	0.896
148	3024	3024–3022	Stable	Stable	Stable	0.896
149	3022	3020–3022	Stable	Stable	Stable	0.896
150	3020	3020–3022	Stable	Stable	Stable	0.896
151	3522	3020–3522	Stable	Stable	Stable	0.896
152	3020	3020–3522	Stable	Stable	Stable	0.896
153	3524	3015–3524	Stable	Stable	Stable	0.896
154	3015	3015–3524	Stable	Stable	Stable	0.896
155	3024	3015–3024	Stable	Stable	Stable	0.896
156	3015	3015–3024	Stable	Stable	Stable	0.896

Table 24.4 Energy Values at the Fault Clearing Time, the Critical Energy by the BCU Method, and the Critical Energy Values Determined by the Group-Based BCU Method for a Group of 156 Coherent Contingencies

Case	Faulted bus	Open line from–to	Energy value at the fault clearing time	Critical energy value by the BCU method	Critical energy value by the group-based BCU method
1	3020	3016–3020	0.377	0.601	0.574
2	316	3016–316	0.203	0.698	0.579
3	3016	3016–316	0.274	0.697	0.579
4	4139	116–4139	0.193	0.706	0.579
5	116	116–4139	0.207	0.706	0.579
6	4139	316–4139	0.197	0.698	0.579
7	316	316–4139	0.203	0.698	0.579
8	4525	4021–4525	0.175	0.698	0.579
9	4026	4026–4027	0.198	0.712	0.579
10	4033	4031–4033	0.185	0.708	0.579
11	4525	4024–4525	0.172	0.705	0.579
12	4026	4026–4029	0.198	0.709	0.579
13	4032	421–4032	0.182	0.69	0.579
14	116	3016–116	0.207	0.703	0.579
15	3016	3016–116	0.27	0.703	0.579
16	3021	3021–421	0.22	0.688	0.579
17	4139	216–4139	0.193	0.703	0.579
18	216	216–4139	0.207	0.703	0.579
19	216	3016–216	0.207	0.703	0.579
20	3016	3016–216	0.27	0.703	0.579
…	…	…	…	…	…
…	…	…	…	…	…
…	…	…	…	…	…
141	4050	4547–4050	0.169	0.685	0.583
142	4547	4547–4050	0.174	0.688	0.583
143	4898	4047–4898	0.17	0.683	0.583
144	4047	4047–4898	0.175	0.684	0.583
145	3522	3524–3522	0.291	0.68	0.583
146	3524	3524–3522	0.379	0.681	0.583
147	3022	3024–3022	0.291	0.68	0.584
148	3024	3024–3022	0.379	0.681	0.584
149	3022	3020–3022	0.291	0.669	0.584
150	3020	3020–3022	0.278	0.668	0.584
151	3522	3020–3522	0.291	0.669	0.584
152	3020	3020–3522	0.278	0.668	0.584
153	3524	3015–3524	0.387	0.683	0.585
154	3015	3015–3524	0.357	0.682	0.585
155	3024	3015–3024	0.387	0.682	0.585
156	3015	3015–3024	0.357	0.682	0.585

We examine another group of coherent contingencies that contains 19 contingencies. They all satisfy the boundary property. Hence, the critical energy given by the BCU method is exactly the same as the one by the group-based BCU method. All the 19 contingencies are stable according to the time–domain simulation method, the BCU method, and the group-based BCU method. The PEBS method classifies these contingencies as stable; however, the critical energy values based on the PEBS method tend to be very large as compared with those by the BCU method and by the group-based BCU method. Further numerical study indicates that the critical energy values determined by the PEBS method are much larger than the exact critical energy values determined by the time–domain simulation method. This study confirms the observation that the PEBS method can give overestimation results on critical energy. This implies that the PEBS method can classify unstable contingencies as stable, making the PEBS method unsuitable for practical applications.

The overall DSA results are shown in Table 24.5, while the energy values at fault clearing time, the critical energy values by the PEBS method, the critical energy values by the BCU method, and the critical energy by the group-based BCU method are summarized in Table 24.6.

Table 24.5 DSA Results by the BCU Method, Time–Domain Assessment, Group-Based BCU Method, PEBS Method, and Boundary Property for Group 17 of the Coherent Contingencies

Case	Faulted bus	Open line from–to	BCU method	Time–domain	G-BCU method	PEBS method	Boundary property
1	3028	3028–3029	Stable	Stable	Stable	Stable	Exact
2	4134	4134–5108	Stable	Stable	Stable	Stable	Exact
3	4134	4134–5607	Stable	Stable	Stable	Stable	Exact
4	3028	3028–5607	Stable	Stable	Stable	Stable	Exact
5	4134	4133–4134	Stable	Stable	Stable	Stable	Exact
6	4134	4134–5107	Stable	Stable	Stable	Stable	Exact
7	3028	3028–5107	Stable	Stable	Stable	Stable	Exact
8	4055	4055–7084	Stable	Stable	Stable	Stable	Exact
9	4055	4055–4056	Stable	Stable	Stable	Stable	Exact
10	3028	3028–3030	Stable	Stable	Stable	Stable	Exact
11	4558	4056–4558	Stable	Stable	Stable	Stable	Exact
12	4134	228–4134	Stable	Stable	Stable	Stable	Exact
13	228	228–4134	Stable	Stable	Stable	Stable	Exact
14	4134	128–4134	Stable	Stable	Stable	Stable	Exact
15	128	128–4134	Stable	Stable	Stable	Stable	Exact
16	128	3028–128	Stable	Stable	Stable	Stable	Exact
17	3028	3028–128	Stable	Stable	Stable	Stable	Exact
18	228	3028–228	Stable	Stable	Stable	Stable	Exact
19	3028	3028–228	Stable	Stable	Stable	Stable	Exact

Table 24.6 Energy Values at the Fault Clearing Time, the Critical Energy Values by the PEBS, the BCU Method and the Group-Based BCU Method for Group 17 of the Coherent Contingencies

Case	Faulted bus	Open line from–to	Energy at fault clearing time	Critical energy by the PEBS method	Critical energy by BCU method	Critical energy by the G-BCU method
1	3028	3028–3029	0.242	1.569	0.31	0.31
2	4134	4134–5108	0.188	0.991	0.31	0.31
3	4134	4134–5607	0.188	0.989	0.307	0.307
4	3028	3028–5607	0.242	1.594	0.308	0.308
5	4134	4133–4134	0.188	0.987	0.308	0.308
6	4134	4134–5107	0.188	0.988	0.308	0.308
7	3028	3028–5107	0.242	1.585	0.304	0.304
8	4055	4055–7084	0.176	3.466	0.316	0.316
9	4055	4055–4056	0.177	3.449	0.314	0.314
10	3028	3028–3030	0.244	1.543	0.305	0.305
11	4558	4056–4558	0.181	3.353	0.302	0.302
12	4134	228–4134	0.189	0.96	0.276	0.276
13	228	228–4134	0.193	0.932	0.276	0.276
14	4134	128–4134	0.189	0.96	0.275	0.275
15	128	128–4134	0.193	0.932	0.275	0.275
16	128	3028–128	0.193	0.93	0.271	0.271
17	3028	3028–128	0.243	1.565	0.271	0.271
18	228	3028–228	0.193	0.93	0.271	0.271
19	3028	3028–228	0.243	1.565	0.271	0.271

24.5 CONCLUDING REMARKS

An integrated group-based BCU method, composed of the BCU method, the group-based verification scheme, the group-based BCU–exit point method, and the group-based BCU–CUEP method has been presented in this chapter. The group-based BCU method does not require either the one-parameter transversality condition or the boundary property, and it computes the correct controlling UEP for each contingency of a contingency list. The group-based BCU method improves the computed controlling UEP or the computed critical energy by the BCU method. Furthermore, the group-based BCU method can reduce the conservativeness of the BCU method.

The integrated group-based BCU method is distinguished in two aspects: it computes both the correct controlling UEP and an accurate yet slightly conservative critical energy value for each contingency. The critical energy value computed by the integrated group-based BCU method is always less than the exact critical energy value determined by the time–domain method. Hence, the integrated group-based

BCU method is more reliable, if not completely reliable, and is less conservative than the BCU method in direct stability assessments of large-scale power systems. Compared to the BCU method, the computational burden of the group-based BCU method is only slightly increased.

The group-based BCU method addresses the issue of rigorously verifying the correctness of DSA results. In the past, this issue has been neglected due to the great computational efforts and difficulty involved. The group-based BCU method has several advantages over the existing direct stability methods: Unlike traditional methods for stability control, the group-based BCU method provides much more information for stability enhancement control design. These advantages can be summarized as follows:

- Traditional stability analysis provides information in a case-by-case manner, while the group-based BCU method provides information in a group-by-group manner. The group-based BCU method offers an effective means of stability enhancement control design.
- The identification of coherent contingencies is useful in several applications such as corrective control and preventive control. The coordinates of the controlling UEP in each group of coherent contingencies provide useful information on how to design controls to stabilize a power system or how to enhance stability with respect to a set of contingencies.
- The property that coherent contingencies are usually located in the same geographic area has several potential applications. This property can lead to stability enhancement control design, with respect to coherent contingencies, by focusing on control devices of the generators and loads in the relevant geographic area.

Chapter 25

Perspectives and Future Directions

25.1 CURRENT DEVELOPMENTS

After decades of research and development in energy function-based direct methods, it has become clear that the capabilities of the controlling unstable equilibrium point (controlling UEP) method and those of the time–domain simulation approach complement each other. The current direction of development is to incorporate the BCU-based controlling UEP method and time–domain simulation programs into the body of overall online transient stability assessments (TSAs) (Chiang et al., 1995, 2007; Chiang 1999; Ernst et al., 2001; IEEE Committee Report, 1988; Kaye and Wu, 1982; Morison et al., 2004; Tada et al., 2002; Yu et al., 2002). Online TSA is designed to provide power system operators with critical information including the following:

1. assessment of the transient stability of a power system subject to a list of contingencies and
2. available (power) transfer limits at key interfaces subject to transient stability constraints.

An integrated architecture for online TSA and control is presented in Figure 25.1. In this architecture, there are two major components in the module of online TSA: fast, dynamic contingency screening and a time–domain stability program for performing detailed stability analysis. When a new cycle of online TSA is warranted, a list of credible contingencies, along with information from the state estimator and topological analysis, is applied to the dynamic contingency screening program whose basic function is to screen out contingencies that are definitely stable or potentially unstable. Contingencies classified to be definitely stable are eliminated from further analysis. Contingencies classified to be potentially unstable are sent to fast time–domain simulation for detailed analysis. Online TSA is feasible due to its ability to perform dynamic contingency screening on a large number of contingencies and to filter out a much smaller number of contingencies requiring further analysis. Contingencies that are either unclassified or identified as unstable are then

Direct Methods for Stability Analysis of Electric Power Systems, by Hsiao-Dong Chiang

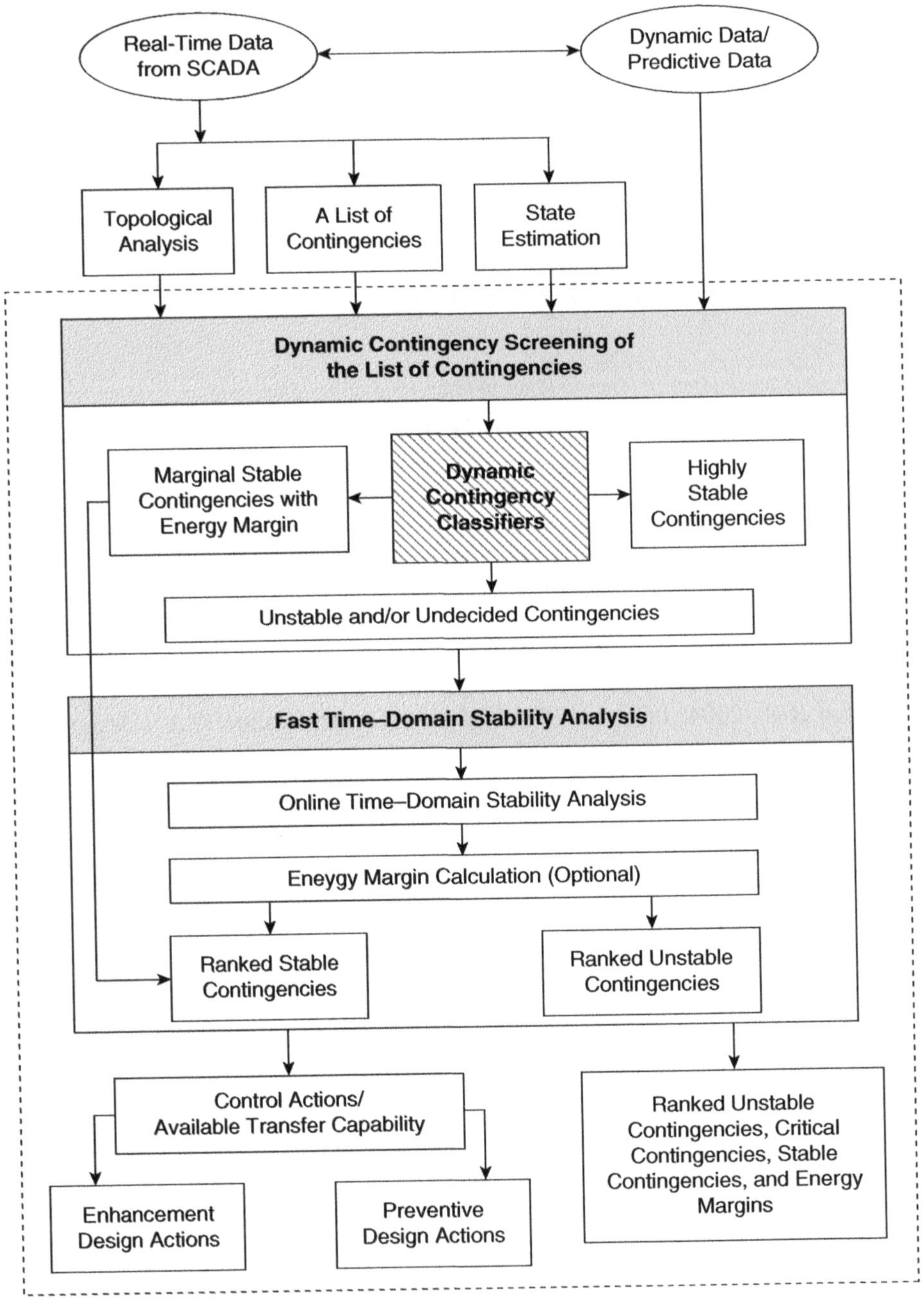

Figure 25.1 An architecture for online transient stability assessment and control.

sent to the time–domain transient stability simulation program for detailed stability analysis.

The block function of control action decisions determines if timely postfault contingency corrective action, such as automated remedial action, is feasible to steer the system away from unacceptable conditions towards a more acceptable state (Li and Bose, 1998; Ota et al., 1996). If appropriate corrective action is not available, the block function of preventive action determines the required precontingency preventive controls should the contingency occur (Faucon and Dousset, 1997; Kaba et al., 2001; Matsuzawa et al., 1995; Scala et al., 1998). These controls include real power redispatches or line switching to maintain the system stability. If the system is marginally stable (i.e., critically stable), the block function of enhancement action determines the required precontingency enhancement controls thereby increasing the degree of system stability should the contingency occur. In this architecture, a fast yet reliable method for performing dynamic contingency screening plays a vital role in the overall process of online TSA.

Suppose a complete online TSA assessment cycle will be completed within, say, 15 min. This cycle starts when all of the necessary data are available to the system and ends when the system is ready for the next cycle. Depending on the size of the underlying power systems, it is estimated that for a large-size power system such as a 15,000-bus power system, the number of contingencies in a contingency list lies between 1000 and 3000. The contingency types will include both three-phase faults with primary clearance and single line-to-ground faults with backup clearance. Several systems have been developed intended for online TSA (El-kady et al., 1986; Mansour et al., 1995; Riverin and Valette, 1998; Takazawa et al., 2006; Vaahedi et al., 1998; Wang and Morison, 2006).

The outputs of online TSA in a given cycle include the following:

- overall status of the system (secure or insecure and the operating margin);
- unstable contingencies (contingency details such as fault type, fault location, and circuits lost);
- stability margin in terms of the energy margin, operating margin in MW, MVAR for each unstable contingency;
- detailed time–domain responses (swing curves) of user-specified quantities for potentially unstable contingencies;
- critical contingencies (contingency details such as fault type, fault location, and circuits lost);
- stability margin in terms of the energy margin, operating margin in MW, MVAR for each critical contingency;
- detailed time–domain responses (swing curves) of user-specified quantities for critical contingencies; and
- power transfer limits at key interfaces and the limiting contingencies.

In addition to the above main functions, the online TSA system needs to have the following functions:

- a study mode with which the users, such as reliability engineers, can analyze various scenarios using cases archived from real-time system models or created from operational planning studies;
- software and hardware failover protection;
- an interface with energy management system (EMS) functions; and
- a definition of the contingency list and creation of the necessary data for stability analysis, data validation and correction, and output visualization.

25.2 ONLINE DYNAMIC CONTINGENCY SCREENING

The strategy of using an effective scheme to screen out a large number of stable contingencies, capture critical contingencies, and apply detailed simulation programs only to potentially unstable contingencies is well recognized. This strategy has been successfully implemented in online static security assessments (Balu et al., 1992). The ability to screen several hundred contingencies to capture tens of the critical contingencies has made the online SSA feasible. This strategy can be applied to online TSA. Given a set of credible contingencies, this strategy would break the task of online TSA into two assessment stages:

Stage 1. Perform the task of dynamic contingency screening to quickly screen out contingencies that are definitely stable from a set of credible contingencies.

Stage 2. Perform a detailed assessment of dynamic performance for each contingency remaining in Stage 1.

Dynamic contingency screening is a fundamental function of an online TSA system. The overall computational speed of an online TSA system depends greatly on the effectiveness of the dynamic contingency screening, the objective of which is to identify contingencies that are definitely stable and thereby to avoid further stability analysis for these contingencies. It is due to the definite classification of stable contingencies that considerable speedup can be achieved for TSA. Contingencies that are either undecided or identified as critical or unstable are then sent to the time–domain transient stability simulation program for further stability analysis.

Hence, it is imperative that a dynamic contingency screening program satisfies the following five requirements (Chiang and Wang, 1998; Chiang et al., 1999):

1. **Reliability Measure.** Absolute capture of unstable contingencies as fast as possible; that is, no unstable (single swing or multiswing) contingencies are missed. In other words, the ratio of the number of captured unstable contingencies to the number of actual unstable contingencies is 1.
2. **Efficiency Measure.** High yield of screening out stable contingencies as fast as possible; that is, the ratio of the number of detected stable contingencies to the number of actual stable contingencies is as close to 1 as possible.

3. **Online Computation.** Little need for off-line computations and/or adjustments in order to meet with the constantly changing and uncertain operating conditions.
4. **Speed Measure.** High speed, that is, fast classification for each contingency case.
5. **Performance Measure.** Robust performance with respect to changes in power system operating conditions.

The requirement of the absolute capture of unstable contingencies is a reliability measure for dynamic contingency screening. This requirement is extremely important for online TSA. However, it is due to the nonlinear nature of the dynamic contingency screening problem that this requirement can best be met by a reliable method such as one with a strong analytical basis. The third requirement asserts that a desired dynamic contingency classifier is one that relies little to none on off-line information, computations, and/or adjustments. This requirement arises because under current and near-future power system operating environments, the correlation between online operational data and presumed off-line analysis data can be minimal or, in extreme cases, the two can be irrelevant to one another. In other words, in a not-too-extreme case, presumed off-line analysis data may become unrelated to online operational data. This uncorrelated relationship is partly attributed to the imminent bulk power transactions resulting from deregulation. The first four requirements should be persistent over all of the operating conditions, as dictated by the requirement for robust performance.

Several methods developed for online dynamic contingency screening have been reported in the literature. These methods can be categorized as follows: the energy function approach (Chadalavada et al., 1997; Chiang and Wang, 1998; Chiang et al., 1999, 2002, 2005), the time–domain approach, and the artificial intelligence (AI) approach (Djukanovic et al., 1994; Jensen et al., 2001; Mansour et al., 1997b; Sobajic and Pao, 1989). The time–domain approach involves the step-by-step simulation of each contingency for a few seconds say 2 or 3 s, to filter out the very stable or very unstable contingencies. This approach may suffer from an accuracy problem in identifying multiswing stable or unstable contingencies. The AI approaches, such as the pattern recognition technique, the expert system technique, the decision tree technique, and the artificial neural network approach, all first perform extensive off-line numerical simulations aiming to capture the essential stability features of the system's dynamic behavior. They then construct a classifier attempting to correctly classify new, unseen online contingencies. As such, the AI approach is likely to become ineffective for online application to current or near-future power systems if little correlation exists between online operational data and presumed off-line analysis data. Unfortunately, the existing AI-based methods fail to meet the online computation requirement and cannot guarantee the reliability requirement.

The BCU-based controlling UEP method presented in this book is an ideal candidate to serve the function of dynamic contingency screening and provides the following advantages:

- fast computational speed to identify definite stable contingencies and to detect potential unstable contingencies;
- computation of energy margins for the degree of stability/instability of each contingency;
- fast computation of transient stability-constrained available transfer capability (ATC) based on energy margins;
- computation of controlling UEPs and their related information, which leads to the development of preventive control schemes against unstable contingencies; and
- computation of controlling UEPs and their related information, which leads to the development of enhancement control schemes for critically stable contingencies.

We emphasize that the deregulated electricity market has resulted in rather rapid changes in operating conditions. System operators now face more new, unknown power flow patterns than ever before (McCalley et al., 1997). At the same time, economic pressure on the electricity market and on grid operators, coupled with limited investment in new generation and transmission networks, pushes power systems close to their stability limits. The traditional tool of off-line dynamic security assessments is inadequate and may ineffectively meet the online requirements. As society grows to be increasingly dependent on reliable electrical supply, the consequences of blackouts are becoming more costly. Thus, any violation of the stability limits will have huge impacts (financially and physically) on society.

25.3 FURTHER IMPROVEMENTS

Further improvements in the modeling capability of direct methods are desirable, for instance the inclusion of appropriate dynamic load models, renewable energy sources, and modern control devices (Hiskens and Hill, 1992; Jing et al., 1995; Ni and Fouad, 1987; Nunes et al., 2004; Tada et al., 2000). The development of controlling UEP methods and BCU methods for AC/DC power systems is needed (DeMarco and Canizares, 1992; Susuki and Hikihara, 2005; Susuki et al., 2004, 2008; Vovos and Galanos, 1985). It is well-known that inaccurate load modeling can lead a power system to operate in modes that result in actual system collapse or separation. Accurate load models capturing load behaviors during disturbances are therefore necessary for more precise assessments of power system stability and calculation of power system stability limits. Most of the load models used in direct methods are limited to the so-called static load models. These static load models may not adequately capture transient load behaviors. Hence, it may prove rewarding to extend direct methods to handle sufficiently detailed load models (more accurate static load models or even adequate dynamic load models) (Chiang et al., 1997; Davy and Hiskens, 1997; Hill, 1993; Lesieutre et al., 1995; Sauer and Lesieutre, 1995).

In recent years, practical experience has shown that a power system may become unstable 10–30 s after a disturbance, even if it is stable in transient states. In other words, the power system is transiently (i.e., short-term) stable and midterm unstable. One mechanism that may contribute to this phenomenon is that the initial state of the postfault system lies inside the stability region of the power system transient stability model but outside the stability region of the power system midterm stability model. This prompts the need for conducting power system midterm stability analysis and emphasizes the necessity for extending the controlling UEP method to power system midterm stability models. In this regard, generalized energy functions for midterm stability models and development of the associated generalized energy function theory are required. It may prove useful to develop generalized energy functions (more general than the energy function) such that any nonlinear system having such a function has the following dynamic properties: (1) the ω-limit set of any of its bounded trajectories consists only of equilibrium points and limit cycles, and (2) the function is nonincreasing along its trajectories. Such a development will allow for the application of the extended controlling UEP method to detailed power system midterm stability models whose stability boundaries contain equilibrium points and limit cycles (Alberto and Chiang, 2008, 2009).

Two crucial factors in power system operations and planning are reliability and efficiency. An adequate return on the high investment costs of power networks can only be achieved if the power network is used efficiently. Optimal efficiency is reachable only if the stability margins of a power network are known precisely. The stability characteristics of a power network must be respected to ensure reliable operation of the power network. In the next section, the application of online TSA to the task of online determination of the ATC is described.

25.4 PHASOR MEASUREMENT UNIT (PMU)-ASSISTED ONLINE ATC DETERMINATION

Due to the upcoming practice of wheeling large amounts of green power over long transmission systems, utilities are more often concerned with determining the power transfer limits across a boundary or the limits of real power generation sent out of a power station. This concern, coupled with the need to share full utilization of transmission among utilities for economic transfer, has made the problem associated with real-time determination of transient stability-constrained power transfer limits a pressing issue. The determination of power transfer limits among various areas and zones of power networks will become more important than ever. A tool, termed the online transient stability-constrained transfer limiter, can be developed using a combination of the family of BCU methods, the conventional time–domain simulation approach, and the continuation power flow method. This tool will provide the information of power transfer limits across a boundary more timely and readily.

ATC has been used to guide power system operations for setting transfer limits on transmission corridors and key tie-lines (Sauer et al., 1983; Wang and Morison, 2006). Currently, most ATC calculations are based on off-line planning studies. The ATC is highly related to a system's operating conditions. To ensure operational security and reliability, the ATC is currently calculated using worst-case scenarios, and it results in very conservative calculations of power transfer limits. Hence, there is a critical need to calculate the ATC based on actual operating conditions.

The task of computing the ATC subject to dynamic security constraints is very challenging due to the nonlinear nature of interconnected power systems and the tremendous computation requirements of the stability analysis of credible contingencies. The dynamic security limit to be considered is the transient stability limit. To put this dynamic security limit into the framework of the ATC, we next define the transient stability limit load margin.

Definition: The transient stability limit load margin is the (minimum) distance (in terms of MW and/or MVAR) from the current operating point to the state vector of the P-V curve of the contingency where the transient instability occurs. Here, the load margin is of a power system dynamic model under a contingency, with respect to a load-generation vector.

We note that the transient stability limit load margin should be smaller than the nose-point load margin of the base case power system since the transient stability limit load margin is not defined when its value is greater than the nose-point load margin of the base case power system. The task of computing the transient stability limit load margin with respect to a set of credible contingencies is challenging.

We suggest a tool, termed the BCU limiter, which can quickly compute the ATC limited by the transient stability of credible contingencies. This tool computes, given a proposed power transaction, the amount of power transfers a power system can withstand before its transient stability limit is reached. In addition, the BCU limiter can rank a given list of contingencies, in terms of their load margins, to transient stability limits and can compute the corresponding ATC. This BCU limiter is an integration of the BCU methods, the continuation power flow method, and a time–domain simulation method. Given an operating point, the BCU limiter not only performs power system dynamic security assessments and ranking but also computes the ATC limited by the transient stability of credible contingencies.

To provide a real-time ATC, it is necessary to have some real-time information regarding a system's operating conditions. For more than a decade, power engineers and researchers have tried to deploy PMUs into power systems and to build the infrastructure for wide area measurement and monitoring. Significant investment in the installation of PMUs is expected, and wide area measurement systems in both the eastern and western interconnections are taking shape. One central topic in the area of wide area measurement is the utilization of this new type of measurement (as opposed to traditional supervisory control and data acquisition [SCADA] measurements). Phasor data, precisely time-synchronized data at a high data rate, provide a wide area view of current power system conditions. To fill the gap between phasor measurements and operation applications, we propose to develop a tool to combine this wide area view with ATC calculation techniques such that an accurate determination of the online ATC with actual operating conditions can be achieved.

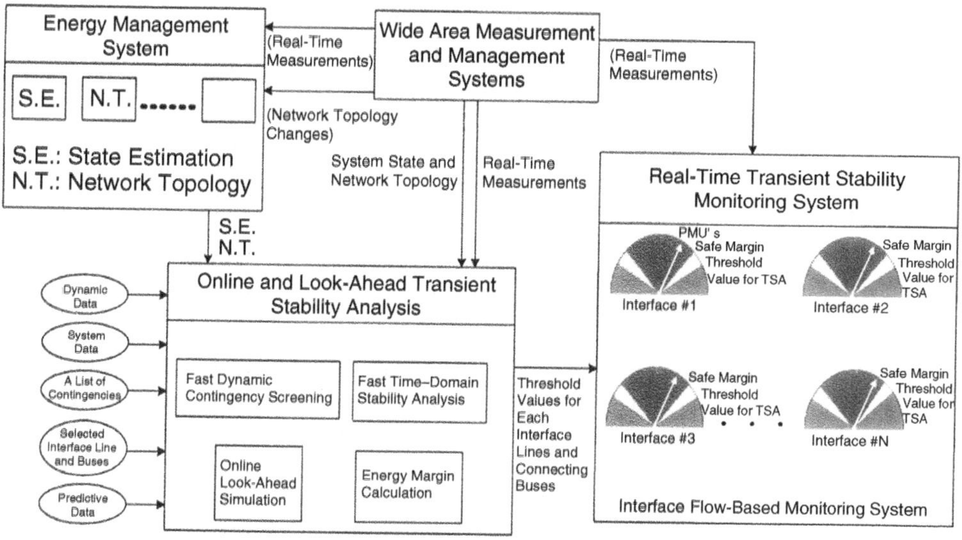

Figure 25.2 Architecture of a wide-area measurement system (WAMS)-assisted, BCU-based, real-time available transfer capability determination.

This online ATC estimation will lead to better power system asset utilization and economic benefits while still ensuring security and reliability.

Given a set of predetermined interfaces and the contingency list associated with each interface, we propose a PMU-assisted, BCU-based, real-time ATC determination system for each interface shown in Figure 25.2. There are two key components to this architecture: the BCU limiter and the real-time measurement units, PMUs.

1. The BCU limiter engine determines the transient stability-constrained power transfer limit across each interface under the contingency list associated with each interface.
2. The real-time measurement of real power flow across each interface from the installed PMUs, the difference between the limit value of power transfer, and the current power transfer is the ATC of the system expressed as the flow across each interface.

25.5 EMERGING APPLICATIONS

During the transient period following the clearing of a large disturbance, voltages and currents can undergo significant excursions, resulting in unexpected switching events. For example, large rotor swings between generators following a large disturbance can result in the tripping of distance relays and can possibly lead to system separation and islanding. Hence, the inclusion of protection operations and viability limits in online dynamic security assessments is desirable. This inclusion would require an estimate of the conditions that induce protection operation or nonviable transients. It is hence required to develop techniques that offer a fast and reliable way of (1) assessing whether a disturbance will initiate an unexpected protection system and (2) adjusting protection and control schemes to avoid such undesirable events. The concept of using relay margins to provide a measure of the closeness of a relay, issuing a trip command, was applied to assess system vulnerability along the postfault system trajectory (Dobraca et al., 1990). Energy margins were used for assessing the out-of-step protection operation in Singh and Hiskens (2001). This work was based on a heuristic relationship between the maximum angular deviation of a separating group of generators and the minimum apparent impedance seen by the out-of-step relays.

Although current power grids cannot completely prevent disastrous cascading, their ability to manage power system disturbances can be considerably enhanced. Power systems have been relying mostly on protection systems, special protection systems, and discrete supplementary control schemes to manage disturbance and to prevent disastrous cascading. The design of these systems and schemes is often based on passive and static considerations. Their parameter settings are not adaptive to system operating conditions and network configurations. In addition, the design and parameter settings do not take into account system dynamic behaviors. Consequently, several adverse behaviors of protection systems and discrete supplementary control systems have occurred, causing service interruption of electricity and system blackouts. These behaviors include unnecessary relay trippings (relays overacting to stable swings) and unnecessary distance relay trippings due to system low voltage and heavy loading conditions.

Traditionally, there is no linkage between EMSs, protection systems, and/or discrete supplementary control systems. A linkage can be established using online measurements, online static and dynamic security assessment results available at the EMS, plus additional required functions to develop hierarchical and adaptive protection systems and adaptive discrete controllers. By periodically broadcasting updated information on both static and dynamic security assessment results from the EMS to selected protection systems and sites of discrete controllers, the rules residing in the relays and control schemes of the discrete supplementary controllers can be updated. It is the building of these hierarchical and adaptive protection systems and discrete controllers that helps the overall power system to better manage disturbances and prevent disastrous cascading.

Another application of online security assessments is the reliability-based maintenance. It is well recognized that the failure of power system equipment has varying degrees of impact on power system security and power system reliability. The failure of some crucial equipment may cause power system insecurity, while the failure of other equipment may have little effect on power system security margins. This fact, in addition to the difficulties in implementing condition-based maintenance, has prompted reliability-based maintenance (Beehler, 1997). One important issue in implementing reliability-based maintenance is a proper ranking scheme for power system equipment on power system security. Again, online dynamic security screening and ranking can provide such vital information.

25.6 CONCLUDING REMARKS

This book has demonstrated, through an exposition of theoretical foundations of direct methods, the family of BCU methods, and the group-based BCU methods, that analytical results can lead to the development of reliable yet efficient solution methodologies for solving practical problems. The occasional failure of the BCU method due to the violation of the boundary property was analyzed, and an integrated verification and remedy scheme for the BCU method, termed the BCU–exit point method, for correcting this problem was developed. In order to reduce the intensive computational efforts associated with the BCU–exit point method, a family of group-based BCU methods was developed to reliably compute the controlling UEP and the accurate critical energy for every contingency in the list of credible contingencies. This further enhances the author's belief that efficiently solving challenging practical problems can be accomplished through a thorough understanding of the underlying theory, in conjunction with an exploration of the special features of the practical problem under study, to develop effective solution methodologies.

The author believes that the following stages of research and development will lead to fruitful and practical applications:

Stage 1. Development of theoretical foundations

Stage 2. Development of the solution methodology

Stage 3. Development of reliable methods to numerically implement the solution methodology

Stage 4. Software implementation and evaluation

Stage 5. Industry user interactions

Stage 6. Practical system installation

The first three stages are suitable for university and research institution application, while the last four stages are more suitable for commercial entities. There is a significant difference between the research criteria of an academic institution and that of a commercial institution. For example, the divergence of Newton methods in solving nonlinear algebraic equations is acceptable in the academic environment. However, in practical applications, this divergence is unacceptable and more reliable numerical methods are needed to safeguard the Newton method. This text focuses on Stages 1 and 2 and touches upon Stage 3. In the following volume, Stage 3 will be more thoroughly explored along with Stages 4 through 6.

Appendix

A1.1 MATHEMATICAL PRELIMINARIES

We shall require some definitions and notation from set theory. Most of our problems are posed in d-**dimensional Euclidean space**, R^d; in particular, $R^1 = R$ is just the set of real numbers or the real line, R^2 is the (Euclidean) plane, and R^3 is the usual (Euclidean) space. Points in R^d are printed in bold type $\boldsymbol{x}$, $\boldsymbol{y}$, and so on, and we will sometimes use the coordinate form $\boldsymbol{x} = (x_1, \ldots, {}_\lambda x_d)$. If $\boldsymbol{x}$ and $\boldsymbol{y}$ are points of R^d, the **distance** between them is

$$|\boldsymbol{x}-\boldsymbol{y}| = \left(\sum_{i=1}^{d} |x_i - y_i|^2\right)^{1/2}.$$

Sets, which will generally be subsets of R^d, are denoted by capital letters (e.g., E, F, and K). In the usual way, $\boldsymbol{x} \in E$ means that the point $\boldsymbol{x}$ is a member of the set E, and $E \subset F$ means that E is a subset of F. We write $\{\boldsymbol{x}\text{: condition}\}$ for the set of $\boldsymbol{x}$ for which the condition is true. The **empty set**, which contains no elements, is written as Ø. We sometimes use a superscript, $^+$, to denote the positive elements of a set (e.g., R^+ is the set of positive real numbers).

The **closed ball** of center $\boldsymbol{x}$ and radius r is defined by $B_r(\boldsymbol{x}) = \{\boldsymbol{y}:|\boldsymbol{y}-\boldsymbol{x}| \le r\}$. Similarly, the **open ball** is $\{\boldsymbol{y}:|\boldsymbol{y}-\boldsymbol{x}| < r\}$. Thus, the closed ball contains its bounding sphere, but the open ball does not. Of course, in R^2, a ball is a disk, and in R^1, a ball is just an interval. If $a < b$, we write $[a, b]$ for the **closed interval** $\{x:a \le x \le b\}$ and (a, b) for the **open interval** $\{x:a < x < b\}$.

We write $E \cup F$ for the **union** of the sets E and F (i.e., the set of points belonging to either E or F). Similarly, we write $E \cap F$ for their **intersection** (i.e., the points in both E and F). More generally, $\cup_i E_i$ denotes the **union** of an arbitrary collection of sets $\{E_i\}$ (i.e., those points in at least one E_i), and $\cap_i E_i$ denotes their **intersection**, consisting of the points common to all of the sets E_i. A collection of sets is **disjoint** if the intersection of any pair is the empty set. The **difference** $E\backslash F$ consists of those points in E that are not in F, and $R^d\backslash E$ is called the **complement** of E.

If E is any set of real numbers, the **supremum**, $\sup E$, is the least number m, such that $x \le m$ for every x in E. Similarly, the **infimum**, $\inf E$, is the greatest number m, such that $m \le x$ for every x in E. Roughly speaking, we think of $\inf E$ and $\sup E$ as the minimum and maximum of the numbers in E, though it should be emphasized that $\inf E$ and $\sup E$ need not themselves be in E.

The **diameter**, diam E, of a subset E of R^d is the greatest distance apart of pairs of points in E; thus, diam $E = \sup\{|\boldsymbol{x}-\boldsymbol{y}|:\boldsymbol{x}, \boldsymbol{y} \in E\}$. A set A is **bounded** if it has a finite diameter or, equivalently, is contained in some sufficiently large ball.

Direct Methods for Stability Analysis of Electric Power Systems, by Hsiao-Dong Chiang

We have already used the terms open and closed in connection with intervals and balls, but these notions extend to much more general sets. Intuitively, a set is closed if it contains its boundary and is open if it contains none of its boundary points. More precisely, a subset E of R^d is **open** if, for every $\boldsymbol{x}$ in E, there is some ball $B_r(\boldsymbol{x})$ of positive radius r, centered at $\boldsymbol{x}$ and contained in E. A set E is **closed** if its complement is open; equivalently, for every sequence $\boldsymbol{x}_r$ in E that is convergent to a point $\boldsymbol{x}$ of $\boldsymbol{R}^d$, we have $\boldsymbol{x}$ in $\boldsymbol{E}$. The empty set Ø and $\boldsymbol{R}^d$ are regarded as both open and closed. The union of any collection of open sets is open, as is the intersection of a ***finite*** collection of open sets. The intersection of any collection of closed sets is closed, as is the union of a ***finite*** number of closed sets.

The smallest closed set containing a set $\boldsymbol{E}$, more precisely, the intersection of all closed sets that contain $\boldsymbol{E}$, is called the **closure** of E. Similarly, the **interior** of a set E is the largest open set contained in E, that is, the union of all open subsets of E. The **boundary** of E is defined as the set of points in the closure of E but not in its interior.

For our purposes, a subset of R^d is **compact** if it is closed and bounded. A set E is thought of as connected if it consists of just one "piece"; formally, E is **connected** if there do not exist open sets U and V such that $U \cup V$ contains E and, with $E \cap U$ and $E \cap V$, is disjoint and nonempty. A subset E of R^2 is termed **simply connected** if both E and $R^2 \backslash E$ are connected.

Occasionally, we need to indicate the degree of smoothness of a curve or surface. We say that such a set is $C^k (k = 1, 2 \ldots)$ if it may be defined locally, with respect to suitable coordinate axes, by a function that is k times differentiable with a continuous kth derivative. A curve or surface is C^∞ if it is C^k for every positive integer k.

Definition: Let $(x, \|.\|)$ be a normed vector space. All points and sets mentioned below are understood to be elements and subsets of x.

1. A *neighborhood* of a point p is a set $B_r(p) = \{x \in X: \|x - p\| < r\}$.
2. A point p is a *limit point* of the set E if every neighborhood of p contains a point $q \neq p$, such that $q \in E$.
3. E is *closed* if every limit point of E is a point of E.
4. E is *open* if, for every point p of E, there is a neighborhood N of p such that $N \subset E$.
5. E is *bounded* if there is a real number M and a point $q \in X$, such that $\|q - p\| < M$ for all $p \in E$.

A1.2 PROOFS OF THEOREMS IN CHAPTER 9

Appendix 9A: Proof of Theorem 9.12

Let $S_c(r)$ denote the set $\{(\delta, \omega): V(\delta, \omega) < r\}$ containing $(\delta_s, 0)$ and let $S_p(r)$ denote the connected component of the set $\{\delta: V_p(\delta) < r\}$ containing δ_s.

Proof of Theorem 9.12

"⇒" Observing $V(.)$ and $V_p(.)$ in Equation 9.10, it follows that $V(\delta, \omega) = V_p(\delta) + V_k(\omega)$, here $V_k(\omega) = \frac{1}{2}\omega_g^T M_g \omega_g$. It is easily verified that

$$S_c\left(V\left(\hat{\delta}, 0\right)\right) \cap \left\{(\delta, \omega) : \omega = 0, \delta \in R^n\right\} = S_p\left(V_p\left(\hat{\delta}\right)\right). \tag{A1.1}$$

Since $(\hat{\delta}, 0)$ is the closest unstable equilibrium point (UEP) of $d(M_g, D)$, condition (iii) of the energy function implies that $S(V(\hat{\delta}, 0))$ is a bounded set containing no other equilibrium points but the stable equilibrium point $(\delta_s, 0)$. It is also noted that $(\hat{\delta}, 0)$ is an equilibrium point of $d(M_g, D)$ if and only if (δ) is an equilibrium point of $d(I)$. Thus, by Equation A1.1, $S_p(V_p(\delta))$ is a bounded set and contains no other equilibrium points but the stable equilibrium point δ_s. On the other hand, condition (i) of the energy function assures that $S_p(V_p(\hat{\delta}))$ is a positive invariant set of system $d(I)$. Conditions (i) and (ii) of the energy function imply that the ω-limit set of system $d(I)$ consists entirely of equilibrium points. With these two facts plus the fact that the ω-limit set of every bounded trajectory exists, we conclude that every trajectory in the set $S_p(V_p(\hat{\delta}))$ converges to δ_s (the only ω-limit set in $S_p(V_p(\hat{\delta}))$). Hence, $S_p(V_p(\hat{\delta})) \subseteq A(\delta_s)$. By the continuity property of energy functions, $S_p(V_p(\hat{\delta})) \subseteq A(\delta_s)$ means that $\hat{\delta}$ is in the closure of the stability region of δ_s. However, $A(x_s)$ cannot contain any equilibrium points other than x_s. Thus, $\hat{\delta}$ must be lying on the stability boundary of δ_s and consequently possesses the minimum value of energy function $V_p(.)$ among all the equilibrium points on the stability boundary.

"⇐" Suppose $\hat{\delta}$ is the closest UEP of $d(I)$. From Proposition 9.3 and condition (iii) of the energy function, it follows that the set $S_p(V_p(\hat{\delta}))$ is bounded and contains no other equilibrium points but δ_s. In addition, the following is true:

$$\begin{aligned}
\bar{S}_c\left(V\left(\hat{\delta}, 0\right)\right) :=\ & \text{the connected component of} \\
& \left\{(\delta, \omega) := V(\delta_s, 0) \le V(\delta, \omega) \le V\left(\hat{\delta}, 0\right)\right\} \\
& \text{containing } (\delta_s, 0) \\
=\ & \left\{(\delta, \omega) : V(\delta_s, 0) \le V_p(\delta) + V_k(\omega) \le V\left(\hat{\delta}, 0\right)\right\} \\
& \text{containing } (\delta_s, 0) \\
\subseteq\ & \left\{(\delta, \omega) : \delta \in S_p\left(V_p\left(\hat{\delta}\right)\right), V(\delta_s, 0) - V_p\left(\hat{\delta}\right) \right. \\
& \left. \le V_k(\omega) \le V_p(\delta) + V\left(\hat{\delta}, 0\right)\right\}.
\end{aligned} \tag{A1.2}$$

Since the map $V_k(\omega) := \omega^T M_g \omega$: $R^n \rightarrow R$ is a proper map (a map f: $X \rightarrow Y$ is called proper if the preimage of every compact set in Y is compact in S); thus, the set $\left\{\omega : V(\delta_s, 0) - V_p\left(\hat{\delta}\right) < V_k(\omega) < V_p\left(\hat{\delta}\right) + V\left(\hat{\delta}, 0\right)\right\}$ is bounded. Because the product of compact spaces is compact, it is from Equation A1.2 that the set $S_c\left(V\left(\hat{\delta}, 0\right)\right)$ is also bounded. Since the only equilibrium point in the set $S_p\left(V_p\left(\hat{\delta}\right)\right)$ is δ_s, it is from Equation A1.1 that the set $S_c\left(V\left(\hat{\delta}, 0\right)\right)$ contains no other equilibrium point but (δ_s, 0). Since the ω-limit set of the system $d(M_g, D)$ consists entirely of equilibrium points and the ω-limit set of every bounded trajectory exists, every trajectory in $S_c\left(V\left(\hat{\delta}, 0\right)\right)$ converges to (δ_s, 0). In other words, $S_c\left(V\left(\hat{\delta}, 0\right)\right) \subseteq A(\delta_s, 0)$. Since equilibrium points cannot lie inside the stability region, $\left(\hat{\delta}, 0\right)$ must be on the stability boundary of (δ_s, 0) and must possess the minimum value of energy function $V(.)$ among all the equilibrium points on the stability boundary. This completes the proof.

Appendix 9B: Proof of Theorem 9.13

The following lemmas are useful in the proof of this theorem.

Lemma B1

Let $\hat{\delta}$ be the closest UEP of the stable equilibrium point δ_s of the system $d(I)$ with respect to $V_p(\delta)$. Then, for any ε' with $V_p(\hat{\delta}) - V_p(\delta_s) > \varepsilon' > 0$, there exists a number $\eta > 0$, such that for the system (Eq. 9.17) satisfying $|P - \bar{P}| < \eta$, we have $S_p(V_p(\hat{\delta}) - \varepsilon') \subseteq A(\hat{\delta}_s)$, where $\hat{\delta}_s \in S_p(V_p(\hat{\delta}) - \varepsilon')$ is a stable equilibrium point of the system (Eq. 9.17).

Proof: Because $\hat{\delta}$ is the closest UEP of $d(I)$, by Proposition 9.3, we have $S_p(V_p(\hat{\delta})) \subseteq A(\delta_s)$. It is obvious that $S_p(V_p(\hat{\delta}) - \varepsilon') \leq S_p(V_p(\delta))$ for $V_p(\hat{\delta}) - V_p(\delta_s) > \varepsilon' > 0$. Hence, we have $S_p(V_p(\hat{\delta}) - \varepsilon') \subseteq A(\delta_s)$.

We consider the level surface $\partial S_p(V_p(\hat{\delta} - \varepsilon'))$, which is the boundary of the set $S_P(V_p(\hat{x}) - \varepsilon')$. At every point of level surface $\partial S_p(V_p(\hat{\delta}) - \varepsilon)$,

$$\frac{d}{dt}(V_p(\delta)) = \left\langle \nabla V_p(\delta), \dot{\delta} \right\rangle < 0, \text{ for all } \delta \in \partial S_p\left(V_p\left(\hat{\delta}\right) - \varepsilon'\right), \tag{A1.3}$$

where <.,.> is the usual inner product. $\nabla V_p(\delta)$ denotes the gradient of $V_p(.)$ at δ. Since the level surface $\partial S_p(V_p(\hat{\delta}) - \varepsilon')$ is a compact set (It is a closed set because the boundary of any open set is a closed set; it is a bounded set because of Proposition 9.3 and condition (iii) of the energy function.), it follows that

$$\left\langle \nabla V_p(\delta), \dot{\delta} \right\rangle \leq \alpha < 0, \text{ for all } \delta \in \partial S_p\left(V_p\left(\hat{\delta}\right) - \varepsilon'\right), \tag{A1.4}$$

where $\alpha = \text{maximum}_{\delta \in \partial S_p\left(V_p\left(\hat{\delta}\right) - \varepsilon'\right)} \left\langle \nabla V_p(\delta), \dot{\delta} \right\rangle$. It is noted that system $d(I)$ is described by $\dot{\delta} = -\nabla V_p(\delta)$, which means that the vector field of $d(I)$ is normal to the level surface at δ. So, Equation A1.4 imports that all the trajectories of $d(I)$ starting from the level surface $\partial S_p(V_p(\hat{\delta})-\varepsilon')$ leave the level surface normally and are in the direction of decreasing $V_p(\delta)$.

Next, let $|.|$ be a norm in R^n and let $s = \max_{\delta \in \partial S_p(V_p(\hat{\delta})-\varepsilon')} |\nabla V_p(\delta)|$; it follows that, for all $\delta_1 \in \partial S_p\left(V_p\left(\hat{\delta}\right)-\varepsilon'\right)$, the following holds:

$$\begin{aligned}
\left\langle \nabla V_p(\delta_1), \dot{\delta} \right\rangle &= \left\langle \nabla V_p(\delta_1), \dot{\delta} - (P - \bar{P}) \right\rangle \\
&= \left\langle \nabla V_p(\delta_1), \dot{\delta} \right\rangle - \left\langle \nabla V_p(\delta_1), P - \bar{P} \right\rangle \\
&\leq \left\langle \nabla V_p(\delta_1), \dot{\delta} \right\rangle + \left| \left\langle \nabla V_p(\delta_1), P - \bar{P} \right\rangle \right| \\
&\leq \left\langle \nabla V_p(\delta_1), \dot{\delta} \right\rangle + \left| \nabla V_p(\delta_1) \right| \left| P - \bar{P} \right| \left(\text{if } \left| P - \bar{P} \right| < \eta \right) \\
&\leq \alpha + s\eta < 0, \text{ if } \eta < \frac{-\alpha}{2}.
\end{aligned} \tag{A1.5}$$

Equation A1.5 indicates that the vector field of the system (Eq. 9.17) also points inward along the level surface $\partial S_p\left(V_p\left(\hat{\delta}\right)-\varepsilon'\right)$ (but may not be normal to the level surface) and, hence, all the trajectories of the system (Eq. 9.17) starting from the level surface $\partial S_p(V_p(\hat{\delta})-\varepsilon')$ leave the level surface transversely and are in the direction of decreasing $V_p(\delta)$. This implies that the set $S_p(V_p(\hat{\delta})-\varepsilon')$ is also a positive invariant set for the system (Eq. 9.17) satisfying $|P - \bar{P}| < \eta$. On the other hand, because of the hyperbolicity of the equilibrium points of $d(I)$, given any neighborhood u of δ_s, there exists a positive number r_2, such that there is only one equilibrium point, say, $\delta_s' \in u$, and δ_s' is a stable equilibrium point of the system (Eq. 9.17) if $|P - \bar{P}| < r_2$. Since every bounded trajectory of the system (Eq. 9.17) converges to one of its equilibrium points, the trajectories in the set $S_p(V_p(\hat{\delta})-\varepsilon')$ of the system (Eq. 9.17) will converge to the stable equilibrium point δ_s' if $|P - \bar{P}| < \eta$, where $\eta = \min\{r_1, r_2\}$. Consequently, the set $S_p(V_p(\hat{\delta})-\varepsilon')$ belongs to the stability region $A(\delta_s')$.

Lemma B2

Let $\hat{\delta}$ be the closest UEP of the system $d(I)$. Then, for any given neighborhood u_1 of $\hat{\delta}$, there exists a real number $r_3 > 0$ such that, for the system (Eq. 9.17) satisfying $|P - \bar{P}| < r_3$, the unique equilibrium point $\hat{p} \in u_1$ of the system (Eq. 9.17) is also on the stability boundary $\partial A(\delta_s')$, where $\delta_s' \in S_p(V_p(\delta))$ is a stable equilibrium point of the system (Eq. 9.17).

Proof: First, according to Theorem 9.5, we have $W^u\left(\hat{\delta}\right) \cap S_p\left(V_p\left(\hat{\delta}\right)-\varepsilon'\right) \neq \phi$. Second, because of the hyperbolicity of $\hat{\delta}$, for any given neighborhood u_1, there exists a real number, $c_1 > 0$, such that $|P - \bar{P}| < c_1$, and there exists a unique equilibrium point, $\hat{p} \in u_1$, of the system (Eq. 9.17) satisfying $|P - \bar{P}| < c_1$ (Cate et al., 1984). Next, applying the continuity result of the local unstable manifold, which asserts that the local unstable manifold of an equilibrium point depends continuously on its vector field in the C^k topology, to the equilibrium point $\hat{\delta}$, we have shown that there exists a real number, $c_2 > 0$, with $c_2 < c_1$ such that, for the system (Eq. 9.17) satisfying $|P - \bar{P}| < c_2$, the unstable manifold $W^u(\hat{p}) \cap S_p(V_p(\hat{\delta}) - \varepsilon') \subseteq \phi$. On the other hand, Lemma B1 asserts that, if system (Eq. 9.17) satisfies $|P - \bar{P}| < \eta$, then $S_p(V_p(\hat{\delta}) - \varepsilon^1) \subseteq A(\hat{\delta}_s)$, where $\hat{\delta}_s \in S_p(V_p(\hat{\delta}))$ is a stable equilibrium point of the system (Eq. 9.17). Combining the above results, it follows that if the system (Eq. 9.17) satisfies $|P - \bar{P}| < r_3$, then $W^u(\hat{p}) \cap A(\hat{\delta}_s) \neq \phi$, where $r_3 = \min\{c_2, \eta\}$. Hence, $\hat{p} \in u$ is also on the stability boundary $\partial A(\hat{\delta}_s)$ of the system (Eq. 9.17) satisfying $|P - \bar{P}| < r_3$.

We now return to the proof of Theorem 9.13. To show $\hat{p}$ is the closest UEP of $\hat{\delta}_s$ of the system (Eq. 9.17) with respect to the energy function $\bar{V}(.)$ in Equation 9.18, we need to show that (i) $\hat{p} \in \partial A(\hat{\delta}_s)$ and (ii) $\hat{V}(\hat{p}) = \min_{x \in \partial A(\hat{\delta}_s) \cap \hat{E}} \bar{V}(x)$, where $\hat{E}$ is the set of equilibrium points of the system (Eq. 9.17). In Lemma B2, we have shown that $\hat{p} \in \partial_A(\hat{\delta}_s)$ if the system (Eq. 9.17) satisfies $|P - \bar{P}| < r_3$, and $\hat{p}$ is in the neighborhood u_1 of $\hat{x}$. For later use, we set $\bar{c} := V(\hat{\delta}) + \max_{\delta \in \bar{u}_1} \langle P - \bar{P}, \delta \rangle$, where $|P - \bar{P}| < r_3$ and $\bar{u}_1$ is the closure of u_1. Part (ii) remains to be verified. First, we consider all the equilibrium points $\delta_i \in \{\partial A(\delta_s) - \hat{\delta}\}$ of system $d(I)$. By the hyperbolicity of the equilibrium points of system $d(I)$, it follows that there exists a number $r' > 0$ such that, for the system (Eq. 9.17) satisfying $|P - \bar{P}| < r'$, there exists a unique equilibrium point $\bar{\delta}_i$ of the system (Eq. 9.17) in a given neighborhood u of δ_i. So, by a proper selection of the given neighborhood u, there exists a $r_4 > 0$ such that, for the system (Eq. 9.17) satisfying $|P - \bar{P}| < r_4$, we have $\bar{V}(\bar{\delta}_i) > \bar{c}$, where r_3 can be reduced if necessary. Next, we consider the equilibrium points $\delta_i \in \left\{\partial A(\delta_s) - \hat{\delta}\right\}$ of system $d(I)$. Let $c = \bar{c} + 1$. From Proposition 9.11 and condition (iii) of the energy function, we have shown that the intersection between the level set $\{\delta: V_p(\delta) \leq c\}$ and $\partial A(\delta_s)$ is a bounded and closed set. So, the sets $\upsilon_1 := \{\delta: V_p(\delta) \leq c\} \cap \partial A(\delta_s)$ and $\upsilon_2 := \{\delta: V_p(\delta) = c\} \cap \bar{A}(\delta_s)$ are compact. By compactness argument, there exists a $r_5 > 0$ such that, if $|P - \bar{P}| < r_5$, then the function $V_p(.)$ in Equation 9.18 has the property $\bar{V}_p(\delta) > \bar{c}$, for $\delta \in \upsilon_2$. On the other hand, there exists a neighborhood υ_3 of υ_1 in the set $\{\delta: V_p(\delta) \leq c\}$; that is, $\upsilon_3 := \{\delta: V_p(\delta) \leq c, |\delta - \upsilon_1| < \varepsilon\}$, such that υ_3 contains no other equilibrium points except those on the stability boundary $\partial A(\delta_s)$. Since every bounded trajectory of $d(I)$ converges to one of the equilibrium points of $d(I)$, we have $\cap_{t \in R^+} \Phi(\upsilon_3, t) = \upsilon_1$. Thus, it follows from Chiang and Thorp (1989) and from Chiang and Chu (1995) that there exists a neighborhood υ_4 of υ_1 in the set $\{\delta: V_p(\delta) \leq c\}$, such that

$$\langle -(f(\delta) - P), n_\delta \rangle > 0 \tag{A1.6}$$

for every $\delta \in \partial\upsilon_4$, where $\partial\upsilon_4$ is the boundary of υ_4 and n_δ is the outward normal vector to $\partial\upsilon_4$ at δ. From Equation A1.6 and the continuity property of the vector $f(\delta) - P$, we have, by using the same procedure as we did in Equation A1.5, $r_6 > 0$, such that

$$\langle -(f(\delta) - \bar{P}), n_\delta \rangle > 0 \tag{A1.7}$$

for $\delta \in \partial\upsilon_4$ and $|P - \bar{P}| < r_6$. Equation A1.7 implies that for any equilibrium point $\delta_i \notin \partial A(\delta_s)$ of system $d(I)$, if its corresponding equilibrium point $\bar{\delta}_i$ of the system

(Eq. 9.17) satisfying $|P - \bar{P}| < r_6$ has the property that its unstable manifold $W^u(\bar{\delta}_i)$ converges to $S_p(V_p(\hat{\delta}) - \varepsilon')$, then $W^u(\bar{\delta}_i)$ cannot intersect υ_1 and must intersect the set υ_2. So, $\bar{V}(\bar{\delta}_i) > \bar{c}$. Hence $\bar{\delta}_i$ cannot be the closest UEP of the system (Eq. 9.17) satisfying $|P - \bar{P}| < \min\{r_3, r_5, r_6\}$. Consequently, taking $r = \min\{r_3, r_4, r_5, r_6\}$, we have $\hat{p}$ as the closest UEP of the stable equilibrium point $\hat{\delta}_s \in S_p(V_p(\hat{\delta}))$ of the system (Eq. 9.17) if $|P - \bar{P}| < r$. Now, applying Theorem 9.12 to the closest UEP, $\hat{p}$, we have $p := (\hat{p}, 0)$ as the closest UEP of the stable equilibrium point $(\hat{\delta}_s, 0) \in S(V_p(\hat{\delta}))$ of the system (Eq. 9.16) if $|P - \bar{P}| < r$, and this proof is completed.

A1.3 PROOFS OF THEOREMS IN CHAPTER 10

Proof of Theorem 10.3

First, we prove the following.

Lemma A1

Suppose x_s is a stable equilibrium point of $d(D)$. (Here, D is nonsingular, not necessarily positive diagonal.) Then, there exist $r > 0$ and $\varepsilon_1 > 0$ such that $B_r(x_s) \subset A_D(x_s)$, and for every matrix $\bar{D}$ satisfying $\|D - \bar{D}\| < \varepsilon_1$, we have $B_r(x_s) \subset A\bar{D}(x_s)$.

Proof: Since x_s is a stable equilibrium point of $d(D)$, it follows that all of the real parts of eigenvalues of the linearized vector field at x_s are smaller than a negative number:

$$\sigma(-DJ_{x_s}) < -c, \tag{A1.8}$$

where $\sigma(A)$ denotes the real part of eigenvalue of A,

$$J_{x_s} = del\left[\frac{\partial f}{\partial x}\right]_{x_s} \quad \text{and } c > 0.$$

It has been shown in Beehler (1997) that there exists a basis, say, F, whose corresponding norm $|.|_F$ and inner product satisfying

$$\langle -DJ_{x_s}x, x\rangle \le -c|x|^2, \text{ for all } x \in \Re^n. \tag{A1.9}$$

(Note that we shall neglect the subscript F in the following derivation.) It follows from the definition of the derivative that

$$\lim_{x \to x_s} \frac{|-Df(x) + DJ_{x_s}(x - x_s)|}{|x - x_s|} = 0 \tag{A1.10}$$

or, by Cauchy inequality,

$$\lim_{x \to x_s} \frac{\langle -Df(x) + DJ_{x_s}(x - x_s), x - x_s \rangle}{|x - x_s|^2} = 0. \tag{A13.11}$$

Then, for any small number $\varepsilon_2 > 0$, there exists $r_1 > 0$ such that if $|x - x_s| \leq r_1$ (or $x \in B_{r_1}(x_s)$) implies

$$\langle -Df(x) + DJ_{x_s}(x - x_s), x - x_s \rangle \leq \varepsilon_2 |x - x_s|^2 \tag{A1.12}$$

or

$$\begin{aligned} \langle -Df(x), x - x_s \rangle &\leq \langle -DJ_{x_s}(x - x_s), x - x_s \rangle + \varepsilon_2 |x - x_s|^2 \\ &\leq -(c - \varepsilon_2)|x - x_s|^2 \\ &= -b^2 |x - x_s|^2, \end{aligned} \tag{A1.13}$$

where we choose $\varepsilon_2 < c$, so $b_2 > 0$.

It follows from Equation A1.13 that there exists a "neighborhood" of D such that for every matrix $\bar{D}$ in this neighborhood, the following inequality holds:

$$\langle -\bar{D}f(x), x - x_s \rangle \leq -b_3 |x - x_s|^2, \text{ for } x \in B_{r_1}(x_s), \tag{A1.14}$$

where $b_3 > 0$ and $|D - \bar{D}|_E < \varepsilon_1$, $|.|_E$ denotes the norm corresponding to the Euclidean basis.

We define the following positive definite function:

$$V(x) = \langle x - x_s, x - x_s \rangle. \tag{A1.15}$$

Then, the derivative of $V(x)$ along the trajectory of $\dot{x} = -\bar{D}f(x)$ is

$$\dot{V}_{\bar{D}}(x) = 2\langle -\bar{D}f(x), x - x_s \rangle \leq 0, \text{ for } x \in B_{r_1}(x_s). \tag{A1.16}$$

Now, consider the level set

$$S(r_1) = \{x \in \Re^n : V(x) \leq r_1^2\}. \tag{A1.17}$$

It is clear that the set $S(r_1)$ is bounded for any bounded r_1. From Equation A1.16, it follows that $S(r_1)$ is an invariant set of $d(D)$ and that x_s is the largest invariant set of $d(D)$ contained in $S(r_1)$. Consequently, all the trajectories of $x^{\cdot} = -\bar{D}f(x)$ starting from $x \in S(r_1)$ approach x_s as $t \to \infty$. Thus, $B_{r_1}(x_s) \subseteq A_{\bar{D}}(x_s)$. Note that $B_{r_1}(x_s)$ is defined in terms of a norm corresponding to the basis F, which may be different from the Euclidean basis. However, $\Re^n$ is a normal space ; it follows that there exists an $r > 0$ such that $B_r(x_s) \subseteq B_r(x_s)$, where $B_r(x_s)$ is defined in terms of the Euclidean norm corresponding to the Euclidean basis.

Proof of Theorem 10.3: Let $U \subset \Re^n$ be a neighborhood of the UEP x_i of $d(D)$, and let us define the local stable and unstable manifolds of $d(D)$ as

$$\hat{W}_D^s(x_i) = \{x \in U | \phi_D(x,t) \to x_i \text{ as } t \to \infty\}, \text{and } \phi_D(x,t) \in U \text{ for } t \geq 0.$$

$$\hat{W}_D^u(x_i) = \{x \in U | \phi_D(x,t) \to x_i \text{ as } t \to -\infty\} \text{ and } \phi_D(x,t) \in U \text{ for } t \geq 0.$$

From a characterization of UEPs lying on the stability boundary, for each UEP x_i on the stability boundary $\partial A(D)$ and for any $r > 0$ such that $B_r(x_s) \subseteq A\bar{D}(x_s)$, we have $\hat{W}_D^u(x_i) \cap B_r(x_s) \neq \varnothing$, $i = 1, 2, \ldots, N$. According to Lemma A1, there exist $r > 0$ and $\hat{\varepsilon}_1 > 0$, such that $\bar{D}$ satisfying $\|D - \bar{D}\| < \hat{\varepsilon}_1$ implies that $B_r(x_s) \subseteq A\bar{D}(x_s)$.

Since local unstable and stable manifolds depend continuously on the vector field $Df(x)$ in the C^k topology, we have shown that, for each UEP $x_i \in \partial A(D)$, there exists an $\varepsilon_i > 0$ such that for every matrix D_i satisfying $\|D_i - D\| < \hat{\varepsilon}_i$,

$$\hat{W}_{D_i}^u(x_i) \cap B_r(x_s) \neq \varnothing. \tag{A1.18}$$

Let $\hat{\varepsilon}_2 = \min\{\varepsilon_1, \varepsilon_2, \ldots, \varepsilon_N\}$. Then, for every matrix $\hat{D}$ satisfying $\|\hat{D} - D\| \leq \hat{\varepsilon}_2$, we have

$$\hat{W}_{\hat{D}}^u(x_i) \cap B_r(x_s) \neq \varnothing, \text{ for } i = 1, 2, \ldots, N. \tag{A1.19}$$

Let us take $\varepsilon_D = \min\{\hat{\varepsilon}_1, \hat{\varepsilon}_2\}$. Then, for every matrix $\bar{D}$ satisfying $\|D - \bar{D}\| \leq \varepsilon_D$, we have

$$\hat{W}_{\bar{D}}^u(x_i) \cap B_r(x_s) \neq \varnothing, \text{ for } i = 1, 2, \ldots, N.$$

In particular, we have

$$\hat{W}_{\bar{D}}^u(x_i) \cap A_{\bar{D}}(x_s) \neq \varnothing, \text{ for } i = 1, 2, \ldots, N.$$

By Theorem 3.12, it follows that the equilibrium points $x_1, x_2, \ldots, x_N$ are also on the stability boundary $\partial A(\bar{D})$.

Proof of Theorem 10.4

This theorem is proved through a series of lemmas.

Lemma B1

Let $\hat{u}$ be a neighborhood of the UEP $\hat{x}$ of $d(D_n)$ with the matrix $D(n) \in D(\lambda)$. Let $\Phi_{D(n)}(p, t)$ denote the trajectory of the system $d(D_n)$ starting from $p \in \hat{u}$ and converging to the stable equilibrium point x_s. Then, $\Phi_{D(n)}(p, t)$ is uniformly bounded for all n.

Proof: Along the trajectory $\Phi_{D(n)}(p, t)$, we define

$$T_{\eta_1}^n = \{t : \Phi_{D(n)}(p, t) \notin B_{\eta_1}(x_i), x_i \in E\}$$

and let $L(T_{\eta_1}^n)$ denote the Lebesque measure of $T_{\eta_1}^n$. It follows that, given $\eta_1 > 0$, there exists an $\eta_2 > 0$ such that $|f(x)| > \eta_2$, for $x \notin B_{\eta_1}(x_i)$, $x_i \in E$. Because $f(x)$ is a gradient vector, let $f(x) = \nabla V(x)$. The derivative of $V(x)$ along the trajectory $\Phi_{D(n)}(p, t)$ is

$$\dot{V}(x)\big|_{d(D_n)} = \langle f(x), -D(n) f(x) \rangle \leq 0. \tag{A1.20}$$

It follows that

$$\int_{T^n_{\eta_1}} \langle f(x), -D(n)f(x)\rangle \left\langle \int_0^\infty < f(x), D(n)f(x) \right\rangle < V_{\max} - V(x_s), \quad (A1.21)$$

where V_{max} is the maximum value of $V(.)$ over $\hat{u}$.

But

$$\int_{T^n_{\eta_1}} \langle f(x), D(n)f(x)\rangle \geq d_n \eta_2^2 L(T^n_{\eta_1}), \quad (A1.22)$$

where $d_n = \min\{D^1(n), D^2(n), \ldots, D^n(n)\}$ and $D(n) = \text{diag}\{D^1(n), D^2(n), \ldots, D^n(n)\}$.

Since $D(n) \in D(\lambda)$, we have

$$d_n \geq d := min\{D_1^1, D_1^2, \ldots, D_1^n, D_2^1, D_2^2, \ldots, D_2^n\} \quad (A1.23)$$

and

$$\int_{T^n_{\eta_1}} \langle f(x), D(n)f(x)\rangle > d\eta_2^2 L(T^n_{\eta_1}), \quad (A1.24)$$

where matrix $D_1 := \text{diag}\{D_1^1, D_1^2, \ldots, D_1^n\}$ and matrix $D_2 := \text{diag}\{D_2^1, D_2^2, \ldots, D_2^n\}$. Thus,

$$L(T^n_{\eta_1}) < \frac{V_{max} - V(x_s)}{d\eta_2^2} := T. \quad (A1.25)$$

Recall that the vector field $D(n)f(x)$ is uniformly bounded. From Equation A1.25, it follows that the time interval, within which the trajectory $\Phi_{D(n)}(p, t)$ lies outside the set $\{B_{\eta_1}(x_j), x_j \in E\}$, is uniformly bounded. Consequently, the trajectory $\Phi_{D(n)}(p, t)$ is uniformly bounded for all $D(n) \in D(\lambda)$.

Lemma B2 below follows from Lemma B1.

Lemma B2

Let p be the ω-limit set of a trajectory of $d(D(n))$. Then, p only consists of equilibrium points of $d(D(n))$.

Lemma B3

Let C be a nonempty interior compact set and $C \cap \partial A_D(x_s) \neq \emptyset$. Let N be a closed neighborhood of $\partial A_D(x_s)$ and $N - \partial A_D(x_s)$ contains no other equilibrium point. Let $C_p = C \cap N$. Then, $\bigcap_{t\in\Re^+} \Phi_D(C_p, t) = \partial A_D(x_s) \cap C_p$.

Proof: By contradiction, if this were not true, then there exists a point $x \in C_p - \partial A_D(x_s)$ such that $\bigcap_{t\in\Re^+} \Phi_D(x, t) \subset C_p$. We want to show that this is impossible. Since C_p is compact, by Lemma B2, the trajectory $\Phi_D(x, t)$ must converge to an equilibrium point. However, by construction, the set $C_p - \partial A_D(x_s)$ contains no equilibrium point.

So, $\Phi_D(x, t)$ must converge to an equilibrium point, say, $\hat{x}$, and $\hat{x} \in \partial A_D(x_s)$. In other words, $x \in W_D^s(\hat{x})$. According to Theorem 3.12, this is a contradiction.

Lemma B4

Let $\hat{x} \in \partial A_D(x_s)$ be an equilibrium point. Let $\{D(n); n = 1, 2, \ldots\} \subseteq \{D_\lambda: \lambda \in [0, 1]\}$ and satisfy the following:

(1) $\|D(n) - D\| < \varepsilon_D$.

(2) $D(n) \to \bar{D}$ and $\|\bar{D} - D\| = \varepsilon_D$.

Suppose that $\hat{x} \in \partial A_D(n)(x_s)$, for $n = 1, 2, \ldots$. Then, $\hat{x} \in \partial A\bar{D}(x_s)$.

Proof: Suppose that $\hat{x} \notin \partial A\bar{D}(x_s)$. Then, $\hat{x} \in [\partial A\bar{D}(x_s)]^c$; since a UEP cannot lie inside a stability region. In Lemma B1, we have shown that any trajectory of $d(D(n))$ starting from a neighborhood $\hat{u}$ of $\hat{x}$ and converging to x_s is uniformly bounded; in other words, $\bigcup_{t \in \Re^+} \Phi_{D(n)}(x, t) \subset B_b(\hat{x})$, for all $x \in \hat{u} \cap A_{D(n)}(x_s), B_b(\hat{x}) := \{x: \|x - \hat{x}\| < b\}$, $b > 0$ is constant. In Lemma B3, we have shown that for any nonempty interior compact set C, $C \cap \partial A\bar{D}(x_s) \neq \emptyset$, there exists a neighborhood of u of $\partial A\bar{D}(x_s)$ such that $\bigcap_{t \in \Re^+} \Phi(C_p, t) = \partial A_{\bar{D}}(x_s) \cap C_p$, where $C_p := C \cap u$. It therefore follows from Cheng and Ma (2003) that there exists a neighborhood $\tilde{u}$ of $\partial A\bar{D}(x_s)$ such that $\langle -\bar{D}f(x), n_x\rangle > 0$ for every $x \in \partial\tilde{u} \cap C$, where $\partial\tilde{u}$ denotes the boundary of $\tilde{u}$ and n_x is the outward normal vector to $\partial\tilde{u}$ at x with respect to $\partial A\bar{D}(x_s)$. Consequently, by the continuity property of $-\bar{D}f(x)$, it follows that there exists $\varepsilon > 0$ such that $\langle -D^*f(x), n_x\rangle > 0$ if $\|D^* - \bar{D}\| < \varepsilon$ and $x \in \partial u \cap C$. Now, we take $C = B_b(\hat{x})$. It follows that there exists an $\varepsilon_1 > 0$ such that, if $\|D^* - \bar{D}\| < \varepsilon$ and $\hat{D} \in D(n)$, it implies that the trajectory $\Phi_{\hat{D}}(p, t)$ starting from $p \in \hat{u}$ cannot transverse the set $\partial u \cap C$. Thus, $\Phi\hat{x}(p, t)$ cannot converge to x_s, for $p \in \hat{u}$. This is a contradiction to our assumption that $\hat{x} \in \partial A_{D(n)}(x_s)$. Thus, this lemma is true.

Proof of Theorem 10.4: Since $d(D_1)$ (or $d(D_2)$) satisfies Assumptions (C1) and (C2), it follows from Theorem 10.3 that there exists $\varepsilon_{D_1} > 0$ (or $\varepsilon_{D_2} > 0$) such that, for every matrix D satisfying $\|D - D_1\| < \varepsilon_{D_1}$ (or $\|D - D_2\| < \varepsilon_{D_2}$, every equilibrium point on $\partial A(D_1)$ (or $\partial A(D_2)$) is also on $\partial A(D)$. Therefore, if $\|D_1 - D_2\| < \min\{\varepsilon_{D_1}, \varepsilon_{D_2}\}$, then

$$\{x : x \in \partial A(D_1) \cap E\} = \{x : x \in \partial A(D_2) \cap E\}, \tag{A1.26}$$

and we complete this proof. Hence, without loss of generality, we assume that $\|D_1 - D_2\| > \varepsilon_{D_1}$. By Lemma B4, every equilibrium point on $\partial A(D_1)$ is also on $\partial A(D')$ if $\|D' - D_1\| = \varepsilon_{D_1}$ and $D' \in D_\lambda$. Similarly, because $d(D')$ satisfies Assumptions (C1) and (C2), it follows that there exists an $\varepsilon_{D'} > 0$ such that for every system $d(D'')$ satisfying $\|D'' - D'\| < \varepsilon_{D'}$, every equilibrium point on $\partial A(D'')$ is also on $\partial A(D')$. Since the convex hull $C_0(D_1, D_2)$ generated by D_1 and D_2 is compact, there exists a finite number of open covers in $C_0(D_1, D_2)$. Thus, repeating the above procedure, we conclude that every equilibrium point on $\partial A(D_1)$ is also on $\partial A(D_2)$; that is,

$$\{x : x \in \partial A(D_1) \cap E\} \subseteq \{x : x \in \partial A(D_2) \cap E\}. \tag{A1.27}$$

Next, we start from the dynamical system $d(D_2)$ and follow an argument similar to the above, we can obtain the conclusion that every equilibrium point on $\partial A(D_2)$ is also on $\partial A(D_1)$; that is,

$$\{x : x \in \partial A(D_2) \cap E\} \subseteq \{x : x \in \partial A(D_1) \cap E\}. \tag{A1.28}$$

Combining Equations A1.27 and A1.28, we complete this proof.

Proof of Theorem 10.8

We first state the following lemma whose proof is similar to that of Lemma A1.

Lemma C1

Suppose $(x_s, 0)$ is a stable equilibrium point of $d(M, D)$. Then, there exists $r_1 > 0$ and $\varepsilon_1 > 0$, such that $B_{r_1}(x_s, 0) \subset A_{(M,D)}(x_s, 0)$. Moreover, if $d(\bar{M}, \bar{D})$ is in the ε_1-ball neighborhood of $d(M, D)$, it follows that $B_{r_1}(x_s, 0) \subset A_{(\bar{M},\bar{D})}(x_s, 0)$.

We then use the same technique as that used in the proof of Theorem 10.3 to conclude this theorem. The details are therefore omitted.

Proof of Theorem 10.9

The following lemma is useful in this proof.

Lemma D1

Let $\Phi_{(M(n),D)}(p, t)$ be the trajectory of system $d(M(n), D)$ starting from $p \in \hat{u}$ and converging to the stable equilibrium point $(x_s, 0)$, where $\hat{u}$ is a neighborhood of the equilibrium point $(\hat{x}, 0)$. Then, $\Phi_{(M(n),D)}(p, t)$ is uniformly bounded for all $M(n) \in M_\lambda$.

Proof: Let $(x_n(t), y_n(t))$ denote the trajectory $\Phi_{(M(n),D)}(p, t)$. From Theorem 10.3, it follows that the component $y_n(t)$ is uniformly bounded. Therefore, it follows from Equation 10.27 and from Brenan et al. (1989) that the time required for the component $x_n(t)$ to leave one ball, $B_\varepsilon(\hat{x}_i)$, and to enter another ball, $B_\varepsilon(\hat{x}_1)$, is bounded below, say, by $\hat{c}$, $\hat{c} = c_1\delta/m\alpha$, where $c_1 > 0$ and where $\hat{x}_i, \hat{x}_1 \in E$:

$$m = \max\{M_1^1 + M_2^1, M_1^2 + M_2^2, \ldots, M_1^n + M_2^n\}$$

$$M = \operatorname{diag}\{M_i^1, M_i^2, \ldots, M_i^n\},$$

and α is the maximum value of $\hat{x}(t)$ over the dynamical systems $d(M_\lambda, D)$. The existence of the maximum value is guaranteed due to the compactness of the convex hull generated by (M_1, D) and (M_2, D).

Without loss of generality, we assume that $x_n(t)$ passes through a sequence of balls, $B_\varepsilon(\hat{x}_i)$. Let us define an increasing sequence, $\{s'_{i(n)}\}$, where $s'_{i(n)}$ is the time that $x_n(t)$ leaves the ball $B_\varepsilon(\hat{x}_i)$. In the same manner, we define another increasing sequence, $\{t'_{i(n)}\}$, where $t'_{i(n)-1}$ is the time that $x_n(t)$ enters the ball $B_\varepsilon(\hat{x}_{i(n)})$. Two situations are possible:

(1) $t'_{i(n)} - s'_{i(n)} < \hat{c}$. This means that $x_n(t)$ is coming back to the same ball after traveling a distance smaller than ε.

(2) $t'_{i(n)} - s'_{i(n)} \geq \hat{c}$. This indicates that either $x_n(t)$ leaves one ball and enters another ball or $x_n(t)$ is coming back to the same ball.

Since we are interested in determining whether $x_n(t)$ is bounded or not, we only need to consider Case 2. Take two subsequences, $\{s_i(n)\} \subset \{s'_{i(n)}\}$ and $\{t_{i(n)}\} \subset \{t'_{i(n)}\}$, such that $t_{i(n)} - s_{i(n)} \geq \hat{c}$; in other words, $t_{i(n)} - s_{i(n)} = n_{i(n)}\hat{c} + \alpha_{i(n)}\hat{c}$, where $n_{i(n)} \geq 1$ is an integer and $0 \leq \alpha_{i(n)} < 1$. We set

$$T_\varepsilon^n \equiv \{t : t \in [S_{i(n)}, t_{i(n)}], i(n) = 1, 2, \ldots\}.$$

Let $L(T_\varepsilon^n)$ denote the Lebesque measure of T_ε^n. It can be shown (Brenan et al., 1989) that

$$L(T_\varepsilon^n) \leq \frac{4}{d} \frac{[V_{max} - V(x_s, 0)]}{\left[\frac{1}{2\alpha}\left(\frac{\delta}{m}\right)^2\right]^2}, \text{ if } c_1 \geq 2$$

or

$$L(T_\varepsilon^n) \leq \frac{4}{d} \frac{[V_{max} - V(x_s, 0)]}{\left[\frac{c_1(4 - c_1)}{8\alpha}\left(\frac{\delta}{m}\right)^2\right]^2}, \text{ if } 0 < c_1 \leq 2,$$

where $d = \min\{D^1, D^2, \ldots, D^n\}$, $D = \text{diag}\{D^1, D^2, \ldots, D^n\}$, and $V_{max} = \max\{V(x, y):(x, y) \in \hat{u}\}$.

Now, because the vector field $y_n(t)$ is uniformly bounded and $L(L(T_\varepsilon^n))$ is uniformly bounded, it follows that $x_n(t)$ is also uniformly bounded. Thus, this lemma is true.

Lemma D2

Every bounded trajectory of the system (Eq. 10.27) converges to one of the equilibrium points.

Proof: Apply Theorem 3.1 in Brenan et al. (1989) to obtain this result.

Using a proof similar to that in Lemmas B3 and D2, we derive Lemma D3 below.

Lemma D3

Let $\tilde{C}$ be a nonempty interior compact set and $\tilde{C} \cap \partial A_{(\tilde{M},D)}(x_s, 0) \neq 0$. Let $\tilde{N}$ be a closed neighborhood of $\partial A_{(\tilde{M},D)}(x_s, 0)$ and suppose that $\tilde{N} - \partial A_{(\tilde{M},D)}(x_s, 0)$ contains no other equilibrium points. Let $\tilde{C}_p \equiv \tilde{C} \cap \tilde{N}$. Then, $\bigcap_{t \in R^+} \phi(\tilde{C}_p, t) = \partial A_{(\tilde{M},D)} \cap \tilde{C}_p$.

Lemma D4

Let $(\tilde{x}, 0) \in \partial A_{(M,D)}(x_s, 0)$ be an equilibrium point. Let $\{M(n): n = 1, 2, \ldots\} \subseteq \{M_\lambda: \lambda \in [0, 1]\}$ and satisfy the following conditions:

(1) $d(M(n), D)$ is in the ε-ball neighborhood of $d(M, D)$.

(2) $M(n) \to \tilde{M}$ and $d(\tilde{M}, D)$ are on the ε-ball neighborhood of $d(M, D)$.

Suppose $(\tilde{x}, 0) \in \partial A_{(M(n),D)}(x_s, 0)$, $n = 1, 2, \ldots$. Then, $(\tilde{x}, 0) \in \partial A_{(\tilde{M},D)}(x_s, 0)$.

Proof: This lemma follows from Lemmas D1 and D3, and the arguments are similar to those used for Lemma B4.

Theorem 10.8 asserts that there exists $\varepsilon_{M_1} > 0$ (or $\varepsilon_{M_2} > 0$) such that, for every matrix M satisfying $\|M - M_1\| < \varepsilon_{M_1}$ (or $\|M - M_2\| < \varepsilon_{M_2}$), every equilibrium point on $\partial A(M_1, D)$ (or $\partial A(M_2, D)$) is also on $\partial A(M, D)$. Therefore, if $\|M_1 - M_2\| < \min\{\varepsilon_{M_1}, \varepsilon_{M_2}\}$, then

$$\{x : x \in \partial A(M_1, D) \cap E\} = \{x : x \in \partial A(M_2, D) \cap E\}, \qquad \text{(A1.29)}$$

and we complete this proof. Hence, without loss of generality, we assume that $\|M_1 - M_2\| > \varepsilon_{M_1}$. By Lemma D4, every equilibrium point on $\partial A(M_1, D)$ is also on $\partial A(M', D)$ if $\|M' - M_1\| = \varepsilon_{M_1}$ and $M' \in M_\lambda$. Similarly, because $d(M', D)$ satisfies Assumptions (C1) and (C2), it follows from Theorem 10.8 that there exists an $\varepsilon_{M'} > 0$ such that for every system $d(M'', D)$ satisfying $\|M'' - M'\| < \varepsilon_{M'}$, every equilibrium point on $\partial A(M'', D)$ is also on $\partial A(M', D)$. Since the convex hull $C_0(M_1, M_2)$ generated by M_1 and M_2 is compact, there exists a finite number of open covers in $C_0(M_1, M_2)$. Thus, repeating the above procedure, we conclude that every equilibrium point on $\partial A(M_1, D)$ is also on $\partial A(M_2, D)$; that is,

$$\{x : x \in \partial A(M_1, D) \cap E\} \subseteq \{x : x \in \partial A(M_2, D) \cap E\}. \qquad \text{(A1.30)}$$

Next, we start from the dynamical system $\partial A(M_2, D)$ and follow an argument similar to the above; we can obtain the conclusion that every equilibrium point on $\partial A(M_2, D)$ is also on $\partial A(M_1, D)$; that is,

$$\{x : x \in \partial A(M_1, D) \cap E\} \supseteq \{x : x \in \partial A(M_2, D) \cap E\}. \qquad \text{(A1.31)}$$

Combining Equations A1.30 and A1.31, we complete this proof.

Bibliography

E. H. Abed and P. P. Varaiya (1984). "Nonlinear oscillations in power systems," *Int. J. Electr. Power Energ. Syst.* 6: 37–43.

L. F. C. Alberto (1997). *Transient Stability Analysis: Studies of the BCU Method; Damping Estimation Approach for Absolute Stability in SMIB Systems* (in Portuguese). Escola de Engenharia de São Carlos, Universidade de São Paulo.

L. F. C. Alberto and H. D. Chiang (2007). "Uniform approach for stability analysis of fast subsystem of two-time scale nonlinear systems," *Int. J. Bifurcat. Chaos* 17(11): 4195–4203.

L. F. C. Alberto and H. D. Chiang (2009). "Controlling unstable equilibrium point theory for stability assessment of two-time scale power system models," IEEE Power and Energy Society General Meeting, July 26–30, Calgary, Canada.

L. F. C. Alberto and H. D. Chiang (2009). "Theoretical foundation of CUEP method for two-time scale power system models," IEEE Power and Energy Society General Meeting, July 26–30, Calgary, Canada.

J. C. Alexander (1986). "Oscillatory solutions of a model system of nonlinear swing equations," *Int. J. Electr. Power Energ. Syst.* 8: 130–136.

P. M. Anderson and A. A. Fouad (2003). *Power System Stability and Control*, 2nd edition. Piscataway, NJ: IEEE Press.

A. Arapostathis, S. S. Sastry, and P. Varaiya (1982). "Global analysis of swing dynamics," *IEEE Trans. Circuits Syst.* CAS-29:673–679.

T. Athay et al. (1984). "Contribution to power system state estimation and transient stability analysis," DOE/et/29362-1, U.S. Department of Energy, February, 1984.

T. Athay, R. Podmore, and S. Virmani (1979). "A practical method for the direct analysis of transient stability," *IEEE Trans. Power Appar. Syst.* PAS-98(2): 573–584.

P. D. Aylett (1958). "The energy integral-criterion of transient stability limits of power systems," *Proc. IEEE* 105c(8): 527–536.

N. Balu, T. Bertram, A. Bose, V. Brandwajn, G. Cauley, D. Curtice, A. Fouad, L. Fink, M. G. Lauby, B. Wollenberg, and J. N. Wrubel (1992). "On-line power system security analysis," *Proc. IEEE* 80(2): 262–280.

M. E. Beehler (1997). "Reliability centered maintenance for transmission systems," *IEEE Trans. Power Deliv.* 12: 1023–1028.

A. R. Bergen and D. J. Hill (1981). "A structure preserving model for power system stability analysis," *IEEE Trans. Power Appar. Syst.* PAS-100: 25–35.

A. R. Bergen, D. J. Hill, and C. L. DeMarco (1986). "Lyapunov function for multimachine power systems with generator flux decay and voltage dependent loads," *Int. J. Electr. Power Energ. Syst.* 8(1): 2–10.

K. E. Brenan, S. L. Campbell, and L. R. Petzold (1989). *Numerical Solution of Initial-Value Problems in Differential-Algebraic Equations*. Amsterdam, Elsevier Science Publishers.

U. D. Caprio (1986). "Accounting for transfer conductance effects in Lyapunov transient stability analysis of a multimachine power system," *Int. J. Electr. Power Energ. Syst.* 8: 27–41.

E. G. Cate, K. Hemmaplardh, J. W. Manke, and D. P. Gelopulos (1984). "Time frame notion and time response of the models in transient, mid-term and long-term stability programs," *IEEE Trans. Power Appar. Syst.* PAS-103(1): 143–151.

Direct Methods for Stability Analysis of Electric Power Systems, by Hsiao-Dong Chiang

V. Chadalavada et al. (1997). "An on-line contingency filtering scheme for dynamic security assessment," *IEEE Trans. Power Syst.* 12(1): 153–161.

L. Chen and K. Aihara (2001). "Stability and bifurcation analysis of differential-difference-algebraic equations," *IEEE Trans. Circuits Syst. I Fundam. Theory Appl.* 48(3): 308–326.

D. Z. Cheng and J. Ma (2003). "Calculation of stability region," in *Proceedings of the 42nd IEEE Conference on Decision and Control, 2003*, Vol. 6, pp. 5615–5620.

H. D. Chiang (1989). "Study of the existence of energy functions for power systems with losses," *IEEE Trans. Circuits Syst.* CAS-36(11): 1423–1429.

H. D. Chiang (1991). "Analytical results on the direct methods for power system transient stability analysis," in C. T. Leondes (ed.), *Control and Dynamic Systems: Advances in Theory and Application*, Vol. 43, pp. 275–334. New York, Academic Press.

H. D. Chiang (1995). "The BCU method for direct stability analysis of electric power systems: pp. theory and applications," in J. Chow, P. V. Kokotovic, and R. J. Thomas (eds.), *Systems Control Theory for Power Systems*, Vol. 64 of IMA Volumes in Mathematics and Its Applications, pp. 39–94. New York, Springer-Verlag.

H. D. Chiang (1996). "On-line method for determining power system transient stability," U.S. Patent 5,483,462, January 9, 1996.

H. D. Chiang (1999). "Power system stability," in J. G. Webster (ed.), *Wiley Encyclopedia of Electrical and Electronics Engineering*, pp. 104–137. New York, John Wiley & Sons.

H. D. Chiang and C. C. Chu (1995). "Theoretical foundation of the BCU method for direct stability analysis of network-reduction power system model with small transfer conductances," *IEEE Trans. Circuits Syst.* CAS-42(5): 252–265.

H. D. Chiang and C. C. Chu (1996). "A systematic search method for obtaining multiple local optimal solutions of nonlinear programming problems," *IEEE Trans. Circuits Syst. I Fundam. Theory Appl.* 43(2): 99–109.

H. D. Chiang and L. Fekih-Ahmed (1992). "On the direct method for transient stability analysis of power system structure preserving models," in *IEEE International Symposium on Circuits and Systems*, pp. 2545–2548. San Diego, IEEE.

H. D. Chiang and L. Fekih-Ahmed (1996a). "Quasi-stability regions of nonlinear dynamical systems: Theory," *IEEE Trans. Circuits Syst. I Fundam. Theory Appl.* 43(82): 627–635.

H. D. Chiang and L. Fekih-Ahmed (1996b). "Quasistability regions of nonlinear dynamical systems: Optimal estimation," *IEEE Trans. Circuits Syst. I Fundam. Theory Appl.* CAS-43: 636–642.

H. D. Chiang and J. S. Thorp (1989a). "The closest unstable equilibrium point method for power system dynamic security assessment," *IEEE Trans. Circuits Syst.*, 36(9): 1187–1200.

H. D. Chiang and J. S. Thorp (1989b). "Stability regions of nonlinear dynamical systems: A constructive methodology," *IEEE Trans. Automat. Contr.*, 34(12): 1229–1241.

H. D. Chiang and C-S. Wang (1998). "Method for on-line dynamic contingency screening of electric power systems," U.S. Patent 5,719,787, February 17, 1998.

H. D. Chiang and F. F. Wu (1988). "Stability of nonlinear systems described by a second-order vector differential equation," *IEEE Trans. Circuits Syst.* 35: 703–711.

H. D. Chiang, F. F. Wu, and P. P. Varaiya (1987). "Foundations of direct methods for power system transient stability analysis," *IEEE Trans. Circuits Syst.* CAS-34(2): 160–173.

H. D. Chiang, F. F. Wu, and P. P. Varaiya (1988). "Foundations of the potential energy boundary surface method for power system transient stability analysis," *IEEE Trans. Circuits Syst.* CAS-35: 712–728.

H. D. Chiang, M. W. Hirsch, and F. F. Wu (1988). "Stability region of nonlinear autonomous dynamical systems," *IEEE Trans. Automat. Contr.* 33(1): 16–27.

H. D. Chiang, F. F. Wu, and P. P. Varaiya (1994). "A BCU method for direct analysis of power system transient stability," *IEEE Trans. Power Syst.* 8(3): 1194–1208.

H. D. Chiang, C. C. Chu, and G. Cauley (1995). "Direct stability analysis of electric power systems using energy functions: Theory, applications and perspective," *Proc. IEEE* 83(11): 1497–1529.

H. D. Chiang, J. C. Wang, C. T. Huang, Y. T. Chen, and C. H. Huang (1997). "Development of a dynamic ZIP-motor load model from on-line field measurements," *Int. J. Electr. Power Energ. Syst.* 19(7): 459–468.

H. D. Chiang, C. S. Wang, and H. Li (1999). "Development of BCU classifiers for on-line dynamic contingency screening of electric power systems," *IEEE Trans. Power Syst.* 14(2): 660–666.

H. D. Chiang, Y. Zheng, Y. Tada, H. Okamoto, K. Koyanagi, and Y. C. Zhou (2002). "Development of an on-line BCU dynamic contingency classifiers for practical power systems," 14th Power System Computation Conference (PSCC), Spain, June 24–28, 2002.

H. D. Chiang, A. Kurita, H. Okamoto, R. Tanabe, Y. Tada, K. Koyanagi, and Y. Zhou (2005). "Method and system for online dynamical screening of electric power system," U.S. Patent 6,868,311, March 15, 2005.

H. D. Chiang, Y. Tada, H. Li, and T. Takazawa (2006). "TEPCO-BCU for online dynamic security assessments of 12,000-bus power systems," X SEPOPE, May 21–25, Florianópolis, Brazil.

H. D. Chiang, Y. Tada, and H. Li (2007). "Power system on-line transient stability assessment" (an invited chapter), in *Wiley Encyclopedia of Electrical and Electronics Engineering*. New York, John Wiley & Sons.

H. D. Chiang, H. Li, Y. Tada, T. Takazawa, T. Yamada, A. Kurit, and K. Koyanagi (2009). "Group-based BCU methods for online dynamical security assessments and energy margin calculations of practical power systems," U.S. Patent 7,483,826, January 27, 2009.

B. K. Choi, H. D. Chiang, Y. H. Li, H. Li, Y. T. Chen, and D. H. Huang (2006). "Measurement-based dynamic load models: Derivation, comparison and validation," *IEEE Trans. Power Syst.*, 21(3): 1276–1283.

J. Chow (1982). *Time-Scale Modeling of Dynamic Networks with Applications to Power Systems*. Berlin, Springer-Verlag.

C. C. Chu and H. D. Chiang (1999). "Constructing analytical energy functions for network-reduction power systems with models: Framework and developments," *Circuits Syst. Signal Process.* 18(1): 1–16.

C. C. Chu and H. D. Chiang (2005). "Constructing analytical energy functions for network-preserving power system models," *Circuits Syst. Signal Process.* 24(4): 363–383.

L. O. Chua and A. C. Deng (1989). "Impasse points-part II: Analytic aspects," *Int. J. Circuit Theory* 17(3):271–289.

CIGRE Task Force 38.02.05. (1990). "Load modeling and dynamics," *Electra*.

CIGRE Task Force Report (1992). "Assessment of practical fast transient stability methods," Convener S. Geeves, June 1992.

R. H. Craven and M. R. Michael (1983). "Load characteristic measurements and representation of loads in the dynamic simulation of the queensland power system," in *1983 CIGRE and IFAC Symposium*, pp. 39–83. Florence, Italy, IEE Proc. C. Gener. Transm..

P. A. Cook and A. M. Eskicioglu (1983). "Transient stability analysis of electric power systems by the method of tangent hypersurface," *IEE Proc. C. Gener. Transm. Distrib. UK* 130: 183–193.

R. J. Davy and I. A. Hiskens (1997). "Lyapunov functions for multimachine power systems with dynamic loads," *IEEE Trans. Circuits Syst. I Fundam. Theory Appl.* 44(9): 796–812.

R. A. DeCarlo, M. S. Branicky, S. Pettersson, and B. Lennartson (2000). "Perspectives and results on the stability and stabilizability of hybrid systems," *Proc. IEEE* 88(7): 1069–1082.

C. L. DeMarco and A. R. Bergen (1984). "Application of singular perturbation techniques to power system transient stability analysis," in *IEEE International Symposium on Circuits and Systems*, pp. 597–601. Piscataway, NJ: IEEE Press.

C. L. DeMarco and A. R. Bergen (1987). "A security measure for random load disturbances in nonlinear power system models," *IEEE Trans. Circuits Syst.* CAS-34: 1546–1557.

C. L. DeMarco and C. A. Canizares (1992). "A vector energy function approach for security analysis of ac/dc systems," *IEEE Trans. Power Syst.*, 7(3): 1001–1011.

M. Djukanovic, D. Sobajic, and Y-H. Pao (1994). "Neural-net based tangent hypersurfaces for transient security assessment of electric power systems," *Int. J. Electr. Power Energ. Syst.* 16(6): 399–408.

F. Dobraca, M. A. Pai, and P. W. Sauer (1990). "Relay margins as a tool for dynamical security analysis," *Int. J. Electr. Power Energ. Syst.* 12: 226–234.

G. C. Ejebe and J. Tong (1995). "Discussion of clarifications on the BCU method for transient stability analysis," *IEEE Trans. Power Syst.* 10: 218–219.

G. C. Ejebe, C. Jing, B. Gao, J. G. Waight G. Pieper, F. Jamshidian, and P. Hirsch (1999). "On-line implementation of dynamic security assessment at Northern States Power Company," in *IEEE PES Summer Meeting*, Edmonton, Alberta, Canada, July 18–22, pp. 270–272.

A. H. El-Abiad and K. Nagappan (1966). "Transient stability regions for multi-machine power systems," *IEEE Trans. Power Appar. Syst.* PAS-85: 169–179.

Electric Power Research Institute (EPRI) (1992). *Extended Transient/Midterm Stability Program Package* (ETMSP Version 3.0). Palo Alto, CA.

Electric Power Research Institute (EPRI) (1995). *User's Manual for DIRECT 4.0*, EPRI TR-105886s. Palo Alto, CA.

M. A. El-kady, C. K. Tang, V. F. Carvalho, A. A. Fouad, and V. Vittal (1986). "Dynamic security assessment utilizing the transient energy function method," *IEEE Trans. Power Syst.* 1(3): 284–291.

D. Ernst, D. Ruiz-Vega, M. Pavella, P. Hirsch, and D. Sobajic (2001). "A unified approach to transient stability contingency filtering, ranking and assessment," *IEEE Trans. Power Syst.* 16(3): 430–443.

D. N. Ewart (1978). "Whys and wherefores of power system blackouts," *IEEE Spectr.* 15: 36–41.

F. Fallside and M. R. Patel (1965). "Step-response behavior of a speed-control system with a back-e.m.f. nonlinearity," *Proc. IEE (London)* 112: 1979–1984.

O. Faucon and L. Dousset (1997). "Coordinated defense plan protects against transient instabilities," *IEEE Comput. Appl. Power* 10(1): 22–26.

L. Fekih-Ahmed (1991). Theoretical results on the stability regions and bifurcations of nonlinear dynamical systems and their applications to electric power system analysis. PhD dissertation, Ithaca, NY, Cornell University.

R. Fischl, F. Mercede, F. F. Wu, and H. D. Chiang (1988). "A comparison of dynamic security indices based on direct methods," *Int. J. Electr. Power Energ. Syst.* 10: 210–232.

A. A. Fouad and S. E. Stanton (1981). "Transient stability of a multimachine power systems. Part I: Investigation of system trajectories," *IEEE Trans. Power Appar. Syst.* PAS-100: 3408–3414.

A. A. Fouad and V. Vittal (1988). "The transient energy function method," *Int. J. Electr. Power Energ. Syst.* 10(4): 233–246.

A. A. Fouad and V. Vittal (1991). Power System *Transient Stability Analysis: Using the Transient Energy Function Method*. Englewood Cliffs, NJ, Prentice-Hall.

A. A. Fouad, V. Vittal, Y-X. Ni, H. M. Zein-Eldin, and E. Vaahedi (1989). "Direct transient stability assessment with excitation control," *IEEE Trans. Power Syst.* 4(1): 75–82.

J. M. Franks (1980). "Homology and dynamical systems," in *CBMS Regional Conference Series in Mathematics*, Vol. 49. Providence, RI: American Mathematical Society.

R. Genesio and A. Vicino (1984a). "Some results on the asymptotic stability of second-order nonlinear systems," *IEEE Trans. Automat. Contr.* AC-29: 857–861.

R. Genesio and A. Vicino (1984b). "New techniques for constructing asymptotic stability regions for nonlinear systems," *IEEE Trans. Circuits Syst.* CAS-31: 574–581.

R. Genesio, M. Tartaglia, and A. Vicino (1985). "On the estimation of asymptotic stability regions: State of the art and new proposals," *IEEE Trans. Automat. Contr.* AC-30: 747–755.

M. Gibescu, C. C. Liu, H. Hashimoto, and H. Taoka (2005). "Energy-based stability margin computation incorporating effects of ULTCs," *IEEE Trans. Power Syst.* 20(2): 843–851.

G. E. Gless (1966). "Direct method of Lyapunov applied to transient power system stability," *IEEE Trans. Power Appar. Syst.* PAS-85: 164–179.

C. G. Groom, K. W. Chan, R. W. Dunn, and A. R. Daniels (1996). "Real-time security assessment of electric power systems," *IEEE Trans. Power Syst.* 11(2): 1112–1117.

J. Guckenheimer and P. Holmes 1983. *Nonlinear Oscillations, Dynamical Systems, and Bifurcation of Vector Fields*. New York: Springer-Verlag.

V. Guillemin and A. Pollack (1974). *Differential Topology*. Englewood Cliffs, NJ, Prentice-Hall.

U. Guadra (1975). "A general Lyapunov function for multi-machine power systems with transfer conductances," *Int. J. Contr.* 21(2): 333–343.

M. E. Guindi and M. Mansour (1982). "Transient stability of a power system by the Lyapunov function considering the transfer conductances," *IEEE Trans. Power Appar. Syst.* PAS-101: 1088–1094.

W. Hahn (1967). *Stability of Motion*. New York, Springer-Verlag.

L. N. Hannett, G. Jee, and B. Fardanesh (1995). "A governor/turbine model for a twin-shaft combustion turbine," *IEEE Trans. Power Syst.* 10(1): 133–140.

P. Hartman (1973). *Ordinary Differential Equations*. New York, Wiley.

R. He, J. Ma, and D. J. Hill (2006). "Composite load modeling via measurement approach," *IEEE Trans. Power Syst.* 21(2): 663–672.

V. E. Henner (1976). "Comments on a general Lyapunov function for multimachine power systems with transfer conductance," *Int. J. Contr.* 23: 143.

D. J. Hill (1993). "Nonlinear dynamic load models with recovery for voltage stability analysis," *IEEE Trans. Power Syst.* 8(1): 166–176.

D. J. Hill and I. M. Y. Mareels (1990). "Stability theory for differential/algebraic systems with application to power systems," *IEEE Trans. Circuits Syst.* CAS-37: 1416–1422.

M. W. Hirsch (1976). *Differential Topology*. New York, Springer-Verlag.

M. W. Hirsch and C. C. Pugh (1970). "Stable manifolds and hyperbolic sets," *Proc. Symp. Pure Math.* 14: 133–163.

M. W. Hirsch and S. Smale (1974). *Differential Equations, Dynamical Systems and Linear Algebra*. New York, Academic Press.

I. A. Hiskens (2001). "Nonlinear dynamic model evaluation from disturbance measurement," *IEEE Trans. Power Syst.* 16(4): 702–710.

I. A. Hiskens and D. J. Hill (1989). "Energy functions, transient stability and voltage behaviour in power systems with nonlinear loads," *IEEE Trans. Power Syst.* 4(4): 1525–1533.

I. A. Hiskens and D. J. Hill (1992). "Incorporation of SVCs into energy function methods," *IEEE Trans. Power Syst.* 7(1): 133–140.

F. Hoppensteadt (1974). "Asymptotic stability in singular perturbation problems II: Problems have matched asymptotic expansion solutions," *J. Differ. Equ.* 15: 510–521.

W. Hurewicz and H. Wallman (1948). *Dimension Theory*. Princeton, NJ, Princeton University Press.

IEEE Committee Report (1988). "Dynamic security assessment practices in North America," *IEEE Trans. Power Syst.* 3(3): 1310–1321.

IEEE Committee Report (1992). "Transient stability test systems for direct methods," *IEEE Trans. Power Syst.* 7: 37–43.

IEEE Standard 421.1 (1986). *IEEE Standard Definitions for Excitation Systems and Synchronous Machines*. New York: IEEE.

IEEE Standard 421.5 (1992). *IEEE Recommended Practice for Excitation System Models for Power System Stability Studies*. New York: IEEE.

IEEE Task Force on Load Representation for Dynamic Performance (1993). "Load representation for dynamic performance analysis," *IEEE Trans. Power Syst.* 8(2): 472–482.

IEEE TF Report (1982). "Proposed terms and definitions for power system stability," *IEEE Trans. Power Appar. Syst.* PAS-101: 1894–1897.

J. L. Jardim, C. S. Neto, and W. T. Kwasnicki (2004). "Design features of a dynamic security assessment system," IEEE Power System Conference and Exhibition, New York, October 13–16, 2004.

C. A. Jensen, M. El-Sharkawi, and R. J. Marks (2001). "Power system security assessment using neural networks: Feature selection using Fisher discrimination," *IEEE Trans. Power Syst.* 16: 757–763.

C. Jing, V. Vittal, G. C. Ejebe, and G. D. Irisarri (1995). "Incorporation of HVDC and SVC models in the Northern State Power Co. (NSP) network; for on-line implementation of direct transient stability assessment," *IEEE Trans. Power Syst.* 10(2): 898–906.

P. Ju, E. Handschin, and D. Karlsson (1996). "Nonlinear dynamic load modeling: Model and parameter estimation," *IEEE Trans. Power Syst.* 11(4): 1689–1697.

M. Kaba, D. Karlsson, J. McDaniel, and K. Vu (2001). "Wide area protection and optimization," in *International Conference on Bulk Power System Dynamics and Control*, August 26–31, Onomichi, Japan, pp. 85–90.

N. Kakimoto, N. Y. Ohsawa, and M. Hayashi (1978). "Transient stability analysis of electric power system via lure-type Lyapunov function," *Trans. IEE Japan* 98: 566–604.

N. Kakimoto, N. Y. Ohsawa, and M. Hayashi (1984). "Transient stability analysis of large-scale power systems by Lyapunov's direct method," *IEEE Trans. Power Appar. Syst.* PAS-103: 160–167.

R. J. Kaye and F. F. Wu (1982). "Dynamic security regions for power systems," *IEEE Trans. Circuits Syst.* CAS-29: 612–623.

H. K. Khalil (2002). *Nonlinear Systems*. New York, Macmillan.

J. Kim (1994). "On-line transient stability calculator," Final Report RP2206–1, EPRI, Palo Alto, CA, March, 1994.

P. Kundur (1981). "A survey of utility experience with power plant response during partial load rejection and system disturbances," *IEEE Trans. Power Appar. Syst.* PAS-100: 2471–2475.

P. Kundur (1994). *Power System Stability and Control*. New York, McGraw–Hill.

D. H. Kuo and A. Bose (1995). "A generation rescheduling method to increase the dynamics security of power systems," *IEEE Trans. Power Syst.* 10(1): 68–76.

A. Kurita et al. (1993). "Multiple time-scale power system dynamic simulation," *IEEE Trans. Power Syst.* 8: 216–223.

H. G. Kwatny, L. Y. Bahar, and A. K. Pasrija (1985). "Energy-like Lyapunov function for power systems stability analysis," *IEEE Trans. Circuits Syst.* CAS-32: 1140–1148.

J. P. La Salle and S. Lefschetz (1961). *Stability by Lyapunov's Direct Method*. New York, Academic Press.

J. Lee (2003). "Dynamic gradient approaches to compute the closest unstable equilibrium point for stability region estimate and their computation limitations," *IEEE Trans. Automat. Contr.* 48(2): 321–324.

J. Lee and H. D. Chiang (2001). "Convergent regions of the Newton homotopy method for nonlinear systems: Theory and computational applications," *IEEE Trans. Circuits Syst. I Fundam. Theory Appl.* 48(1): 51–66.

J. Lee and H. D. Chiang (2004). "A singular fixed-point homotopy method to locate the closest unstable equilibrium point for transient stability region estimate," *IEEE Trans. Circuits Syst. II Express Briefs* 51(4): 185–189.

W. W. Lemmon, K. R. C. Mamandur, and W. R. Barcelo (1989). "Transient stability prediction and control in real-time by QUEP," *IEEE Trans. Power Syst.* 4(2): 627–642.

B. C. Lesieutre, P. W. Sauer, and M. A. Pai (1995). "Development and comparative study of induction machine based dynamic P, Q load models," *IEEE Trans. Power Syst.* 10(1): 182–191.

W. P. Li and A. Bose (1998). "A coherency based rescheduling method for dynamic security," *IEEE Trans. Power Syst.* 13: 810–815.

Y. Liang, C. O. Nwankpa, R. Fischl, A. Devito, and S. C. Readinger (1998). "Dynamic reactive load model," *IEEE Trans. Power Syst.* 13(4): 1365–1372.

A. Llamas, J. De La Ree Lopez, L. Mili, A. G. Phadke, and J. S. Thorp (1995). "Clarifications on the BCU method for transient stability analysis," *IEEE Trans. Power Syst.* 10: 210–219.

M. Loccufier and E. Noldus (2000). "A new trajectory reversing method for estimating stability regions of autonomous dynamic systems," *Nonlinear Dyn.* 21: 265–288.

K. A. Loparo and G. L. Blankenship (1978). "Estimating the domain of attraction of nonlinear feedback systems," *IEEE Trans. Automat. Contr.* AC-23: 602–608.

L. Luyckx, M. Loccufier, and E. Noldus (2004). "Computational methods in nonlinear stability analysis: Stability boundary calculations," *J. Comput. Appl. Math.* 168(1–2): 289–297.

A. V. Machias (1986). "Analysis of transient stability of a multimachine power system using a variable gradient Lyapunov function," *Proc. Inst. Elect. Eng.* 133(Pt C): 81–86.

P. C. Magnusson (1947). "Transient energy method of calculating stability," *AIEE Trans.* 66: 747–755.

Y. Mansour, E. Vaahedi, A. Y. Chang, B. R. Corns, B. W. Garrett, K. Demaree, T. Athay, and K. Cheung (1995). "B.C. Hydro's on-line transient stability assessment (TSA): Model development, analysis and post-processing," *IEEE Trans. Power Syst.* 10(1): 241–253.

Y. Mansour, E. Vaahedi, A. Y. Chang, B. R. Corns, J. Tamby, and M. A. El-Sharkawi (1997a). "Large scale dynamic security screening and ranking using neural networks," *IEEE Trans. Power Syst.* 12(2): 954–960.

Y. Mansour, E. Vaahedi, and M. El-Sharkawi (1997b). "Dynamic security contingency screening and ranking using neural networks," *IEEE Trans. Neural Netw.* 8(4): 942–950.

W. S. Massey (1967). *Algebraic Topology: An introduction*. New York: Harcourt Brace Jovanovich.

K. Matsuzawa, K. Yanagihashi, J. Tsukita, M. Sato, T. Nakamura, and A. Takeuchi (1995). "Stabilizing control system preventing loss of synchronism from extension and its actual operating experience," *IEEE Trans. Power Syst.* 10: 1606–1613.

R. M. May (1973). *Stability and Complexity in Model Ecosystems*. Princeton, NJ, Princeton University Press.

J. D. McCalley, S. Wang, Q. Zhao, G. Zhou, R. T. Treinen, and A. D. Papalexopoulos (1997). "Security boundary visualization for systems operation," *IEEE Trans. Power Syst.* 12(2): 940–947.

F. P. de Mello, J. W. Feltes, T. F. Laskowski, and L. J. Oppel (1992). "Simulating fast and slow dynamic effects in power systems," *IEEE Comput. Appl. Power* 5(3): 33–38.

A. N. Michel and R. K. Miller (1977). *Qualitative Analysis of Large Scale Dynamical Systems*. New York, Academic Press.

A. N. Michel, N. R. Sarabudla, and R. K. Miller (1982). "Stability analysis of complex dynamical systems some computational methods," *Circuits Syst. Signal Process.* 1: 171–202.

A. N. Michel, A. A. Fouad, and V. Vittal (1983). "Power system transient stability using individual machine energy functions," *IEEE Trans. Circuits Syst.* CAS-26: 266–276.

A. N. Michel, B. H. Nam, and V. Vittal (1984). "Computer generated Lyapunov functions of interconnected systems: Improve results with applications to power systems," *IEEE Trans. Circuits Syst.* CAS-31: 189–198.

J. Milnor (1965). *Topology from the Differential Viewpoint*. Charlottesville: University of Virginia Press.

J. Milnor (1973). *Morse Theory*, Annals of Math. Stud. Princeton University Press.

H. Miyagi and K. Yamashita (1986). "Stability studies of control systems using a non-Lure type Lyapunov function," *IEEE Trans. Automat. Contr.* 31(10): 970–972.

S. Mokhtari et al. (1994). "Analytical methods for contingency selection and ranking for dynamic security assessment," Final Report RP3103-3, EPRI, Palo Alto, CA, May.

Y. H. Moon, B. H. Cho, T. H. Rho, and B. K. Choi (1999). "The development of equivalent system technique for deriving an energy function reflecting transfer conductances," *IEEE Trans. Power Syst.* 14(4): 1335–1341.

K. Morison, L. Wang, and P. Kundur (2004). "Power system security assessment," *IEEE Power Energ. Mag.* 2(5): 30–39.

J. R. Munkres (1975). *Topology - A First Course*. Englewood Cliffs, NJ: Prentice-Hall.

N. Narasimhamurthi and M. R. Musavi (1984). "A general energy function for transient stability analysis of power systems." *IEEE Trans. Circuits Syst.* CAS-31: 637–645.

T. B. Nguyen, M. A. Pai, and I. A. Hiskens (2002). "Sensitivity approaches for direct computation of critical parameters in a power system," *Int. J. Electr. Power Energ. Syst.* 24: 337–343.

Y. X. Ni and A. A. Fouad (1987). "A simplified two-terminal HVDC model and its use in direct transient stability assessment," *IEEE Trans. Power Syst.* 2(4): 1006–1012.

Z. Nitecki and M. Shub (1975). "Filtration, decompositions, and explosions," *Am. J. Math.* 107: 1029.

M. V. A. Nunes, J. A. P. Lopes, H. H. Zurn, U. H. Bezerra, and R. G. Almeida (2004). "Influence of the variable-speed wind generators in transient stability margin of the conventional generators integrated in electrical grids," *IEEE Trans. Energ. Conv.* 19(4): 692–701.

H. Ota, Y. Kitayama, H. Ito, N. Fukushima, K. Omata, K. Morita, and Y. Kokai (1996). "Development of transient stability control system based on on-line stability calculation," *IEEE Trans. Power Syst.* 11: 1463–1472.

K. R. Padiyar and K. K. Ghosh (1989). "Direct stability evaluation of power systems with detailed generator models using structure-preserving energy functions," *Int. J. Electr. Power Syst.* 11(1): 47–56.

K. R. Padiyar and H. S. Y. Sastry (1987). "Topological energy function analysis of stability of power systems," *Int. J. Electr. Power Syst.* 9(1): 9–16.

F. Paganini and B. C. Lesieutre (1997). "A critical review of the theoretical foundations of the BCU method," Technical Report TR97-005, MIT Laboratory for Electromagnetic and Electronic Systems, July, 1997.

F. Paganini and B. C. Lesieutre (1999). "Generic properties, one-parameter deformations, and the BCU method," *IEEE Trans. Circuits Syst.* 46(6), pp. 760–763.

M. A. Pai (1989). *Energy Function Analysis for Power System Stability*. Boston, Kluwer Academic Publishers.

M. A. Pai and P. G. Murthy (1973). "On Lyapunov functions for power systems with transfer conductances," *IEEE Trans. Automat. Contr.* AC-18: 181–183.

M. A. Pai and S. D. Varwandkar (1977). "On the inclusion of transfer conductances in Lyapunov functions for multimachine power systems," *IEEE Trans. Automat. Contr.* AC-22: 983–985.

J. Palis (1969). "On Morse–Smale dynamical systems," *Topology*, 8: 385–405.

J. Palis and W. de Melo, Jr. (1981). *Geometric Theory of Dynamical Systems: An introduction*. New York: Springer-Verlag.

M. Pavella and P. G. Murthy (1994). *Transient Stability of Power Systems: Theory and Practice*. New York, Wiley.

M. Pavella, D. Ernst, and D. Ruiz-Vega (2000). *Transient Stability of Power Systems - A Unified Approach to Assessment and Control*. Kluwer Academic Publishers.

R. Podmore (1978). "Identification of coherent generators for dynamic equivalents," *IEEE Trans. Power Appar. Syst.* PAS-97(4): 1344–1354.

F. S. Prabhakara and A. H. El-Abiad (1975). "A simplified determination of stability regions for Lyapunov method," *IEEE Trans. Power Appar. Syst.* PAS-94: 672–689.

K. L. Praprost and K. A. Loparo (1996). "A stability theory for constrained dynamic systems with applications to electric power systems," *IEEE Trans. Automat. Contr.* 41: 1605–1617.

W. W. Price, K. A. Wirgau, A. Murdoch, J. V. Mitsche, E. Vaahedi, and M. A. El-Kady (1988). "Load modeling for power flow and transient stability computer studies," *IEEE Trans. Power Syst.* 3(1): 180–187.

C. C. Pugh and M. Shub (1970). "Ω-Stability theorem for flows," *Inv. Math.* 11: 150.

J. N. Qiang and S. W. Zhong (2005). "Clarifications of the integration path of transient energy function," *IEEE Trans. Power Syst.* 20(2): 883–887.

J. Qiu, S. M. Shahidehpour, and Z. Schuss (1989). "Effect of small random perturbations on power systems dynamics and its reliability evaluation," *IEEE Trans. Power Syst.* 4: 197–204.

F. A. Rahimi (1990). "Evaluation of transient energy function method software for dynamic security analysis," Final Report RP 4000-18, EPRI, Palo Alto, CA, December, 1990.

F. A. Rahimi, M. G. Lauby, J. N. Wrubel, and K. L. Lee (1993). "Evaluation of the transient energy function method for on-line dynamic security assessment," *IEEE Trans. Power Syst.* 8(2): 497–507.

M. Ribbens-Pavella and F. J. Evans (1985). "Direct methods for studying dynamics of large-scale electric power systems - A survey," *Automatica* 32: 1–21.

M. Ribbens-Pavella and B. Lemal (1976). "Fast determination of stability regions for on-line power system studies," *Proc. IEEE* 123: 689–969.

M. Ribbens-Pavella, P. G. Murthy, and J. L. Horward (1981). "The acceleration approach to practical transient stability domain estimation in power systems," in *IEEE Control and Decision Conference*, Vol. 1, pp. 471–476. San Diego, CA: IEEE Press.

L. Riverin and A. Valette (1998). "Automation of security assessment for Hydro-Quebec's power system in short-term and real-time modes," CIGRE-1998, 39–103, Paris.

H. M. Rodrigues, L. F. C. Alberto, and N. G. Bretas (2000). "On the invariance principle, generalizations and applications to synchronization," *IEEE Trans. Circuits Syst. I Fundam. Theory Appl.* 47(5): 730–739.

H. M. Rodrigues, L. F. C. Alberto, and N. G. Bretas (2001). "Uniform invariance principle and synchronization: robustness with respect to parameter variation," *J. Differ. Equ.* 169(1): 228–254.

M. Saeki, M. Araki, and B. Kondo (1983). "A Lure-type Lyapunov function for multimachine power systems with transfer conductances," *Int. J. Contr.* 37: 333–343.

S. Saha, A. A. Fouad, W. H. Kliemann, and V. Vittal (1997). "Stability boundary approximation of a power system using the real normal form of vector fields," *IEEE Trans. Power Syst.* 12(2): 797–802.

H. Sasaki (1979). "An approximate incorporation of fields flux decay into transient stability analysis of multi-machine power systems by the second method of Lyapunov," *IEEE Trans. Power Appar. Syst.* PAS-98(2), pp. 473–483.

S. S. Sastry (1999). *Nonlinear Systems: Analysis, Stability, and Control*. New York, Springer-Verlag.

S. Sastry and C. A. Desoer (1981). "Jump behavior of circuits and systems," *IEEE Trans. Circuits Syst.* CAS-28(12): 1109–1124.

S. Sastry and P. P. Varaiya (1980). "Hierarchical stability and alert state steering control of interconnected power systems," *IEEE Trans. Circuits Syst.* CAS-27: 1102–1112.

P. W. Sauer and B. C. Lesieutre (1995). "Power system load modeling," in J. Chow, P. V. Kokotovic, and R. J. Thomas (eds.), *Systems Control Theory for Power Systems*, Vol. 64 of IMA Volumes in Mathematics and Its Applications, pp. 283–314. New York, Springer-Verlag.

P. W. Sauer and M. A. Pai (1998). *Power System Dynamics and Stability*. Upper Saddle River, NJ, Prentice-Hall.

P. W. Sauer, K. D. Demaree, and M. A. Pai (1983). "Stability limited load supply and interchange capability," *IEEE Trans. Power Appar. Syst.* PAS-102: 3637–3643.

P. W. Sauer, A. K. Behera, M. A. Pai, J. R. Winkelman, and J. H. Chow (1989). "Trajectory approximation for direct energy methods that use sustained faults with detailed power system models," *IEEE Trans. Power Syst.* PWRS-4(2): 499–506.

M. L. Scala, M. Trovato, and C. Antonelli (1998). "On-line dynamic preventive control: An algorithm for transient security dispatch," *IEEE Trans. Power Syst.* 13: 601–610.

R. Seydel (1988). *From Equilibrium to Chaos: Practical Bifurcation and Stability*, New York, Elsevier.

M. Shub (1987). *Global Stability of Dynamical Systems*. New York, Springer-Verlag.

F. H. J. R. Silva, L. F. C. Alberto, J. B. A. London, and N. G. Bretas (2005). "Smooth perturbation on a classical energy function for lossy power system stability analysis," *IEEE Trans. Circuits Syst.* 52(1): 222–229.

C. Singh and A. Hiskens (2001). "Direct assessment of protection operation and non-viable transients," *IEEE Trans. Power Syst.* 16: 427–434.

S. Smale (1967). "Differentiable dynamical systems," *Bull. Am. Math. Soc.* 73: 747–817.

D. Sobajic and Y. Pao (1989). "Artificial neural-net based dynamic security assessment for electric power systems," *IEEE Trans. Power Syst.* 4: 220–228.

B. Stott (1979). "Power system dynamic response calculations," *Proc. IEEE* 67: 219–241.

M. Stubbe, A. Bihain, J. Deuse, and J. C. Baader (1989). "Euro-Stag - A new unified software program for the study of the dynamic behavior of electrical power systems," *IEEE Trans. Power Syst.* 4: 129–138.

Y. Susuki and T. Hikihara (2005). "Transient dynamics of electric power system with dc transmission: Fractal growth in stability boundary," *IEE Proc. Circ. Dev. Syst.* 152(2): 159–164.

Y. Susuki, T. Hikihara, and H. D. Chiang (2004). "Stability boundaries analysis of electric power system with DC transmission based on differential algebraic equation system," *IEICE Trans. Fund. Electr.* E87-A(9): 2339–2346.

Y. Susuki, T. Hikihara, and H. D. Chiang (2008). "Discontinuous dynamics of electric power system with DC transmission: A study on DAE system," *IEEE Trans. Circuits Syst.* 55(2): 697–707.

Y. Tada and H-D. Chiang (2008). "Design and implementation of on-line dynamic security assessment," *IEEJ Trans. Electr. Electron. Eng.* 4(3): 313–321.

Y. Tada, A. Kurita, M. Masuko, Y. Takahara, and K. Koyanagi (2000). "Development of an integrated power system analysis package," Power System Technology, 2000. Proceedings of PowerCon 2000 International Conference, Perth, Australia, December 4–7, 2000.

Y. Tada, A. Kurita, Y. C. Zhou, K. Koyanagi, H. D. Chiang, and Y. Zheng (2002). "BCU-guided time-domain method for energy margin calculation to improve the BCU-DSA system," IEEE/PES Transmission and Distribution Conference and Exhibition, April, Yokohama, Japan.

Y. Tada, A. Ono, A. Kurita, Y. Takahara, T. Shishido, and K. Koyanagi (2004). "Development of analysis function for separated power system data based on linear reduction techniques on integrated power system analysis package," 15th Conference of the Electric Power Supply, Shanghai, China, October, 2004.

Y. Tada, T. Takazawa, H. D. Chiang, H. Li, and J. Tong (2005). "Transient stability evaluation of a 12,000-bus power system data using TEPCO-BCU," 15th Power System Computation Conference (PSCC), Belgium, August 24–28, 2005.

T. Takazawa, Y. Tada, H. D. Chiang, and H. Li (2006). "Development of parallel TEPCO-BCU and BCU screening classifiers for on-line dynamic security assessment," The 16th Conference of the Electric Power Supply Industry, Mumbai, India, November 6–10, 2006.

T. Tanaka, T. Nagao, and K. Takahashi (1994). "Integrated analysis software for bulk power system stability," IV Symposium of Specialists in Electric Operational and Expansion Planning, Brazil, May 23–27, 1994.

C. K. Tang, C. E. Graham, M. El-Kady, and R. T. H. Alden (1994). "Transient stability index from conventional time domain simulatin," *IEEE Trans. Power Syst.* 9: 1524–1530.

C. W. Tang, M. Varghese, P. Varaiya, and F. F. Wu (1995). "Bifurcation, chaos, and voltage collapse in power systems," *Proc. IEEE*, 83(11): 1484–1496.

J. S. Thorp and S. A. Naqavi (1989). "Load flow fractals," Proceedings of the 28th IEEE Conference on Control and Decision, December, Los Angeles.

J. Tong, H. D. Chiang, and T. P. Conneen (1993). "A sensitivity-based BCU method for fast derivation of stability limits in electric power systems," *IEEE Trans. Power Syst* PWRS-8: 1418–1437.

R. T. Treinen, V. Vittal, and W. Klienman (1996). "An improved technique to determine the controlling unstable equilibrium point in a power system," *IEEE Trans. Circuits Syst. I Fundam. Theory Appl.* CAS-43(4): 313–323.

N. Tsolas, A. Arapostathis, and P. P. Varaiya (1985). "A structure preserving energy function for power system transient stability analysis," *IEEE Trans. Circuits Syst.* CAS-32: 1040–1049.

K. Uemura, J. Matuski, J. Yamada, and T. Tsuji (1972). "Approximation of an energy function in transient stability analysis of power systems," *Elect. Eng. Japan* 92(6): 96–100.

S. Ushiki (1980). "Analytic expressions of the unstable manifolds," *Proc. Japan Acad. Ser. A* 56: 239–243.

E. Vaahedi, Y. Mansour, and E. K. Tse (1998). "A general purpose method for on-line dynamic security assessment," *IEEE Trans. Power Syst.* 13: 243–249.

P. P. Varaiya, F. F. Wu, and R-L. Chen (1985). "Direct methods for transient stability analysis of power systems: Recent results," *Proc. IEEE* 73: 1703–1715.

M. Varghese and J. S. Thorp (1988). "An analysis of truncated fractal growths in the stability boundaries of three-node swing equations," *IEEE Trans. Circuits Syst.* 35: 825–834.

V. Venkatasubramanian, H. Schattler, and J. Zaborszky (1991). "A taxonomy of the dynamics of the large power system with emphasis on its voltage stability," in *Proceedings, Bulk Power System Voltage Phenomena II: Voltage Stability and Security: An International Seminar*, Deep Lake, MD, pp. 9–52.

V. Venkatasubramanian, H. Schattler, and J. Zaborszky (1995a). "Local bifurcations and feasibility regions in differential-algebraic systems," *IEEE Trans. Automat. Contr.* AC-40(12): 1992–2013.

V. Venkatasubramanian, H. Schattler, and J. Zaborsky (1995b), "Dynamic of large constrained nonlinear systems - A taxonomy theory," *Proc. IEEE* 83(11): 1530–1560.

M. Vidyasagar (2002). Nonlinear Systems Analysis, 2nd edition, *Classics in Applied Math*, Vol. 42. Philadelphia: SIAM.

V. Vittal and A. N. Michel (1986). "Stability and security assessment of a class of systems governed by Lagrange's equation with application to multi-machine power systems," *IEEE Trans. Circuits Syst.* CAS-33: 623–636.

N. A. Vovos and G. D. Galanos (1985). "Enhancement of the transient stability of integrated AC/DC systems using active and reactive power modulation," *IEEE Trans. Power Appar. Syst.* PAS-104(7): 1696–1702.

K. T. Vu and C. C. Liu (1992). "Shrinking stability regions and voltage collapse in power systems," *IEEE Trans. Circuits Syst.* 39(4): 271–289.

L. Wang and K. Morison (2006). "Implementation of on-line security assessment," *IEEE Power Energ. Mag.* 4(5): 46–59.

H. K. Wimmer (1974). "Inertia theorems for matrices, controllability, and linear vibrations," *Linear Algebra Appl.* 8: 337–343.

F. F. Wu and N. N. Murlhi (1983). "Coherency identification for power system dynamic equivalents," *IEEE Trans. Circuits Syst.* CAS-30(3): 140–147.

W. Xu and Y. Mansour (1994). "Voltage stability analysis using generic dynamic load models," *IEEE Trans. Power Syst.* 9(1): 479–493.

Y. Xue et al. (1992). "Extended equal area criterion revised," *IEEE Trans. Power Syst.* PS-7: 1012–1022.

H. Yee and B. D. Spading (1997). "Transient stability analysis of multi-machine systems by the method of hyperplances," *IEEE Trans. PAS* PAS-96: 276–284.

N. Yorino, Y. Kamei, and Y. Zoka (2005). "A new method for transient stability assessment based on critical trajectory," 15th Power Systems Computation Conference, Liege, Belgium, August 22–26, 2005.

Y. X. Yu, Y. Zeng, and F. Feng (2002). "Differential topological characteristics of the DSR on injection space of electrical power system," *Sci. China Ser. E* 45(6): 576–584.

J. Zaborszky, G. Huang, B. Zheng, and T. C. Leung (1988a). "On the phase portrait of a class of large nonlinear dynamic systems such as the power system," *IEEE Trans. Automat. Contr.* AC-32(1): 4–15.

J. Zaborszky, G. Huang, and B. Zheng (1988b). "A counterexample on a theorem by Tsolas *et al.*, and an independent result by Zaborszky *et al.*," *IEEE Trans. Automat. Contr.* 33: 316–317.

Y. Zou, M. H. Yin, and H. D. Chiang (2003). "Theoretical foundation of controlling UEP method for direct transient stability analysis of network-preserving power system models," *IEEE Trans. Circuits Syst. I Fundam. Theory Appl.* 50: 1324–1356.

Index

Note: Page numbers in *italics* refer to figures, those in **bold** to tables.

analytical energy functions for transient stability models, 80–81
existence of, 84–89, **90**
existence of local, 92–93
for lossless network-preserving models, 82–89, **90**, 93
for lossless network-reduction models, 81–82, 93
nonexistence of, for lossy models, 81, 89–92, 93
Ω-stable theorem, 92
angle stability, 19, 24
approximation
numerical evaluations of, 108, 110, **112**
ray, 94, 96–97, *98*, 108, 110, **112**
of stability boundary, and controlling UEP method, 182, *183*, 184, *185*
of stability boundary, and PEBS method, 148, 150, 172, 173
trapezoidal, 97–98, *98*, 110, **112**, 112–114, *113*, **114**, **115**, **116**, 116, 118
asymptotic stability, 30–31, *31*, 32, 34
ATC (available transfer capability), determination of, 453–455, *455*
available transfer capability, determination of, 453–455, *455*

BCU limiter, 454, 455
BCU method, 9, 26, 27, 213
BCU exit-point method, 365–371, *367*, *369*, **370**, **372**, **373**, 373–382, **374**, **378**, **380**, **381**
group-based BCU method (integrated), 430–431, *432*, 433–434, *435*, 436, *437*, 438, *439*, 440–441, **440**, **441**, **442**, **443**, 444–446, **444**, **445**
group-based BCU-CUEP methods, 420–425, *425*, *426*, **426**, 427–429, **427**, **428**, **429**
group-based BCU-exit point method, 401–404, *404*, **405–407**, 407, **408**, **409**, *409*, 410–414, *412*, **414–417**, *417*, **418–419**, *419*
network-preserving BCU method, 279–297, *286*, *298*, 299
network-reduction BCU method, 235–253, *240*, *249*, *250*, *252*, *253*
numerical network-preserving BCU method, 300–317, *303*, *304*, *311*, *313*, **317**, **318**, **319**, 319, **320**, **321**, **322**, 322–325, **323**, **324**
numerical network-reduction BCU method, 254–263, *258*, *259*, *260*, *261*, *262*, *263*, *264*, 265–270, **268**, *269*, **269**, *270*, **271**, **272**, 272, *273*, 274–275, **274**, **276**, **277**, 278
numerical studies of, from stability boundary perspectives, 326–336, **328**, **329**, *330*, *332*, *333*, *334*, **335**, **336**, *337*, *338*, 338–344, *340*, *341*, *342*, *343*
transversality conditions, 345–356, *347*, *348*, *349*, **350**, *350*, **355**, *355*, **357**, *357*, 358–362, *361*, *362*, *363*, 364
BCU-exit point, 366, 368. *See also* BCU-exit point distance; BCU-exit point method
computation of, 373–376
and critical energy, 376–377, **378**, 379
BCU-exit point distance, 366
BCU-exit point method, 365, 378–382, **380**, **381**. *See also* group-based BCU-exit point method
boundary property, 365–371, *367*, *369*, **370**, **372**, **373**, 373, **374**, 379–381

Direct Methods for Stability Analysis of Electric Power Systems, by Hsiao-Dong Chiang

BCU-exit point method (*cont'd*)
verification scheme for boundary property, 368–370, *369*, **370**
boundary distance
definition, 366
and groups of coherent contingencies, 395, **398**, 401
boundary of stability region-based controlling unstable equilibrium point. *See* BCU limiter; BCU method; BCU-exit point; BCU-exit point distance; BCU-exit point method; group-based BCU method (integrated); group-based BCU-CUEP methods; group-based BCU-exit point method; network-preserving BCU method; network-reduction BCU method; numerical network-preserving BCU method; numerical network-reduction BCU method
boundary property, 367
analytical investigation of, 351–353
and BCU-exit point method, 365–371, *367*, *369*, **370**, **372**, **373**, 373, **374**, 379–381
and damping, 371, **372**, **373**, 373, **374**
as group property, 383, 420
and groups of coherent contingencies, 388, 395, **396**, **397**, 401
and invariant property, 360
and structural stability, 353
and transversality conditions of BCU method, 351–353, 358–359, 360–362, *361*, *362*, *363*, 364
verification scheme for, 368–370, *369*, **370**
bounded stability boundary, 44
bounded stability regions, 44

center of inertia (COI) framework, and numerical network-preserving BCU method, 315–317
characterization-based method of computing controlling UEP, 205–206
classical model of power system, 62–63, 67–68
closed orbit, characterization of on stability boundary, 40–41
closest UEP, 70, 75, 132–135, 146
dynamic characterization of, 134–135
existence of, 133–134, 138
invariant property, 143
relationship between, 139
robustness of, 129, 140–143, 146
robustness to network topology, 142–143
and stability boundary, 132–133, 138
topological characterization of, 135
uniqueness of, 133–134, 138
closest UEP method, 25, 122–123, *122*, *123*, 129, 135–136, 194
improved method, 129, 136–143, 146
network-preserving model, 129–132
numerical studies, 144–146, **144**, *145*, *146*
optimal estimation, 136, 146
relationship between original system and new system, 141–142
and stability boundary, 138
static relationship, 138–139
closest unstable equilibrium point. *See* closest UEP; closest UEP method
coherent contingencies, groups of, 383, 384–386, 399
boundary distance, 395, **398**, 401
boundary property, 388, 395, **396**, **397**, 401
critical energy values, 397, **399**
dynamic group properties, 395, 397, **399**, 399, **400**
identification of, 386, **387**
locational relationship, 387–388, **389–390**, *391*, *392*, **393–394**, *394*, *395*
multiswing stability/instability, 397, **400**
static group properties, 387–388, **389–390**, *391*, *392*, **393–394**, *394*, *395*, 395, **396**, **397**, **398**, 399
coherent generators, 383–384
conceptual methods for computing controlling UEP, 205–207
conceptual network-preserving BCU method, 295–297, *298*, 299
conceptual network-reduction BCU method, 250–251
constant energy surface, and stability region, 70, *71*
constrained nonlinear equations, for equilibrium points, 199–201

construction of numerical energy functions for lossy transient stability models, 93, 94–95, 117–118
  corrected functions, 116–117
  first integral-based procedure, 98–105
  ill-conditioned numerical problems, 105–106, *106*, **106**, **107**, **108**, 108, **109**, **110**, **111**
  multistep trapezoidal scheme, 110, **112**, 112–114, *113*, **114**, **115**, **116**, 116, 118
  numerical evaluations of approximation schemes, 108, 110, **112**
contingencies
  dynamic behaviors after, 17–18
  group properties of. *See* coherent contingencies, groups of
contingency list, 385
controlling UEP, 8–9, 177, 178–179, 178, *179*
  characterization-based method, 205–206
  computing, challenges of, 8, 299
  computing, in numerical network-preserving BCU method, 310–313, *311*, *313*, 325
  conceptual methods for computing, 205–207
  convergence regions of equilibrium points, 197, 203–205, *204*
  for DAE systems, 222–224
  dynamic characterization, 191–192, *193*
  dynamic property, 339–342, *340*, *341*, *342*, *343*
  exact, method for computing, 421–423
  existence of, 180–181, 223
  finding, 196, 197–198
  geometric characterization, 192–193, *193*
  integrated group-based BCU method for, 434, 436, *437*, 438, *439*
  numerical examples, 188–189, **189**, *190*, 191
  and postfault stability boundary, 180
  for singularly perturbed model, 223–224
  time-domain-based method for computing, 206–207, 422–423
  uniform, 223, 224
  uniqueness of, 180–181, 223
  verification procedure for, in group-based BCU-exit point method, 402–403
controlling UEP method, 7, 8, 25–26, 27–28, 123–125, *124*, *125*, 177, 181–183, *181*, *183*, 193–195, *194*, 196–197
  for accurate critical energy, 185–188, *186*, *187*, *188*
  analysis of, 183–188, *185*, *186*, *187*, *188*
  computational challenges, 197–198, 212–213
  constrained nonlinear equations, for equilibrium points, 199–201
  critical energy values, 182, 185–188, *186*, *187*, *188*
  for DAE systems, 224–226
  direct stability assessment, 182
  equilibrium equations, 200–201
  excitation system and power system stabilizer, 200
  for network-preserving transient stability models, 215–231, **228**, *229*, **230**, **231**, **232**, **233**, **234**
  numerical studies, 207–209, *208*, **209**, *209*, **210**, *211*, 211–212, *212*, *213*, *214*, 226–230, **228**, *229*, **230**
  numerical techniques for computing equilibrium points, 201–203
  singular perturbation approach, for network-preserving models, 216, 219–221, 222, 223, 224, 226
  stability boundary, relevant, approximation of, 182, *183*, 184, *185*
controlling unstable equilibrium point. *See* controlling UEP; controlling UEP method
convergence regions, 203
  of equilibrium points, 197, 203–205, *204*
  of exit points, 197
  fractal property of, 198, *204*, 204–205
  geometric definition of, 203
  size of, 197, 198, *204*, 204, 205
  of UEP, 197–198
corrected numerical energy functions, and lossy transient stability models, 116–117, 118
critical elements, 53
  characterization of on ∂Ã, 54–55
  characterization of on stability boundary, 42–43
  on quasi-stability boundary, 53–55, 57

critical energy values, 397, **399**, 403, 404, **405–407**, 407, **408–409**, *409*, 410, 411, 412–413, **414–417**, *417*, **418–419**, *419*
  and BCU-exit point, 376–377, **378**, 379
  and controlling UEP method, 185–188, *186*, *187*, *188*
  determination of, 182
  and groups of coherent contingencies, 397, **399**
  and integrated group-based BCU method for, 431, *432*, 433–434, *435*
CUEP. *See* controlling UEP; controlling UEP method

DAE systems
  controlling UEP for, 222–224
  controlling UEP method for, 224–226
  stability boundary of, 219
damping, and boundary property, 371, **372**, **373**, 373, **374**
differential and algebraic equations systems. *See* DAE systems
direct methods of transient stability analysis, 6–9, 25–26, **27**, 27–28, 60–61, 119–120, 126–128, *127*, *128*. *See also* BCU method; closest UEP method, controlling UEP method; PEBS method
  challenges of, 6–9
  and energy function theory, 60–70, *71*, 72–75, **74**, *75*, *76*, 77–79
  limitations of, 6–9
  and network-preserving models, 80
  and network-reduction models, 80
  simple system, 120–122, *121*
  and stability regions of nonlinear dynamical systems, 29
  steps, 60–61
direct stability assessment
  and controlling UEP method, 182
  and numerical network-preserving BCU method, 302–303
dynamic contingency screening
  online, 4, 447, *448*, 450–452
  in transient stability assessment, 4, 447, *448*, 450–452
dynamic data, for stability analysis, 24–25
dynamic properties
  of controlling UEP, 339–342, *340*, *341*, *342*, *343*
  and groups of coherent contingencies, 395, 397, **399**, 399, **400**
  in network-preserving BCU method, 281–288, *286*, 289, 292–295
  in network-reduction BCU method, 237–241, *240*, 243, 244–245, 246–250, *249*, *250*

energy function theory, 60–61, 64–69. *See also* energy functions
  and direct methods of transient stability analysis, 60–70, *71*, 72–75, **74**, *75*, *76*, 77–79
  trajectories, global behavior of, 65
  trajectories on stability boundary, 65
energy function-based direct methods of transient stability analysis. *See* direct methods of transient stability analysis
energy functions, 61–64, 143
  analytical, 80–93, **90**
  and bounded property, in relation to quasi-stability region, 75, 77
  corrected numerical functions, and lossy transient stability models, 116–117
  and equilibrium points, 66, 143
  local, existence of, 92–93
  for lossless network-preserving models, 82–89, **90**, 93
  for lossless network-reduction models, 81–82, 93
  for lossy models, nonexistence of, 81, 89–92, 93
  for lossy network-preserving models, 221–222
  and multimachine power systems, 91–92
  numerical, construction of, for lossy transient stability models, 93, 94–106, *98*, *106*, **106**, **107**, **108**, 108, **109**, **110**, 110, **111**, **112**, 112–114, *113*, **114**, **115**, **116**, 116–118
  and optimal estimation of stability regions, 73–75, **74**, *75*
  and quasi-stability region, 75, *76*, 77–78
  for singularly perturbed network-preserving model, 220

and stability boundary, 65
use in estimating stability region, 69–70, *71*, 72–75, **74**, *75*
equilibrium equations, 200–201
equilibrium points, 30, 133
and asymptotic stability, 30–31
characterization, on stability boundary, 39–40, 41–42, *42*
computation of, in numerical network-preserving BCU method, 308–313, *311*, *313*
constrained nonlinear equations for, 199–201
convergence regions of, 197, 203–205, *204*
and energy functions, 66, 143
finite number on stability boundary, 155
hyperbolic, 36, 37, 67–68, 132
and Lyapunov stability, 30–32
number of on stability boundary, 43
numerical techniques for computing, 201–203
same, with same inertia, 154–155
stability of, in controlling UEP method for network-preserving models, 218
structure of on stability boundary, 44, 66
type of, 36
event disturbances, 2, 15
exact controlling UEP, method for computing, in group-based BCU-CUEP methods, 421–423
existence
of analytical energy functions, 84–89, **90**
of closest UEP, 133–134, 138
of controlling UEP, 180–181, 223
of local energy functions, 92–93
exit points, 178, *178*
computing, in numerical network-reduction BCU method, 256–257
convergence regions of, 197
numerical scheme to detect, in numerical network-preserving BCU method, 305–307
of original model, method for computing, in group-based BCU-CUEP methods, 421–422
time-domain method for computing, 207, 421–422

fault-on stage, 15–17, **16**, *17*, *18*, 24, 26, **27**

generalized gradient systems, 153–154
characterization of stability boundary of *d(D)*, 155
global invariant property, 156, 157, *157*
local invariant property, 156
same equilibrium points with same inertia, 154–155
generators, coherent, 383–384
generic property, 66
geometric characterization of controlling UEP, 192–193, *193*
geometric definition of convergence region, 203
geometry, of PEBS, 165–166
global behavior of trajectories, 65
global characterization of stability boundary, 43–45
global invariant property
and generalized gradient systems, 156, 157, *157*
and second-order dynamical systems, 160
golden bisection method, 373–376
group properties of contingencies. *See* coherent contingencies, groups of
group-based BCU method (integrated), 430, 445–446
for accurate critical energy, 431, *432*, 433–434, *435*
for controlling UEPs, 434, 436, *437*, 438, *439*
numerical studies, 438, 440–441, **440**, **441**, **442**, **443**, 444, **444**, **445**
group-based BCU-CUEP methods, 420–421, 423–424, *425*, *426*, 428–429, 438
computing exact controlling UEP, 421–423
computing exit point of original model, 421–422
numerical studies, 424–425, **426**, 427–428, **427**, **428**, **429**
group-based BCU-exit point method, 401–402, 410–412, *412*, 413–414, 433–434
group-based procedure to verify computed controlling UEPs, 402–403
group-based verification scheme, 402–403, 413–414

group-based BCU-exit point method (*cont'd*)
linear/nonlinear relationships, 403–404, *404*, **405–407**, 407, **408–409**, *409*, 410
numerical studies, 412–413, **414–417**, *417*, **418–419**, *419*

Hartman-Grobman theorem, 32
hyperbolic equilibrium points, 36, 37, 67–68, 132

IEEE test system, in numerical network-reduction BCU method, 274–275, **274**, **276**, **277**
ill-conditioned numerical problems, and construction of energy functions, 105–106, *106*, **106**, **107**, **108**, **109**, **110**, **111**, 117–118
improved closest UEP method, 129, 136–143, 146
and existence of closest UEP, 138
invariant property of closest UEP, 143
relationship between closest UEPs, 139
relationship between original system and new system, 141–142
robustness of closest UEP, 140–143
robustness to network topology, 142–143
and stability boundary, 138
static relationship, 138–139
and uniqueness of closest UEP, 138
inertia, same, with same equilibrium points, 154–155
instabilities in power systems, 1, 24
integrated group-based BCU method. *See* group-based BCU method (integrated)
invariant property
and boundary property, 360
of closest UEP, 143
global, and generalized gradient systems, 156, 157, *157*
global, and second-order dynamical systems, 160
local, and generalized gradient systems, 156
local, and second-order dynamical systems, 160
local, on stability boundary, in network-reduction BCU method, 245
and transversality conditions of BCU method, 358–359, 360
and two-machine infinite bus (TMIB) system, 358–359
invariant sets, 34, 36, 159–160

large-disturbance stability. *See* general stability
limit sets, 34–37
limiter,
BCU, 454, 455
online transient stability-constrained, 453
linear/nonlinear relationships, in group-based BCU-exit point method, 403–404, *404*, **405–407**, 407, **408–409**, *409*, 410
load modeling, 21–22
component-based approach, 22–23
measurement-based approach, 22–23, *23*
need for accurate models, 452
static load model structure, 22
load variations, 15
local characterizations of stability boundary, 38–43, *42*
local energy functions, existence of, 92–93
local invariant property
and generalized gradient systems, 156
and second-order dynamical systems, 160
on stability boundary, in network-reduction BCU method, 245
locational relationship, and groups of coherent contingencies, 387–388, **389–390**, *391*, *392*, **393–394**, *394*, *395*
lossless models, 80, 81
energy functions for network-preserving models, 82–89, **90**, 93
energy functions for network-reduction models, 81–82, 93
lossy models, 80, 81
and corrected numerical energy functions, 116–117, 118
energy functions for network-preserving models, 221–222
nonexistence of energy functions for, 81, 89–92, 93

numerical energy functions, construction of for, 93, 94–106, *98*, *106*, **106**, **107**, **108**, 108, **109**, **110**, 110, **111**, **112**, 112–114, *113*, **114**, **115**, **116**, 116–118
Lyapunov function theory, 32–33, 50, 80, 81
Lyapunov stability, 30, *30*, 33
and equilibrium points, 30–32
and limit sets, 34

manifolds, stable/unstable, 34–37, *36*
measurement-based modeling, 20, 21–23
frequency-domain methods, 21
and load modeling, 22–23, *23*
time-domain methods, 21
MGP. *See* minimum gradient point
minimum gradient point
in numerical network-preserving BCU method, 306, 307–308
in numerical network-reduction BCU method, 255, 257–263, *258*, *259*, *260*, *261*, *262*, *263*, *264*
modeling, load, 22–23, *23*
modeling, measurement-based. *See* measurement-based modeling
modeling, system, and stability problems. *See* system modeling and stability problems
models, power system, 18–19
for calculating system dynamic response relative to disturbance, 14–15
for direct transient stability analysis, 80
long-term stability models, 19
midterm stability models, 19
short-term stability models, 19
multimachine power systems, and energy functions, 91–92
multistep trapezoidal approximation scheme, 110, **112**, 112–114, *113*, **114**, **115**, **116**, 116, 118
multiswing stability/instability, and groups of coherent contingencies, 397, **400**

network equations
in controlling UEP method, 200
in numerical network-preserving BCU method, 309
network-preserving BCU method, 279–280, 299. *See also* numerical network-preserving BCU method
analytical results, 288–292
conceptual method, 295–297, *298*, 299
dynamic properties, 281–288, *286*, 289, 292–295
persistence property, 290–291
reduced-state model, 280–285
static properties, 280, 281–289, *286*, 291–293
network-preserving models, 80, 216–218
advantages of use for direct stability analysis, 215
and BCU method, 334–336, **335**, **336**, *337*, *338*, 338–339
and closest UEP method, 129–132
and controlling UEP methods, 215–231, **228**, *229*, **230**, **231**, **232**, **233**, **234**
energy functions for lossless models, 82–89, **90**, 93
energy functions for lossy models, 221–222
singular perturbation approach, in controlling UEP method, 216, 219–221, 222, 223, 224, 226
singularly perturbed systems in, 83, 216, 220
network-reduction BCU method, 235–236, 250–252, *252*, *253*. *See also* numerical network-reduction BCU method
analytical results, 237–246
conceptual method, 250–251
dynamic properties, 237–241, *240*, 243, 244–245, 246–250, *249*, *250*
local invariant property on stability boundary, 245
persistence property, 242
reduced-state system, 236–237, *240*, 328, **328**, 329, **329**
robustness of equilibrium points on stability boundary, 242–243
static properties, 237–241, *240*, 243–244, 246
network-reduction models, 80
energy functions for lossless models, 81–82, 93
stability boundary of, 328–334, **328**, **329**, *330*, *332*, *333*, *334*
nonexistence of energy functions for lossy models, 81, 89–92, 93

numerical energy functions
construction, for lossy transient stability models, 93, 94–106, *98*, *106*, **106**, **107**, **108**, 108, **109**, **110**, 110, **111**, **112**, 112–114, *113*, **114**, **115**, **116**, 116–118
correcting, 116–117, 118
in direct methods, 93
first integral-based procedure, 94, 98–105
and ill-conditioned numerical problems, 105–106, *106*, **106**, **107**, **108**, **109**, **110**, **111**, 117–118
multistep trapezoidal approximation scheme, 110, **112**, 112–114, *113*, **114**, **115**, **116**, 116, 118
path-dependent (analytical) terms in, 94, 95, 96–98, *98*, 117
path-independent terms in, 94
ray approximation scheme, 94, 96–97, *98*, 117–118
two-step procedure, 94, 95–98, *98*
numerical evaluations of approximation schemes, 108, 110, **112**
numerical examples
of controlling UEP, 188–189, **189**, *190*, 191
and numerical network-preserving BCU method, 313–317, **317**, **318**, **319**, 319, **320**, **321**, **322**
numerical illustrations, in numerical network-reduction BCU method, 270, **271**, **272**, 272, *273*
numerical network-preserving BCU method, 300–304, *303*, *304*, 325
COI (center of inertia) framework, 315–317
computational considerations, 304–305
controlling UEP, computing, 310–313, *311*, *313*, 325
direct stability assessment steps, 302–303
dot product, 305–306
equilibrium points, computing, 308–313, *311*, *313*
excitation system and power system stabilizer, 309
exit points, numerical scheme to detect, 305–307
generator electric dynamic equations, 309
generator motion dynamic equations, 308–309
minimum gradient point (MGP), 306, 307–308
network equations, 309
numerical examples, 313–317, **317**, **318**, **319**, 319, **320**, **321**, **322**
numerical scheme to detect exit points, 305–307
SEP, computing, 310, *311*, 312, *313*, 325
stability-boundary-following procedure, 306–307, 308
test systems, 319, 322–325, **323**, **324**
numerical network-reduction BCU method, 254–256, 275, 278
dot product, 256–257
exit points, computing, 256–257
IEEE test system, 274–275, **274**, **276**, **277**
minimum gradient point (MGP), 255, 257–263, *258*, *259*, *260*, *261*, *262*, *263*, *264*
numerical illustrations, 270, **271**, **272**, 272, *273*
stability-boundary-following procedure, 257–263, *258*, *259*, *260*, *261*, *262*, *263*, *264*
numerical studies
of BCU method, from stability boundary perspectives, 326–336, **328**, **329**, *330*, *332*, *333*, *334*, **335**, **336**, *337*, *338*, 338–344, *340*, *341*, *342*, *343*
of closest UEP method, 144–146, **144**, *145*, *146*
of controlling UEP method, 207–209, *208*, **209**, *209*, **210**, *211*, 211–212, *212*, *213*, *214*, 226–230, **228**, *229*, **230**
of group-based BCU-CUEP methods, 424–425, **426**, 427–428, **427**, **428**, **429**
of group-based BCU-exit point method, 412–413, **414–417**, *417*, **418–419**, *419*
of integrated group-based BCU method, 438, 440–441, **440**, **441**, **442**, **443**, 444, **444**, **445**
of transversality conditions of BCU method, 360–362, *361*, *362*, *363*
numerical techniques for computing equilibrium points, 201–203

off-line transient stability assessment, 3, 6, 451

Ω-stable theorem, and existence of local energy functions, 92
online dynamic contingency screening, 4, 447, *448*, 450–452
online transient stability assessment, 3, 4–5, *5*, 6, 447, *448*, 449–452
  dynamic contingency screening, 4, 447, *448*, 450–452
  financial benefits of, 4–5, *5*
  outputs of, 449
online transient stability-constrained limiter, 453
operating environment, trends of, 2–3
optimal estimation
  and closest UEP method, 136, 146
  and stability regions, 73–75, **74**, *75*
optimal schemes for estimating stability regions, 73–75, **74**, *75*
oscillatory behaviors, 24

parameter values for model structures, 20–21. *See also* measurement-based modeling
parametric study of transversality conditions of BCU method, 346–351, *347*, *348*, *349*, **350**, *350*
PEBS, 126, 165, 168, *168*, 169, *169*
  dynamic relationships, 167–168, 175–176
  geometry of, 165–166
  static relationships, 167, 175
PEBS method, 25, 125–126, 148, 444
  analysis of, 164–175, *168*, *169*, *170*, *171*, *172*, *173*, *174*
  and approximation of stability boundary, 148, 150, 172, 173
  artificial model, 150–153, *152*, 160–164, 175–176
  dynamic relationships, 162–163, 167–168, 175–176
  generalized gradient systems, 153–157, *157*
  original model, 150–153, *152*, 160–164, 175–176
  procedure, 149–150
  relation between original and artificial models, 160–164
  second-order dynamical systems, 157–160
  static relationship, 161–162, 167, 175
persistence property
  in network-preserving BCU method, 290–291
  in network-reduction BCU method, 242
phasor measurement unit-assisted online ATC determination, 453–455, *455*
PMU-assisted online ATC determination, 453–455, *455*
postfault stability boundary, and controlling UEP, 180
postfault stage, **16**, 17, *17*, *18*, 24, 26, **27**
potential energy boundary surface. *See* PEBS; PEBS method
power flow data, for stability analysis, 24
power flow equation, energy function associated with, 100–102
power outages/interruptions, 1, 2
power system stabilizer and excitation system
  and controlling UEP method, 200
  and numerical network-preserving BCU method, 309
power systems
  classical model of, 62–63, 67–68
  dispersed generation systems, effect of, 3
  disturbances, types of, 2
  dynamic behaviors of, after contingencies, 17–18
  group properties of contingencies in, 9
  increasing load, 3
  instabilities in, 1, 24
  nonlinear nature of, 14
  stability analysis, 2
  stability problems, 15–19, **16**, *17*, *18*, 23–25
prefault stage, 15, **16**, 17, *17*, *18*, 24, 26, **27**
PSS (power system stabilizer) and excitation system
  and controlling UEP method, 200
  and numerical network-preserving BCU method, 309

quasi-stability boundary, 51–59, *52*, *56*, *192*
  characterization of, 57, 58
  critical elements on, 53–55, 57
quasi-stability regions, 51–59, *52*, *56*
  and bounded property, in relation to energy function, 75, 77

quasi-stability regions (*cont'd*)
characterization of, 56–58
characterization of critical elements on $\partial\bar{A}$, 54–55
connected but not simply connected property, in relation to energy function, 77–78
and energy function, 75, *76*, 77–78
necessary condition, 53–54
relation between ∂A_p and $\partial\bar{A}$, 54, 55
topological property, 56, 58

ray approximation scheme, 94, 96–97, *98*, 108, 110, **112**
reduced-state system
in network-preserving BCU method, 280–285
in network-reduction BCU method, 236–237, *240*, 328, **328**, 329, **329**
regions of attraction. *See* stability regions
reliability-based maintenance, 456
robustness
of closest UEP, 129, 140–143, 146
of equilibrium points on stability boundary, in network-reduction BCU method, 242–243
to network topology, improved closest UEP method, 142–143
rotor circuit equations, energy function associated with, 102–105

saturation effects of control system internal variables, 15
scenario limitation of direct methods of transient stability analysis, 7–8
second-order dynamical systems, 157, 160
characterization of stability boundary, 158–159
global invariant property, 160
local invariant property, 160
and PEBS method, 157–160
positive invariant set *Q*(*D*), 159–160
static relationship, 158
security, North American Electric Reliability Council definition, 2
SEP separation, 385
separatrix. *See* stability boundary
SEPs
computing, in numerical network-preserving BCU method, 310, *311*, 312, *313*, 325
global, 37
short-term stability models, 19
singular perturbation approach, in controlling UEP method for network-preserving models, 216, 219–221, 222, 223, 224, 226
singular surfaces, 216, 217–218
singularly perturbed systems
controlling UEP for, 223–22
in network-preserving power system models, 83, 216, 220
sink, 31
small-disturbance stability, 24
small-signal stability, 24
SSA (static security assessment) and control, 3, 4
stability, 29
stability analysis, 2, 19–20, 25–26, **27**. *See also* transient stability analysis
approaches for, 25–26, **27**
data needed, 24–25
dynamic data, 24–25
long-term, 19
midterm, 452–453
static data, 24
time-domain approach, 5–6, 26, **27**, 27, 119
stability, angle, 19, 24
stability, asymptotic, 30–31, *31*, 32, 34
stability boundary, 37, 38, 47–49, *47*, *48*, *49*, 50, 51, 52, *52* *See also* numerical studies: of BCU method, from stability boundary perspectives
algorithm to determine, 45–49, *47*, *48*, *49*
approximation of, and PEBS method, 148, 150, 172, 173
bounded, 44
characterization of, 44, 68
characterization of, and second-order dynamical systems, 158–159
characterization of closed orbit on, 40–41
characterization of critical elements on, 42–43
characterization of equilibrium points on, 39–40, 41–42, *42*

and closest UEP, 132–133, 138
of DAE systems, 219
of $d(D)$, in generalized gradient systems, 155
energy functions and, 65
finite number of equilibrium points on, 155
global characterization, 43–45
local characterizations, 38–43, *42*
of network-reduction models, 328–334, **328**, **329**, *330*, *332*, *333*, *334*
number of equilibrium points on, 43
relevant, approximation of, 182, *183*, 184, *185*
structure of equilibrium points on, 44, 66
topological properties, 37–38
trajectories on, 65
stability, general, 24
stability, Lyapunov, 30–32, *30*, 33, 34
stability problems, 15–19, **16**, *17*, *18*, 23–25
fault-on stage, 15–17, **16**, *17*, *18*, 24, 26, **27**
postfault stage, **16**, 17, *17*, *18*, 24, 26, **27**
prefault stage, 15, **16**, 17, *17*, *18*, 24, 26, **27**
and system modeling, 14–28, **16**, *17*, *18*, *23*, **27**
transient, 2, 119
stability regions, 37–38, *38*, 50, 51, 52, *52*. *See also* quasi-stability regions
bounded, 44
and constant energy surface, 70, *71*
and controlling UEP method for network-preserving models, 218–219
determining, 46–49, *47*, *48*, *49*
estimation of, using energy functions, 69–70, *71*, 72–75, **74**, *75*
of nonlinear dynamical systems, 29
optimal estimation, 73–75, **74**, *75*
topological properties, 37–38
unbounded, 45, 66, 68–69
stability, small-disturbance, 24
stability, small-signal, 24
stability, structural
and analytical investigation of boundary property, 353
and two-machine infinite bus (TMIB) system, 359–360
stability-boundary-following procedure
in numerical network-preserving BCU method, 306–307, 308
in numerical network-reduction BCU method, 257–263, *258*, *259*, *260*, *261*, *262*, *263*, *264*
stable equilibrium points. *See* SEPs
stable limit set, 34
stable manifolds, 34–37, *36*
static data, for stability analysis, 24
static load model structure, 22
static properties
and groups of coherent contingencies, 387–388, **389–390**, *391*, *392*, **393–394**, *394*, *395*, 395, **396**, **397**, **398**, 399
in network-preserving BCU method, 280, 281–289, *286*, 291–293
in network-reduction BCU method, 237–241, *240*, 243–244, 246
static relationship
and closest UEP method, 138–139
and PEBS method, 161–162, 167, 175
and second-order dynamical systems, 158
static security assessment and control, 3, 4
structural stability
and analytical investigation of boundary property, 353
and two-machine infinite bus (TMIB) system, 359–360
structure-preserving models. *See* network-preserving models
swing equation, energy function associated with, 100
system modeling and stability problems, 14–15, 27–28
measurement-based modeling, 20, 21–23, *23*
model structures/parameters, 19–21
power system stability problems, 15–19, **16**, *17*, *18*, 23–25
stability analysis, approaches for, 25–26, **27**

test systems
IEEE, in numerical network-reduction BCU method, 274–275, **274**, **276**, **277**
and numerical network-preserving BCU method, 319, 322–325, **323**, **324**

time-domain methods
of computing controlling UEP, 206–207, 422–423
of computing exit points, 207, 421–422
of measurement-based modeling, 21
of online dynamic contingency screening, 451
of stability analysis, 5–6, 26, **27**, 27, 119
TMIB (two-machine infinite bus) system, and transversality conditions of BCU method, 354–356, **355**, *355*, **357**, *357*, 358–360
topological characterization
of closest UEP, 135
and estimating stability region using energy functions, 70
topological properties
of quasi-stability region, 56, 58
in singular perturbation approach, in controlling UEP method for network-preserving models, 220
of stability boundary, 37–38
of stability regions, 37–38
trajectories, 65
trajectory function, system, 30
transfer capability, 4–5, *5*
transient angle stability, 19
transient (short-term) dynamics, 19
transient stability assessment, 3, 4–5, *5*, 19. *See also* direct methods of transient stability analysis
dynamic contingency screening in, 4, 447, *448*, 450–452
off-line, 3, 6, 451
online, 3, 4–5, *5*, 6, 447, *448*, 449–452
and short-term stability model, 19
transient stability limit load margin, 454
transient stability models, analytical energy functions for. *See* analytical energy functions for transient stability models
transient stability problems, 2, 119
transient voltage stability, 19
transmission systems. *See* power systems
transversality, 37, 41, 42, *42*
transversality conditions of BCU method, 345–346
and boundary property, 351–353, 358–359, 360–362, *361*, *362*, *363*, 364
and invariant property, 358–359, 360
numerical studies, 360–362, *361*, *362*, *363*
parametric study, 346–351, *347*, *348*, *349*, **350**, *350*
structural stability, 359–360
two-machine infinite bus (TMIB) system, 354–356, **355**, *355*, **357**, *357*, 358–360
transverse intersection, 155
trapezoidal approximation schemes, 97–98, *98*, 110, **112**, 112–114, *113*, **114**, **115**, **116**, 116, 118
TSA. *See* transient stability assessment
type-k equilibrium points, 36
type-one equilibrium points, 36

UEP
convergence regions, 197–198
distance, 366
unbounded stability regions, 45, 66, 68–69
uniform controlling UEP, 223, 224
uniqueness
of closest UEP, 133–134, 138
of controlling UEP, 180–181, 223
unstable manifolds, 34–37, *36*

verification procedure for controlling UEP, in group-based BCU-exit point method, 402–403
verification scheme for boundary property, in BCU-exit point method, 368–370, *369*, **370**
voltage stability, 19, 24

Printed and bound by CPI Group (UK) Ltd, Croydon, CR0 4YY
16/09/2026
14963412-0006